Fungal Biology

Series Editors

Vijai Kumar Gupta
Department of Chemistry and Biotechnology
ERA Chair of Green Chemistry
School of Science
Tallinn University of Technology
Tallinn
Estonia

Maria G. Tuohy
Head of the Molecular Glycobiotechnology Group
Biochemistry
School of Natural Sciences
National University of Ireland Galway
Galway
Ireland

About the Series

Fungal biology has an integral role to play in the development of the biotechnology and biomedical sectors. It has become a subject of increasing importance as new fungi and their associated biomolecules are identified. The interaction between fungi and their environment is central to many natural processes that occur in the biosphere. The hosts and habitats of these eukaryotic microorganisms are very diverse; fungi are present in every ecosystem on Earth. The fungal kingdom is equally diverse, consisting of seven different known phyla. Yet detailed knowledge is limited to relatively few species. The relationship between fungi and humans has been characterized by the juxtaposed viewpoints of fungi as infectious agents of much dread and their exploitation as highly versatile systems for a range of economically important biotechnological applications. Understanding the biology of different fungi in diverse ecosystems as well as their interactions with living and non-living is essential to underpin effective and innovative technological developments. This series will provide a detailed compendium of methods and information used to investigate different aspects of mycology, including fungal biology and biochemistry, genetics, phylogenetics, genomics, proteomics, molecular enzymology, and biotechnological applications in a manner that reflects the many recent developments of relevance to researchers and scientists investigating the Kingdom Fungi. Rapid screening techniques based on screening specific regions in the DNA of fungi have been used in species comparison and identification, and are now being extended across fungal phyla. The majorities of fungi are multicellular eukaryotic systems and therefore may be excellent model systems by which to answer fundamental biological questions. A greater understanding of the cell biology of these versatile eukaryotes will underpin efforts to engineer certain fungal species to provide novel cell factories for production of proteins for pharmaceutical applications. Renewed interest in all aspects of the biology and biotechnology of fungi may also enable the development of "one pot" microbial cell factories to meet consumer energy needs in the 21st century. To realize this potential and to truly understand the diversity and biology of these eukaryotes, continued development of scientific tools and techniques is essential. As a professional reference, this series will be very helpful to all people who work with fungi and should be useful both to academic institutions and research teams, as well as to teachers, and graduate and postgraduate students with its information on the continuous developments in fungal biology with the publication of each volume.

More information about this series at http://www.springer.com/series/11224

Marcela C. Pagano • Mónica A. Lugo
Editors

Mycorrhizal Fungi in South America

Springer

Editors
Marcela C. Pagano
Federal University of Minas Gerais
Belo Horizonte, Brazil

Mónica A. Lugo
Biological Sciences
National University of San Luis, Grupo MICODIF (Micología, Diversidad e Interacciones Fúngicas)/IMIBIO (Instituto Multidisciplinario de Investigaciones Biológicas)-CONICET-CCT SL
San Luis, San Luis, Argentina

ISSN 2198-7777 ISSN 2198-7785 (electronic)
Fungal Biology
ISBN 978-3-030-15227-7 ISBN 978-3-030-15228-4 (eBook)
https://doi.org/10.1007/978-3-030-15228-4

This Springer imprint is published by the registered company Springer Nature Switzerland AG
The registered company address is: Gewerbestrasse 11, 6330 Cham, Switzerland

January 6, 2019
This document formally acknowledges the participation of Matilde Maria Crespo, ID 39621950, in the revision of many of the chapters from the book Mycorrhizal Fungi in South America, edited by M.C. Pagano and M.A. Lugo and included in the book series Fungal Biology published by Springer Editorial, New York. Her skills as an English translator were immensely helpful for the correction of the chapters and are consequently deeply appreciated.

Contents

About the Editors

Marcela C. Pagano was born in La Plata, Argentina. She taught for some years at high school and universities. In 2018, she was appointed as technical analyst in Health Promotion and Management at the National Sanitary Surveillance Board, Santa Catarina, Brazil. Prior to that, she spent over 10 years in mycorrhizal research, completing her PhD at the Federal University of Minas Gerais, Brazil, and four postdoctoral fellowships.

Education: Dr. Pagano completed her DSc (Sciences – Applied Botany) and her MSc in Ecology, Conservation, and Wild Life Management from the Federal University of Minas Gerais, Brazil.

Experience: She has vast experience in different areas of plant-soil-fungal biology: to name a few, rhizobial and mycorrhizal symbiosis, nanomaterials focusing on anthropogenic soils and biochar, terrestrial ecology, etc.

Publication: Since 1998, she has published 28 papers in national and international peer-reviewed journals like *Ecological Indicators*, *Applied Soil Ecology*, *Symbiosis*, *Soil & Tillage Research*, etc. and from different journals of Elsevier, Springer, etc.

International Collaboration: Dr. Pagano also has collaborated in research projects in Brazil and with other research groups, which has resulted in 27 book chapters and two edited books.

Editorial Experience: She is editorial board member of two international journals (*European Journal of Soil Biology*, *Frontiers in Microbiology*).

Specialization Keywords: Plant-soil interactions/soil ecology, mycorrhizal fungi, plant symbiosis, nanomaterials.

Additional Information:
Research Gate: https://www.researchgate.net/profile/Marcela_Pagano/?ev=hdr_xprf
http://scholar.google.com/citations?user=I4sFFysAAAAJ

Mónica A. Lugo MICODIF-IMIBIO-CONICET-UNSL, Facultad de Química, Bioquímica y Farmacia, 5700, San Luis, Argentina. E-mail lugo@unsl.edu.ar/monicalugo63@gmail.com

Mónica Lugo was born in Tigre, Argentina. She taught for 28 years at high school and universities such as the National University of Buenos Aires, National University of Córdoba, and National University of San Luis; and her current positions are adjunct professor (Biology Department, FQByF, National University of San Luis) to Bachelor of Biological Sciences and professor in Biological Sciences, Vegetal Diversity I, and Plant-Fungi Interactions: Mycophyllas and Mycorrhizas, Biology of Protist and Fungi, and Fungi and Plant Systematics. She won three doctoral scholarships (1994–1999) and a postdoc fellowship (1999–2001). Further, she is the director of the Mycology, Diversity and Fungi Interaction Herbarium (MICODIF) of the National University of San Luis-National Biological Data System (SNDB), Argentina.

Education: Dr. Lugo completed her DSc (Biological Sciences) from the National University of Córdoba, Argentina.

Experience: She has worked in diverse research topics such as plant-fungi associations (stem fungal endophytes/mycophyllas, arbuscular mycorrhizas and dark septate endophytes, ectomycorrhiza ecology), environmental microbiology (plant and soil microorganism interactions in highlands; stressed, arid, and semiarid native and agronomic ecosystems; secondary communities), mycology, fungal diversity, etc.

Publication: Since 1995, she has published 24 papers in national and international peer-reviewed journals, like *Microbial Ecology*, *Mycologia*, *Symbiosis*, *Mycorrhiza*, *Pedobiologia*, *Journal of Arid Environments*, etc., 3 book chapters, 1 book of research methodologies, and 2 didactic books from different editorials such as Nova Science Publishers, Springer, etc.

International Collaboration: Dr. Lugo also has collaborated in research projects in Argentina, Colombia, and Italy and with other research groups.

Editorial Experience: She has reviewed research works as expert reviewer of international journals (*Applied Soil Ecology*, *Fungal Ecology*, *Microbial Ecology*, *Mycorrhiza*, *Symbiosis*, etc.) and national journals.

Specialization Keywords: Plant-soil interactions/soil ecology, mycorrhizal fungi, plant symbiosis/arid, semiarid, highlands, stressed environments.

Additional Information:
http://www.sanluis-conicet.gob.ar/imibio-sl/ www.bosquesnativossl.wixsite.com/misitio

https://es-la.facebook.com/Micodif-302821103462298/
https://southmycorrhizas.org/

Chapter 1
Overview of the Mycorrhizal Fungi in South America

Mónica A. Lugo and Marcela C. Pagano

1.1 Introduction

The advances in cataloging the flora and the employment of new technology has led to an integrated assessment of all the known native species of vascular plants in the Americas (Ulloa Ulloa et al. 2017), 38 years after the publication in Spanish of *Biogeografía de América Latina* by Cabrera and Willink (1980), who compiled the vegetation units within the phytogeographic provinces. It is known that countries in the tropical Andes such as Colombia, Ecuador, and Perú share a large number of plant species (between 6799 and 9226, Ulloa Ulloa et al. 2017) and that the temperate Southern Cone has the most commonality with Brazil and Bolivia. In their work, Ulloa Ulloa et al. (2017) have cited 143,903 native plant species in South America (SA), ca. 51,380 of which are endemic of this continent. In this book, there have been reported and analyzed published and unpublished data of 169 taxa woody plant species such as native and exotic trees and shrubs in relation to their mycorrhizal associations and the biological invasions of hosts along SA (see chapter by Urcelay et al.); many hosts in tropical ecosystems of Colombia (Peña Venegas and Vasco-Palacios); 16 native plant species in the Chaco region (see chapter by Grilli et al.); 205 native, endemic and exotic plant taxa in the Highlands (Lugo and Menoyo); 45 plant taxa in the Mountain forests (see chapter by Soteras et al.); 44 vascular plant taxa in the Salt-flats (see chapter by Becerra et al.); 1576 plant species from the Mediterranean Chilean "Matorral" (see chapter by Silva-Flores et al.), 245 vascular plant species in the Chilean temperate rainforests (see chapter by

M. A. Lugo
Biological Sciences, National University of San Luis, Grupo MICODIF
(Micología, Diversidad e Interacciones Fúngicas)/IMIBIO (Instituto Multidisciplinario de Investigaciones Biológicas)-CONICET-CCT SL, San Luis, San Luis, Argentina

M. C. Pagano (✉)
Federal University of Minas Gerais, Belo Horizonte, Brazil

M. C. Pagano, M. A. Lugo (eds.), *Mycorrhizal Fungi in South America*,
Fungal Biology, https://doi.org/10.1007/978-3-030-15228-4_1

Godoy and Marín), and some crop species (Pagano et al. Chap. 17; Aguilera et al. Chap. 7).

Thus, in this book, a total of ca. 2300 vascular plant taxa were analyzed considering its mycorrhizal and fungal root endophytes. This high number of host plants only represents 1.6% of the native plants existing in SA. Further, it has been shown that plant cover is the most important vegetation feature to control soil erosion as well as for the provisioning of plant products, invasion resistance, pathogen and pest regulation, soil fertility regulation, and interaction with communities of microorganisms (Pagano et al. 2017). Therefore, there is an urgent need to consider the microbial interactions in the plant-soil system as most plant species associate with microorganisms in a benefic way (legumes associate with rhizobial bacteria, and the majority of vascular plants associate with mycorrhizas and/or endophytic microorganisms) (Kumar et al. 2017). Moreover, mycorrhizas provide several ecosystem services. Among the papers published between 1990 and 2015 on ecosystem services, the largest number is from Europe (38%), with SA representing only 6% (Adhikari and Hartemink 2016).

Of particular concern are the increasing interest to study the unique vegetation types and the reduction of the environmental impacts on the protected native vegetation. Moreover, there was reported a new map of vegetation units in Argentina within the phytogeographic provinces described by Cabrera (1976) in SA using descriptions of the vegetation published in recent decades and physiognomic-floristic maps (Oyarzabal et al. 2018). In this book, a huge spectrum of phytogeographic provinces, biomes, ecoregions, environment types, and biogeographic domains were analyzed considering latitudinal, altitudinal, and salinity ranges among others environments in SA such us the Caatinga, Cerrado, Chaco region, High-Central-Magenallinic-Patagonian Andes, Páramo, Puna, Prepuna, Tropical dry and rainforest, and Valdivian temperate rainforest together with their anthropized areas. As a general result of this book's studies, the mycorrhizas and endophytic fungal root associations have shown first the arbuscular mycorrhizas (AM) and second the ectomycorrhizas (ECM) in order of predominance in SA.

In the beginning of the twenty-first century, new alternatives for the study of plants increased research in diversity, ecology and plant microbiome. As part of the microbiome, colonizing plant roots and soil biota, the arbuscular mycorrhizal fungi (AMF), links the biotic and geochemical components of the ecosystems providing varied ecosystem services. Research on mycorrhizas has gone through several stages (Fig. 1.1) (Pagano et al. 2016); however, the present period has revolutionized research on this fungal group. In 2014, from 300 to 1000 AMF species were molecularly described (Öpik et al. 2014). Most AMF are disseminated worldwide (Davison et al. 2015). Individual host plants may associate from 1 to 60 taxa (Davison et al. 2015), and different plant species can present different AMF richness (Lekberg et al. 2013), maybe due to differential resource supply by plants, as AMF richness correlates positively with fungal biomass (Antoninka et al. 2011). Further, plant species can vary in both AMF richness and composition (Meadow and Zabinski 2012; Lekberg et al. 2013), being preferentially related to different AMF (e.g. Pagano et al. 2011) due to dissimilar phenology, root architecture and other

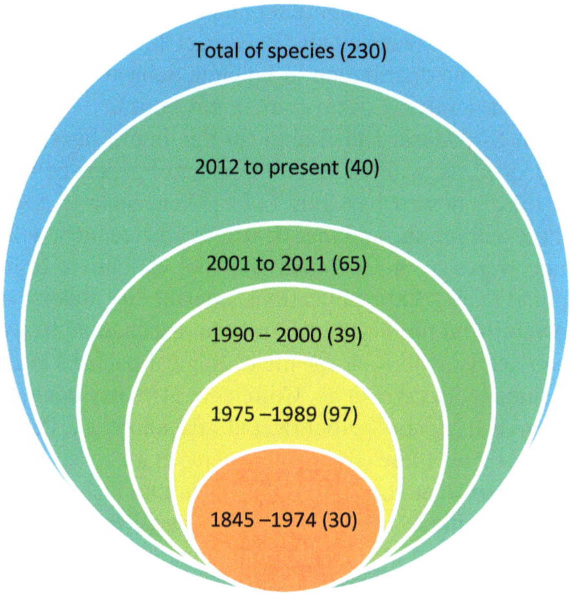

Fig. 1.1 Number of arbuscular mycorrhizal species reported in the different periods of the history of AMF (Oehl et al. 2011a, b, 2014; Goto et al. 2012; Pagano et al. 2016). Morphologically described until 2016. (References: Species number in brackets)

factors affecting the distribution, colonization strategy or function (Hart and Reader 2002; Oehl et al. 2005; Maherali and Klironomos 2007).

Recent studies reported AMF as belonging to Mucoromycota and Glomeromycotina (Spatafora et al. 2016). Arbuscular mycorrhizal fungi develop mutual symbiotic associations with most terrestrial plants and with some macrophytes from lakes, streams (Kai and Zhiwei 2006), rivers (Marins et al. 2009) and marshy plants (Radhika and Rodrigues 2007) species. It is expected that the number of known fungal species will increase since numerous potential fungal habitats and areas remain understudied (Hawksworth 2001; Hawksworth and Lücking 2017); moreover, the distribution of fungal species, phyla, and functional groups as well as the determinants of fungal diversity and biogeographic patterns are still poorly understood (Tedersoo et al. 2014; Marín and Bueno in this book).

Mycorrhizas are symbiotic associations between soil fungi and plant roots, as some structures occur in the soil compartment and others in the plant roots. The importance of AMF for soil quality is nowadays more recognized (Bradford 2014) and the association of trees with different mycorrhizal fungi is highlighted to understand the biotic interactions in global carbon dynamics (Averill et al. 2014). In spite of the urgently recommended study of AMF function under global change (Kivlin et al. 2013), few projects have initiated such research in SA. The diverse soil functions of natural and managed terrestrial ecosystems support the delivery of complex ecosystem services, that are not fully understood as they occur at the different interfaces of the lithosphere, hydrosphere, biosphere and atmosphere (reviewed by Adhikari and Hartemink 2016). Moreover, the global assessment of AMF diversity revealed very low endemism (Davison et al. 2015). This wide distribution range of

AMF also has been reported in this book in SA (Cofré et al.), and in salt flats, natural and crop environments (Becerra et al.); in contrast, it has also been demonstrated the endemic distribution in La Gran Sabana of Venezuela of the species belonging to Gigasporales (Lovera et al. in this book).

In SA, Marín and Bueno (in this book) have found 797 articles published on mycorrhizas, with the largest number of papers being from Brazil (44%), Argentina (21%), followed by Chile (12%), Venezuela (6.5%), and Ecuador (5%); further, these authors showed that Brazil was the country with the highest number of scientific works on mycorrhizal and root endophyte research. Moreover, in this book more than 2000 publications considering diverse issues such as mycorrhizas, mycorrhizal fungi and other root fungal endophytic associations in SA were revised and analyzed to look for the ecological patterns at different scales in the chapters written by Becerra et al., Cofré et al., Urcelay et al., Godoy and Marín, Grilli et al., Lugo and Menoyo, Nouhra et al., Pagano et al., Peña-Venegas and Vasco-Palacios, Silva-Flores Flores et al. and Soteras et al.

Tedersoo et al. (2010) pointed out that ECM-dominated habitats and plant hosts in SA continue undersampled compared to the north temperate regions, together with AMF and most of the fungi that inhabit soils (Tedersoo et al. 2014); further, there is a need for improving the molecular studies in the region. As regards mycorrhizal information, Varga (2015) has pointed out that most of the available publications usually lack important details about the descriptions of the experimental and methodological procedures, materials used, applied techniques and treatments, and the methodology for data collection. Thus, the authors and researchers of this book have put special emphasis on the detailed description of the methodology of data collection and analysis, completely and clearly illustrating the results obtained. This chapter discusses and analyzes the advances on mycorrhizal fungi in native ecosystems of SA and the state of conservation of mycorrhizal fungi and associations among the South American protected areas.

1.2 Mycorrhizal Symbioses in South America

In this book, different researchers show the occurrence of different types of mycorrhizas and mycorrhizal fungi species among the different biogeographic regions of SA. Arbuscular mycorrhizas and ECM were the most frequent type of fungal root symbiotic associations, but also dark septate endophytes (DSE), orchid and ericoid mycorrhizas were recorded and studied in SA. Thus, in the Highlands, from the northern of SA to the southernmost cone of the continent, AM were the most prevailing plant symbiosis followed by DSE, a few orchid and ericoid mycorrhizas and dual and triple associations (Lugo and Menoyo). These multiple associations in the Highlands of SA could be related to the recent evolutionary plant—fungal symbiont processes as it has been suggested by Brundrett and Tedersoo (2018). The chapter by Godoy and Marín provides a view of the state of the art of mycorrhizal research in the old-growth temperate rainforests, located in the Patagonian and Valdivian regions of southern Chile and Argentina (exclusive for SA), characterized by elevated precipitation levels without atmospheric pollution. The mycorrhizal traits of the dominant flora are special: *Nothofagus* spp. associates with ECM fungi, while the native conifer species associate with AMF, which is an opposite pattern to that of the northern hemisphere. These authors show an overview of the different mycorrhizal types of 245 vascular plant species and they also show the role of mycorrhizal fungi on crucial ecosystem processes such as biogenic weathering or potential use as ecological restoration tools for the re-establishment of native flora. Godoy and Marín found that the co-inoculation of two ECM species significantly increases the growth of *Nothofagus* spp. when compared to singular inoculations, highlighting the key role on nutrient cycling, maintenance of biodiversity and ecosystem productivity. They conclude that mycorrhizas of southern Chile temperate rainforests are affected by the mountain system (the Andes and Coastal mountain ranges), the mycorrhizal dominance of the forest (either ECM or AM), soil chemistry and altitude. However, how the abiotic and biotic factors interact and affect mycorrhizal communities -and the mycorrhizal symbiosis- remains to be thoroughly studied in the South American temperate rainforests (Bueno et al. 2017, Godoy and Marín in this book) and Mountain forest (Soteras et al. in this book). They stressed that the diversity and function of soil biota under climatic change provides essential information about the ecosystem processes that take place over long periods; and that traditional approaches commonly restricted to a few years due to funding and logistical restrictions (Amano and Sutherland 2013) need scientific collaboration to better understand the role of soil biota, particularly mycorrhizal fungi, in future studies of biogeochemical cycles in pristine temperate rainforests of SA (Truong et al. 2017; Oeser et al. 2018). Meanwhile, the biogeographic analysis of ECM fungi in SA was conducted by Nouhra et al. in this book (see paragraphs forward).

The chapter by Clara P. Peña-Venegas and Aída M. Vasco-Palacios provides a view of the state of the art of mycorrhizal research from Colombia, including a case study of AMF and ECM in the Amazonian tropical rainforest. They explain the restrictions on mycorrhizal research to root colonization and AMF spore quantification in commercial-plants, or fungal fruiting bodies near reported ectomycorrhizal plant species for inventories. They stress that few studies included mycorrhizal associations in natural ecosystems or the use of molecular tools in Colombia, which limited the understanding of mycorrhizal symbiosis that could be useful for agriculture, timber production and soil bioremediation. They found a total of 97 reports related to ECM fungi in Colombia, showing that 172 species of ECM fungi have been reported mainly in *Quercus*-dominated montane forests.

Silva-Flores et al. (see Chap. 14) stress the little research regarding mycorrhizal symbiosis on the South American Mediterranean-type ecosystem (Chilean matorral), which is a biodiversity hotspot. They highlight and compile the knowledge of mycorrhizal symbiosis in these environments showing the lack of studies on ericoid mycorrhiza (also at a national level) and the existence of scarce reports. However, they point out an emerging interest of several researchers mainly in studies of AM, ECM and orchid mycorrhizal (OM).

In this book, Cofré et al. (see Chap. 3) address the cumulative information that has increased considerably over the last years, particularly in SA. They present the published literature of AMF morphological richness for SA to evaluate richness patterns across the ecological divisions of the region. They compiled evidence of an increasing interest in the study of these fungi in SA; however, an uneven distribution among the ecodivisions show the main research focus is in the Amazonia, Atlantic forest, Caatinga and Chaco while many regions remain unstudied and others poorly sampled (e.g. Guianan lowlands and Patagonia respectively). They also highlight that the soybeanization accompanied by monocultures, clearing, fumigation and movement of peasants in the ecoregions of SA affect the unexplored biodiversity.

The chapter by Carlos Urcelay et al. addresses how the biological invasions constitute a global environmental threat that fast alters natural communities and ecosystem functioning. A way to understand the success of alien trees in novel ecosystems is by comparing their ecological strategies with those of native ones. As it happens on a global scale, the different types of mycorrhizas are not randomly distributed across the biomes and are related to environmental variables instead. They examine the patterns of mycorrhizal distribution in native and alien tree species occurring in contrasting montane ecosystems across a broad latitudinal gradient in SA. They point out that the effects of mycorrhizal fungi on the growth, nutrition, and then expansion and dominance of most native and alien trees in the ecosystems of southern SA still remain to be determined; however, it seems that mycorrhizal associations have an unambiguous role in tree invasions in the montane forests across different climates but the relative importance of each mycorrhizal type in each ecosystem remains unknown.

Interestingly, some native AM plants from SA (*Flaveria bidentis*, Zhang et al. 2017; *Bidens pilosa*, Song et al. 2011) are invasive in other countries/ continents. For example, *F. bidentis*, a native of SA, is an aggressive invader in North China. Moreover, the invasion of the exotic Pinaceae in SA is also revised in this book (see Chap. 2, Urcelay et al.; Chap. 6, Peña-Venegas and Vasco-Palacios). Special attention should be given to exotic pastures such as *Brachiaria*, which is a species largely used for Brazilian agropecuary and in the integrating cropping livestock systems. Species of *Brachiaria* and *Panicum* are mainly included in tropical agro-systems. These tropical pastures (*Urochloa decumbens* syn. *Brachiaria decumbens* Stapf) in no-till cultivation are commonly used as pastures due to their adaptation capacity and large root system, together with intercropped grasses and legumes to increase the production, forage quality, and the profitability and sustainability of these systems in tropical regions (reviewed by Pagano et al. 2017).

The high diversity of Orchidaceae, which is present in two biodiversity hotspots in the northwestern region of SA: the tropical Andean region and the Chocó-Darien hotspot (Myers et al. 2000), is subject to constraint due to plant extraction for commercialization, and was investigated in the pioneer studies carried out by Dr. Joel T. Otero in Colombia (Otero and Bayman 2009; Otero et al. 2013), which reported *Rhizoctonia*-like fungi being isolated from several orchids and evaluated for biocontrol potential against the pathogenic *Rhizoctonia solani* in rice (Mosquera et al. 2010). They found discrete groups of mycorrhizas, including fungi from tropical epiphytic orchids; from plant pathogenic *Rhizoctonia* species (*Thanatephorus* spp.) used as a positive control; from terrestrial orchids; and from *Vanilla* species (Mosquera et al. 2013). Others studies on epiphytic orchids in Oncidiinae have revealed moderate to high levels of preference for specific clades of *Ceratobasidium* spp. (Otero et al. 2002, 2004, 2005). Other reports on Orchidaceae are scarce and distributed in a few countries such as Brazil, with the research leaders being Maria Catarina M Kasuya and Marlon C Pereira (e.g. Pereira et al. 2011; Detmann et al. 2018). As Brazil is the home of approximately 2500 species of orchids of which 1627 are endemics (Barros et al. 2012), new projects on OM led by MC Pereira and Prof. MC Kasuya are improving these studies. Also in Argentina, some reports of OM are mentioned (see chapter by Lugo and Menoyo), Brazil (Pagano et al., see Chap. 9) and Chile (see chapters by Silva-Flores et al. and Godoy and Marín).

Finally, it has been stressed that most published reports (74%) commonly lack any important detail when describing the study such as those about the experimental treatment, the abiotic growing conditions, the soil nutrient concentrations, the duration of the study or a better description of the methodology for data collection (Varga 2015). In SA, the most published papers have shown the AMF association with native and also with some agronomical plant species predominate over the rest of the mycorrhizal and other root fungal endosymbiotic associations (see also chapters by Grilli et al., Lugo and Menoyo, Marín and Bueno, and Silva-Flores et al. in this book). These results are also in agreement with the global distribution of presence or absence of the diverse mycorrhizal and root fungal endophytic associations under an evolutionary approach (Brundrett and Tedersoo 2018).

1.3 The Mycorrhizal Fungal Species in South America

A growing worldwide attention on fungi is being noticed, as of 120,000 known fungal species (Hawksworth and Lücking 2017), more than one million (Schueffler and Anke 2014) or 2.2–3.8 million (Hawksworth and Lücking 2017) are predictable to exist, but also the number of fungal species estimated ranges from 500,000 to ca. ten million species. However, the range of 1.5–5 million species is the estimation most accepted by the mycologists (Hawksworth and Lücking 2017 and references therein). Taking into account that the number of native plants for SA would be approximately 143,903 species (Ulloa Ulloa et al. 2017), and if approximately 70–80% would be colonized forming AM (Brundrett and Tedersoo 2018), this yields a possible number of associated plants forming AM in SA of ca. 100,000 putative host plants. Therefore, if 1–60 taxa of AMF could be associated by a host (Davison et al. 2015), a simple calculation has yielded a putative number of AMF in SA that surpasses the worldwide total number recorded or virtually proposed until now by morphological and molecular tools.

Further, for Fungi in general, it has been proposed (Hawksworth and Lücking 2017 and references therein) that the possible sources or sites where fungal, yet undescribed species, could be found are "in biodiversity hot spots in the tropics, little-explored habitats, and material in collections awaiting to be studied". Thus, in SA exist all possible sites and conditions where the undiscovered fungal mycorrhizal species are available and need to be studied and described.

1.3.1 The ECM Fungal Species in South America

South America has about 6–7% of the total number of papers on mycorrhizas worldwide. As regard to ECM fungi and its ECM associations, they are commonly widespread in forests and woodlands of temperate and cold regions in both Hemispheres (Tedersoo et al. 2012), but also in tropical and subtropical regions worldwide (Moyersoen et al. 1998a, b, 2001; Founoune et al. 2002; Onguene and Kuyper 2002); particularly in SA, where ECM fungi can be diverse (Henkel et al. 2012; Kennedy et al. 2011; Riviere et al. 2007; Smith et al. 2013, 2017; Tedersoo et al. 2007; Vasco-Palacios et al. 2018 among others). Recently, Brundrett and Tedersoo (2018) have updated the global biogeographical patterns of plant and fungi associated forming mycorrhizas; however, the fungal diversity of tropical and subtropical habitats in the Southern Hemisphere is understudied in contrast to Northern Hemisphere forests (Corrales et al. 2018; Hawksworth and Lücking 2017; Tedersoo et al. 2007). In addition, many recent reports in SA have reported high fungal diversity in Andean Nothofagaceae forests (Truong et al. 2017), in the Yungas forests (Geml et al. 2014), and fungal diversity in Caesalpiniaceae legume-dominated Neotropical forests (Henkel et al. 2012).

In this book, ECM fungi have been also studied along the whole SA (Chap. 4 by Nouhra et al.), in Tropical forest of Colombia (Chap. 6 by Peña-Venegas and Vasco-Palacios), Mediterranean-type Matorral (Chap. 14 by Silva-Flores et al.) and Temperate rainforest of Chile (Chap. 16 by Godoy and Marín). In the chapter by Eduardo Nouhra et al., it is provided an overview of the wide range of ECM habitats and EMC fungi lineages from South America. These authors have found that the Patagonian forest dominated by Nothofagaceae could harbor a high ECM diversity and the largest amount of ECM fungi lineages in contrast to the Neotropical sites considered by Nouhra et al. Further, in Patagonia forests, ECM fungi belonging to Pezizales (Ascomycota) were highly diverse. In Chap. 4, the analysis performed by Nouhra et al. has shown one unique lineage at the global scale (/guyanagarika) in Guiana Region; instead, the ECM linages /cortinarius, /russula-lactarius, /amanita and /clavulina are present in all of the areas treated in SA; the /cortinarius lineage was rich in Patagonia and, three lineages were notably rich in the Guiana Shield.

In Colombia, a total of 172 species of ECM fungi have been reported in Tropical forests of *Quercus*, *Pinus*, *Colombobalanus*, *Dicymbe* and *Aldina*, and *Pseudomonotes tropenbosii* (Peña-Venegas and Vasco-Palacios, see Chap. 6). The fungal ECM taxa reported has been included into typical ECM lineages of Basidiomycota such as Amanitaceae, Cantharellaceae, and genera of the families Russulaceae and Boletaceae; in addition, some particular taxa such as *Polyporoletus sublividus* (Albatrellaceae), *Tremellogaster surinamensis* (Diplocystidiaceae) and four endemic species of *Sarcodon* (Bankeraceae) have been also registered in Chap. 6. The genus *Sarcodon* was considered to be distributed in the Northern Hemisphere, but now 10 species are known from lowland areas in tropics, to be expanding the range of distribution and the knowledge about plants host-associated to this genera (Grupe et al. 2016). The EM lineages *Tomentella* and *Sebacina* are commonly detected on root analysis (Vasco-Palacios 2016; Vasco-Palacios et al. 2014). However, those have not been reported in Colombia yet, probably, because the basidiomata of these genera can be easily overlooked, as they occur erratic and are resupinate and/or cryptic (Moyersoen 2006). In Mediterranean-type Chilean Matorral, 43 species of ECM fungi have been recorded in four localities which belong to the genera *Amanita, Austropaxillus, Boletus, Cortinarius* (represented with the greatest species number), *Dermocybe, Descolea, Inocybe, Laccaria, Paxillus, Russula, Stephanopus, Thaxterogaster, Tricholoma*, and *Zelleromyces* (Chap. 14 by Silva-Flores et al.). Along the temperate rainforest of Chile, Godoy and Marín (Chap. 16) have cited previous reports registering a huge number of 651 ECM fungi taxa exclusive to *Nothofagus* spp. forests (Garrido 1988) and, considered as the most abundant ECM fungal orders on *Nothofagus* forests to Boletales, Cortinariales, Gautieriales, and Russulales (Palfner and Godoy 1996; Flores et al. 1997; Godoy and Palfner 1997; Palfner 2001; Nouhra et al. 2013).

1.3.2 The AMF Species in South America

With regard to AMF, there are currently applied three main worldwide and different systematic points of view; one of them is the AMF morphological and molecular classification proposed by Redecker et al. (2013) and Schüßler and Walker (2010) (which is online updated in http://www.amf-phylogeny.com/amphylotaxonomy.html). This classification system recognizes the AMF within the phylum Glomeromycota and the class Glomeromycetes, 4 orders (Glomerales, Diversisporales, Paraglomerales and Archaeosporales), 12 families, 34 genera and approximately 316 AMF morphospecies correctly described and the actually excluded species *Glomus tenue* that was changed to *Planticonsortium tenue* in the subphylum Mucoromycotina (Walker et al. 2018). Another AMF classification system based in morphological and developmental features of AMF has been proposed by Błaszkowski (2012), Błaszkowski et al. (2018 among others), in this systematic classification of AMF, 3 classes, 5 orders, 16 families, 41 genera were included and 300 species have recently been reported (Oehl et al. 2011a, 2014; Goto et al. 2012; Stürmer 2012; Stürmer et al. 2018; Błaszkowski 2012; Błaszkowski et al. 2018). Finally, the third AMF classification involves virtual taxa (VT), which are putative taxa defined by Öpik et al. (2010) and that are part of the most important MaarjAM database of AMF molecular diversity; under this systematic point of view, AMF comprise ca. 350–1000 molecularly defined taxa (Davison et al. 2015 and references therein). Although in SA, molecular characterization of AMF communities is poorly known (Grilli et al. 2015; Senés-Guerrero and Schüßler 2016; Soteras et al.

2016), the AMF diversity prospection by means of VT analyses was carried out in native (see chapters Grilli et al., Peña-Venegas and Vargas-Palacios, Silva-Flores et al., Soteras et al.) and agronomic (see chapter Pagano et al. in this book) environments. Besides, the AMF high tolerance to Al and higher P efficiency to host vs. non-host plants were studied in acidic and anthropized soils of Chilean farming systems (see chapter of Aguilera et al.).

In the 1990s only a few papers were published, but, in the following two decades, there has been many studies mainly using morphological identification compared to genetical one. Between 2011 and 2014, studies on mycorrhizal occurrence and diversity exceeded the number of studies on ecology. Moreover, the number of studies on mycorrhizal ecology outperformed all other reports in the last years until now (Fig. 1.1).

In SA, the largest number of described species is from Brazil followed by Argentina. The importance of checklists has been highlighted for plant and fungi diversity knowledge by Ulloa Ulloa et al. (2017) and Hawksworth and Lücking (2017), respectively. In SA, Brazilian researchers have conducted, published and maintained over time these biodiversity useful tool of checklists for AMF (Goto et al. 2010; Jobim et al. 2016, 2018 among others); furthermore, a complete revised list of publications of AMF morphological diversity in SA is addressed in Cofré et al. (Chap. 3), salt flats (see Becerra et al., Chap. 15) and anthropized environments (Schalamuk et al. 2013; Aguilera et al., Chap. 7; Pagano et al., Chaps 9 and 17).

In this book, Cofré et al. (see Chap. 3) have shown that most globally distributed taxa of Glomeromycota are present in SA, including 62% of the worldwide currently known AMF; however, a huge amount of SA is still unstudied. These authors have concluded that AMF communities of the ecodivisions of Guianan (Uplands, Highlands, and Lowlands), Peruvian-Chilean (Atacama) desert, and Caribbean need to be studied. In addition, Patagonia has been poorly sampled. Moreover, Cofré et al. have found represented in SA 186 different AMF morphospecies which surpass the 131 identified AMF morphospecies registered until 2012 in a recent review (Stürmer et al. 2018).

In Chap. 8, Lovera et al. shows in detail how patterns of AMF in savannas intermixed with forests, shrublands, meadows and palm swamps, are crucial for ecosystem functioning by improving plant nutrition and resistance to environmental stress. They observed a high diversity of Gigasporaceae and four new species (*Scutellospora spinosissima*, *S. crenulata*, *S. striata*, *S. tepuiensis*) and, some undescribed morphotypes, are considered endemic, suggesting the region as a center of diversification for *Scutellospora* or even for the Gigasporaceae. Similarly, due to the several species of Gigasporaceae that have been described from Northeast Brazil, and to the recorded species of the family in this region (ca. 60% of the Gigasporaceae species diversity), Brazil has been proposed as the center of diversification for Gigasporaceae (Marinho et al. 2014; de Souza et al. 2016). Thus, Lovera et al. (in this book) suggest that the Guayana Shield, in addition to the Atlantic Shield in the Northeast Brazil, can be hotspots of diversification for Gigasporaceae, shrublands hosting the greatest diversity and endemism. Further, Lovera et al. (see Chap. 8) have shown the presence of endemic AMF in La Gran Sabana (LGS) of Venezuela in the biome of shrublands, which also included different vegetation types such as the tropical grasslands, savannas that host the greatest diversity and endemism of Gigasporaceae and that are considered an evolutionary hotspot for AMF (Pärtel et al. 2017). Therefore, these surpassing results are evidencing an increasing interest in the study of these fungi in the region in recent years. Thus, probably diversity of AMF from SA in relation to worldwide diversity will increase, as has been proposed by Veblen et al. (2015).

In SA, many soil properties are relevant to AMF community composition and diversity (Table 1.1.). Soteras et al. (see Chap. 13) have analyzed the AMF diversity in the high mountain ecosystems, one of the main hotspots of biodiversity of South America. The AMF taxa in South American mountain forests have been reviewed and the richness of morphospecies and structure of AMF communities have been analyzed in relation to microscale (host species, pH, N, P) and macroscale factors (latitude, temperature, precipitation). In this book, Soteras et al. has shown that AMF communities differed in both scales being associated with sampling site, vegetation type or host identity. The families Glomeraceae and Gigasporaceae were related to micro- and macro-scale factors, while Acaulosporaceae was not significantly related with neither micro- nor with macro-scale factors. The AMF community composition at higher scales of tropical and temperate ecosystems differed due to latitude, precipitation and temperature. However, at lower scales soil characteristics

Table 1.1 List of key soil properties related to AMF in natural ecosystems from South America

Key soil properties	Mycorrhizal mycelium length	Changes in AMF community structure and /or diversity
Soil organic carbon	X	
Soil pH		X
Potassium		X
Calcium		X
Magnesium		X
Phosphorus availability		X
Iron		X
Sulfur		X
Zinc		X
Sodium		X
Base saturation		X
Clay mineralogy		X
Silt		X
Water stable macroaggregates	X	
Coarse and total sand		X
Water in soil		X
Boron		X
CEC (Cation exchange capacity).		X

and host species were the most relevant factors in differentiating AMF sites composition. Thus, Soteras et al. have found that AMF communities of high mountain forests of SA are differentially affected by the particular characteristics of these environments, different from the cosmopolitan pattern. Meanwhile, in Salt flats environments, Becerra et al. (see Chap. 15) has registered that the Glomeraceae was the dominant AMF family in stressful habitats in Argentinean Pampas and in Northwestern and Central of Argentina.

1.4 New Insights for AMF Conservation

1.4.1 The Mycorrhizas in Protected Areas of South America

In SA, there are very important hotspots areas (Myers et al. 2000; Cardoso da Silva and Bates 2002; Madriñán et al. 2013; Young et al. 2015) such as Caatinga, Cerrado, Dry tropical forest, Highlands of Central High Andes Mountain forest, Mediterranean Central Chilean matorral, Páramo, Patagonia steppe, Puna, Tropical and Temperate rainforests. All of these megadiverse and/or unique diverse environments have been analyzed throughout this book. However, for the huge importance of these areas as reserves of wild biodiversity, there are very scant conservatory politics and efforts conducted. Moreover, the settlement of National Parks and Reserves and mycorrhizal research in these areas are still undeveloped.

As regards the AMF occurrence in native ecosystems, more information on indigenous species were compiled by Turrini and Giovanetti (2012). They analyzed the AMF occurrence in protected areas worldwide, showing the lists of AMF species present in ten South American habitats, namely, Canaima National Park – La Gran Sabana (Venezuela), Reserva Biológica San Francisco and Podocarpus National Park (Ecuador), Manu National Park (Perú), Brasilia National Park, Vale do Catimbau National Park, Serra do Cipó National Park, State Park of Campo de Jordão, State Park of Alto Ribeira, in Brazil. Moreover, other preserved sites are mentioned in this book for Chile (see chapter by Silva-Flores), and El Palmar National Park, Quebrada del Condorito, Sierra de las Quijadas and Nahuel Huapi National Park in Argentina (Fig. 1.2). Further AMF diversity and occurrence revision is showed in the chapters by Cofré et al., Lugo and Menoyo, Soteras et al. in this book. With regard to ECM, they were investigated in *Nothofagus* forests in Patagonia (boundaries of the Lanin National Park), Argentina (Nouhra et al. 2013), showing that the composition was influenced by altitude. In this book, ECM associations have been revised along a latitudinal gradient in SA which has shown a similar pattern of frequency of colonization and presence versus AM and other mycorrhizal associations than at a global scale (Urcelay et al.), in the Valdivian forests ECM were predominant in native trees (Godoy and Marín), in the Mediterranean Chaparral of Chile (Silva-Flores et al.) and they were not still found in South American Highlands (Lugo and Menoyo). The ECM fungi in SA have been

Fig. 1.2 Some views of National Parks or preserved areas studied for AMF. Clockwise, from upper left: National Park El Palmar in Argentina, native trees from Patagonia region, Argentina and sand dunes (Protected preserved area) in Brazil (Photo-credit: M. Pagano); down, national Park, Sierra de las Quijadas, San Luis province, Argentina (Photo-credit: M. Lugo)

also studied and revised (Nouhra et al.) and some new records were also mentioned in the Mediterranean Chaparral (Silva-Flores et al.); further, some exotic ECM and its relationship with exotic host plants invasions in SA have been explored by Urcelay et al.

In their chapter on ECM fungi biogeographic research revision in SA, Nouhra et al. compile data on various aspects since the first appearance of studies in the region (areas of interest along the Andes, Guiana, Amazonian Basin and the north-eastern coast of SA). They stress the wide variety of unique biomes in SA, the ectotrophic Nothofagaceae forest in Patagonia, which presents the highest ECM diversity. However, substantial ECM diversity remains to be discovered in all regions of SA. The highest richness present in Patagonia is consistent with global patterns of ECM distribution (Tedersoo et al. 2012). They analyze the different lineages of ECM by regions and biomes, indicating that the Guiana Region also harbors a higher diversity of ECM, being home of at least one unique lineage, and that the tropical and subtropical Andes regions with their Northern Hemisphere-derived ECM hosts and the Amazon basin with its widely dispersed Neotropical hosts are apparently the least diverse regions.

Recently, most important protocols for studying the fungi including AMF were compiled by Lugo et al. (2018), in order to help researchers to investigate the interactions with AMF, fungi and endophytes. The manual is available in Spanish and contains methodologies applied in Argentinian environments, in many studies of AM and AMF and fungal endophytes in roots. Some protocols have been used in protected areas of Argentina, such as Sierra de las Quijadas National Park (Fig. 1.2.), where diversity of AMF in Bromeliaceae rhizosphere, terrestrial and epiphytic Bromeliaceae root colonization by AMF and DSE, and AM association in relation with plants functional types were studied (Rivero Mega et al. 2014; Lugo et al. 2009, 2015).

Moreover, new reports for protected areas have been published. For example, in Brazil, permanent plots (100m × 100m) in three phytophysiognomies of the Atlantic Forest, established in the Biota-Program from the São Paulo State Research Foundation, in the Serra do Mar State Park, Brazil (Joly et al. 2012) were studied for mycorrhizal symbioses (Duarte et al. 2018), after previous reports from Aidar et al. (2004) at the Tourist State Park of the High Ribeira Valley (PETAR) also in São Paulo. Nevertheless, more studies are needed as well as enough understanding of the native mycorrhizal biodiversity (Fig. 1.3). At the level of plant-AMF community interactions, Lekberg and Waller (2016) showed that plant species harbored distinct AMF communities in 25% of the studied sites of different biomes worldwide, suggesting host plant identity a weak driver for AMF community assemblage. They call for obtaining more samples with more plant species and replicates within communities to expand the understanding of plant-AMF community interactions.

In this book, mycorrhizal fungi and associations have been addressed in protected areas of Argentina, Bolivia, Brazil, Chile, Colombia and Venezuela throughout the most of the chapters, and those that have not been included are registered in the Table 1.2.

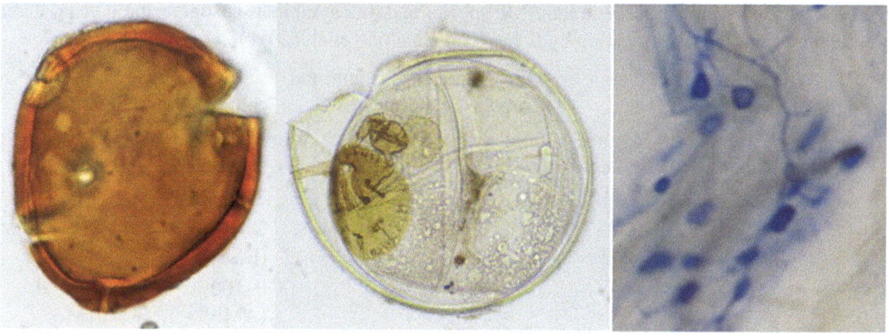

Fig. 1.3 Native AMF from sites in Brazil. Clockwise, from left: AMF spores of Glomerales and Gigasporales, colonized roots of a native tree (Photo-credit: M. Pagano)

In addition, new reports show increased number of identifiable and unidentifiable AMF species in SA. For example, Freitas et al. (2014) have identified 41 spore morphotypes in rhizosphere of three legume trees in Amazonian forest; Carvalho et al. (2012) reported 49 AMF species for six different natural habitats (highland fields, bogs, Cerrado, etc.) and Silva et al. (2014) 50 AMF species for three semiarid sites (dry forest, a transitional zone and a moist forest). The information of mycorrhizal fungi studies performed in the last decade in the protected areas of SA is summarized in the Table 1.2. However, these reports are still very few, considering the great extension of SA.

Among the conservation units, none National Park is under continuous research in SA and few ones have been investigated for the mycorrhizal and root fungal endophytic symbioses. There is a high number of unidentified morphotypes of AMF, which are annotated as Genus + sp. 1 and to cope with this problem, different strategies can be performed such us AMF culture using trap culture method or transformed root culture (see paragraphs forward).

1.4.2 *The Mycorrhizal Species in the Soil Profile*

In 2005, Oehl et al. stressed that the deep soil layers should be included in studies to get a complete representation of AMF diversity, as they can show different composition than the topsoil. Despite that time, there are few reports of research on AMF with soil depth, among them there are two reports from South America. One from Terra Preta de Índio (Pagano et al. 2016) and other from mixed cultivation *Acacia mangium* with *Eucalyptus grandis* (Pereira et al. 2018). Moreover, recently it was confirmed that different AMF communities can occur in subsoils of in agricultural field, being also pointed as a potential biodiversity reservoir (Sosa-Hernández et al. 2018). However, we do not know the functional traits of those communities, thus the subsoil biology needs to be included in agricultural management (Sosa-Hernández et al. 2018).

Table 1.2 Total number of identified AMF species and root symbiosis in some natural ecosystems in the biomes of SA classification by Echeverría-Londoño et al. (2018)

Country	Biome/Vegetation type/State/Region	AMF species/ Morphotypes	Unidentified species	% Root colonization	Reference
Argentina	Arid Chaco, Monte/ San Luis/Sierra de las Quijadas National Park	7	I (3)	NI	Rivero Mega et al. (2014)
				7–37 (AMF)[b] 27–91% (DSE)[b]	Lugo et al. (2009)
				0–100% (AMF)[b] (65, 13% AMF)[b]	Lugo et al. (2015)
Argentina	Espinal, Palmar/Entre Ríos/El Palmar National Park	46	I	NI	Velázquez et al. (2008, 2013), Velázquez and Cabello (2011)
Argentina	Temperate Rainforest/ Río Negro/Patagonian Andes/National Park Nahuel Huapi	27	I	NI	Velázquez et al. (2016)
Brazil	Amazonia/Terra firme forest	39	I (1)	NI	Freitas et al. (2014)
Brazil	Atlantic rain Forest/ Mature forest, Paraná state,	47–68	I (22)	~20–80 (AMF)[b] (52% AMF)[b]	Zangaro et al. (2013)
Brazil	Southern Pernambuco state/	34–44	I (6)	NI	Pereira et al. (2014)
Brazil	Northern São Paulo state	23–25	I (6)	~46–80 (AMF)[b] (63.12% AMF)[b]	Aidar et al. (2004)
Brazil	Serra do Mar State Park, São Paulo	5–13	I (8)	50–53 (AMF)[b]	Duarte et al. (2018)
Brazil	Atlantic rain Forest — Cerrado/Riparian vegetation	27	–	NI	Pagano and Cabello (2012)
Brazil	Araucaria Forest/*Araucaria angustifolia*	18	NI	NI	Patreze et al. (2009)
		13	I (5)	(42.5% AMF)[b]	Moreira et al. (2016)
Brazil	Pantanal/Semi-deciduous forest, Cerrado, Cerradão, grasslands	19–25	I (18)	NI	Gomide et al. (2014)

(continued)

Table 1.2 (continued)

Country	Biome/Vegetation type/State/Region	AMF species/ Morphotypes	Unidentified species	% Root colonization	Reference
Brazil	Cerrado/Natural Cerrado forest, Northern	29–33 (57)[ω]	I (6)	NI	Pontes et al. (2017)
Brazil	Cerrado/Murundu fields[a], Goiás	24–27	I (3)	NI	Assis et al. (2014)
Brazil	Cerrado/Highland fields[a], Minas Gerais	51–75	I (18)	NI	Oki et al. (2016)
Brazil	Cerrado/Ferruginous fields, Iron mining areas[a], Minas Gerais	24	I (7)	NI	Teixeira et al. (2017)
Brazil	Cerrado†	NI	NI	NI	Lemes et al. (2016)
Brazil	Pantanal/Seasonal flooding – Grasslands, Cerrado	21–37	I (18)	NI	Gomide et al. (2014)
Brazil	Caatinga/Dry forest, moist forest	27–42 (50)[ω]	I (14)	NI	Da Silva et al. (2014)
Brazil	Caatinga, Pernambuco	16	–	NI	Mello et al. (2012)
Brazil	Caatinga/Deciduous Forest	13–15	I (2)	~10–30 (AMF)[b]	Pagano et al. (2013)
Brazil	Caatinga/Carrasco	16–18	I (2)	~20–50 (AMF)[b]	Pagano et al. (2013)
	Woody caatinga	9–23	I (3)	~10–40 (AMF)[b]	Pagano et al. (2013)
Brazil	Coastal ecosystems Restinga	25	I (2)	(36% AMF)[b]	Stürmer et al. (2013)
Brazil	Mangrove forest, Restinga forest	17–22	I (3)	~2–74 (AMF) (36.75% AMF)[b]	Silva et al. (2017)
Chile†	Chilean temperate rainforests	59	I	NI	Marín et al. (2017)
Chile†	Evergreen primary forest, secondary forest and natural grassland	6–29 (58)[ω]	I (7)	NI	Castillo et al. (2016)
Colombia	Orchidaceae	NI	–	I (OMF)[b]	Otero et al. (2013)

References: [a]Reported as AMF hotspots sites; † Checklist or review; ω = maximum number of AMF reported; I = informed, number of AMF unidentified species between parentheses; NI = not informed; [b] = mean value of root colonization % and/or type of fungal symbionts involved are showed between parentheses; AMF = arbuscular mycorrhizal fungi; DSE = dark septate endophytes; OMF = orchid mycorrhizal fungi

Edaphic factors such as the soil pH is related with species richness and with the diversity index. However, the height above sea level can also modulate the AMF community composition which conducts to a diverse distribution of species in patches with little influence of the type of cultural traits showing the importance of developing specific biofertilizers for crops that contain AMF naturally adapted to the different characteristics of the multiple soil types present in agriculture (Mahecha-Vásquez 2017).

More recently, Araujo et al. (2018) have checked for distribution of AMF along the soil profile in pure and mixed *Eucalyptus grandis* and *Acacia mangium* plantations in Brazil. *Acacia mangium* in the mixed cultivation system stimulates greater root colonization rates in *Eucalyptus grandis* plants in the 0–20 cm and 20–50 cm layers. AMF spores are present in all the management systems studied, in all soil layers down to 800 cm of depth. The same applies to AMF root colonization. Evaluation of sampled spores identified the following six AMF genera: *Acaulospora, Gigaspora, Glomus, Intraornatospora, Scutellospora* and *Racocetra*, distributed among 16 species.

As it was pointed out before, mycorrhizal fungi and its associations have been scantly studied in protected areas in SA. The few conservation units studied are inhabited by a high number of unidentified morphotypes of AMF. Thus, to resolve this gap in the AMF species recorded, two main strategies can be performed. One option is to propagate the new species using the roots transformed by *Agrobacterium rhizogenes* which are also effective as inocula which generally utilized carrot and is used as experimental model systems for research purposes (see Giovannetti and Avio 2002). But these inoculation procedures are highly expensive and only utilized in agriculture of high value products.

In vivo cultures of species from different regions are maintained in ex-situ collections worldwide (Giovannetti and Avio 2002). For that purpose, the spores are inoculated nearby the roots of a host plant cultivated in soil, sand, expanded clay, peat or other substrates (after sterilization by steam, fumigation or irradiation). In general, new spores are produced in the pot cultures 3 months after inoculation (see Giovannetti and Avio 2002). Observations under microscopes of stained fungal structures in the roots can confirm the percentage of mycorrhizal colonization.

The selection of appropriate fungal endophytes plays a fundamental role in preventing growth after transplant (Requena et al. 2001) and plant micropropagation, which can presently benefit from AM biotechnology in SA. The applications of mycorrhizas in restoration and environmental issues are still incipient. AMF inoculant for farm application requires large-scale multiplication fungi, which is generally carried out in substrate-based or in vitro systems (Ijdo et al. 2011). Commercial inocula exist, but often these inoculants do not work nor contribute satisfactorily, especially under field conditions (Wetzel et al. 2014). Infective propagules of AMF (spores, hypha and colonized roots) can be used as inocula (Sieverding 1991). Some experts have tested the production of AMF inoculum together with biofertilizers using the on-farm method in SA (Czerniak and Stürmer 2015). Moreover, those authors have confirmed the probable use of Gigasporaceae as inoculants.

In addition to the application of these techniques to complete the knowledge in the protected areas previously sampled, it becomes necessary to reinforce the research of the fungal symbiosis and the mycorrhizal fungi in the rest of the protected areas of SA. This would be possible with the joint effort not only of the South American researchers, but also with the reinforcement of policies of joint work with groups of countries more developed in the subject, and with statal policies that promote this activity in protected areas by financially supporting these topics.

1.5 Conclusion

In this chapter, the needs for more information to understand native ecosystems in South America under different vegetation types have been highlighted. The examination and use of arbuscular mycorrhizas in different biomes from SA have been mentioned. Throughout the chapter, the applications or economic importance of mycorrhizas have been shown as still incipient. Morphological identification procedure of AMF continues to be important, although it requires a specific training and experience.

Technology for commercial mycorrhizal inoculum has been developed in a few countries, mostly in Brazil. Finally, native ecosystems present high AMF diversity; however, rain forests and Cerrado are less studied ecosystems. Further, two important research gap areas have been detected in SA: first, the mycorrhizal and fungal endophyte associations studies and its followings, along the time in Protected Parks and Reserves, and second, the researches on mycorrhizal ecosystem services provision, considering that among papers published between 1990–2015 on ecosystem services, the largest number is from Europe (38%), SA represents only 6% of the total number. Consequently, further research is necessary on this field, especially regarding the new species of mycorrhizas and its functions in the ecosystem.

Acknowledgements Mónica A. Lugo and Marcela C. Pagano wish to express their deepest gratitude to all the researchers who are the authors and co-authors of the chapters of this book and whose scientific contributions have made this book possible, as well as to all the researchers whose contributions were a starting point for this book, and which have complemented and broadened the knowledge of mycorrhizal fungi in South America. In addition, the authors are grateful for the help provided by the critical reading of English to Matilde M. Crespo and to Dr. Lucía V. Risio for her collaboration in the preparation of the Table on protected areas of SA.

References

Adhikari K, Hartemink AE (2016) Linking soils to ecosystem services — A global review. Geoderma 262:101–111

Aidar MPM, Carrenho R, Joly CA (2004) Aspects of arbuscular mycorrhizal fungi in an Atlantic Forest chronosequence in Parque Estadual Turístico do Alto Ribeira (PETAR), SP. Biota Neotropica 4:1–15

Amano T, Sutherland WJ (2013) Four barriers to the global understanding of biodiversity wealth, language, geographical location and security. Proc R Soc B 2013 280: 2012–2649

Antoninka A, Reich PB, Collins Johnson N (2011) Seven years of carbon dioxide enrichment, nitrogen fertilization and plant diversity influence arbuscular mycorrhizal fungi in a grassland ecosystem. New Phytol 192:200–214

Araujo APP, Santana MC, Bonfim JA, Mescolotti DL, Cardoso EJBN (2018) Digging deeper to study the distribution of mycorrhizal arbuscular fungi along the soil profile in pure and mixed *Eucalyptus grandis* and *Acacia mangium* plantations. Appl Soil Ecol in press

Assis PCR, Saggin Júnior OJ, Paulino HB et al. (2014) Arbuscular mycorrhizal fungi in "murundu fields" after conversion to farm systems in the cerrado. Rev Bras Ciênc Solo 38:1703–1711

Averill C, Turner BL, Finzi AC (2014) Mycorrhiza-mediated competition between plants and decomposers drives soil carbon storage. Nature 505: 543–545

Barros F, Vinhos F, Rodrigues VT, Barberena FFVA, Fraga CN, Pessoa EM (2012) *Orchidaceae*. In: Forzza RC et al. (eds). Lista de Espécies da Flora do Brasil. Jardim Botânico do Rio de Janeiro

Błaszkowski J (2012) Glomeromycota. W. Szafer Institute of Botany, Polish Academy of Sciences

Błaszkowski J, Kozłowska A, Niezgoda P, Goto BT, Dalpé Y (2018) A new genus, *Oehlia* with *Oehlia diaphana* comb. nov. and an emended description of *Rhizoglomus vesiculiferum* comb. nov. in the Glomeromycotina. Nova Hedwigia 107:501–518

Bradford MA (2014) Good dirt with good friends. Nature 505: 486–487

Brundrett MC, Tedersoo L (2018) Evolutionary history of mycorrhizal symbioses and global host plant diversity. New Phytol 220:1108–1115

Bueno CG, Marín C, Silva-Flores P, Aguilera P, Godoy R (2017) Think globally, research locally: emerging opportunities for mycorrhizal research in South America. New Phytol 215(4):1306–1309

Cabrera AL (1976) Regiones Fitogeográficas Argentinas. In: Enciclopedia Argentina de Agricultura y Ganadería, 2nd ed. ACME S.A.C.I., Argentina, 80 p

Cabrera AL, Willink A (1980) Biogeografía de América Latina. OEA, Washington, DC (In Spanish)

Cardoso da Silva JM and Bates JM (2002) Biogeographic Patterns and Conservation in the South American Cerrado: A tropical savanna hotspot. BioScience 52:225–233

Carvalho F, Souza FA, Carrenho R, Moreira FMS, Jesus EC, Fernandes GW (2012) The mosaic of habitats in the high-altitude Brazilian rupestrian fields is a hotspot for arbuscular mycorrhizal fungi. Appl Soil Ecol 52:9–19

Castillo CG, Borie F, Oehl F, Sieverding E (2016) Arbuscular mycorrhizal fungi biodiversity: prospecting in Southern-Central zone of Chile. A review. J Soil Science Plant Nutrition 16 (2):400–422

Czerniak MJ, Stürmer SL (2015) On-farm production of mycorrhizal inoculum using residues from the forestry industry. Rev Bras Ciênc Solo 38(6):1712–1721

Corrales A, Henkel TW, Smith ME (2018) Ectomycorrhizal associations in the tropics – biogeography, diversity patterns and ecosystem roles. New Phytol online, https://doi.org/10.1111/nph.15151

da Silva IR, de Mello CMA, Neto RAF, da Silva DKA, de Melo AL, Oehl F, Maia LC (2014) Diversity of arbuscular mycorrhizal fungi along an environmental gradient in the Brazilian semiarid. Appl Soil Ecol 84:166–175

de Souza TAF, Rodriguez-Echeverría S, de Andrade LA, Freitas H (2016) Arbuscular mycorrhizal fungi in *Mimosa tenuiflora* (Willd.) Poir from Brazilian semi-arid. Brazilian J Microbiol 47 (2):359–366

Davison J, Moora M, Öpik M, Adholeya A, Ainsaar L, Bâ A, Burla S, Diedhiou AG, Hiiesalu I, Jairus T, Johnson NC, Kane A, Koorem K, Kochar M, Ndiaye C, Pärtel M, Reier Ü, Saks Ü, Singh R, Vasar M, Zobel M (2015) Global assessment of arbuscular mycorrhizal fungus diversity reveals very low endemism. Science 349: 6251

Detmann KSC, Leite TS, Oliveira Neto RR (2018) Arbuscular mycorrhizae and absence of cluster roots in the Brazilian Proteaceae *Roupala montana* Aubl. Symbiosis https://doi.org/10.1007/s13199-018-0581-0

Duarte LM, Bertini SCB, Stürmer SL, Lambais MR, Azevedo LCB (2018) Arbuscular mycorrhizal fungal communities in soils under three phytophysiognomies of the Brazilian Atlantic Forest. Acta Bot Brasilica. https://doi.org/10.1590/0102-33062018abb0236

Echeverría-Londoño S, Enquist BJ, Neves DM, Violle C et al. (2018) Plant Functional Diversity and the Biogeography of Biomes in North and South America. Front. Ecol. Evol. 6: 11–12, https://doi.org/10.3389/fevo.2018.00219

Flores R, Godoy R, Palfner G (1997) Morfo – anatomía de la ectomicorriza *Cenococcum geophilum* Fr. en *Nothofagus alessandrii* Esp. Gayana Bot 54(2):157–167

Founoune H, Duponnois R, Bâd AM, El Bouamib F (2002). Influence of the dual arbuscular endomycorrhizal / ectomycorrhizal symbiosis on the growth of Acacia holosericea (A. Cunn. ex G. Don) in glasshouse conditions. Ann For Scie 59: 93–98

Freitas RO, Buscardo E, Nagy L, Maciel ABS, Carrenho R, Luizão RCC (2014) Arbuscular mycorrhizal fungal communities along a pedo-hydrological gradient in a Central Amazonian terra firme forest. Mycorrhiza 24:21–32

Garrido N (1988) Agaricales sl und ihre Mykorrhizen in den *Nothofagus*-Wäldern Mittelchiles

Geml J, Pastor N, Fernández L, Pacheco S, Semenova T, Becerra AG, Wicaksono CY, Nouhra ER (2014) Large-scale fungal diversity assessment in the Andean Yungas forests reveals strong community turnover among forest types along an altitudinal gradient. Molecular Ecology 23:2452–2472

Giovannetti M, Avio L (2002) Biotechnology of Arbuscular Mycorrhizas. Appl Mycol Biotechnol 2:275–310

Godoy R, Palfner G (1997) Ectomicorrizas en *Nothofagus alpina* (Poepp. & Endl.) Oerst. y *N. dombeyi* (Mirb. Oerst.) del sur de Chile. Boletín Micológico 12(1–2):55–61

Gomide PHO, Silva MLN, Soares CRFS et al. (2014) Arbuscular mycorrhizal fungi in vegetation types in the Pantanal of Nhecolândia, Mato Grosso do Sul, Brazil. Rev Bras Ciênc Solo 38:1114–1127

Goto BT, da Silva GA, Yano-Melo AM, Maia LC (2010) Checklist of the arbuscular mycorrhizal fungi (Glomeromycota) in the Brazilian semiarid. Mycotaxon 113:251–254

Goto BT, Silva GA, Assis DMA, Silva DKA, Souza RG, Ferreira ACA, Jobim K, Mello CMA, Vieira HEE, Maia LC, Oehl F. (2012) *Intraornatosporaceae* (*Gigasporales*), a new family with two new genera and two new species. Mycotaxon 119:117–132

Grilli G, Urcelay C, Galetto L, Davison J, Vasar M, Saks Ü, Jairus T, Öpik M (2015) The composition of arbuscular mycorrhizal fungal communities in the roots of a ruderal forb is not related to the forest fragmentation process. Environ Microbiol 17:2709–2720

Grupe AC, Vasco-Palacios AM, Boekhout T, Smith M, Henkel T (2016) *Sarcodon* in the Neotropics: four new species from Colombia and a key to selected species. Mycologia 108:791–805, https://doi.org/10.3852/15-254

Hart MM, Reader RJ (2002) Taxonomic basis for variation in the colonization strategy of arbuscular mycorrhizal fungi. New Phytol 153(2):335–344

Hawksworth DL (2001) The magnitude of fungal diversity: the 1.5 million species estimate revisited. Mycol Res 105:1422–1432

Hawksworth DL, Lücking R (2017) Fungal diversity revisited: 2.2 to 3.8 million species. Microb Spectrum 5(4):FUNK-0052-2016. https://doi.org/10.1128/microbiolspec.FUNK-0052-2016.

Henkel TW, Aime MC, Chin MML, Miller S, Vilgalys R, Smith M (2012) Ectomycorrhizal fungal sporocarp diversity and discovery of new taxa in *Dicymbe* monodominant forests of the Guiana Shield. Biodiversity and Conservation 21:2195–2220

Ijdo M, Cranenbrouck S, Declerck S (2011) Methods for large-scale production of AM fungi: past, present, and future. Mycorrhiza 21:1–16

Jobim K, Santos Oliveira BI, Goto BT (2016) Checklist of the Glomeromycota in the Brazilian Savanna. Mycotaxon 131:255

Jobim K, Vista XM, Goto BT (2018) Updates on the knowledge of arbuscular mycorrhizal fungi (Glomeromycotina) in the Atlantic Forest biome – an example of very high species richness in Brazilian biomes. Mycotaxon 133 (1)

Joly CA, Assis MA, Bernacci LC, et al. (2012) Florística e fitossociologia em parcelas permanentes da Mata Atlântica do sudeste do Brasil ao longo de um gradiente altitudinal. Biota Neotropica 12:123–145

Kai W, Zhiwei Z (2006) Occurrence of arbuscular mycorrhizas and dark septate endophytes in hydrophytes from lakes and streams in southwest China. Int Rev Hydrobiol 91:29–37

Kennedy P, Garibay-Orijel R, Higgins L, Angeles-Arguiz R (2011) Ectomycorrhizal fungi in Mexican *Alnus* forests support the host co-migration hypothesis and continental-scale patterns in phylogeography. Mycorrhiza 21(6):559–568

Kivlin SN, Emery SM, Rudgers JA (2013) Fungal symbionts alter plant responses to global change. Am J Bot 100 (7):1445–1457

Kumar P, Pagano M, O'Donovan A (2017) Mycosphere Essay 18: Biotechnological advances of beneficial fungi for plants. Mycosphere 8(3):445–455, https://doi.org/10.5943/mycosphere/8/3/6

Lekberg Y, Gibbons SM, Rosendahl S, Ramsey PW et al. (2013) Severe plant invasions can increase mycorrhizal fungal abundance and diversity. ISME J 7(7):1424–1433

Lekberg Y, Waller LP (2016) What drives differences in arbuscular mycorrhizal fungal communities among plant species? Fungal Ecol 24 B: 135–138

Lemes EM, Rodrigues GI, de Paula ADM, de Lima DT, Torres JLR (2016) Mycorrhizal associations in Cerrado soils: A review of benefits and management. Australian J Crop Science 10(11):1504–1510

Lugo MA, Molina MG, Crespo EM (2009) Arbuscular mycorrhizas and dark septate endophytes in bromeliads from South American arid environment. Symbiosis 47:17–21

Lugo MA, Reinhart KO, Menoyo E, Crespo EM, Urcelay C (2015) Plant functional traits and phylogenetic relatedness explain variation in associations with root fungal endophytes in an extreme arid environment. Mycorrhiza 25:85–95

Lugo MA et al. (2018) MICODIF: Manual de metodologías para el trabajo con hongos y sus simbiosis /– 1ª ed. – San Luis: Nueva Editorial Universitaria – U.N.S.L.

Madriñán S, Cortés AJ, Richardson JE (2013) Páramo is the world's fastest evolving and coolest biodiversity hotspot. Frontiers in Genetics 4:1–6

Mahecha-Vásquez G, Sierra S, Posada R (2017) Diversity indices using arbuscular mycorrhizal fungi to evaluate the soil state in banana crops in Colombia Appl Soil Ecol 109:32–39

Maherali H, Klironomos JN (2007) Influence of phylogeny on fungal community assembly and ecosystem functioning. Science 316:1746–1748

Marín C, Aguilera P, Oehl F, Godoy R (2017) Factors affecting arbuscular mycorrhizal fungi of Chilean temperate rainforests. J Soil Science Plant Nutrit 17 (4):966–984

Marinho F, Silva GA, Ferreira ACA, Veras JSN, Sousa NMF, Goto BT, Maia L, Oehl F (2014) *Bulbospora minima*, a new genus and a new species in the Glomeromycetes from semi-arid Northeast Brazil. Sydowia 66:313–323

Marins JF, Rosilaine Carrenho R, Sidinei Magela Thomaz SM (2009) Occurrence and coexistence of arbuscular mycorrhizal fungi and dark septate fungi in aquatic macrophytes in a tropical river–floodplain system. Aquatic Botany 91(1):13–19

Meadow JF, Zabinski CA (2012) Linking symbiont community structures in a model arbuscular mycorrhizal system. New Phytol 194:800–809

Mello CMA, da Silva IR, Pontes JS, Goto BT, da Silva GA, Maia LC (2012) Diversity of arbuscular mycorrizal fungi in an area of Caatinga, PE, Brazil. Acta Botanica Brasilica 26(4):938–943

Moreira M, Zucchi MI, Gomes JE, Tsai SM, Alves-Pereira A, Cardoso EJBN (2016) Araucaria angustifolia aboveground roots presented high arbuscular mycorrhizal fungal colonization and diversity in the Brazilian Atlantic Forest. Pedosphere 26(4):561–566

Mosquera-Espinosa AT, Bayman P, Otero JT (2010) *Ceratobasidium* como hongo micorrízico de orquídeas en Colombia. Acta Agronom 59:316–326

Mosquera-Espinosa AT, Bayman PP, Prado GA, Gomez-Carabali A, Otero JT (2013) Pathogenicity of orchid mycorrhizal fungi (*Ceratobasidium* sp.) on rice and biocontrol of *Rhizoctonia solani* sheath blight. Mycologia 105:141–150

Moyersoen B (2006) *Pakaraimaea dipterocarpacea* is ectomycorrhizal, indicating an ancient Gondwanaland origin for the ectomycorrhizal habit in Dipterocarpaceae. New Phytol 172 (4):589–794, https://doi.org/10.1111/j.1469-8137.2006.01860.x

Moyersoen B, Fitter A, Alexander I (1998a) Spatial distribution of ectomycorrhizas and arbuscular mycorrhizas in Korup National Park rainforest, Cameroon, in relation to edaphic parameters. New Phytol 139(2):311–320

Moyersoen B, Alexander I, Fitter A (1998b) Phosphorus nutrition of ectomycorrhizal and arbuscular mycorrhizal tree seedlings from a lowland tropical rain forest in Korup National Park, Cameroon. J Trop Ecol 14(1):47–61

Moyersoen B, Becker P, Alexander I (2001). Are ectomycorrhizas more abundant than arbuscular mycorrhizas in tropical heath forests? New Phytol (2001) 150:591–599

Myers N, Mittermeier RA, Mittermeier CG, da Fonseca GAB, Kent J (2000) Biodiversity hotspots for conservation priorities. Nature 403:853–858

Nouhra E, Urcelay C, Longo S, Tedersoo L (2013) Ectomycorrhizal fungal communities associated to *Nothofagus* species in Northern Patagonia Mycorrhiza 23:487–496

Oehl F, Sieverding E, Ineichen K, Ris EA, Boller T, Wiemken A (2005) Community structure of arbuscular mycorrhizal fungi at different soil depths in extensively and intensively managed agroecosystems. New Phytol 165:273–283

Oehl F, Sieverding E, Palenzuela J, Ineichen K, Silva GA (2011a) Advances in *Glomeromycota* taxonomy and classification. IMA Fungus 2:191–199

Oehl F, Silva DKA, Maia LC, Sousa NMF, Vieira HEE, Silva GA (2011b) *Orbispora* gen. nov., ancestral in the *Scutellosporaceae* (*Glomeromycetes*). Mycotaxon 116:161–169

Oehl F, Sánchez–Castro I, Palenzuela J, Silva GA (2014) *Palaeospora spainii*, a new arbuscular mycorrhizal fungus from Swiss agricultural soils. Nova Hedwigia 101:1–2

Oeser RA, Stroncik N, Moskwa LM, Bernhard N, Schaller M, Canessa R et al (2018) Chemistry and microbiology of the Critical Zone along a steep climate and vegetation gradient in the Chilean Coastal Cordillera. Catena 170:183–203

Oki Y, Goto BT, Jobim K, Rosa LH, Costa Ferreira M, Silva Coutinho E, de Azevedo Xavier JH, Fernanda Carvalho, de Souza Moreira FM, de Souza FA, Louro Berbara RL, Fernandes GW (2016) Arbuscular Mycorrhiza and Endophytic Fungi in Ruspestrian Grasslands. In: GW Fernandes (ed) Ecology and Conservation of Mountain top Grasslands in Brazil, Springer, Switzerland. https://doi.org/10.1007/978-3-319-29808-5-8, p 157–179

Onguene N, Kuyper T (2002) Importance of the ectomycorrhizal network for seedling survival and ectomycorrhiza formation in rain forests of south Cameroon. Mycorrhiza 12: 13–17

Otero JT, Bayman P (2009) Symbiotic vs. asymbiotic seed germination in epiphytic orchids. Acta Agronom 58:270–276

Otero JT, Ackerman JD, Bayman P (2002) Diversity and host specificity of mycorrhizal fungi from tropical orchids. Amer J Bot 89:1852–1858

Otero JT, Ackerman JD, Bayman P (2004) Differences in mycorrhizal specificity between two tropical orchids. Molec Ecol 13:2393–2404

Otero JT, Bayman P, Ackerman JD (2005) Individual variation in plant and fungus for mycorrhizal seed germination in an epiphytic orchid. Evol Eco. 19:29–43

Otero JT, Mosquera AT, Flanaga NS (2013) Tropical orchid mycorrhizae: potential applications in orchid conservation, commercialization, and beyond. Lankesteriana 13(1–2):57–63

Oyarzabal M, Clavijo J, Oakley L, Biganzoli F, Tognetti P, Barberis I, Maturo HM, Aragón R, Campanello PI, Prado D, Oesterheld M, León RJ.C (2018) Vegetation units of Argentina. Ecología Austral 28:040–063

Öpik M, Vanatoa A, Vanatoa E, Moora M, Davison J, Kalwij JM, Reier U, Zobel M (2010) The online database MaarjAM reveals global and ecosystemic distribution patterns in arbuscular mycorrhizal fungi (Glomeromycota). New Phytol 188(1) 223–241

Öpik, M., J. Davison, M. Moora, M. Zobel (2014) DNA-based detection and identification of Glomeromycota: the virtual taxonomy of environmental sequences Botany 92:135–147

Pagano MC, Cabello MN (2012) Mycorrhizas in natural and restored riparian zones. In: Pagano MC, editor Mycorrhiza: Occurrence and Role in Natural and Restored Environments. Nova Science Publishers, Hauppauge, NY. ISBN: 978-1-61209-226-3.
Pagano MC, Utida MK, Gomes EA, Marriel IE, Cabello MN, Scotti MR (2011) Plant-type dependent changes in arbuscular mycorrhizal communities as soil quality indicator in semi-arid Brazil. Ecological Indicators 11 (2):643–650
Pagano MC, Zandavalli RB, Araújo FS (2013) Biodiversity of arbuscular mycorrhizas in three vegetational types from the semiarid of Ceará State, Brazil. Applied Soil Ecology 67: 37–46
Pagano MC, Oehl F, Silva GA, Maia LC, Silva DK, Cabello MN (2016) Advances in Arbuscular mycorrhizal taxonomy. In: Pagano MC (ed) Recent Advances on Mycorrhizal Fungi. Springer Nature, Switzerland, p 15–21
Pagano MC, Correa EJA, Duarte NF, Yelikbayev B, O'Donovan A, Gupta VK (2017) Advances in Eco-Efficient Agriculture: The Plant-Soil Mycobiome. Agriculture 7:14 https://doi.org/10.3390/agriculture7020014
Palfner G (2001) Taxonomische Studien an Ektomykorrhizen aus den-*Nothofagus*-Wäldern Mittelsüdchiles
Palfner G, Godoy R (1996) "Nothofagirhiza vinicolor" + *Nothofagus pumilio* (Poepp. et Endl.) Kraser. Descriptions of Ectomycorrhizae 1, 65–70
Pärtel M, Öpik M, Moora M, Tedersoo L, Szava-Kovats R, Rosendahl S, Rillig MC, Lekberg Y, Kreft H, Helgason T, Eriksson O, Davison J, de Bello F, Caruso T, Zobel M (2017) Historical biome distribution and recent human disturbance shape the diversity of arbuscular mycorrhizal fungi. New Phytol 216 (1):227–238
Patreze CM, De Paulo EN, Martinelli AP, Cardoso EJB, Tsai SM (2009) Characterization of fungal soil communities by F-RISA and arbuscular mycorrhizal fungi from *Araucaria angustifolia* forest soils after replanting and wildfire disturbances. Symbiosis 48(1–3):164–172
Pereira MC, Moreira Vieira N, Tótola MR, Kasuya MCM (2011) Total fatty acid composition in the characterization and identification of orchid mycorrhizal fungi *Epulorhiza* spp. Rev Bras Ciênc Solo 35(4):1159–1165
Pereira CMR, da Silva DKA, Ferreira ACA, Goto BT, Maia LC (2014) Diversity of arbuscular mycorrhizal fungi in Atlantic forest areas under different land uses. Agriculture, Ecosystems and Environment 185:245–252
Pereira CMR, Silva DKA, Goto BT, Rosendahl S, Maia L (2018) Management practices may led to loss of arbuscular mycorrhizal fungal diversity in protected areas of the Brazilian Atlantic Forest. Fungal Ecol 34:50–58
Pontes JS, Oehl F, Pereira CD, Machado CTT, Coyne D, da Silva DKA, Maia LC (2017) Diversity of arbuscular mycorrhizal fungi in the Brazilian's Cerrado and in soybean under conservation and conventional tillage Applied Soil Ecology:178–189
Radhika KP, Rodrigues BF (2007) Arbuscular Mycorrhizae in association with aquatic and marshy plant species in Goa, India. Aquatic Botany 86(3):291–294
Redecker D, Schüßler A, Stockinger H, Stürmer SL, Morton JB, Walker C (2013) An evidence-based consensus for the classification of arbuscular mycorrhizal fungi (Glomeromycota). Mycorrhiza 23:515–531
Requena N, Perez-Solis E, Azcón-Aguilar C, Jeffries P, Barea JM (2001) Management of indigenous plant-microbe symbioses aids restoration of desertified. Appl Environm Microbiol 67:495–498
Rivero Mega MS, Crespo EM, Molina MG, Lugo MA (2014) Diversidad diferencial de esporas de Glomeromycota en la rizosfera de bromeliáceas nativas del Parque Nacional Sierra de las Quijadas (San Luis, Argentina). Bol Soc Argent Bot 49:317–325
Riviere T, Diedhiou A, Diabate M, Senthilarasu G, Natarajan K, Verbeken A, Buyck B, Dreyfus B, Bena G, Ba A (2007) Genetic diversity of ectomycorrhizal Basidiomycetes from African and Indian tropical rain forests. Mycorrhiza 17:415–428
Schalamuk S, Druille M, Cabello MN (2013) Capítulo 3. Hongos formadores de micorrizas arbusculares: Influencia de las prácticas agronómicas sobre su diversidad y dinámica de colonización.

Arbuscular mycorrhizal fungi: Influence of agronomic practices on diversity and dynamics of colonization. In (García de Salomone E, Vázquez S, Penna C, Cassan F (eds) Rizosfera, biodiversidad y agricultura sustentable, TIRBAS, Río Cuarto, Argentina, pp 47–71

Schueffler A, Anke T (2014) Fungal natural products in research and development. Nat Prod Rep 31:1425–1448

Schüßler A, Walker C (2010) The Glomeromycota: a species list with new families and new genera. Gloucester, in libraries at The Royal Botanic Garden Edinburgh, The Royal Botanic Garden Kew, Botanische Staatssammlung Munich and Oregon State University

Senés-Guerrero C, Schüßler A (2016) A conserved arbuscular mycorrhizal fungal core-species community colonizes potato roots in the Andes. Fungal Divers 77:317–333

Sieverding E (1991) Vesicular-arbuscular mycorrhiza management in tropical agrosystems. Eschborn: Deutsche Gesellschaft für Technische Zusammenarbeit (GTZ) GmnBH, Esch boran

Silva IR, de Mello CMA, Ferreira Neto RA, da Silva DKA, de Mello AL, Oehl F, Maia LC (2014) Diversity of arbuscular mycorrhizal fungi along an environmental gradient in the Brazilian semiarid. Appl. Soil Ecol. 84:166–175

Silva IR, da Silva, DKA, de Souza FA, Oehl F, Maia LC (2017) Changes in arbuscular mycorrhizal fungal communities along a river delta island in northeastern Brazil. Acta Oecologica 79:8–17

Smith ME, Henkel TW, Uehling JK, Fremier AK, Clarke HD, Vilgalys R. (2013) The Ectomycorrhizal Fungal Community in a Neotropical Forest Dominated by the Endemic Dipterocarp *Pakaraimaea dipterocarpacea*. PLoS ONE 8(1): e55160

Smith ME, Henkel TW, Williams GC, Aime MC, Fremier AK, Vilgalys R (2017) Investigating niche partitioning of ectomycorrhizal fungi in specialized rooting zones of the monodominant leguminous tree *Dicymbe corymbosa*. New Phytol 215(1):443–453

Song H, Zhong Z, Yang W, Chen Q (2011) Analysis of the activities of protective enzymes in *Bidens pilosa* L. inoculated with *Glomus mosseae* under drought stress. Shengtai Xuebao/ Acta Ecologica Sinica 31:2471–2477

Sosa-Hernández MA, Roya J, Hempel S, Kautz T, Köpke U, Uksad M, Schloter M, Caruso T, Rillig MC (2018) Subsoil arbuscular mycorrhizal fungal communities in arable soil differ from those in topsoil Soil Biol Biochem 117:83–86

Soteras F, Coutinho Moreira B, Grilli G, Pastor N, Carneiro Mendes F, Ruela Mendes D, Renison D, Megumi Kasuya MC, de Souza FA, Becerra A (2016) Arbuscular mycorrhizal fungal diversity in rhizosphere spores versus roots of an endangered endemic tree from Argentina: is fungal diversity similar among forest disturbance types? Appl Soil Ecol 98:272–277

Spatafora JW, Chang Y, Benny GL, Lazarus K, Smith ME, Berbee ML, Bonito G, Corradi N, Grigoriev I, Gryganskyi A, James TY, O'Donnell K, Roberson RW, Taylor TN, Uehling J, Vilgalys R, White MM, Stajich JE (2016) A phylum-level phylogenetic classification of zygomycete fungi based on genome-scale data. Mycologia 108:1028–1046

Stürmer SL (2012) A history of the taxonomy and systematics of arbuscular mycorrhizal fungi belonging to the phylum Glomeromycota. Mycorrhiza 22:247–258

Stürmer SL, Stürmer R, Pasqualini D (2013) Taxonomic Diversity and community structure of arbuscular mycorrhizal fungi (Phylum *Glomeromycota*) in three maritime sand dunes in Santa Catarina State, South Brazil. Fungal Ecology 6:27–36

Stürmer SL, Bever JD, Morton JB (2018) Biogeography of arbuscular mycorrhizal fungi (Glomeromycota): a phylogenetic perspective on species distribution patterns. Mycorrhiza 28:587–603

Tedersoo L, May TW, Smith ME (2010) Ectomycorrhizal lifestyle in fungi: Global diversity, distribution, and evolution of phylogenetic lineages. Mycorhiza 20 (4):217–263

Tedersoo L, Suvi T, Beaver K, Kõljalg U (2007) Ectomycorrhizal fungi of the Seychelles: diversity patterns and host shifts from the native *Vateriopsis seychellarum* (Dipterocarpaceae) and *Intsia bijuga* (Caesalpiniaceae) to the introduced *Eucalyptus robusta* (Myrtaceae), but not *Pinus caribea* (Pinaceae). New Phytol 175:321–333

Tedersoo L, Bahram M, Toots M, Diédhiou A, Henkel t, Kjøller R, Morris MH, Nara K, Nouhra E, Peay K, Põlme, Ryberg M, Smith ME, Kõljalg U (2012) Towards global patterns in the diversity and community structure of ectomycorrhizal fungi. Mol Ecol 21:4160–4170

Tedersoo L, Bahram M, Põlme S, Kõljalg U, Yorou NS, Wijesundera R, Ruiz L. V, Vasco-Palacios AM, Thu PQ, Suija A, Smith ME, Sharp C, Saluveer E, Saitta A, Rosas M, Riit T, Ratkowsky D, Pritsch K, Põldmaa K, Piepenbring M, Phosri C, Peterson M, Parts K, Pärtel K, Otsing E, Nouhra E, Njouonkou AL, Nilsson RH, Morgado LN, Mayor J, May TW, Majuakim L, Lodge DJ, Lee SS, Larsson K-H, Kohout P, Hosaka K, Hiiesalu I, Henkel TW, Harend H, Guo L, Greslebin A, Grelet G, Geml J, Gates G, Dunstan W, Dunk, Drenkhan R, Dearnaley J, De Kesel A, Dang T, Chen X, Buegger F, Brearley FQ, Bonito G, Anslan S, Abell S, Abarenkov K (2014) Global diversity and geography of soil fungi. Science 346(6213):1256688–1–10

Teixeira AFS, Kemmelmeier K, Marascalchi MN, Stürmer SL, Carneiro MAC, Moreira FMS (2017) Arbuscular mycorrhizal fungal communities in an iron mining area and its surroundings: Inoculum potential, density, and diversity of spores related to soil properties. Ciência e Agrotecnologia, 41(5):511–525

Truong C, Mujic A, Healy R, Kuhar F, Furci G, Torres D, Niskanen T, Sandoval-Leiva P, Fernández N, Escobar J, Moretto A, Palfner G, Pfister D, Nouhra E, Swenie R, Sánchez-García M, Matheny B, Smith M (2017). How to know the fungi: combining field inventories and DNA-barcoding to document fungal diversity. New Phytol 214:913–919

Turrini A, Giovanetti M (2012) Arbuscular mycorrhizal fungi in national parks, nature reserves and protected areas worldwide: a strategic perspective for their in situ conservation. Mycorrhiza 22:81–97, https://doi.org/10.1007/s00572-011-0419-6

Ulloa Ulloa C, Acevedo-Rodríguez P, Beck S, Belgrano MJ, Bernal R, Berry PE, Brako L, Celis M, Davidse G, Forzza RC, S. Gradstein R, Hokche O, León B, León-Yánez S, Magill RE, Neill DA, Nee M, Raven PH, Stimmel H, Strong MT, Villaseñor JL, Zarucchi JL, Zuloaga FO, Jørgensen PM (2017) An integrated assessment of the vascular plant species of the Americas. Science 358:1614–1617

Vasco-Palacios AM (2016) Ectomycorrhizal fungi in Amazonian tropical forests in Colombia. Doctoral Thesis, Utrecht University. Editorial Panamericana, Bogotá. 201 p

Vasco-Palacios AM, López-Quintero C, Franco-Molano AE, Boekhou, T (2014) *Austroboletus amazonicus* sp. nov. and Fistulinella campinaranae var. scrobiculata, two commonly occurring boletes from a forest dominated by *Pseudomonotes tropenbosii* (Dipterocarpaceae) in Colombian Amazonia. Mycologia 106: 1004–1014. Vasco-Palacios AM, Hernandez J, Peñuela-Mora MC, Franco-Molano AE, Boekhout T (2018) Ectomycorrhizal fungi diversity in a white sand forest in western Amazonia. Fungal Ecol 31:9–18

Vasco-Palacios AM, Hernandez J, Peñuela-Mora MC, Franco-Molano AE, Boekhou, TE (2018) Ectomycorrhizal fungi diversity in a white sand forest in western Amazonia. Fungal Ecology 31:9–18

Varga S (2015) On the importance of details in arbuscular mycorrhizal research. Appl Soil Ecol 87:87–90

Veblen TT, Young KR, Orme AR (eds) (2015) The physical geography of South America. Oxford University Press

Velázquez MS, Cabello M, Irrazabal G, Godeas A (2008) Acaulosporaceae from El Palmar National Park, Entre Ríos, Argentina. Mycotaxon 103:171–187

Velázquez S, Cabello M (2011) Occurrence and diversity of arbuscular mycorrhizal fungi in trap cultures from El Palmar National Park soils. Eur J Soil Biol 47:230–235

Velázquez M, Cabello M, Barrera M (2013) Composition and structure of arbuscular-mycorrhizal communities in El Palmar National Park, Argentina. Mycologia 105(3):509–520

Velázquez MS, Stürmer SL, Bruzone C, Fontenla S, Barrera M, Cabello M (2016) Occurrence of arbuscular mycorrhizal fungi in high altitude sites of the Patagonian Altoandina region in Nahuel Huapi National Park (Argentina). Acta Bot Bras 30(4):521–531

Walker C, Gollotte A, Redecker D (2018) A new genus, *Planticonsortium* (Mucoromycotina), and new combination (*P. tenue*), for the fine root endophyte, *Glomus tenue* (basionym *Rhizophagus tenuis*). Mycorrhiza 28:213–219

Wetzel K, Silva GA, Matczinski U, Oehl F, Fester T (2014) Superior differentiation of arbuscular mycorrhizal fungal communities from till and no-till plots by morphological spore identification when compared to T-RFLP. Soil Biol Biochem 72:88–96

Young BE, Josse C, Stern M, Vasconez S, Olander J, Smyth R, Zador M, Sánchez de Lozada A, Comer PJ, Moull K, Echavarría M, Hak J (2015) Hotspot de biodiversidad de los Andes tropicales. In: Resumen técnico del perfil del ecosistema. Washington, D.C.: NatureServer & EcoDecisi6n

Zhang F-J, Qiao Li Q, Chen F-X, Xu H-Y, Inderjit, Wan F-H (2017) Arbuscular mycorrhizal fungi facilitate growth and competitive ability of an exotic species Flaveria bidentis. Soil Biol Biochem 115:275–284

Zangaro W, Rostirola LV, de Souza PB, de Almeida Alves R, Lescano LEAM, Rondina ABL, Nogueira MA, Carrenho R (2013) Root colonization and spore abundance of arbuscular mycorrhizal fungi in distinct successional stages from an Atlantic rainforest biome in southern Brazil. Mycorrhiza 23(3):221–233

Chapter 2
Latitudinal Distribution of Mycorrhizal Types in Native and Alien Trees in Montane Ecosystems from Southern South America

Carlos Urcelay, Paula A. Tecco, Valentina Borda, and Silvana Longo

2.1 Introduction

Biological invasions constitute a global environmental threat that rapidly alters natural communities and ecosystem functioning (Mack et al. 2000; MA 2005). The changes caused by alien plant invasions into novel ecosystems are accompanied by economic losses and environmental and social problems (Pimentel et al. 2000, 2005; Charles and Dukes 2007; Pejchar and Mooney 2009). For these reasons, it is extremely important to know the mechanisms that make an exotic plant to expand into new ecosystems, particularly tree species that are known to profoundly alter biological communities and ecosystem processes (Richardson et al. 2014).

One way to understand the success of alien trees in novel ecosystems is by comparing their ecological strategies with those of natives (Pyšek and Richardson 2007; Van Kleunen et al. 2010). In the case of plants, contrasting strategies to successfully invade novel ecosystems can be expected: (a) those that share attributes with natives and (b) those that differ from native communities (converging and diverging functional strategies, respectively) (e.g. Cleland 2011; Leishman et al. 2007, 2010; Pyšek and Richardson 2007). Whether alien strategies tend to converge or diverge from those of natives depend on geographical scale, climatic conditions, land uses, plant life form, and suite of biological attributes selected (e.g. Tecco et al. 2010, 2013; Zeballos et al. 2014; Funk et al. 2017). However, the comparisons between native and alien species rarely include symbiotic interactions (but see Tecco et al. 2013).

Biological interactions such as belowground symbiosis between plant and fungi, known as mycorrhizas, have also shown to influence the success of alien species in novel ecosystems (Richardson et al. 2000). Mycorrhizas are associations between

C. Urcelay (✉) · P. A. Tecco · V. Borda · S. Longo
Instituto Multidisciplinario de Biología Vegetal (CONICET) and Facultad de Ciencias Exactas, Físicas y Naturales, Universidad Nacional de Córdoba, Córdoba, Argentina
e-mail: curcelay@imbiv.unc.edu.ar

M. C. Pagano, M. A. Lugo (eds.), *Mycorrhizal Fungi in South America*, Fungal Biology, https://doi.org/10.1007/978-3-030-15228-4_2

plant and fungi at the root level and are one of the most widespread symbioses. In exchange for carbon, mycorrhizal fungi provide plants the access to limiting nutrients, among other benefits (Smith and Read 2008). According to the anatomy, morphology, and functional attributes of the symbiosis (including the phylogenetic identity of the plant and fungal symbionts), three basic and widespread types of mycorrhizas can be recognized dominating terrestrial ecosystems (Smith and Read 2008; Brundrett 2009): Arbuscular mycorrhizas (AM), Ectomycorrhizas (ECM), and Ericoid mycorrrhizas (ERM). There also exist other types of mycorrhizas but they do not dominate in any ecosystem.

Despite the phylogenetic imprints that often characterize mycorrhizal distribution among plants (Brundrett 2009), models on mycorrhizal distribution across environmental gradients, vegetation types (Read 1991; Read 1993) and plant life forms (Brundrett 1991, 2009) have been proposed. At the global scale, the one proposed by Read (1991) is still the best proxy to distribution of mycorrhizal types among biomes. Recent results at the continental scale (Europe), support Read's model and found that distribution of mycorrhizal types is mainly driven by mean annual temperature, soil pH and net primary productivity (Bueno et al. 2017).

Arbuscular mycorrhizas are formed by more than 80% of the terrestrial plants and dominate soils with high mineral N but low P availability such as temperate grasslands, savannahs and subtropical deciduous forests. Ectomycorrhizas are formed by several plant lineages, mainly trees growing in acidic soils with litter accumulation and seasonal N and P availability. They typically dominate in temperate forests and taiga. Instead, Ericoid mycorrrhizas are restricted to Ericales and dominate in acidic soils with low N and P contents such as heathlands and the arctic tundra. These mycorrhizal types are associated with different patterns of carbon and nutrient cycling, thus have pivotal role in ecosystem functioning (Cornelissen et al. 2001; Read and Perez-Moreno 2003). For example, Cornelissen et al. (2001) observed that among British plant species, those with ERM mycorrhizas show low growth rates, low foliar N and P concentration, and poor decomposition rates. In contrast, AM plants show comparatively higher growth rates, N and P foliar concentrations, and decomposition rates. While ECM plants show intermediate levels of these attributes. It is worth mentioning that nearly 20% of plant species are Non-mycorrhizal (Brundrett 2009) but they are rarely considered in models of mycorrhizal distribution (but see Brundrett 2017; Bueno et al. 2017).

Some studies aimed to answer whether patterns of mycorrhizal distribution in alien plant species tend to converge or diverge with those in natives. For example, the majority of naturalized pants in Great Britain belong to AM families (Fitter 2005). Menzel et al. (2017) found that mycorrhizal aliens inhabit a wider geographical range when compared with non-mycorrhizal ones in Germany. This trend was more marked when only woody species were compared. Conversely, invasive aliens from non-mycorrhizal plant families are higher in number than those from mycorrhizal ones in California (Pringle et al. 2009). These differences suggest that mycorrhizal types may play different roles in plant invasions in different ecosystems. Moreover, they show that woody species behave differently than non-woody.

In this chapter we examine the patterns of distribution of mycorrhizal types among the most abundant native and alien trees in montane forest ecosystems along a latitudinal gradient in Argentina (Fig. 2.1). We aim to answer two general questions: (1) Do patterns of mycorrhizal distribution in contrasting montane ecosystems behave as is predicted by models on mycorrhizal distribution across biomes, and (2) Do patterns of mycorrhizal distribution in alien species tend to converge or diverge with those in observed in natives?

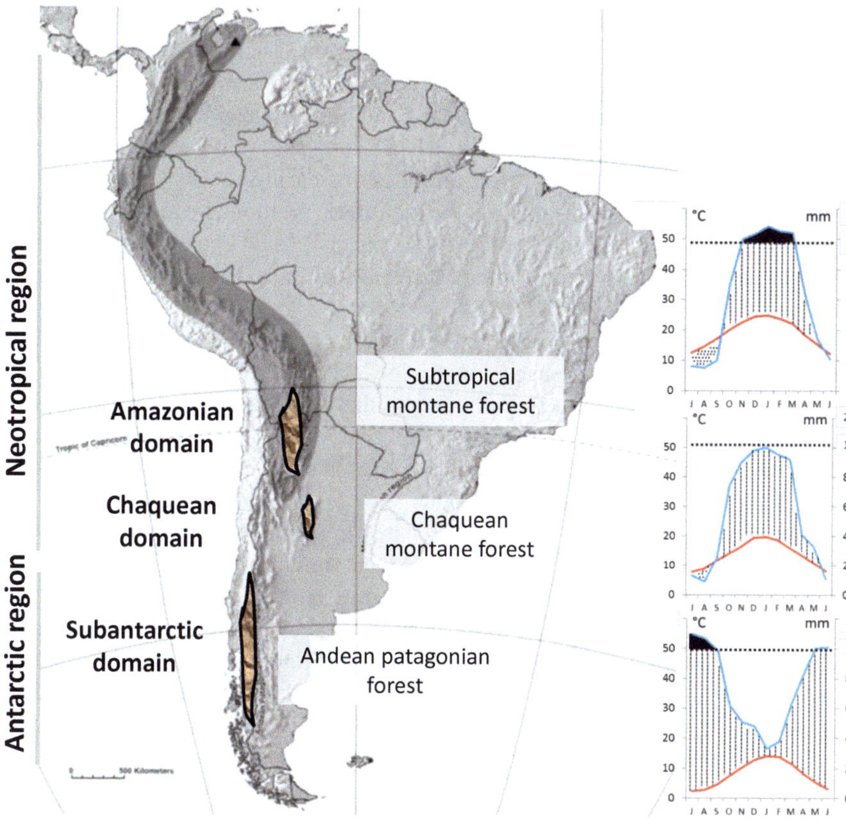

Fig. 2.1 Distribution of montane forests in Argentina. In brown the areas that were subjected to analyses in this chapter. In the right margin representative climate diagrams of these forest ecosystems (26°51′S/65°21′W, 31°6′S/64°27′W and 41°0′S/71°23′W, respectively; WorldClim – Global Climate Data)

2.2 Data Sources

Argentina includes an extensive land area that corresponds to two main regions: Neotropical and Antarctic. These regions are crossed by important montane ranges in north-south direction. The montane ranges correspond to the amazonian, chaquean, and sub-antarctic domains and are mostly covered by forests (Cabrera 1971) (Fig. 2.1). These forests are known as: Subtropical montane forests (also known as Yungas), Chaquean montane forests (also known as Chaco Serrano), and Andean-Patagonian forests (Morello et al. 2012; Oyarzabal et al. 2018). Data from soils were extracted from Rubio et al. (2019).

The **subtropical montane forests** occupy the east slopes in subandean and pampean mountains in northwest Argentina (Catamarca, Tucumán, Salta and Jujuy provinces), between 400 and 3000 m asl. The climate is warm and humid to subhumid with variable mean annual precipitation ranging between 800–3000 mm, 80% concentrated in summer (Fig. 2.1). There is a strong seasonal variation. The mean annual temperature is 22 °C at lower altitudes decreasing to 13 °C at higher altitudes. Three main vegetation types can be identified: premontane rain forests (400–700 m asl), montane rain forests (700–1500 m asl), and upper montane forests (1500–3000 m asl). Soils belong to the Mollisolls, Alfisols, Entisols and Inceptisols orders with a pH 5–7.

The **chaquean montane forests** are seasonally dry forest ecosystems located at central Argentina mountain ranges, mainly in Córdoba Province, between 500 and 2790 m asl. The climate is subxerophytic and the mean annual precipitation range between 500 and 900 mm, concentrated in warm months (Fig. 2.1). Mean annual temperature range from 15 °C in lower altitudes and 7.4 °C at the highest points. Soils belong mainly to Entisols, pH 6–7.

The **Andean-patagonian forests** are located in southwest Argentina and Chile. The climate is cool temperate and humid. Mean annual precipitations range from 750 mm in the eastern areas while reach 4000 mm in some western areas known as Valdivian rain forests, albeit most areas average 1800 mm concentrated in winter as rain and/or snow (Fig. 2.1). The mean annual temperature is 8 °C decreasing with increasing latitude and altitude. Andisols, Molisols, Inceptisols and Entisols are the main soil orders represented in the area, pH 4.5–6.

Each of these montane forests consists in different subunits of vegetation. Due to the scale of analysis, each of them is considered as one ecological unit here.

The basic sources of literature for selecting the most abundant native tree species in each region were: Morello et al. (2012) and Oyarzabal et al. (2018), but also Cabido et al. (2018) for the Chaquean region.

Naturalized alien species are those foreign species that have successfully invaded any ecosystem: they have self-sustaining populations that do not require repeated reintroduction (Fitter 2005). For selecting naturalized alien tree species we used: Grau and Aragón (2000) and Sirombra and Meza (2010) for Subtropical montane forests, Giorgis and Tecco (2014) for Chaquean montane forests, and Simberloff et al. (2002, 2003), Kutschker et al. (2015), Datri et al. (2015), and Calviño et al. (2018) for Andean-Patagonian forests.

For assesing the mycorrhizal types in native and alien trees we gathered data from own field surveys and data available from literature such as Wang and Qiu (2006), Brundrett (2009), Fracchia et al. (2009), Tecco et al. (2013), Godoy et al. (1994), Castillo et al. (2006), among other specific resources. Some species form both ECM and AM. They were considered as ECM (except for *Juniperus communis*, see below) because this type has been shown to be more important in terms of mycorrhizal colonization rates (e.g. Van der Heijden 2001) and their nutrient and carbon cycling traits are more similar to those of ECM trees.

A total of 169 cases were analyzed. In the 41 cases of species for which information on mycorrhizal type was not available, we assigned the mycorrhizal type corresponding to congeneric species because there is a strong phylogenetic conservatism in mycorrhizal symbiosis (Brundrett and Tedersoo 2018, but also see Brundrett 2017). In some few cases (4) for which information from congeneric species was also unavailable, we assigned the mycorrhizal type corresponding to the majority of the species in that family. We did not consider mycorrhizal status (i.e. facultative -somes cases colonized by mycorrhizal fungi, others not- or obligate mycorrhizal -always colonized-) because in our experience mycorrhizal plants from these regions are consistently colonized by mycorrhizal fungi in the field. Moreover, such status categories are subjected to a high probability of erroneous assignment (Bueno et al. 2019).

2.3 Mycorrhizal Distribution in Native Trees of Montane Forests from Argentina

In South America, the first approach to mycorrhizal distribution was made by Singer and Morello (1960). They postulated that "a completely (ecto)-mycorrhizal community is characteristic of strongly contrasted thermoperiodical climates" (p. 549), excluding those with excessive dryness or humidity. In other words, in terms of richness and abundance, the importance of ECM trees increases with increasing altitude and latitude but excluding arid or highly humid ecosystems.

More recently, Read (1991) postulated that montane forests from Southern South America would shift from those dominated by AM trees in the subtropics to those dominated by ECM trees in temperate regions. Accordingly, we found that the proportion of ECM native species is greater in temperate Andean-patagonian forests in comparison to subtropical and subxerophytic chaquean montane forests types ($x^2 = 17.7, p = 0.0014$) (Fig. 2.2).

For seasonal tropical and subtropical forests from South America, Read (1991) postulated that they are dominated by AM species with some ECM. Accordingly, among the 48 native trees species in subtropical montane forests in northwest Argentina surveyed here, 46 are AM while two are ECM (Fig. 2.2). These two belong to the genus *Alnus* (Betulaceae) and *Salix* (Salicaceae). *Alnus* is a holartic genus that migrated southward through the Andes from North America. *Alnus acu-*

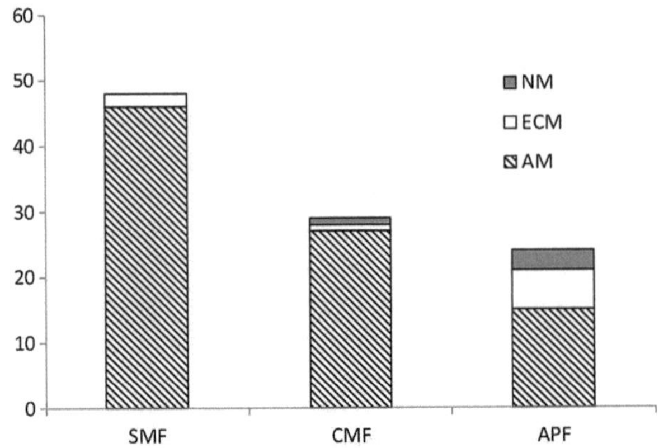

Fig. 2.2 Frequency of mycorrhizal types in native tree species from Subtropical (SMF), Chaquean (CMF), and Andean Patagonian (APF) montane forests. NM, non-mycorrhizal; ECM, ectomycorrhizal; AM, arbuscular mycorrhizal. Surveyed species in SMF: **AM**, *Allophylus edulis, Anadenanthera colubrina, Blepharocalyx salicifolius, Bocconia integrifolia, Calycophyllum multiflorum, Cedrela angustifolia, Chloroleucon tenuiflorum, Cordia americana, Cordia trichotoma, Enterolobium contortisiliquum, Eugenia uniflora, Ficus maroma, Handroanthus impetiginosus, Heliocarpus popayanensis, Ilex argentina, Inga edulis, Inga marginata, Inga saltensis, Jacaranda mimosifolia, Juglans australis, Myracrodruon urundeuva, Myrcianthes callicoma, Myrcianthes pseudomato, Myrcianthes pungens, Myroxylon peruiferum, Nectandra cuspidata, Ocotea porphyria, Ocotea puberula, Parapiptadenia excelsa, Phyllostylon rhamnoides, Podocarpus parlatorei, Polylepis australis, Prunus tucumanensis, Pterogyne nitens, Sambucus nigra (var. Peruviana), Schinus areira, Senna spectabilis, Solanum riparium, Tecoma stans, Tessaria integrifolia, Tipuana tipu, Trema micrantha, Urera baccifera, Urera caracasana, Vachellia albicorticata, Zanthoxylum coco* – **ECM,** *Alnus acuminata, Salix humboldtiana*; in CMF: **AM,** *Acacia aroma, Acacia caven, Acacia gilliesii, Acacia praecox, Aspidosperma quebracho-blanco, Celtis ehrenbergiana, Condalia buxifolia, Condalia montana, Geoffroea decorticans, Jodina rhombifolia, Kageneckia lanceolata, Lithrea molleoides, Maytenus boaria, Myrcianthes cisplatensis, Parkinsonia aculeata, Polylepis australis, Porliera microphylla, Prosopis alba, Prosopis caldenia, Prosopis nigra, Prosopis torquata, Ruprechtia apétala, Schinopsis marginata, Schinus fasciculatus, Sebastiania commersoniana, Zanthoxylum coco, Ziziphus mistol* – **ECM,** *Salix humboldtiana* – **NM,** *Bougainvillea stipitata*; and in APF: **AM,** *Aextoxicon punctatus, Araucaria araucana, Austrocedrus chilensis, Dasyphyllum diacanthoides, Drimys winteri, Fitzroya cupressoides, Laureliopsis philippiana, Luma apiculata, Maytenus boaria, Persea lingue, Pilgerodendron uviferum, Podocarpus nubigenus, Saxegothaea conspicua, Schinus patagonicus, Weinmannia trichosperma,* – **ECM,** *Nothofagus alpina, Nothofagus Antarctica, Nothofagus betuloides, Nothofagus dombeyi, Nothofagus obliqua, Nothofagus pumilio* – **NM,** *Embothrium coccineum, Lomatia hirsuta, Gevuina avellana*

minata forms monospecific stands in the upper montane forest (1700–2500 m asl) but also occupies riparian areas at lower altitudes. The other ECM species is *Salix humboldtiana* that occurs at riparian areas in the lower altitudinal belts. It has been suggested that this species also migrated from the northern hemisphere to the south through riparian corridors (*see* Tedersoo 2017). Both species also form AM (Becerra et al. 2005a, b, 2009). Metagenomic analyses of soils show the presence of several ECM fungal lineages in *A. acuminata* forests (Geml et al. 2014; Wicaksono et al. 2017). It was also observed different ECM lineages in soils from the lower altitudinal belts, although the diversity was lower than in *Alnus* forests (Geml et al. 2014). This lower diversity of ECM fungal lineages could be attributed to the fact that there is no dominant ECM tree species in these forests. The only species surveyed is *S. humboldtiana* that is restricted to certain riparian ecosystems. It is also possible that these fungi are associated with ECM tree species not listed in our survey because they are represented in low abundance.

Besides those two exceptions, the other native tree species in these forests (95.8%) form AM and belong to different plant families among which Fabaceae and Myrtaceae are the most numerous (Table 2.1). It is worth mentioning that this region is characterized for the scarcity of mycorrhizal studies.

For subxerophytic Chaquean montane forests, the dominance of AM trees is also predicted (Read 1991). In line, we found that among 29 native tree species, 27 were AM while one was ECM and the other one was Non-mycorrhizal (Fig. 2.2). Here, the only ectomycorrhizal tree is *S. humboldtiana* that, as in the subtropical montane forests, is restricted to certain riparian habitats without forming extensive forests.

Table 2.1 Number of native tree species surveyed in each family in Subtropical montane forests in north-west Argentina

Family	No of species	Mycorrhizal type
Fabaceae	12	AM
Myrtaceae	5	AM
Bignoniaceae	3	AM
Lauraceae	3	AM
Anacardiaceae	2	AM
Boraginaceae	2	AM
Rosaceae	2	AM
Urticaceae	2	AM
Adoxaceae	1	AM
Aquifoliaceae	1	AM
Asteraceae	1	AM
Betulaceae	1	AM-ECM
Cannabaceae	1	AM
Juglandaceae	1	AM
Meliaceae	1	AM
Moraceae	1	AM
Papaveraceae	1	AM
Podocarpaceae	1	AM
Rubiaceae	1	AM
Rutaceae	1	AM
Salicaceae	1	AM-ECM
Sapindaceae	1	AM
Solanaceae	1	AM
Tiliaceae	1	AM
Ulmaceae	1	AM

The non-mycorrhizal species is *Boungainvillea stipitata* (Nyctaginaceae) for which no study has assessed its mycorrhizal status. We assigned the mycorrhizal status according to information of the congeneric species *B. spectabilis* (Wang and Qiu 2006). This is also supported by the fact that Nyctaginaceae includes numerous non-mycorrhizal species (Brundrett 2017). The AM tree species of this seasonally dry montane forests (93%) belong to different families among which Fabaceae is also the most numerous (Table 2.2).

In turn, ECM trees are predicted to dominate in temperate forests (Read 1991). Among the 24 native species surveyed in Temperate Andean Patagonian forests, 6 are ECM. However, in terms of species number, AM species show higher values (15 species). In addition, 3 species correspond to NM type (Table 2.3).

The ECM species correspond to the Gondwanic genus *Nothofagus* (Nothofagaceae). Despite being lower in terms of species number, in comparison to AM, the ECM *Nothofagus* spp. species are dominant trees in these forests (e.g Veblen et al. 1992). The non-mycorrhizal species belong to the Proteaceae, a typical non-mycorrhizal gondwanic family that forms clusters roots specialized in obtaining nutrient in infertile soils (Brundrett 2017). Among AM families, Cupressaceae and Podocarpaceae show the highest numbers with 3 and 2 species, respectively. *Austrocedrus chilensis* (Cupressaceae) cover important parts of this territory (Veblen et al. 1992).

Altogether, these surveys suggest that the importance of ECM and, to a lesser degree, NM trees in montane forests ecosystems increases from subtropical towards temperate region. The results support the proposed models for mycorrhizal distribution in Southern South America.

Table 2.2 Number of native tree species surveyed in each family in Chaquean montane forests in north-west Argentina

Family	No of species	Mycorrhizal type
Fabaceae	10	AM
Rhamnaceae	3	AM
Anacardiaceae	3	AM
Rosaceae	2	AM
Apocynaceae	1	AM
Cannabaceae	1	AM
Celastraceae	1	AM
Euphorbiaceae	1	AM
Myrtacea	1	AM
Nyctaginaceae	1	NM
Polygonaceae	1	AM
Rutaceae	1	AM
Salicaceae	1	AM-ECM
Santalaceae	1	AM
Zygophyllaceae	1	AM

Table 2.3 Number of native tree species surveyed in each family in Andean patagonian forests in south-west Argentina

Family	No of species	Mycorrhizal type
Nothofagaceae	6	ECM
Cupressaceae	3	AM
Proteaceae	3	NM
Podocarpaceae	2	AM
Aextoxicaceae	1	AM
Anacardiaceae	1	AM
Araucariaceae	1	AM
Asteraceae	1	AM
Atherospermataceae	1	AM
Celastraceae	1	AM
Cunoniaceae	1	AM
Lauraceae	1	AM
Myrtaceae	1	AM
Winteraceae	1	AM

2.4 Mycorrhizal Distribution in Alien Trees Occurring in Montane Forests from Argentina

Our Survey reveal that the Chaquean montane forests show the highest number of invasive alien tree species (34 spp.) when compared with Subtropical montane forests (15 spp.) and Andean patagonian forests (19 spp.). This could be the result of more intense sampling in chaquean forests or because this ecosystem is more susceptible to invasions.

The proportion of mycorrhizal types in alien trees also differ among ecosystems ($x^2 = 19.2$, $p = 0.0007$) and the proportion of ECM type also increases with increasing latitude (Fig. 2.3). Instead, non-mycorrhizal alien trees were only present in subtropical montane forest, represented by only one species (*Grevillea robusta*, Proteaceae).

Among the 15 alien trees species surveyed in subtropical montane forest in northwest Argentina, the majority were AM, albeit two ECM and one NM were also registered. The ECM trees were *Eucalyptus grandis* (Myrtaceae) and *Pinus taeda* (Pinaceae). Both genera are well known by having several ECM invasive tree species (Richardson and Rejmánek 2011). It is worth mentioning that the invasive alien tree species were more evenly distributed among plant families in these forests (Table 2.4) than in the other two regions (see below).

The chaquean montane forests present a wide variety of alien trees species (34 spp, Table 2.5). The majority is AM followed by ECM, with 28 (82%) and 6 (17.6%) species, respectively. Rosaceae is the family with the highest number of alien species. Most of them are AM but *Crataegus monogyna* has been reported to

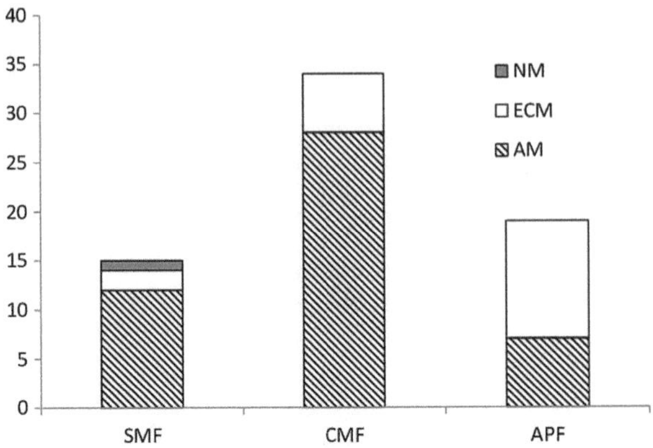

Fig. 2.3 Frequency of mycorrhizal types in alien tree species from Subtropical (SMF), Chaquean (CMF), and Andean Patagonian (APF) montane forests. NM, non-mycorrhizal; ECM, ectomycorrhizal; AM, arbuscular mycorrhizal. Surveyed species in SMF: **AM**, *Bauhinia candicans, Citrus aurantium, Eriobotrya japonica, Gleditsia triacanthos, Ligustrum lucidum, Ligustrum sinense, Morus alba, Morus nigra, Persea americana, Prunus pérsica, Psidium guajava, Pyracantha angustifolia* – **ECM**, *Eucalyptus grandis, Pinus taeda* – **NM**, *Grevillea robusta;* in CMF: **AM**, *Acacia dealbata, Acer negundo, Ailanthus altissima, Bauhinia forficata, Celtis australis, Cotoneaster franchetii, Cotoneaster glaucophyllus, Cotoneaster horizontalis, Gleditsia triacanthos, Jacaranda mimosifolia, Ligustrum lucidum, Ligustrum sinense, Maclura pomífera, Manihot grahamii, Melia azedarach, Morus alba, Olea europea, Phytolacca dioica, Prunus cerasifera, Prunus persica, Pyracantha angustifolia, Pyracantha coccinea, Robinia pseudoacacia, Schinus areira, Tamarix gallica, Tamarix ramosissima, Ulmus pumila, Zanthoxylum armatum* – **ECM**, *Betula pendula, Crataegus monogyna, Eucalyptus camaldulensis, Pinus elliottii, Pinus halepensis, Salix viminalis;* and in APF: **AM**, *Acer pseudo-platanus, Cytisus scoparius, Juniperus communis, Laburnum anagyroides, Malus sylvestris, Rosa rubiginosa, Sambucus nigra* – **ECM**, *Alnus glutinosa, Alnus incana, Alnus rubra, Crataegus monogyna, Pinus contorta, Pinus montícola, Pinus ponderosa, Pinus radiata, Pinus sylvestris, Pseudotsuga menziesii, Salix fragilis, Salix viminalis*

Table 2.4 Number of alien tree species surveyed in each family in Subtropical montane forests in northwest Argentina

Family	No of species	Mycorrhizal type
Rosaceae	3	AM
Fabaceae	2	AM
Moraceae	2	AM
Myrtaceae	2	1 AM, 1 AM-ECM
Oleaceae	2	AM
Lauraceae	1	AM
Pinaceae	1	ECM
Proteaceae	1	NM
Rutaceae	1	AM

Table 2.5 Number of alien tree species surveyed in each family in Chaquean montane forests in central Argentina

Family	No of species	Mycorrhizal type
Rosaceae	8	7 AM, 1 ECM
Fabaceae	4	AM
Oleaceae	3	AM
Moraceae	2	AM
Pinaceae	2	ECM
Tamaricaceae	2	AM
Anacardiaceae	1	AM
Betulaceae	1	ECM
Bignoniaceae	1	AM
Cannabaceae	1	AM
Euphorbiaceae	1	AM
Meliaceae	1	AM
Myrtaceae	1	AM-ECM
Phytolaccaceae	1	AM
Rutaceae	1	AM
Salicaceae	1	AM-ECM
Sapindaceae	1	AM
Simaroubaceae	1	AM
Ulmaceae	1	AM

Table 2.6 Number of alien tree species surveyed in each family in Andean patagonian forests in southwest Argentina

Family	No of species	Mycorrhizal type
Pinaceae	6	1 AM-ECM, 5 ECM
Betulaceae	3	ECM
Rosaceae	3	2 AM, 1 ECM
Fabaceae	2	AM
Salicaceae	2	1 AM-ECM, 1 ECM
Adoxaceae	1	AM
Cupressaceae	1	AM (ECM)
Sapindaceae	1	AM

be ECM in Europe (Maremmani et al. 2003). Then we suggest that the mycorrhizal type of this species should be confirmed. The other ECM trees belong to well-known ECM families such as Pinaceae, Betulaceae, Salicaceae and Myrtaceae.

Unlike both neotropical montane ecosystems (Subtropical and Chaquean), in Andean patagonian forests ECM is represented by higher percentage of species than AM (63 and 37%, respectively) (Table 2.6). Pinaceae and Betulaceae showed the highest species number (6 and 3, respectively) (Simberloff et al. 2002). Among them, *Pseudosuga menziesii* forms both AM and ECM (Salomón et al. 2018). The other ECM families were Rosaceae and Salicaceae. *Juniperus communis* (Cupressaceae) is mainly cited as AM but ECM has been occasionally reported suggesting a facultative relationship with ECM fungi (Thomas et al. 2007). For this reason, here it is considered as AM but further studies would confirm the mycorrhizal status of this alien plant in subantarctic forests.

Two AM shrubby species were also included due to their importance in terms of abundance and distribution: *Cystus scoparius* (Fabaceae) (Simberloff et al. 2002, 2003) and *Rosa rubiginosa* (Rosaceae) (Simberloff et al. 2002; Zimmermann et al. 2010), both AM species.

The patterns of mycorrhizal distribution in alien trees occurring in montane ecosystems across Argentinian territory also show that ECM species frequency increase with increasing latitude. However, this increase in frequency is notably higher in aliens than in natives. Instead, NM aliens are less represented than NM natives.

2.5 Mycorrhizas and Plant Invasions

In the last 20 years, the importance of belowground mutualistic interactions such as "mycorrhizas" in plant invasions has been widely recognized (e.g. Richardson et al. 2000; Callaway et al. 2003; Dickie et al. 2017). Few studies, however, compared patterns of mycorrhizal distribution in native and alien flora at the biome or regional level (Fitter 2005; Pringle et al. 2009; Menzel et al. 2017).

Our surveys include three floras in montane ecosystems from different biomes. Considering the montane forests altogether, the distribution of mycorrhizal types show a greater proportion of ECM in alien (20/67 cases) than in native (9/101) tree species ($x^2 = 12.8$, $p = 0.017$). Inversely, there was a higher proportion of AM and NM in natives (88 and 4, respectively) than in alien trees (46 and 1, respectively). The negligible presence of NM alien trees throughout the montane ranges (i.e. single specie) is in line with Menzel et al. (2017), highlighting the relevance of belowground mutualistic interactions for tree invasion success. The overall prevalence of AM associations among invasive species is in line with Fitter (2005).

The general trend of a higher proportion of ECM in aliens than in natives kept significant within the Temperate Andean Patagonian forests ($x^2 = 8.17$, $p = 0.0169$) when analysing each biome separately (Fig. 2.4). However, there were no differences between aliens and natives within the Subtropical and Chaquean montane forests ($x^2 = 5.02$, $p = 0.0811$; $x^2 = 4.22$; $p = 0.1213$, respectively). These analyses suggest that ECM may have an advantage over AM aliens in expanding their ranges and that this advantage would be higher with increasing latitude (see below).

The non-significant differences in proportion of mycorrhizal types in alien and natives trees in Neotropical region (Subtropical and Chaquean forests) is explained by the majority of AM tree species in both native and aliens in these forests. According to the literature, the most widely studied alien species, and probably more widely distributed, in Subtropical montane forests are AM (*Ligustrum*

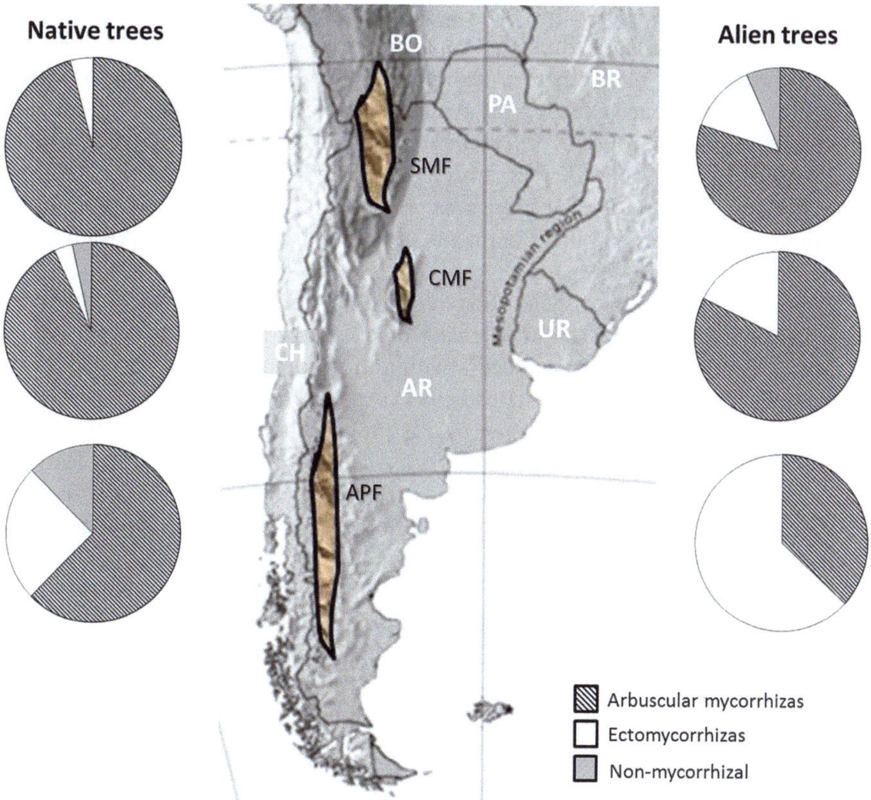

Fig. 2.4 Relative abundance of native and alien tree species forming different mycorrhizal type in Subtropical (SMF), Chaquean (CMF), and Andean Patagonian (APF) montane forests from Southern South America. NM, non-mycorrhizal; ECM, ectomycorrhizal; AM, arbuscular mycorrhizal. AR, Argentina; CH, Chile; UR, Uruguay; BO, Bolivia; PA, Paraguay; BR, Brazil

lucidum, *Morus alba* and *Gleditsia triacanthos*; Aragón and Morales 2003; Fernandez et al. 2017) but the role of mycorrhizal fungi in their success in these forests has not been studied. In Chaquean Montane forest, *Ligustrum lucidum* (Hoyos et al. 2010; Zeballos et al. 2014; Giorgis et al. 2017), *Gleditsia triacanthos* (Giorgis et al. 2011a; Furey et al. 2014; Fernandez et al. 2017; Marcora et al. 2018), and *Pyracantha angustifolia* (Tecco et al. 2007; Zeballos et al. 2014) are the most widely expanded and studied in the region. It has been recently shown that seedling of these alien tree species benefit from AM fungi, mainly for P nutrition, either from already invaded or from non- invaded elevations in these montane ranges (Urcelay et al. 2019). The prevalence of AM association in trees of both neotropical regions could be interpreted as functional convergence in belowground strategies among native and alien trees in these biomes. However, it is worth mentioning that ECM alien trees include some serious invaders such as Pinaceae (Richardson and Rejmánek 2011) that are naturalized in both montane biomes. For example, in

Chaquean mountains *Pinus elliottii* expand outside the afforestations, either to native forests or grasslands, due to its capacity for co-invade with alien ECM symbionts, particularly *Suillus granulatus* and *Rhizopogon pseudoreseolus* (Urcelay et al. 2017). No evidence exists on the symbiotic interactions between *P. elliottii* and native ECM fungi reported for *Salix humboldtiana*, the only ECM native tree in the region (Becerra et al. 2009). The expansion of *Pinus elliottii* is still incipient (Giorgis et al. 2011b; Urcelay et al. 2017) but it is predicted to become a great threat in the near future as occurred in other continents in the southern hemisphere (Richardson 2006).

The higher proportion of ECM alien trees within temperate forests (compared to both the coexisting natives and to aliens of other regions) highlights the advantage of this association for invasion success within this biome. It is further in line with global scale distribution of ECM (Read 1991). Among the invasive aliens in Andean Patagonian forests, *Pseudosuga menziesii* is probably the most widely distributed and studied (e.g. Simberloff et al. 2002; Sarasola et al. 2006; Orellana and Raffaele 2010). Also important are *Pinus ponderosa*, *P. contorta*, *P. radiata*, *P. monticola*, *P. sylvestris*, and *Juniperus communis* (Simberloff et al. 2002; Sarasola et al. 2006; Richardson et al. 2008). *Pseudosuga* and *Pinus* species need to establish ECM symbiosis to succeed in the novel environments. In these species, the specificity for mycorrhizal partners is higher than in AM trees which are assumed to be generalists (Nuñez and Dickie 2014). Thus, ECM trees need alien ECM fungi to expand outside plantations. This was particularly evidenced for *P. menziesii*, *P. contorta*, and *P. ponderosa*. They were found to be associated with the different fungi such as *Suillus luteus, S. lakei, Amphinema* spp., *Melanogaster* sp., *Rhizopogon cf. rogersii, R. cf. arctostaphyli, R. roseolus, R. villosuslus, Hebeloma mesophaeum, Lactarius quieticolor, Cortinarius* spp. *Pseudotomentella tristis, Wilcoxina* spp., among others. These alien ECM fungi are dispersed either by the wind or by exotic mammals (Nuñez et al. 2009, 2013; Salomón et al. 2011, 2018; Hayward et al. 2015a, b). Inferring convergence or divergence in mycorrhizal association between coexisting alien and natives in temperate forests is not conclusive. There is a prevalence of AM in native species in terms of species number but this underestimate the relevance of ECM in the system in terms of dominance since *Nothofagus* spp. are very relevant in terms of structural cover in this biome (Veblen et al. 1992). Thus, we do not infer divergence out of our results. The analysis of patterns of mycorrhizal distribution ponderated by species' abundance will certainly give a better insight to this question.

2.6 Conclusions

Although not explicitly tested, the results gathered here suggest that the distribution of mycorrhizal types in native trees at a broad geographic scale in South America is mainly driven by climate. From the analyses we conclude that patterns of mycorrhizal distribution in alien and native trees occurring in montane ecosystems from subtropical to temperate regions, roughly follow those predicted by models of mycorrhizal distribution (e.g. Read 1991). This is seemingly in line with the idea of broad scale environmental filters driving greater predominance of convergences than divergences in the functional strategies of coexisting tree species along these mountain biomes (e.g. Weiher et al. 1998; Cornwell et al. 2006; Cornwell and Ackerly 2009; Lohbeck et al. 2014). Nonetheless, ECM in aliens is in higher proportion compared to natives, particularly in temperate forests. Further studies should incorporate the analyses of the abundance of mycorrhizal types in a given ecosystem, not only the proportion of plant species, in order to estimate their impacts on ecosystem functioning and biogeochemical cycling (Soudzilovskaia et al. 2017). The evidence suggests that ECM trees co-invade with alien mycorrhizal fungi. In the case of Pinaceae, they mainly co-invade with species belonging to the genus *Suillus* and *Rhizopogon* (Policelli et al. 2018). Arbuscular mycorrhizal type shows the highest proportion in native and alien trees from all the ecosystems except for aliens in the Andean-patagonian region. In contrast to ECM trees, it is presumed that AM alien trees form mycorrhizas with native fungi.

It is worth remarking that mainly trees were included in the analyses. This selection can exclude certain mycorrhizal types. For example, the occurrence of small ericoid shrubs is known for the three montane forests. In some of them, indeed, the ERM structures and the identity of symbionts were studied (Urcelay 2002; Selosse et al. 2007; Bruzone et al. 2015). Then, the proportion mycorrhizal types represented in native and alien species may change if other plant life forms were included. In this scenario, a higher proportion ERM and NM could be expected because these types are mainly represented in small shrubs and herbs, respectively.

Since some of the mycorrhizal status lists consulted here may contain errors (see Dickie et al. 2007; Brundrett and Tedersoo 2019), the type of mycorrhiza in certain tree species still should be confirmed. However, the patterns reported here, based on updated and more precise information, do not differ much from those communicated some years ago based on more limited data sets (Urcelay and Tecco 2006, 2008, 2010). This suggests that the patterns of mycorrhizal distribution are robust in face of the addition and/or refinement of the data.

Finally, the effects of mycorrhizal fungi on growth, nutrition, and then expansion and dominance of most native and alien trees in ecosystems of southern South America still remains to be determined. Altogether, these findings suggest that mycorrhizal associations have an unambiguous role in tree invasions in montane forests across different climates but the relative importance of each mycorrhizal type in each ecosystem remains unknown.

Acknowledgements This work was supported by Secyt (UNC). We thank CONICET and the Universidad Nacional de Córdoba (Argentina), both of which supported the facilities used in this investigation. G. Robledo provided the map. A. Cingolani kindly provided climate data.

References

Aragón R, Morales JM (2003) Species composition and invasion in NW Argentinian secondary forests: effects of land use history, environment and landscape. Journal of Vegetation Science 14: 195–204

Becerra AG, Pritsch K, Arrigo N, Palma M, Bartoloni N (2005a). Ectomycorrhizal colonization of *Alnus acuminata* Kunth in northwestern Argentina in relation to season and soil parameters. Annals of Forest Science 62: 325–332

Becerra A, Zak MR, Horton TR, Micolini J (2005b). Ectomycorrhizal and arbuscular mycorrhizal colonization of Alnus acuminata from Calilegua National Park (Argentina). Mycorrhiza 15: 525–531

Becerra AG, Nouhra ER, Silva MP, McKay D (2009) Ectomycorrhizae, arbuscular mycorrhizae, and dark-septate fungi on Salix humboldtiana in two riparian populations from central Argentina. Mycoscience 50: 343–352

Brundrett MC (1991) Mycorrhizas in natural ecosystems. In: Macfayden A, Begon M, Fitter AH (eds) Advances in ecological research. Academic Press 171–313. London, UK

Brundrett MC (2009) Mycorrhizal associations and other means of nutrition of vascular plants: understanding the global diversity of host plants by resolving conflicting information and developing reliable means of diagnosis. Plant Soil 320: 37–77

Brundrett MC (2017) Distribution and Evolution of Mycorrhizal Types and Other Specialised Roots in Australia. In: Tedersoo L (ed) Biogeography of Mycorrhizal Symbiosis. Springer International 361–394. Cham, Switzerland

Brundrett MC, Tedersoo L (2018) Evolutionary history of mycorrhizal symbiosis and global host plant diversity. New Phytologist 220: 1108–1115

Brundrett MC, Tedersoo L (2019) Misdiagnosis of mycorrhizas and inappropriate recycling of data can lead to false conclusions. New Phytologist 221: 18–24.

Bruzone MC, Fontenla SB, Vohník M (2015). Is the prominent ericoid mycorrhizal fungus *Rhizoscyphus ericae* absent in the Southern Hemisphere's Ericaceae? A case study on the diversity of root mycobionts in Gaultheria spp. from northwest Patagonia, Argentina. Mycorrhiza 25: 25–40

Bueno CG, Moora M, Gerz M, Davison J, Öpik M, Pärtel M, Helm A, Ronk A, Kühn I, Zobel M (2017). Plant mycorrhizal status, but not type, shifts with latitude and elevation in Europe. Global Ecology and Biogeography 26: 690–699

Bueno CG, Gerz M, Zobel M, Moora M (2019) Conceptual differences lead to divergent trait estimates in empirical and taxonomic approaches to plant mycorrhizal trait assignment. Mycorrhiza 29: 1–11

Cabido M, Zeballos SR, Zak M, Carranza ML, Giorgis MA, Cantero JJ, Acosta ATR (2018) Native Woody vegetation in central Argentina: Classification of Chaco and Espinal forests. Applied Vegetation Science 21: 298–311

Cabrera AL (1971) Fitogeografía de la república Argentina. Boletín de la Sociedad Argentina de Botánica 14: 1–42

Callaway RM, Mahall BE, Wicks C, Pankey J, Zabinski C (2003) Soil fungi and the effects of an invasive forb on grasses: neighbor identity matters. Ecology 84: 129–135

Calviño CI, Edwards P, Fernández M, Relva MA, Ezcurra C (2018) Not one but three: undetected invasive Alnus species in northwestern Patagonia confirmed with cpDNA and ITS sequences. Biological Invasions 20: 2715–2722

Castillo CG, Borie F, Godoy R, Rubio R, Sieverding E (2006) Diversity of mycorrhizal plant species and arbuscular mycorrhizal fungi in evergreen forest, deciduous forest and grassland ecosystems of Southern Chile. Journal of Applied Botany and Food Quality 80: 40–47

Charles H, Dukes JS (2007) Impacts of Invasive Species on Ecosystem Services. In: Nentwig W (ed) Biological invasions. Springer 217–237. Berlin, Heidelberg

Cleland EE (2011). Trait divergence and the ecosystem impacts of invading species. New Phytologist 189: 649–652

Cornelissen J, Aerts R, Cerabolini B, Werger M, Van Der Heijden M (2001) Carbon cycling traits of plant species are linked with mycorrhizal strategy. Oecologia 129: 611–619
Cornwell WK, Schwilk DW, Ackerly DD (2006) A trait-based test for habitat filtering: convex hull volume. Ecology 87: 1465–1471
Cornwell WK, Ackerly DD (2009) Community assembly and shifts in plant trait distributions across an environmental gradient in coastal California. Ecological Monographs 79: 109–126
Datri LA, Faggi AM, Gallo LA (2015) Modelo de invasión no lineal y funciones bioingenieras de sauce fragilis en Patagonia (Argentina). European Scientific Journals 1: 265–272
Dickie IA, Thomas MM, Bellingham PJ (2007) On the perils of mycorrhizal status lists: the case of Buddleja davidii. Mycorrhiza 17: 687
Dickie IA, Bufford JL, Cobb RC, Desprez-Loustau ML, Grelet G, Hulme PE, Klironomos J, Makiola A, Nuñez MA, Pringle A, Thrall PH, Tourtellot SG, Waller L, Williams NM (2017) The emerging science of linked plant–fungal invasions. New Phytologist 215: 1314–1332
Fernandez RD, Ceballos SJ, Malizia A, Aragón R (2017) *Gleditsia triacanthos* (Fabaceae) in Argentina: a review of its invasion. Australian Journal of Botany 65: 203–213
Fitter AH (2005) Darkness visible: reflections on underground ecology. Journal of Ecology 93: 231–243
Fracchia S, Aranda A, Gopar A, Silvani V, Fernandez L, Godeas A (2009) Mycorrhizal status of plant species in the Chaco Serrano Woodland from central Argentina. Mycorrhiza 19: 205–214
Funk JL, Larson JE, Ames GM, Butterfield BJ, Cavender-Bares J, Firn J, Laughlin DC, Sutton-Grier AE, Williams L, Wright J (2017) Revisiting the Holy Grail: using plant functional traits to understand ecological processes. Biological Reviews 92: 1156–1173
Furey, C, Tecco PA, Perez-Harguindeguy N, Giorgis MA, Grossi M (2014) The importance of native and exotic plant identity and dominance on decomposition patterns in mountain woodlands of central Argentina. Acta Oecologica 54: 13–20
Geml J, Pastor N, Fernandez L, Pacheco S, Semenova TA, Becerra AG, Wicaksono CY, Nouhra ER (2014) Large-scale fungal diversity assessment in the Andean Yungas forests reveals strong community turnover among forest types along an altitudinal gradient. Molecular Ecology 23: 2452–2472
Giorgis MA, Cingolani AM, Chiarini F, Chiapella J, Barboza G, Ariza Espinar L, Morero R, Gurvich DE, Tecco PA, Subils R Cabido M (2011a) Composición florística del Bosque Chaqueño Serrano de la provincia de Córdoba, Argentina. Kurtziana 36: 9–43
Giorgis MA, Tecco PA, Cingolani AM, Renison D, Marcora P, Paiaro V (2011b) Factors associated with woody alien species distribution in a newly invaded mountain system of central Argentina. Biological Invasions 13: 1423–1434
Giorgis MA, Tecco PA (2014) Árboles y arbustos invasores de la Provincia de Córdoba (Argentina): una contribución a la sistematización de bases de datos globales. Boletín de la Sociedad Argentina de Botánica 49: 581–603
Giorgis MA, Cingolani AM, Gurvich DE, Tecco PA, Chiapella J, Chiarini F, Cabido M (2017) Changes in floristic composition and physiognomy are decoupled along elevation gradients in central Argentina. Applied Vegetation Science 20: 558–571
Godoy R, Romero R, Carrillo R (1994) Status micotrófico de la flora vascular en bosques de coníferas nativas del sur de Chile. Revista Chilena de Historia Natural 67: 209–220
Grau HR, Aragón R (2000) Ecología de árboles exóticos de las Yungas argentinas. Laboratorio de Investigaciones Ecológicas de las Yungas. Universidad Nacional de Tucumán
Hayward J, Horton TR, Pauchard A, Nuñez MA (2015a) A single ectomycorrhizal fungal species can enable a *Pinus* invasion. Ecology 96: 1438–1444
Hayward J, Horton TR, Nuñez MA (2015b) Ectomycorrhizal fungal communities coinvading with Pinaceae host plants in Argentina: Gringos bajo el bosque. New Phytologist 208: 497–506
Hoyos LE, Gavier-Pizarro GI, Kuemmerle T, Bucher EH, Radeloff VC, Tecco, PA (2010) Invasion of glossy privet (*Ligustrum lucidum*) and native forest loss in the Sierras Chicas of Córdoba, Argentina. Biological invasions 12: 3261–3275

Kutschker A, Hechem V, Codesal P, Rafael M, López S, Silva V (2015) Diversidad de plantas exóticas en áreas sometidas a distintos disturbios en el Parque Nacional Los Alerces, Chubut (Argentina). Boletín de la Sociedad Argentina de Botánica 50: 47–59. Córdoba, Argentina
Leishman MR, Haslehurst T, Ares A, Baruch Z (2007) Leaf trait relationships of native and invasive plants: community and global scale comparisons. New Phytologist 176: 635–643
Leishman MR, Thomson V P, Cooke J (2010) Native and exotic invasive plants have fundamentally similar carbon capture strategies. Journal of Ecology 98: 28–42
Lohbeck M, Poorter L, Martínez-Ramos M, Rodriguez-Velázquez J, Breugel, M, Bongers F (2014) Changing drivers of species dominance during tropical forest succession. Functional Ecology 28: 1052–1058
MA (2005) Millennium Ecosystem Assessment. Ecosystems and Human Well-being: Synthesis. Island Press, Washington, DC, USA
Mack RN, Simberloff D, Lonsdale MW, Evans H, Clout M, Bazzaz FA (2000) Biotic invasions: causes, epidemiology, global consequences and control. Ecological Applications 10: 689–710
Maremmani A, Bedini S, Matoševic I, Tomei PE, Giovannetti M (2003) Type of mycorrhizal associations in two coastal nature reserves of the Mediterranean basin. Mycorrhiza 13: 33–40
Marcora PI, Ferreras AE, Zeballos SR, Funes G, Longo S, Urcelay C, Tecco PA (2018) Context-dependent effects of fire and browsing on woody alien invasion in mountain ecosystems. Oecologia 188: 479–490
Menzel A, Hempel S, Klotz S, Moora M, Pyšek P, Rillig M C, Zobel M, Kühn I (2017) Mycorrhizal status helps explain invasion success of alien plant species. Ecology 98: 92–102
Morello J, Matteuci S, Rodriguez A, Silva M (2012). Ecorregiones y complejos ecosistémicos argentinos. Buenos Aires, Orientación Gráfica Editora
Nuñez MA, Horton TR, Simberloff D (2009) Lack of belowground mutualisms hinders Pinaceae invasions. Ecology 90: 2352–2359
Nuñez MA, Hayward J, Horton TR, Amico GC, Dimarco RD, Barrios-Garcia MN, Simberloff D (2013) Exotic mammals disperse exotic fungi that promote invasion by exotic trees. PLoS one 8(6): e66832
Nuñez MA, Dickie IA (2014) Invasive belowground mutualists of woody plants. Biological Invasions 16: 645–661
Orellana IA, Raffaele E (2010) The spread of the exotic conifer Pseudotsuga menziesii in Austrocedrus chilensis forests and shrublands in northwestern Patagonia, Argentina. New Zealand Journal of Forestry Science (New Zealand Forest Research Institute Ltd (trading as Scion)) 40
Oyarzabal M, Clavijo J, Oakley L, Biganzoli F, Tognetti P, Barberis I, Maturo HM, Aragón R, Campanello PI, Prado D, Oesterheld M, León RJC (2018) Unidades de vegetación de la Argentina. Ecología Austral 28: 40–63
Pejchar L, Mooney HA (2009) Invasive species, ecosystem services and human well-being. Trends in ecology & evolution 24: 497–504
Pimentel D, Lach L, Zuniga R, Morrison D (2000) Environmental and Economic Costs of Nonindigenous Species in the United States. BioScience 50: 53–65
Pimentel D, Zuniga R, Morrison D (2005) Update on the environmental and economic costs associated with alien-invasive species in the United States. Ecological Economics 52: 273–288
Policelli N, Bruns TD, Vilgalys R, Nuñez MA (2018) Suilloid fungi as global drivers of pine invasions. New Phytologist 222: 714–725
Pringle A, Bever JD, Gardes M, Parrent JL, Rillig MC, Klironomos JN (2009) Mycorrhizal Symbioses and Plant Invasions. Annual Review of Ecology, Evolution, and Systematics 40: 699–715
Pyšek P, Richardson DM (2007) Traits Associated with Invasiveness in Alien Plants: Where Do we Stand? In: Nentwig W (ed) Biological Invasions. Springer 97–125. Berlin, Heidelberg
Read D J (1991) Mycorrhizas in ecosystems. Experientia 47: 376–391
Read DJ (1993) Mycorrhiza in plant communities. Advances in Plant Pathology 9: 1–31

Read DJ, Perez-Moreno J (2003). Mycorrhizas and nutrient cycling in ecosystems – a journey towards relevance? New Phytologist 157: 475–492
Richardson DM, Pyšek P, Rejmánek M, Barbour MG, Panetta FD, West CJ (2000) Naturalization and invasion of alien plants: concepts and definitions. Diversity & Distributions 6: 93–107
Richardson DM (2006) *Pinus*: a model group for unlocking the secrets of alien plant invasions? Preslia 78: 375–388
Richardson DM, van Wilgen BW, Nuñez MA (2008) Alien conifer invasions in South America: short fuse burning? Biological invasions 10: 573–577
Richardson DM, Rejmánek M (2011) Trees and shrubs as invasive alien species – a global review. Diversity and Distributions 17: 788–809
Richardson DM, Hui C, Nuñez MA, Pauchard A (2014) Tree invasions: patterns, processes, challenges and opportunities. Biological Invasions 16: 473–81
Rubio G, Lavado RS, Pereyra FX (2019) The soils of Argentina. Springer International, Cham Switzerland
Salomón MES, Barroetaveña C, Rajchenberg M (2011) Do pine plantations provide mycorrhizal inocula for seedlings establishment in grasslands from Patagonia, Argentina? New forests 41: 191–205
Salomón MES, Barroetaveña C, Pildain MB, Williams EA, Rajchenberg M (2018) What happens to the mycorrhizal communities of native and exotic seedlings when Pseudotsuga menziesii invades Nothofagaceae forests in Patagonia, Argentina? Acta Oecologica 91: 108–119
Sarasola MM, Rusch VE, Schlichter TM, Ghersa CM (2006). Invasión de coníferas forestales en áreas de estepa y bosques de ciprés de la cordillera en la Región Andino Patagónica. Ecología austral 16: 143–156
Selosse MA, Setaro S, Glatard F, Richard F, Urcelay C, Weiß M (2007). Sebacinales are common mycorrhizal associates of Ericaceae. New Phytologist 174: 864–878
Simberloff D, Relva MA, Nuñez M (2002) Gringos en el bosque: introduced tree invasion in a native Nothofagus/Austrocedrus forest. Biological Invasions 4: 35–53
Simberloff D, Relva MA, Nuñez M (2003). Introduced species and management of a Nothofagus/Austrocedrus forest. Environmental Management 31: 0263–0275
Smith SE, Read DJ (2008) Mycorrhizal symbiosis. 3rd. Academic Press. New York
Singer R, Morello JH (1960) Ectotrophic forest tree mycorrhizae and forest communities. Ecology 41: 549–551
Sirombra MG, Mesa LM (2010) Composición florística y distribución de los bosques ribereños subtropicales andinos del río Lules, Tucumán, Argentina. Revista de Biología Tropical 58: 499–510
Soudzilovskaia NA, Vaessen S, van't Zelfde M, Raes N (2017) Global patterns of mycorrhizal distribution and their environmental drivers. In: Tedersoo L (ed) Biogeography of Mycorrhizal Symbiosis. Springer International 223–235. Cham, Switzerland
Tecco PA, Diaz S, Gurvich DE, Perez-Harguindeguy N, Cabido M, Bertone GA (2007) Facilitation and interference underlying the association between the woody invaders *Pyracantha angustifolia* and *Ligustrum lucidum*. Applied Vegetation Science 10: 211–218
Tecco PA, Díaz S, Cabido M, Urcelay C (2010) Functional traits of alien plants across contrasting climatic and land-use regimes: do aliens join the locals or try harder than them? Journal of Ecology 98: 17–27
Tecco PA, Urcelay C, Díaz S, Cabido M, Pérez-Harguindeguy N (2013) Contrasting functional trait syndromes underlay woody alien success in the same ecosystem. Austral Ecology 38: 443–451
Tedersoo L (2017) Global Biogeography and invasions of Ectomycorrhizal Plants: Past, Present and Future. In: Tedersoo L (ed) Biogeography of Mycorrhizal Symbiosis. Springer International 469–532. Cham, Switzerland
Thomas PA, El-Barghathi M, Polwart A (2007) Biological flora of the British Isles: Juniperus communis L. Journal of Ecology 95: 1404–1440

Urcelay C (2002) Co-occurrence of three fungal root symbionts in Gaultheria poeppiggi DC in Central Argentina. Mycorrhiza 12: 89–92

Urcelay C, Tecco P (2006) Distribución de tipos micorrícicos en especies leñosas exóticas de Dominios fitogeográficos de Argentina. XXII Reunión Argentina de Ecología, Córdoba, Argentina. Libro de resúmenes, p. 344

Urcelay C, Tecco PA (2008) Micorrizas en el cono sur sudamericano; una aproximación micogeográfica y sus implicancias en los procesos de invasiones biológicas. VI Congreso Latinoamericano de Micología, Mar del Plata, Argentina. Libro de resúmenes, p.242

Urcelay C, Tecco PA. (2010) Mycorrhizal types in native and exotic woody species in Southern South America: a biogeographic approach. 9th International Mycological Congress – The Biology of Fungi. Edinburgh, UK. CD room (P3.48)

Urcelay C, Longo S, Geml J, Tecco PA, Nouhra E (2017) Co-invasive exotic pines and their ectomycorrhizal symbionts show capabilities for wide distance and altitudinal range expansion. Fungal Ecology 25: 50–58

Urcelay C, Longo S, Geml J, Tecco PA (2019). Can arbuscular mycorrhizal fungi from non-invaded montane ecosystems facilitate the growth of alien trees? Mycorrhiza 29: 39–49

Van der Heijden EW (2001) Differential benefits of arbuscular mycorrhizal and ectomycorrhizal infection of Salix repens. Mycorrhiza 10: 185–193

Van Kleunen M, Dawson W, Schlaepfer D, Jeschke JM, Fischer M (2010) Are invaders different? A conceptual framework of comparative approaches for assessing determinants of invasiveness. Ecology Letters 13: 947–958

Veblen TT, Kitzberger T, Lara A (1992) Disturbance and forest dynamics along a transect from Andean rain forest to Patagonian shrubland. Journal of Vegetation Science 3: 507–520

Wang B, Qiu YL (2006) Phylogenetic distribution and evolution of mycorrhizas in land plants. Mycorrhiza 16: 299–363

Weiher E, Clarke GP, Keddy PA (1998) Community assembly rules, morphological dispersion, and the coexistence of plant species. Oikos 81: 309–322

Wicaksono CY, Aguirre-Guiterrez J, Nouhra E, Pastor N, Raes N, Pacheco S, Geml, J (2017) Contracting montane cloud forests: a case study of the Andean alder (Alnus acuminata) and associated fungi in the Yungas. Biotropica 49: 141–152

WorldClim – Global Climate Data. Available at: http://www.worldclim.org

Zeballos SR, Giorgis MA, Cingolani AM, Cabido M, Whitworth Hulse JI, Gurvich DE (2014) Do alien and native tree species from Central Argentina differ in their water transport strategy? Austral Ecology 39: 984–991

Zimmermann H, Ritz CM, Hirsch H, Renison D, Wesche K, Hensen I (2010). Highly reduced genetic diversity of Rosa rubiginosa L. populations in the invasive range. International Journal of Plant Sciences 171: 435–446

Chapter 3
Biodiversity of Arbuscular Mycorrhizal Fungi in South America: A Review

M. Noelia Cofré, Florencia Soteras, M. del Rosario Iglesias, Silvana Velázquez, Camila Abarca, Lucía Risio, Emanuel Ontivero, Marta N. Cabello, Laura S. Domínguez, and Mónica A. Lugo

3.1 Introduction

Identification of species is crucial in understanding how diversity changes affect ecosystemic processes. Particularly, soil microbial are key factors of ecosystemic functioning (Copley 2000). Among soil microbes, arbuscular mycorrhizal fungi (AMF, phylum Glomeromycota) are worldwide distributed (Tedersoo et al. 2018) and form symbiotic associations with almost 80% of the vascular plants of the earth, except for one species, *Geosiphon pyriformis*, which associates with the cyanobacteria *Nostoc* (Smith and Read 2008). AMF comprise around 300 morphologically defined or 350–1000 molecularly defined taxa (Davison et al. 2015 and references therein). Since AMF associate with aboveground community, their occurrence and

M. N. Cofré (✉) · L. S. Domínguez
Laboratorio de Micología, IMBIV, CONICET, Universidad Nacional de Córdoba, Córdoba, Argentina

F. Soteras
Laboratorio de Ecología Evolutiva y Biología Floral, IMBIV, CONICET, Universidad Nacional de Córdoba, Córdoba, Argentina

M. del Rosario Iglesias
IMBIV, CONICET, Universidad Nacional de Córdoba, Córdoba, Argentina

S. Velázquez · C. Abarca · M. N. Cabello
Instituto de Botánica Spegazzini, Universidad Nacional de La Plata-CICPBA, La Plata, Buenos Aires, Argentina

L. Risio · E. Ontivero
MICODIF-IMIBIO-CONICET, Universidad Nacional de San Luis, San Luis, Argentina

M. A. Lugo
Biological Sciences, National University of San Luis, Grupo MICODIF (Micología, Diversidad e Interacciones Fúngicas)/IMIBIO (Instituto Multidisciplinario de Investigaciones Biológicas)-CONICET-CCT SL, San Luis, San Luis, Argentina

M. C. Pagano, M. A. Lugo (eds.), *Mycorrhizal Fungi in South America*, Fungal Biology, https://doi.org/10.1007/978-3-030-15228-4_3

composition can influence ecosystemic processes either through affecting plant community composition and thus its processes rates, or soil microbial communities, which are directly involved in nutrient cycling (Rillig 2004). According to Pärtel et al. (2016), soil microorganisms are considered a potentially suitable target for studying regional and local effects on diversity. The symbiosis with AMF not only increases nutrient uptake by the plant of mainly phosphorus (P) and nitrogen (N) in exchange for plant-assimilated carbon (C), but also improves the tolerance of plants to various biotic and abiotic stresses such as pathogens, salinity, and drought (Smith and Read 2008).

External factors (abiotic and biotic) and intrinsic properties of species (dispersal ability, rates of speciation and extinction) affect the AMF geographical distributions (Chaudhary et al. 2008). For instance, the abiotic factors of temperature and precipitation constrain AMF occurrence (Davison et al. 2015) while biotic ones such as host preferences determine the rhizospheric AMF community (Senés-Guerrero and Schüßler 2016; Soteras et al. 2016). Moreover, anthropogenic activities like agricultural practices that alter soil conditions could influence the occurrence of AM fungal taxa (Cofré et al. 2017). At the same time, either external or internal factors may indirectly influence each other, causing changes in AMF taxa occurrence and distribution (Chaudhary et al. 2008). Currently, an increasing number of studies attempt at unravelling the worldwide geographical patterns of AMF (Öpik et al. 2010, 2013; Kivlin et al. 2011; Tedersoo et al. 2014; Davison et al. 2015). These researches reviewed AMF descriptions based on DNA methods and showed contrasting results of AMF biogeographical patterns. For instance, Öpik et al. (2010) found that two-thirds of AMF taxa showed restricted distribution, but Davison et al. (2015) postulated that most of the AMF taxa show a cosmopolitan distribution and that species richness of AMF virtual taxa decreases with latitude at the global scale. However, in South America (SA), molecular characterization of AMF communities is fairly scarce (Grilli et al. 2015; Senés-Guerrero and Schüßler 2016; Soteras et al. 2016). In a recent review, Stürmer et al. (2018) described large-scale patterns of distribution of taxa within the phylum Glomeromycota reported on 7 continents and in 87 countries, including ultramarine country territories like Canary Islands, Kerguelen Island, the Bermudas, and Guadeloupe Islands. They concluded that this phylum mainly comprises cosmopolitan taxa. So far, AMF communities morphologically described have been analyzed from a biogeographical perspective in few studies (Stürmer et al. 2018), while distributional patterns across South American ecological divisions have never been deeply acknowledged. Therefore, in this Chapter we reviewed studies of SA that morphologically described the AMF community. Despite morphologically described AMF species (hereafter morphospecies) being highly intraspecifically variable, not necessarily representing root colonizing taxa, and not always sporulating during sampling, they could give a first approach of the AMF community of a particular place and moment. However, further DNA-based descriptions should be combined with morphospecies approach in order to deeply characterize AMF distribution in SA.

3.2 Arbuscular Mycorrhizal Morphospecies

AMF spores are asexual multinucleate single cells which originate from the differentiation of vegetative hyphae (Smith and Read 2008). Sporulation in the soil depends on several factors such as the fungal and host plant identity, soil fertility, and temperature among others (Smith and Read 2008). Spores represent the genetic unit of fungal species, being responsible for the colonization of new habitats and initiation of new individuals (Morton et al. 1993; Błaszkowski 2012). Since many components of the subcellular structure of the spores such as wall layers are stable under different environmental conditions, they are considered as diagnostic traits for the morphological identification of AMF. Although the identification of species is crucial to understanding geographical patterns of biodiversity (Zak et al. 2003), the taxonomic classification of AMF is under continuous debate due to the high variation of morphological traits, difficulties in spore extraction from soil and their pure culture under controlled conditions (Błaszkowski 2012). Since the definition of species in Glomeromycota is controversial (Rosendahl 2008), the term "morphospecies" has been chosen to refer to AMF species (Robinson-Boyer et al. 2009) when their identification is based on the morphological traits and ontogeny of spores.

In this Chapter we used the AMF morphospecies classification proposed by Redecker et al. (2013) and Schüßler and Walker (2010). We followed the AMF species list that was updated in September 2018 in http://www.amf-phylogeny.com/amphylotaxonomy.html. That last classification recognizes within the phylum Glomeromycota a single class (Glomeromycetes) which includes 4 orders (Glomerales, Diversisporales, Paraglomerales and Archaeosporales), 12 families (Acaulosporaceae, Ambisporaceae, Archaeosporaceae, Claroideoglomeraceae, Diversisporaceae, Geosiphonaecae, Gigasporaceae, Glomeraceae, Pacisporaceae, Paraglomeraceae, Pervetustaceae and Sacculosporaceae), 34 genera and approximately 316 AMF morphospecies validly described. *Glomus tenue* was considered in the species list but it is important to clarify that it was moved to *Planticonsortium* in the subphylum Mucoromycotina (Walker et al. 2018).

3.3 Arbuscular Mycorrhizal Fungi in South America

South America is globally recognized for its vast and incredible biodiversity linked to its unique geology, climate and biogeographic history. The great plant diversity in SA is the result of its complex evolutionary history over the last 250 million years (Fittkau 1969; Lavina and Fauth 2011). The floristic composition of SA (particularly in Chile, Argentina and "Cordillera de los Andes" or Andean Range) turns this area into a vegetation relict (Villagrán and Hinojosa 1997). Moreover, the region presents a great diversity of mycorrhizas that varies from angiosperm and gymnosperm forests dominated by arbuscular mycorrhizal (AM) to ectomycorrhizal

(ECM) *Nothofagus* spp. ones (Fontenla et al. 1998; Palfner 2001; Bueno et al. 2017). In accordance with the particular characteristics of the region, research on mycorrhizal patterns has provided interesting findings, such as the occurrence of new mycorrhizal associations (Bidartondo et al. 2002) and regional differences in the distribution of mycorrhizal fungi (Tedersoo et al. 2014; Davison et al. 2015). Alarming advances in deforestation, desertification and loss of biodiversity mainly due to soybeanization (the expansion of the agricultural frontier due to the planting of soybean) and with consequent processes of marginalization and social persecution, constitute a worrying reality that is increasingly widespread within the South American region (Viglizzo et al. 2011). Therefore, knowing the biotic diversity of the region is of great importance to protect it, so that diversity studies in the different types of ecological divisions of South America are essential for the expansion of knowledge about this ecosystemically diverse and threatened region.

In this Chapter, the available published studies on AMF morphospecies diversity have been compiled. To this end, we searched for Google Scholar articles from 1955 to 2018 containing the term combination "arbuscular mycorrhizal" AND "country name" and grouped them into ecological divisions. Studies on AMF morphological diversity from Paraguay, Guyana, Surinam, French Guiana could not be found. This research process included the review of a total of 110 articles dealing with AMF morphospecies diversity of the following nine countries: Argentina (27), Bolivia (8), Brazil (40), Chile (13), Colombia (6), Ecuador (1), Perú (2), Uruguay (2) and Venezuela (9), see Table 3.1. Particularly, the works of Uruguay only showed species richness number without any other details about the species identified.

3.4 South American Arbuscular Mycorrhizal Morphospecies Diversity

3.4.1 Arbuscular Mycorrhizal Fungi Morphospecies Richness in South America

Considering the 110 articles above mentioned, 186 AMF morphospecies were identified at the species level (Table 3.1) while a large number of taxa were identified at the genus level (608) but were not considered for geographical descriptions. Ordered according to the increasing number of the morphospecies identified: in Brazil, there were recorded 158 morphospecies to species level and 258 to genus level; in Argentina, 83 and 133; Chile, 59 and 36; Venezuela, 38 and 144; Perú, 31 and 21; Colombia, 20 and 6; Bolivia, 15 and 5; and Ecuador, 4 and 4.

In a recent review, Stürmer et al. (2018) described large-scale patterns of taxa distribution within the phylum Glomeromycota. Their results showed 131 identified morphospecies among the 1280 registered until 2012 in South America. In contrast, in our study we found 2187 records until June of 2018 in SA, representing 186

Table 3.1 List of the AMF morphospecies described or cited in the literature of South America analyzed in this Chapter. The total citations of each one and of all the morphospecies, the total of species and the total of research articles are shown for each region[a]

Families	Genus	Species	Biogeographic region														
			Amazonia	Atlantic Forest	Caatinga	Cerrado	Chaco	Dry South central Andes	Guianan Uplands and Highlands	Llanos	Mediterranean Chile	Moist north central Andes	Moist Pacific Mesoamerican	Moist Pacific temperate	Pampas	Patagonia	Total records
Acaulosporaceae	*Acaulospora*	*alpina*					5							4	2	4	15
Acaulosporaceae	*Acaulospora*	*bireticulata*	1	2	3	8	15			1					4	1	35
Acaulosporaceae	*Acaulospora*	*brasiliensis*	1	1		8		1									11
Acaulosporaceae	*Acaulospora*	*cavernata*				1	1							1		3	6
Acaulosporaceae	*Acaulospora*	*colombiana*	3	7	5	4			3	1		1	3				27
Acaulosporaceae	*Acaulospora*	*colossica*		5		4			1	2	1			1			14
Acaulosporaceae	*Acaulospora*	*delicata*	2	1	4	4	2								4	4	21
Acaulosporaceae	*Acaulospora*	*denticulata*		1	3	7	3	1				5			2	1	23
Acaulosporaceae	*Acaulospora*	*dilatata*			1		2				1			1		4	9
Acaulosporaceae	*Acaulospora*	*elegans*	1	2			1										4
Acaulosporaceae	*Acaulospora*	*endographis*		1													1
Acaulosporaceae	*Acaulospora*	*entreriana*					1										1
Acaulosporaceae	*Acaulospora*	*excavata*	1	1	13	2	10					4			4		35
Acaulosporaceae	*Acaulospora*	*foveata*	3	10	8		4		1			1				1	28
Acaulosporaceae	*Acaulospora*	*foveoreticulata*			1												1
Acaulosporaceae	*Acaulospora*	*herrerae*			7												7
Acaulosporaceae	*Acaulospora*	*ignota*		1													1
Acaulosporaceae	*Acaulospora*	*kentinensis*	1	1													2
Acaulosporaceae	*Acaulospora*	*koskei*		2	2	7					1			1			13
Acaulosporaceae	*Acaulospora*	*lacunosa*		4	2		5				1						12
Acaulosporaceae	*Acaulospora*	*laevis*	1	7	5		8		1	1	3	4		5	4	4	43
Acaulosporaceae	*Acaulospora*	*longula*	1	5	4	1				1		2	1	4			19

Table 3.1 (continued)

Families	Genus	Species	Biogeographic region														
			Amazonia	Atlantic Forest	Caatinga	Cerrado	Chaco	Dry South central Andes	Guianan Uplands and Highlands	Llanos	Mediterranean Chile	Moist north central Andes	Moist Pacific Mesoamerican	Moist Pacific temperate	Pampas	Patagonia	Total records
Acaulosporaceae	*Acaulospora*	*mellea*	4	18	9	8	12			2		5		2	4	3	67
Acaulosporaceae	*Acaulospora*	*morrowiae*	3	10	9	7				3		1					33
Acaulosporaceae	*Acaulospora*	*myriocarpa*		2								1		2			5
Acaulosporaceae	*Acaulospora*	*papillosa*		1													1
Acaulosporaceae	*Acaulospora*	*paulinae*	1			3					2			5			11
Acaulosporaceae	*Acaulospora*	*polonica*	1														1
Acaulosporaceae	*Acaulospora*	*punctata*	1								1			2			4
Acaulosporaceae	*Acaulospora*	*reducta*			1												1
Acaulosporaceae	*Acaulospora*	*rehmii*	4	4	9		7			1		5			2	1	33
Acaulosporaceae	*Acaulospora*	*rugosa*		2		2	4								1		9
Acaulosporaceae	*Acaulospora*	*scrobiculata*	5	21	19	6	16			1	1	5		3	3	4	84
Acaulosporaceae	*Acaulospora*	*sieverdingii*		1	3						2			1			7
Acaulosporaceae	*Acaulospora*	*spinosa*	4	11	8	7	11		1			5	1	5	3	3	59
Acaulosporaceae	*Acaulospora*	*spinulifera*		1													1
Acaulosporaceae	*Acaulospora*	*splendida*			1												1
Acaulosporaceae	*Acaulospora*	*thomii*												3			3
Acaulosporaceae	*Acaulospora*	*tuberculata*	3	7	6	1	2		1								20
Acaulosporaceae	*Acaulospora*	*undulata*	1				8							2			11
Ambisporaceae	*Ambispora*	*appendicula*	1	11	8	8	4		1	1		2	2				38
Ambisporaceae	*Ambispora*	*callosa*			1	7											8
Ambisporaceae	*Ambispora*	*fecundispora*		1													1
Ambisporaceae	*Ambispora*	*gerdemannii*	1	3	1		2							1		3	11
Ambisporaceae	*Ambispora*	*jimgerdemannii*		1	1												2
Ambisporaceae	*Ambispora*	*leptoticha*	3	3	7		8			2		4		2	3	1	33

Ambisporaceae	*Ambispora*	*leptoticha*					1										1
Archaeosporaceae	*Archaeospora*	*schenckii*									1			1			2
Archaeosporaceae	*Archaeospora*	*trappei*	5	6	1		3			3		1		9	2		30
Gigasporaceae (unclear)	*Bulbospora*	*minima*		1													1
Gigasporaceae	*Cetraspora*	*auronigra*		1		1											2
Gigasporaceae	*Cetraspora*	*gilmorei*		4	1	1	2				2			1	1	3	15
Gigasporaceae	*Cetraspora*	*pellucida*	2	10	10	3	2								6		33
Gigasporaceae	*Cetraspora*	*striata*							1								1
Claroideoglomeraceae	*Claroideoglomus*	*claroideum*	3	5	2	1					3	4		8	4	2	32
Claroideoglomeraceae	*Claroideoglomus*	*drummondii*	1														1
Claroideoglomeraceae	*Claroideoglomus*	*etunicatum*	2	11	19	8	12			1	3	1		9	5	4	75
Claroideoglomeraceae	*Claroideoglomus*	*lamellosum*		1		3								1			5
Claroideoglomeraceae	*Claroideoglomus*	*luteum*	2	1	1		10								2		16
Claroideoglomeraceae	*Claroideoglomus*	*maculosum simil C. claroideum*		1													1
Claroideoglomeraceae	*Claroideoglomus*	*walkeri*	1														1
Diversisporaceae	*Corymbiglomus*	*corymbiforme*	1														1
Diversisporaceae	*Corymbiglomus*	*globiferum*		1													1
Diversisporaceae	*Corymbiglomus*	*pacificum*												2			2
Diversisporaceae	*Corymbiglomus*	*tortuosum*	1	4	4		1		1	2					1	2	16
Gigasporaceae	*Dentiscutata*	*cerradensis*	2	4	2												8
Gigasporaceae	*Dentiscutata*	*colliculosa*			2												2
Gigasporaceae	*Dentiscutata*	*erythropus*			2												2
Gigasporaceae	*Dentiscutata*	*heterogama*	2	16	3	5	5					1	2		4	1	39
Gigasporaceae	*Dentiscutata*	*nigra*						1									1
Gigasporaceae	*Dentiscutata*	*reticulata*		1										1			2
Gigasporaceae	*Dentiscutata*	*savannicola*		1	2							2					5
Diversisporaceae	*Diversispora*	*eburnea*	1		1		2								1		5

Table 3.1 (continued)

Families	Genus	Species	Biogeographic region														
			Amazonia	Atlantic Forest	Caatinga	Cerrado	Chaco	Dry South central Andes	Guianan Uplands and Highlands	Llanos	Mediterranean Chile	Moist north central Andes	Moist Pacific Mesoamerican	Moist Pacific temperate	Pampas	Patagonia	Total records
Diversisporaceae	*Diversispora*	*insculpta*				1											1
Diversisporaceae	*Diversispora*	*spurca*	4	3	1		5				1			6			20
Diversisporaceae	*Diversispora*	*trimurales*	1														1
Glomeraceae	*Dominikia*	*aurea*		1							1			2	1		5
Glomeraceae	*Dominikia*	*minutum*	1	1	2	1									1		6
	Entrophospora	*baltica*	1											1			2
	Entrophospora	*infrequens*	2	4	12		15	1		3	1	4	1	1	4	1	49
Glomeraceae	*Funneliformis*	*badium*	1				4				1			2			8
Glomeraceae	*Funneliformis*	*caledonium*													1		1
Glomeraceae	*Funneliformis*	*coronatum*					4				1			3	3		11
Glomeraceae	*Funneliformis*	*dimorphicus*					2										2
Glomeraceae	*Funneliformis*	*geosporum*	3	7	3	8	11	1	1					2	3		39
Glomeraceae	*Funneliformis*	*halonatus*		1	3												4
Glomeraceae	*Funneliformis*	*monosporum*		1										1			2
Glomeraceae	*Funneliformis*	*mosseae*		7	7	8	13			2	1			5	9	2	54
Glomeraceae	*Funneliformis*	*verruculosum*				1											1
Gigasporaceae	*Fuscutata*	*heterogama*		1		1											2
Gigasporaceae	*Gigaspora*	*albida*		6	6						1						13
Gigasporaceae	*Gigaspora*	*candida*					1										1
Gigasporaceae	*Gigaspora*	*decipiens*		14	11	7	1								3		36
Gigasporaceae	*Gigaspora*	*gigantea*		6	8	1	4								3		22
Gigasporaceae	*Gigaspora*	*margarita*	1	8	10	7	7								5		38
Gigasporaceae	*Gigaspora*	*ramisporophora*			3				1								4
Gigasporaceae	*Gigaspora*	*rosea*	1	4			7								1		13

Glomeraceae	*Glomus*	*ambisporum*	1	2	3		2							1	1		10
Glomeraceae	*Glomus*	*arborense*		1	1												2
Glomeraceae	*Glomus*	*atrouva*	1														1
Glomeraceae	*Glomus*	*australe*	2														2
Glomeraceae	*Glomus*	*botryoides*		1													1
Glomeraceae	*Glomus*	*brohultii*	2	2	3		10		1					1	2		21
Glomeraceae	*Glomus*	*deserticola*		1											2		3
Glomeraceae	*Glomus*	*formosanum*		1													1
Glomeraceae	*Glomus*	*fuegianum*	1	1			8								2		12
Glomeraceae	*Glomus*	*glomerulatum*	1	4	7	8	1			1		1	1				24
Glomeraceae	*Glomus*	*heterosporum*	1	1													2
Glomeraceae	*Glomus*	*hoi*		2			1				2			3			8
Glomeraceae	*Glomus*	*macrocarpum/F. geosporum*	3	8	15	8					1			4			39
Glomeraceae	*Glomus*	*magnicaule*	1				4										5
Glomeraceae	*Glomus*	*microcarpum*	2	5	7	7	1						1				23
Glomeraceae	*Glomus*	*multicaule*			1												1
Glomeraceae	*Glomus*	*multiforum*				1											1
Glomeraceae	*Glomus*	*nanolumen*	1														1
Glomeraceae	*Glomus*	*pallidum*			1									4			5
Glomeraceae	*Glomus*	*pansihalos*		1													1
Glomeraceae	*Glomus*	*pustulatum*	1														1
Glomeraceae	*Glomus*	*reticulatum*		1													1
Glomeraceae	*Glomus*	*spinuliferum*		1													1
Glomeraceae	*Glomus*	*tenebrosum*	1						1								2
Glomeraceae	*Glomus*	*tenue*								1							1
Glomeraceae	*Glomus*	*trufemii*		1	1												2
Glomeraceae	*Glomus*	*versiforme*	2	1	1									5			9
Pacisporaceae	*Pacispora*	*chimonobambusae*										4					4

Table 3.1 (continued)

Families	Genus	Species	Biogeographic region														
			Amazonia	Atlantic Forest	Caatinga	Cerrado	Chaco	Dry South central Andes	Guianan Uplands and Highlands	Llanos	Mediterranean Chile	Moist north central Andes	Moist Pacific Mesoamerican	Moist Pacific temperate	Pampas	Patagonia	Total records
Pacisporaceae	*Pacispora*	*dominikii*	1			1	4				2			5			13
Pacisporaceae	*Pacispora*	*franciscana*			3												3
Pacisporaceae	*Pacispora*	*patagonica*					4									2	6
Pacisporaceae	*Pacispora*	*robigina*	1			1	3										5
Pacisporaceae	*Pacispora*	*scintillans*				1											1
Gigasporaceae	*Paradentiscutata*	*bahiana*		1		1											2
Gigasporaceae	*Paradentiscutata*	*maritima*		2													2
Pacisporaceae	*Paraglomus*	*bolivianum*		2	2										2		6
Paraglomeraceae	*Paraglomus*	*brasilianum*			1	1											2
Paraglomeraceae	*Paraglomus*	*laccatum*					1							1			2
Glomeraceae	*Paraglomus*	*lacteum*	1									4					5
Paraglomeraceae	*Paraglomus*	*occultum*	2	6	9	10			1		1	2	3	6			40
Paraglomeraceae	*Paraglomus*	*pernambucanum*				1											1
Gigasporaceae	*Racocetra*	*alborosea*		1													1
Gigasporaceae	*Racocetra*	*castanea*	1	1		3											5
Gigasporaceae	*Racocetra*	*coralloidea*		4	2		1										7
Gigasporaceae	*Racocetra*	*fulgida*		4	7	7	1								3		22
Gigasporaceae	*Racocetra*	*gregaria*		1	9		1										11
Gigasporaceae	*Racocetra*	*intraornata*		2	4												6
Gigasporaceae	*Racocetra*	*novaum*			1												1
Gigasporaceae	*Racocetra*	*persica*	1	1	1	2											5
Gigasporaceae	*Racocetra*	*tropicana*		2													2
Gigasporaceae	*Racocetra*	*verrucosa*		4	3												7
Gigasporaceae	*Racocetra*	*weresubiae*		3	2									1			6

Diversisporaceae	*Redeckera*	*fulvum*		1													1
Glomeraceae	*Rhizophagus*	*aggregatus*	1	3	2		5	1		1	1			1	6		21
Glomeraceae	*Rhizophagus*	*antarcticus*															1
Glomeraceae	*Rhizophagus*	*arabicus*			1												1
Glomeraceae	*Rhizophagus*	*clarus*	2	5	3	10			1			4			4	2	31
Glomeraceae	*Rhizophagus*	*diaphanus*	2	5	1	6	2		1		3			5	1		26
Glomeraceae	*Rhizophagus*	*fasciculatus*	2	4	2	6	4					2	2	5	4	1	32
Glomeraceae	*Rhizophagus*	*intraradices*	1	2	5	2	14				3	4		5	4		40
Glomeraceae	*Rhizophagus*	*invermaius*	1	4	1	8								3			17
Glomeraceae	*Rhizophagus*	*irregularis*			1										1		2
Glomeraceae	*Rhizophagus*	*manihotis*		1					1	1		1	2				6
Glomeraceae	*Rhizophagus*	*microaggregatus*	1	3	2	4	5			1					7	2	25
Glomeraceae	*Rhizophagus*	*natalensis*		1													1
Glomeraceae	*Rhizophagus*	*vesiculifer*										1		1			2
Glomeraceae	*Sclerocystis*	*coremioides*	3	3	4		1								2		13
Glomeraceae	*Sclerocystis*	*clavispora*	1	2	5									1			9
Glomeraceae	*Sclerocystis*	*liquidambaris*					1										1
Glomeraceae	*Sclerocystis*	*pachycaulis*								1			1				2
Glomeraceae	*Sclerocystis*	*rubiformis*	4	2	2		3				1			1			13
Glomeraceae	*Sclerocystis*	*sinuosa*		6	7		3			2					3		21
Glomeraceae	*Sclerocystis*	*taiwanensis*	1	2	2												5
Gigasporaceae	*Scutellospora*	*arenicola*	1						1						1		3
Gigasporaceae	*Scutellospora*	*auriglоba*		3	5									1			9
Gigasporaceae	*Scutellospora*	*biornata*	1	1	5	8	12		1			4			2	1	35
Gigasporaceae	*Scutellospora*	*calospora*	1	3	7	6	3				3	1		4	2		30
Gigasporaceae	*Scutellospora*	*crenulata*							2								2
Gigasporaceae	*Scutellospora*	*dipapillosa*		2			3					4			3		12
Gigasporaceae	*Scutellospora*	*dipurpurescens*	1	1	3	3								3			11
Gigasporaceae	*Scutellospora*	*hawaiiensis*		1													1

Table 3.1 (continued)

Families	Genus	Species	Biogeographic region														Total records
			Amazonia	Atlantic Forest	Caatinga	Cerrado	Chaco	Dry South central Andes	Guianan Uplands and Highlands	Llanos	Mediterranean Chile	Moist north central Andes	Moist Pacific Mesoamerican	Moist Pacific temperate	Pampas	Patagonia	
Gigasporaceae	*Scutellospora*	*pernambucana*		5	3	3											11
Gigasporaceae	*Scutellospora*	*rubra*			1	5	3								1		10
Gigasporaceae	*Scutellospora*	*scutata*	1	2	2												5
Gigasporaceae	*Scutellospora*	*spinosissima*	1	1					2								4
Gigasporaceae	*Scutellospora*	*tricalypta*		1													1
Glomeraceae	*Septoglomus*	*constrictum*		3	2	4	11			1	1			2	3		27
Glomeraceae	*Septoglomus*	*titan*			2												2
Glomeraceae	*Septoglomus*	*viscosum*		1											1		2
		Total records/ecoregion	137	416	409	269	355	6	25	36	47	90	20	163	153	60	2187
		Total species/ecoregion	81	120	96	61	71	6	21	24	30	32	12	57	54	26	
		Research articles/ecoregion (110)	5	18	19	4	20	1	6	3	4	7	1	7	9	2	

[a]Research articles' sources: Aguilera et al. (2014, 2017); Aidar et al. (2004); Albuquerque (2008); Angulo-Veizaga and García-Apaza (2014); Becerra and Cabello M (2008); Becerra et al. (2011, 2014); Bonfim et al. (2013); Cabello (1994, 1997); Carrenho et al. (2001); Casanova-Katny et al. (2011); Castillo et al. (2005, 2006, 2010, 2016); Cofré et al. (2017); Colombo et al. (2014); Cordoba et al. (2001); Coutinho et al. (2015); Covacevich et al. (2006); Cuenca et al. (1998); Cuenca and Herrera-Peraza (2008); Cuenca and Lovera (1992); Cuenca and Meneses (1996); da Silva et al. (2005, 2008, 2012, 2014); de Carvalho et al. (2012); de Mello (2011); de Mello et al. (2013); de Oliveira Freitas et al. (2014); Dhillion et al. (1995); Dodd et al. (1990); Escudero and Mendoza (2005); Fernandes and Siqueira (1989); França et al. (2007); de Oliveira Freitas (2006); Frioni et al. (1999); Furrazola et al. (2013); García et al. (2017); Gómez-Carabalí et al. (2011); Goto and Costa Maia (2005); Goto et al. (2009, 2010a, b); Grilli et al. (2012); Herrera-Peraza et al. (2001, 2016); Janos et al. (1995); Jobim et al. (2016, 2018); Krüger (2013); Leal et al. (2009); Lemos (2008); Longo et al. (2014); Lugo and Cabello (1999, 2002); Lugo et al. (2005, 2008); Marín et al. (2016, 2017); Medina et al. (2014, 2015); Meier et al. (2012); Mendoza et al. (2002); Menéndez et al. (2001); Menoyo et al. (2009); Mergulhão (2007); Moreira et al. (2009); Janos et al. (1995); Oehl and Sieverding (2004); Oehl et al. (2010);Oehl et al. (2011a, b); Pagano et al. (2013); Pereira et al. (2014); Pontes et al. (2017); Purin et al. (2006); Rabatin et al. (1993); Rivero-Mega et al. (2014); Rojas-Mego et al. (2014); Schalamuk et al. (2006); Schenck et al. (1984); Schneider et al. (2013); Sieverding and Howeler (1985); Sieverding and Toro (1987); Silva et al. (2007); Siqueira et al. (1987, 1989); Soteras et al. (2012, 2014, 2015); Sousa et al. (2013); Souza et al. (2013); Spain et al. (2006); Stürmer and Bellei (1994); Stürmer and Siqueira (2011); Urcelay et al. (2009); Vasconcellos et al. (2016); Velázquez and Cabello (2011); Velázquez et al. (2013, 2016); Vestberg et al. (1999); Vilcatoma-Medina et al. (2018); Walker et al. (1998); Zangaro et al. (2013)

different AMF morphospecies and evidencing an increasing interest in the study of these fungi in the region in recent years.

All the taxa were included in 9 families: Acaulosporaceae (40 morphospecies), Ambisporaceae (7), Archaeosporaceae (2), Claroideoglomeraceae (7), Diversisporaceae (9), Glomeraceae (62), Gigasporaceae (46), Pacisporaceae (7), Paraglomeraceae (4), and within 24 genera: *Acaulospora* (40 morphospecies), *Ambispora* (7), *Archaeospora* (2), *Bulbospora* (1), *Cetraspora* (4), *Clareoideoglomus* (7), *Corymbiglomus* (4), *Dentiscutata* (7), *Diversispora* (4), *Dominikia* (1), *Entrophospora* (2), *Funneliformis* (8), *Fuscutata* (1), *Gigaspora* (7), *Glomus* (29), *Pacispora* (6), *Paradensticutata* (2), *Paraglomus* (6), *Racocetra* (11), *Redeckera* (1), *Rhizophagus* (13), *Sclerocystis* (7), *Scutellospora* (13), *Septoglomus* (3). In consistency with Stürmer et al. (2018), Glomeraceae was the dominant family in South America according to the number of species per family. The next most dominant family was Acaulosporaceae in SA as well as in other parts of the world like North America, Europe and Antarctica, together with the Gigasporaceae family in Africa, Asia, and Oceania.

3.4.2 *Arbuscular Mycorrhizal Fungi in South America: Ecological Divisions*

We used the primary eighteen ecological divisions of SA described by Kelt and Meserve (2014) and modified by Young et al. (2007) and Josse et al. (2003) (Fig. 3.1). The Amazonia, Atlantic Forest, Caatinga and Chaco were the ecodivisions with the highest species records thus being the main research focus of SA (Table 3.1). Meanwhile, the Cerrado, Moist Pacific Temperate and Pampas showed an intermediate number of species records, while the lowest was observed in the Dry-South Central Andes, Guianan Uplands and Highlands, Llanos, Mediterranean Chile, Moist North-Central Andes, Moist Pacific Mesoamerica and Patagonia.

Morphospecies richness varied from 3 to 68 considering the taxa identified up to genus level. Most of the richest points were located in Brazil, which is considered one of the most biodiverse countries in the world, comprising six biomes with two hotspots: the Cerrado (Brazilian Savanna) and the Atlantic Forest. We observed a general pattern of high AMF richness along the diagonal comprised of the Caatinga, Atlantic Forest, Cerrado (Brazil), Pampas, Chaco (Argentina) and Moist Pacific Temperate (Chile) (Fig. 3.1). Other points of high species richness were located in the Amazonia, Llanos and Moist North Central Andes.

Many authors have discussed evidence about whether the current vegetation of the "savannah corridor" or "diagonal of open formations", which extends across South America from north-northeastern Brazil to the Chaco region of northern Argentina, represents the remnants of a once continuous forests (Prado and Gibbs 1993).

We visualized the similarities of AMF composition among the Atlantic Forest, Caatinga and Chaco in a Venn diagram (Fig. 3.2) using BioVenn (Hulsen et al. 2008).

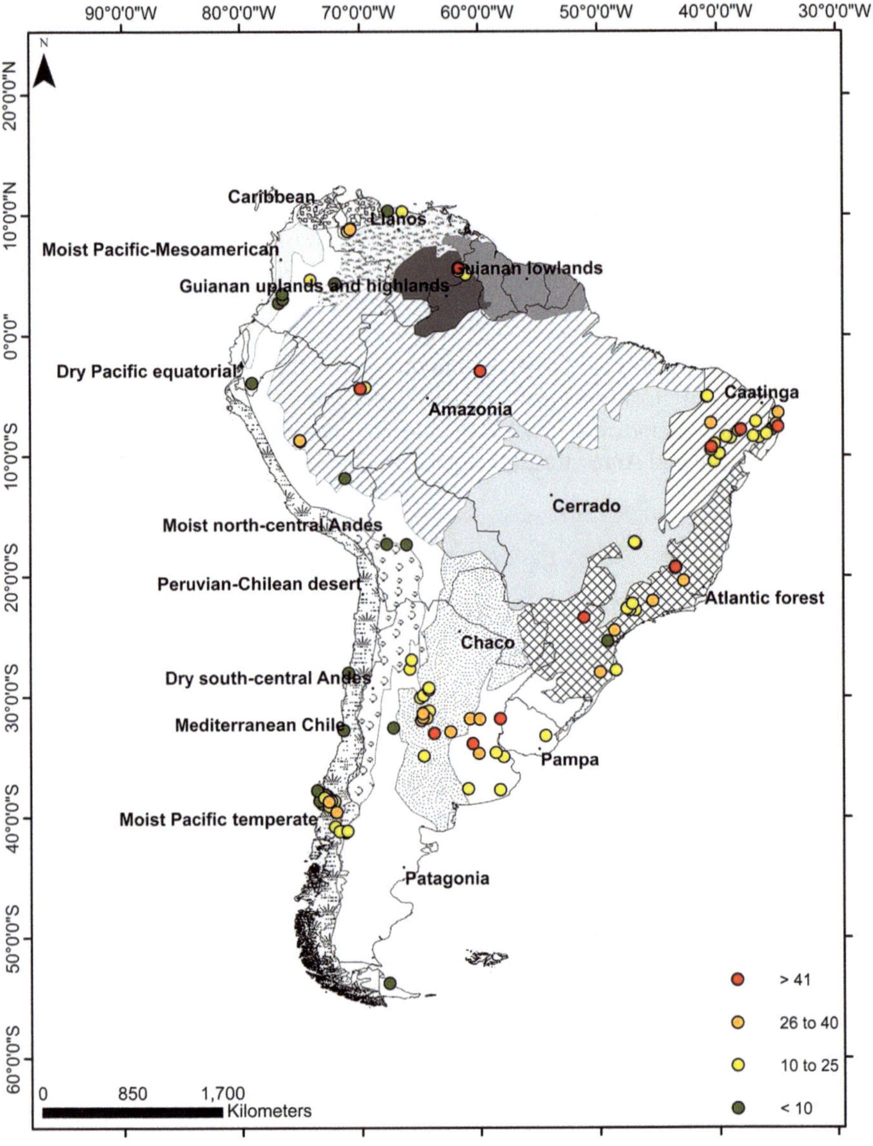

Fig. 3.1 Map with the primary ecological divisions of South America (Kelt and Meserve 2014, modified from Young et al. 2007 and Josse et al. 2003), showing the species richness distribution of AMF cited in the 110 reviewed research articles

The three ecodivisions shared 47 morphospecies of the 154 identified. The Atlantic Forest and Caatinga showed the highest species composition similarities (30), followed by the Atlantic Forest and Chaco (6), and finally by the Caatinga and Chaco (3) which were the most dissimilar ecodivisions. The Atlantic Forest showed 37 unique morphospecies, the Caatinga 16, and Chaco 15.

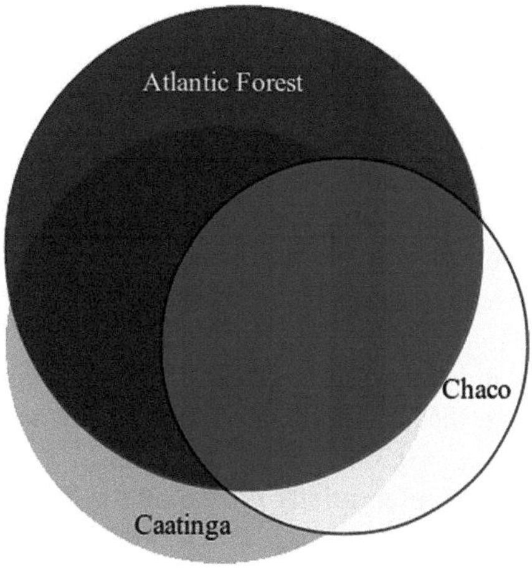

Fig. 3.2 Venn diagram comparing AMF morphospecies occurrence across the diagonal comprised of the Atlantic Forest, Caatinga and Chaco ecodivisions

Acaulospora scrobiculata (84 times recorded), *Claroideoglomus etunicatum* (75), *A. mellea* (67), *A. spinosa* (59), *Funneliformis mosseae* (54), and *Entrophospora infrequens* (49) were the morphospecies more frequently recorded in the studies reviewed. All these morphospecies except for *F. mosseae* occurred in most of the SA ecodivisions (between 11 to 13 of the 15 ecological divisions). Several morphospecies (106) occurred in no more than 3 ecodivisions and were recorded from 1 to 13 times (Table 3.1). These morphospecies could be defined as generalists (e.g. *A. scrobiculata, C. etunicatum* and *A. spinosa*) and specialists (e.g. *A. ignota, A. nicolsonii, A. fecundispora, Bulbospora minima* and *Septoglomus titan*) sensu Oehl et al. (2010). These authors differentiate specialist from generalists considering the number of times a species is found. Specialist morphospecies probably evidence an association with particular niche conditions. In contrast, the generalist species *Acaulospora scrobiculata* was detected in all seven continents (Stürmer et al. 2018) and other generalists such as *A. trappei, C. etunicatum* and *R. intraradices* have been commonly found in previous studies carried out in Central Europe (Wetzel et al. 2014; Säle et al. 2015), hence evidencing the worldwide distribution of some AMF taxa.

To evaluate the strength of the association of ecodivisions with AMF morphospecies, an indicator of species analysis was applied using the *indval()* function of the R package labdsv (Dufrene and Legendre 1997; R Core Team 2018; Roberts 2013). The analyses revealed 28 significant AMF morphospecies associated with 12 of the 14 ecodivisions. *G. australe* and *A. scrobiculata* were significantly associated with the Amazonia ecodivision (indicator value = 0.4, P = 0.025; and indicator value = 0.13, P = 0.001 respectively). *C. etunicatum* (indicator value = 0.13, P = 0.002) predominated in the Caatinga ecodivision. *Ambispora callosa* (indicator value = 0.6510; P = 0.005), *A. brasiliensis* (indicator value = 0.5765; P = 0.003), *R. invermaius* (indicator value = 0.4060, P = 0.028), and *R. clarus* (indicator value = 0.2581, P = 0.036) were significantly associated with the Cerrado, and *A. bireticulata* with the Chaco ecodivision. *Dentiscutata nigra* (indicator value = 1.0000; P = 0.013), *Cetraspora striata* (indicator value: 1.0000; P = 0.013), and *Funneliformis geosporum* (indicator value: 0.2238; P = 0.009) prevailed in the Dry South Central Andes; and both *S. crenulata* (indicator value = 0.6667; P = 0.021) and *S. spinosissima* (indicator value = 0.4861; P = 0.047) in the Guianan Uplands and Highlands. *A. morrowiae* (indicator value = 0.2899, P = 0.002) was indicative of the Llanos, and *S. calospora* (indicator value = 0.2971, P = 0.005), *R. diaphanus* (indicator value = 0.2928, P = 0.046), *A. laevis* (indicator value = 0.2842, P = 0.007), and *R. intraradices* (indicator value = 0.2256, P = 0.008) prevailed in the Mediterranean Chile. *Pac. chimonobambusae* (indicator value = 0.8000, P = 0.012), *Par. lacteum* (indicator value = 0.6400, P = 0.041), *S. dipapillosa* (indicator value = 0.4520, P = 0.050), and *A. rehmii* (indicator value = 0.2697, P = 0.008) were significantly associated with Moist north central Andes. Both *A. thomii* (indicator value = 0.3333, P = 0.043) and *G. pallidum* (indicator value = 0.2879, P = 0.050) were preferentially present in the Moist Pacific Temperate, and *F. mosseae* (indicator value = 0.1862, P = 0.002) in Pampas. *A. dilatata* (indicator value = 0.6165, P = 0.008), *A. alpina* (indicator value = 0.5053, P = 0.003) and *A. delicata* (indicator value = 0.3806, P = 0.006) were indicator of Patagonia.

This chapter revealed, that most taxa of Glomeromycota are present in SA, including 62% of the worldwide currently known AMF (~300). The vast majority of the territory is still poorly studied (Fig. 3.1). As a matter of fact, AMF communities in the ecodivisions of the Guianan Uplands and Highlands, Guianan Lowlands, Peruvian-Chilean desert, and the Caribbean are still unstudied. In addition, the Patagonia has been poorly sampled, which validates the idea that AMF diversity of SA will increase in relation to worldwide diversity (Veblen et al. 2015).

The high richness of AMF observed in the diagonal of Caatinga-Chaco is in concordance with other groups of fungi and plants that were located preferentially along these highly biodiverse ecosystems (Greer 2014). Moreover, the diagonal also matches the most studied ecodivisions thus suggesting that biased research may be overestimating AMF diversity. Studies focused on unstudied ecodivisions of this large area should be carried out in order to have a complete picture of the AMF diversity in SA.

3.5 Conclusion

Research on AMF diversity is still scarce in South America. Given the importance of these soil microorganisms for the functioning of ecosystems, it is essential to know their diversity. South America has a great potential to be explored in terms of AMF diversity due to the great diversity of biomes and its geographical extension. Its diversity of AMF represents 62% of the currently worldwide known diversity, and this information coming from studies concentrated only in few regions. It is crucial to generate inventories of species from all the ecosystems of this great region, to isolate fungi obtained and deposit them in germplasm banks. Soybeanization accompanied by monoculture, clearing, fumigation and displacement of peasants increasingly extends the agricultural frontier in this region, thus promoting the loss of biodiversity of the fauna and flora in South America. Therefore, most of the ecoregions that are affected by soybeanization surely have AMF species not yet described for science, which are undoubtedly of great importance for the maintenance of these systems.

References

Aguilera P, Cornejo P, Borie F, Barea JM, von Baer E, Oehl F (2014) Diversity of arbuscular mycorrhizal fungi associated with *Triticum aestivum* L. plants growing in an Andosol with high aluminum level. Agr Ecosyst Environ 186:178–184

Aguilera P, Marín C, Oehl F, Godoy R, Borie F, Cornejo P (2017) Selection of aluminum tolerant cereal genotypes strongly influences the arbuscular mycorrhizal fungal communities in an acidic Andosol. Agr Ecosyst Environ 246:86–93

Aidar MP, Carrenho R, Joly CA (2004) Aspects of arbuscular mycorrhizal fungi in an Atlantic Forest chronosequence Parque Estadual Turístico do Alto Ribeira (PETAR), SP. Biota Neotrop 4(2):1–13

Albuquerque PP (2008) Diversidade de Glomeromycetes e atividade microbiana em solos sob vegetação nativa do semi-árido de Pernambuco, Doctoral dissertation, PhD Thesis, Universidade Federal de Pernambuco, Recife, Brazil

Angulo-Veizaga WV, García-Apaza E (2014) *Baccharis incarum* and fungus Arbuscular Mycorrhizal symbiotic relationship for land fallow in the Bolivian highland. CienciAgro 3(1):51–58

Becerra A, Cabello M (2008) Hongos micorrícico arbusculares presentes en bosques de *Alnus acuminata* (Betulaceae) de la Yunga Argentina. B Soc Argent Bot 43(3–4):197–203

Becerra A, Bartoloni J, Cofré N, Soteras F, Cabello M (2014) Arbuscular mycorrhizal fungi in saline soils: vertical distribution at different soil depths. Braz J Microbiol 45:585–594

Becerra A, Cabello MN, Bartolini NJ (2011) Native arbuscular mycorrhizal fungi in the Yungas forest, Argentina. Mycologia 103:273–279

Bidartondo MI, Redecker D, Hijri I, Wiemken A, Bruns TD, Domínguez L, Sersic A, Leake JR, Read DJ (2002) Epiparasitic plants specialized on arbuscular mycorrhizal fungi. Nature 419:389–392

Błaszkowski J (2012) Glomeromycota. W. Szafer Institute of Botany, Polish Academy of Sciences

Bonfim JA, Vasconcellos RLF, Stürmer SL, Cardoso EJBN (2013) Arbuscular mycorrhizal fungi in the Brazilian Atlantic forest: A gradient of environmental restoration. Appl Soil Ecol 71:7–14

Bueno CG, Marín C, Silva-Flores P, Aguilera P, Godoy R (2017) Think globally, research locally: emerging opportunities for mycorrhizal research in South America. New Phytol 215:1306–1309

Cabello MN (1994) *Glomus antarcticum* sp. nov., a vesicular-arbuscular mycorrhizal fungus from Antarctica. Mycotaxon 51:123–128

Cabello MN (1997) Hydrocarbon pollution: its effect on native arbuscular mycorrhizal fungi (AMF). FEMS Microbiol Ecol 22 (3):233–236

Carrenho, R, Silva ES, Trufem SFB, Bononi VLR (2001) Successive cultivation of maize and agricultural practices on root colonization, number of spores and species of arbuscular mycorrhizal fungi. Braz J Microbiol 32(4):262–270

Casanova-Katny MA, Torres-Mellado GA, Palfner G, Cavieres LA (2011) The best for the guest: high Andean nurse cushions of *Azorella madreporica* enhance arbuscular mycorrhizal status in associated plant species. Mycorrhiza 21(7):613–622

Castillo C, Borie F, Godoy R, Rubio R, Sieverding E (2005) Diversity of mycorrhizal plant species and arbuscular mycorrhizal fungi in evergreen forest, deciduous forest and grassland ecosystems of Southern Chile. J Appl Bot Food Qual 80:40–47

Castillo CG, Rubio R, Rouanet JL, Borie F (2006) Early effects of tillage and crop rotation on arbuscular mycorrhizal fungal propagules in an Ultisol. Biol Fertil Soils 43(1):83–92

Castillo C, Rubio R, Borie F, Sieverding E (2010) Diversity of arbuscular mycorrhizal fungi in horticultural production systems of southern Chile. J Soil Sci Plant Nut 10(4):407–413

Castillo CG, Oehl F, Sieverding E (2016) Arbuscular mycorrhizal fungal diversity in wheat agro-ecosystems in Southern Chile and effects of seed treatment with natural products. J Soil Sci Plant Nut 16(4):967–978

Chaudhary VB, Lau MK, Johnson NC (2008) Macroecology of Microbes – Biogeography of the Glomeromycota. In: Varma A (eds) Mycorrhiza. Springer, Berlin, Heidelberg

Cofré MN, Ferrari AE, Becerra A, Domínguez L, Wall LG, Urcelay C (2017) Effects of cropping systems under no-till agriculture on arbuscular mycorrhizal fungi in Argentinean Pampas. Soil Use Manage 33:364–378

Colombo RP, Fernandez Bidondo L, Silvani VA, Carbonetto MB, Rascovan N, Bompadre MJ, Pergola M, Cuenca G, Godeas AM (2014) Diversity of arbuscular mycorrhizal fungi in soil from the Pampa Ondulada, Argentina, assessed by pyrosequencing and morphological techniques. Can J Microbiol 60: 819–827.

Copley J. 2000. Ecology goes underground. Nature 406:452–454

Cordoba AS, de Mendonça MM, Stürmer SL, Rygiewicz PT (2001) Diversity of arbuscular mycorrhizal fungi along a sand dune stabilization gradient: A case study at Praia da Joaquina, Ilha de Santa Catarina, South Brazil. Mycoscience 42(4):379–387

Covacevich F, Marino M A, Echeverrica H E (2006) The phosphorus source determines the arbuscular mycorrhizal potential and the native mycorrhizal colonization of tall fescue and wheatgrass. Eur J Soil Biol 42:127–138

Coutinho ES, Fernandes GW, Berbara RLL, Valério HM, Goto BT (2015) Variation of arbuscular mycorrhizal fungal communities along an altitudinal gradient in rupestrian grasslands in Brazil. Mycorrhiza 25(8):627–638

Cuenca G, Herrera-Peraza R (2008) *Scutellospora striata* sp. nov., a newly described glomeromycotan fungus from La Gran Sabana, Venezuela. Mycotaxon 105(6):79–87

Cuenca G, Lovera M (1992) Vesicular-arbuscular mycor¬rhizae in disturbed and revegetated sites from La Gran Sabana, Venezuela. Canadian Journal of Botany 70: 73–79

Cuenca G, Meneses E (1996) Diversity patterns of arbuscular mycorrhizal fungi associated with cacao in Venezuela. Plant Soil 183(2):315–322

Cuenca G, De Andrade Z, Escalante G (1998) Diversity of Glomalean spores from natural disturbed and revegetated communities growing on nutrient-poor tropical soils. Soil Biology and Biochemistry 30: 711–719

de Carvalho F, De Souza FA, Carrenho R, de Souza Moreira FM, da Conçeição Jesus E, Fernandes GW (2012) The mosaic of habitats in the high-altitude Brazilian rupestrian fields is a hotspot for arbuscular mycorrhizal fungi. Appl Soil Ecol 52:9–19

da Silva DKA, de Oliveira Freitas N, Cuenca G, Maia LC, Oehl F (2008) *Scutellospora pernambucana*, a new fungal species in the Glomeromycetes with a diagnostic germination orb. Mycotaxon 106(1):361–370

da Silva DKA, Pereira CMR, de Souza RG, da Silva GA, Oehl F, Maia LC (2012) Diversity of arbuscular mycorrhizal fungi in restinga and dunes areas in Brazilian Northeast. Biodivers Conserv 21(9):2361–2373

da Silva GA, Trufem SFB, Júnior OJS, Maia LC (2005) Arbuscular mycorrhizal fungi in a semi-arid copper mining area in Brazil. Mycorrhiza 15(1):47–53

da Silva IR, de Mello CMA, Neto RAF, da Silva DKA, de Melo AL, Oehl F, Maia LC (2014) Diversity of arbuscular mycorrhizal fungi along an environmental gradient in the Brazilian semiarid. Appl Soil Ecol 84:166–175

de Mello CMA (2011) Fungos micorrízicos arbusculares do núcleo de desertificação de Cabrobó-PE, Doctoral dissertation, Dissertação de mestrado, Programa de Pós-graduação em Biologia de Fungos, Universidade Federal de Pernambuco, Recife p 66

de Mello CMA, da Silva GA, de Assis DMA, de Pontes JS, de Almeida Ferreira AC, Porto Carneiro LM, Evangelista Vieira HE, Costa Maia L, Oehl F (2013) *Paraglomus pernambucanum* sp.

nov. and *Paraglomus bolivianum* comb. nov., and biogeographic distribution of *Paraglomus* and *Pacispora*. J Appl Bot Food Qual 86:113–125

de Oliveira Freitas N (2006) Aspectos da associação de fungos micorrízicos arbusculares (Glomeromycota) em videira (*Vitis* sp.). Dissertação (Mestrado). Programa de Pós-Graduação em Biologia de Fungos, Universidade Federal de Pernambuco, Recife

de Oliveira Freitas R, Buscardo E, Nagy L, dos Santos Maciel AB, Carrenho R, Luizão RC (2014) Arbuscular mycorrhizal fungal communities along a pedo-hydrological gradient in a Central Amazonian terra firme forest. Mycorrhiza 24(1):21–32

Davison J, Moora M, Öpik M, Adholeya A, Ainsaar L, Bâ A, Burla S, Diedhiou AG, Hiiesalu L, Jairus T, Johnson NC, Kane A, Koorem K, Kochar M, Ndiaye C, Pärtel M, Reier Ü, Saks Ü, Singh R, Vasar M, Zobel M (2015) Global assessment of arbuscular mycorrhizal fungus diversity reveals very low endemism. Science 970–973

Dhillion SS, Vidiella PE, Aquilera LE, Friese, CF, De Leon E, Armesto JJ, Zak JC (1995) Mycorrhizal plants and fungi in the fog-free Pacific coastal desert of Chile. Mycorrhiza 5:381–386

Dodd JC, Arias I, Koomen I, Hayman DS (1990) The management of populations of vesicular-arbuscular mycorrhizal fungi in acid-infertile soils of a savanna ecosystem. Plant Soil 122:229–240

Dufrene M, Legendre P (1997) Species assemblages and indicator species: the need for a flexible asymmetrical approach. Ecol Monogr 67(3):345–366

Escudero V, Mendoza R (2005) Seasonal variation of arbuscular mycorrhizal fungi in temperature grassland along a wide hydrologic gradient. Mycorrhiza 15(4):291–299

Fernandes A, Siqueira JO (1989) Micorrizas vesicular-arbusculares em cafeeiros da região Sul do Estado de Minas Gerais. Pesq Agropec Bras 24(12):1489–1498

Fittkau EJ (1969) The fauna of South America. In: Fittkau EJ, Illies J, Klinge H, Schwabe GH, Sioli H (eds) Biogeography and ecology in South America, Junk, La Haya, p 624–650

Fontenla S, Godoy R, Rosso P, Havrylenko M (1998) Root associations in *Austrocedrus* forests and seasonal dynamics of arbuscular mycorrhizas. Mycorrhiza 8:29–33

França SC, Gomes-da-Costa SM, Silveira AP (2007) Microbial activity and arbuscular mycorrhizal fungal diversity in conventional and organic citrus orchards. Biol Agric Horticul 25(2):91–102

Frioni L, Minasian H, Volfovicz R (1999) Arbuscular mycorrhizae and ectomycorrhizae in native tree legumes in Uruguay. Forest Ecol Manag 115(1):41–47

Furrazola E, Goto BT, Silva GAD, Torres-Arias Y, Morais T, Pereira de Lima CE, de Almeida Ferreira AC, Costa ML, Sieverding, E, Oehl F (2013) *Acaulospora herrerae*, a new pitted species in the Glomeromycetes from Cuba and Brazil. Nova Hedwigia 97(3–4):401–413

García S, Pezzani F, Rodríguez-Blanco A (2017) Long-term phosphorus fertilization effects on arbuscular mycorrhizal fungal diversity in Uruguayan grasses. J Soil Sci Plant Nut 17(4):1013–1027

Gómez-Carabalí A, Rao IM, Tupac Otero J (2011) Influence of fertilization, season, and forage species in presence of arbuscular mycorrhizae in a degraded Andisoil of Colombia. Acta Agron 60:84–92

Goto BT, Maia LC (2005) Sporocarpic species of arbuscular mycorrhizal fungi (Glomeromycota), with a new report from Brazil. Acta Bot Brasil 19(3):633–637

Goto BT, Maia LC, da Silva GA, Oehl F (2009). *Racocetra intraornata*, a new species in the Glomeromycetes with a unique spore wall structure. Mycotaxon 109(1):483–491

Goto, BT, Silva GAD, Maia LC, Oehl F (2010a) *Dentiscutata colliculosa*, a new species in the Glomeromycetes from Northeastern Brazil with colliculate spore ornamentation. Nova Hedwigia 90(3–4):383–393

Goto BT, da Silva GA, Yano-Melo AM, Maia LC (2010b) Checklist of the arbuscular mycorrhizal fungi (Glomeromycota) in the Brazilian semiarid. Mycotaxon 113:251–254

Greer FE (ed) (2014) Dry Forests: Ecology, Species Diversity and Sustainable Management. Nova Science Publ, United States of America

Grilli G, Urcelay C, Galetto L (2012) Forest fragment size and nutrient availability: complex responses of mycorrhizal fungi in native–exotic hosts. Plant Ecol 213:155–165

Grilli G, Urcelay C, Galetto L, Davison J, Vasar M, Saks Ü, Jairus T, Öpik M (2015) The composition of arbuscular mycorrhizal fungal communities in the roots of a ruderal forb is not related to the forest fragmentation process. Environ Microbiol 17 (8):2709–2720

Herrera-Peraza RA, Cuenca G, Walker C (2001) *Scutellospora crenulata*, a new species of Glomales from La Gran Sabana, Venezuela. Can J Botany 79(6):674–678

Herrera-Peraza RA, Montilla M, Furrazola E, Ferrer RL, Morales S, Monasterio M (2016) Presencia y distribución de representantes hipógeos de la clase Glomeromycetes (hongos micorrizógenos VA) en ecosistemas andinos venezolanos. Acta Bot Cubana 215(2):196–217

Hulsen T, De Vlieg J, Alkema W (2008) BioVenn – a web application for the comparison and visualization of biological lists using area-proportional Venn diagrams. BMC Genomics 9:488

Janos DP, Sahley CT, Emmons LH (1995) Rodent Dispersal of Vesicular-Arbuscular Mycorrhizal Fungi in Amazonian Peru. Ecology 76(6): 1852–1859

Jobim K, Santos Oliveira BI, Goto BT (2016) Checklist of the Glomeromycota in the Brazilian Savanna. Mycotaxon 131:255

Jobim K, Vista XM, Goto BT (2018) Updates on the knowledge of arbuscular mycorrhizal fungi (Glomeromycotina) in the Atlantic Forest biome – an example of very high species richness in Brazilian biomes. Mycotaxon 133 (1)

Josse C, Navarro G, Comer P, Evans R, Faber-Langendoen D, Fellows M, Kittel G, Menard S, Pyne M, Reid M, Schulz K, Snow K, Teague J (2003). Ecological Systems of Latin America and the Caribbean: A Working Classification of Terrestrial Systems. Nature Serve, Arlington, Virginia

Kelt DA, Meserve P (2014) Status and challenges for conservation of small mammal assemblages in South America. Biol Rev 89:705–722

Kivlin SN, Hawkes CV, Treseder KK (2011) Global diversity and distribution of arbuscular mycorrhizal fungi. Soil Biol Biochem 43:2294–2303

Krüger C (2013) Arbuscular mycorrhizal fungi for reforestation of native tropical trees in the Andes of South Ecuador (Doctoral dissertation, lmu)

Lavina EL, Fauth G (2011) Evolução geológica da América do Sul nos últimos 250 milhões de anos. In: Carvalho JB, Almeida EA (eds) Biogeografia da America do Sul, padrões and processos. Roca, São Paulo, Brasil, p 3–13

Leal PL, Stürmer SL, Siqueira JO (2009) Occurrence and diversity of arbuscular mycorrhizal fungi in trap cultures from soils under different land use systems in the Amazon, Brazil. Braz J Microbiol 40(1):111–121

Lemos IB (2008) Simbiose micorrízica arbuscular em porta enxertos de videira (*Vitis* spp.). Dissertação (Mestrado). Programa de Pós-Graduação em Biologia de Fungos, Universidade Federal de Pernambuco, Recife

Longo S, Nouhra E, Goto BT, Berbara RL, Urcelay C (2014) Effects of fire on arbuscular mycorrhizal fungi in the Mountain Chaco Forest. For Ecol Manag 315:86–94

Lugo MA, Cabello MN (1999) Acaulosporaceae (Glomales, Zygomycetes) en pastizales autóctonos del centro de Argentina. Darwiniana 37(3–4): 323–332

Lugo MA, Cabello MN (2002) Native arbuscular mycorrhizal fungi (AMF) from mountain grassland (Córdoba, Argentina) I. Seasonal variation of fungal spore diversity. Mycologia 94 (4):579–586

Lugo MA, Anton AN, Cabello MN (2005) Arbuscular mycorrhizas in the Larrea divaricata shrubland at arid "Chaco", Central Argentina. J Agric Tech 1:163–178

Lugo MA, Ferrero MA, Menoyo E, Estévez MC, Siñeriz F, Anton AM (2008) Arbuscular mycorrhizal fungi and rhizospheric bacteria diversity along an altitudinal gradient in South American Puna grassland. Microb Ecol 55:705–713

Marín C, Aguilera P, Cornejo P, Godoy R, Oehl F, Palfner G, Boy J (2016) Arbuscular mycorrhizal assemblages along contrasting Andean forests of Southern Chile. J Soil Sci Plant Nut 16(4):916–929

Marín C, Aguilera P, Oehl F, Godoy R (2017) Factors affecting arbuscular mycorrhizal fungi of Chilean temperate rainforests. J Soil Sci Plant Nut 17(4):966–984

Medina J, Cornejo P, Borie F, Meier S, Palenzuela J, Vieira HEE, Ferreira ACA, Silva GAl, Sánchez-Castro I, Oehl F (2014) *Corymbiglomus pacificum*, a new glomeromycete from a saline lakeshore in Chile. Mycotaxon127 (11):173–183

Medina J, Meier S, Rubio R, Curaqueo G, Borie F, Aguilera P, Oehl F, Cornejo P (2015) Arbuscular mycorrhizal status of pioneer plants from the mouth of lake Budi, Araucanía Region, Chile. J Soil Sci Plant Nut 15(1):142–152

Meier S, Borie F, Curaqueo G, Bolan N, Cornejo P (2012) Effects of arbuscular mycorrhizal inoculation on metallophyte and agricultural plants growing at increasing copper levels. Appl Soil Ecol 61:280–287

Mendoza RE, Goldmann V, Rivas J, Escudero V, Pagani E, Collantes M, Marbán L (2002) Poblaciones de hongos micorrícicos arbusculares en relación con las propiedades del suelo y de la planta hospedante en pastizales de Tierra del Fuego. Ecolo Aust 12:105–116

Menéndez AB, Scervino JM, Godeas AM (2001) Arbuscular mycorrhizal populations associated with natural and cultivated vegetation on a site of Buenos Aires province, Argentina. Biol Fertil Soils 33:373–381

Menoyo E, Renison D, Becerra A (2009) Arbuscular mycorrhizas and performance of *Polylepis australis* trees in relation to livestock density. For Ecol Manag 258(12):2676–2682

Mergulhão ACES (2007) Aspectos Ecológicos e Moleculares de Fungos Micorrízicos Arbusculares. Recife, UFP, Brazil

Moreira M, Baretta D, Tsai SM, Cardoso EJBN (2009) Arbuscular mycorrhizal fungal communities in native and in replanted *Araucaria* forest. Sci Agric 66(5):677–684

Morton J, Bentivenga S, Wheeler W (1993). Germ plasm in the International Collection of Arbuscular and Vesicular-Arbuscular Mycorrhizal Fungi (INVAM) and procedures for culture development, documentation and storage. Mycotaxon 48:491–528

Oehl F, Sieverding E (2004) *Pacispora*, a new vesicular-arbuscular mycorrhizal fungal genus in the Glomeromycetes. J Appl Bot Food Qual 78:72–82

Oehl F, Laczko E, Bogenrieder A, Stahr K, Bösch R, van der Heijden M, Sieverding E (2010) Soil type and land use intensity determine the composition of arbuscular mycorrhizal fungal communities. Soil Biol Biochem 42:724–738

Oehl F, da Silva GA, Goto BT, Sieverding E (2011a) Glomeromycota: three new genera and glomoid species reorganized. Mycotaxon 116:75–120

Oehl F, Sieverding E, Palenzuela J, Ineichen K, da Silva GA (2011b) Advances in Glomeromycota taxonomy and classification. IMA Fungus 2:191–199

Öpik M, Vanatoa A, Vanatoa E, Moora M, Davison J, Kalwij JM, Reier Ü, Zobel M (2010) The online database MaarjAM reveals global and ecosystemic distribution patterns in arbuscular mycorrhizal fungi (Glomeromycota). New Phytol 188:223–241

Öpik M, Zobel M, Cantero JJ, Davison J, Facelli JM, Hiiesalu I, Jairus T, Kalwij JM, Koorem K, Leal ME, Liira J, Metsis M, Neshataeva V, Paal J, Phosri C, Põlme S, Reier Ü, Saks Ü, Schimann H, Thiéry O, Vasar M, Moora M (2013) Global sampling of plant roots expands the described molecular diversity of arbuscular mycorrhizal fungi. Mycorrhiza 23:411–430

Pagano MC, Zandavalli RB, Araújo FS (2013) Biodiversity of arbuscular mycorrhizas in three vegetational types from the semiarid of Ceará State, Brazil. Appl Soil Ecol 67:37–46

Palfner G (2001) Taxonomische studien an ektomykorrhizen aus den nothofagus-wáldern mittelsudchiles: mit Tabellen in Text/Götz Palfner. Berlin, Germany: J. Cramer in der Gebrüder Borntraeger Verlagsbuchhandlung, Bibliotheca Mycologica

Pärtel M, Bennett JA, Zobel M (2016) Macroecology of biodiversity: disentangling local and regional effects. New Phytol 211:404–410

Pereira CMR, da Silva DKA, de Almeida Ferreira AC, Goto BT, Maia LC (2014) Diversity of arbuscular mycorrhizal fungi in Atlantic forest areas under different land uses. Agr Ecosyst Environ 185:245–252

Pontes JS, Oehl F, Marinho F, Coyne D, Silva DKAD, Yano-Melo AM, Maia LC (2017) Diversity of arbuscular mycorrhizal fungi in Brazil's Caatinga and experimental agroecosystems. Biotropica 49(3):413–427

Prado DE, Gibbs PE (1993) Patterns of species distribution in the dry seasonal forests of South America. Ann Missouri Bot Gard 80 (4):902–927

Purin S, Klauberg Filho O, Stürmer SL (2006) Mycorrhizae activity and diversity in conventional and organic apple orchards from Brazil. Soil Biol Biochem 38(7):1831–1839

Rabatin SC, Stinner BR, Paoletti MG (1993) Vesicular-arbuscular mycorrhizal fungi, particularly *Glomus tenue*, in Venezuelan bromeliad epiphytes. Mycorrhiza 4(1):17–20

R Core Team (2018) R: A language and environment for statistical computing. R Foundation for Statistical Computing, Vienna, Austria. http://www.R-project.org

Redecker D, Schüßler A, Stockinger H, Stürmer SL, Morton JB, Walker C (2013) An evidence-based consensus for the classification of arbuscular mycorrhizal fungi (Glomeromycota). Mycorrhiza 23:515–531

Rillig MC (2004) Arbuscular mycorrhizae and terrestrial ecosystem processes. Ecol Lett 7(8):740–754

Rivero-Mega MS, Crespo EM, Molina MG, Lugo MA (2014) Diversidad diferencial de esporas de Glomeromycota en la rizosfera de Bromeliáceas nativas del Parque Nacional Sierra de Las Quijadas (San Luis, Argentina) Bol Soc Argen Bot 49 (3):317–325

Roberts DW (2013) labdsv: ordination and multivariate analysis for ecology package. Version 1.6–1. http://cran.r project.org/web/packages/labdsv

Robinson-Boyer L, Grzyb I, Jeffries P (2009) Shifting the balance from qualitative to quantitative analysis of arbuscular mycorrhizal communities in field soils. Fungal Ecol 2:1–9

Rojas-Mego KC, Elizarbe-Melgar C, Gárate-Díaz MH, Ayala-Montejo D, Pedro RC, Sieverding, E (2014) Hongos de micorriza arbuscular en tres agroecosistemas de cacao (*Theobroma cacao* L.) en la amazonía peruana. Folia Amazónica 23(2):149–156

Rosendahl S (2008) Communities, populations and individuals of arbuscular mycorrhizal fungi. New Phytol 178:253–266

Säle V, Aguilera P, Laczko E, Mäder P, Berner A, Zihlmann U, van der Heijden MGA, Oehl F (2015) Impact of conservation tillage and organic farming on the diversity of arbuscular mycorrhizal fungi. Soil Biol Biochem 84:38–52

Schalamuk S, Velazquez S, Chidichimo H, Cabello M (2006) Fungal spore diversity of arbuscular mycorrhizal fungi associated with spring wheat: effects of tillage. Mycologia 98:16–22

Schenck NC, Joyce L, Spain E, Sieverding E, Howeler RH (1984) Several new and unreported vesicular-arbuscular mycorrhizal fungi (Endogonaceae) from Colombia. Mycologia 76:685–699

Schneider J, Stürmer SL, Guilherme LRG, de Souza Moreira FM, de Sousa Soares CRF (2013) Arbuscular mycorrhizal fungi in arsenic-contaminated areas in Brazil. J Hazard Mater 262:1105–1115

Schüßler A, Walker C (2010) The Glomeromycota: a species list with new families and new genera. Gloucester, in libraries at The Royal Botanic Garden Edinburgh, The Royal Botanic Garden Kew, Botanische Staatssammlung Munich and Oregon State University

Senés-Guerrero C, Schüßler A (2016) A conserved arbuscular mycorrhizal fungal core-species community colonizes potato roots in the Andes. Fungal Divers 77:317–333

Sieverding E, Howeler RH (1985) Influence of species of VA mycorrhizal fungi on cassava yield response to phosphorus fertilization. Plant Soil 88:213–221

Sieverding E, Toro TS (1987) *Acaulospora denticulata* sp. nov. and *Acaulospora rehmii* sp. nov. (Endogonaceae) with ornamented spore walls. Angewandte Botanik 61:217–223

Silva LX, Figueiredo MVB, Silva GA, Goto BT, Oliveira JP, Burity HA (2007) Fungos micorrízicos arbusculares em áreas de plantio de leucena e sabiá no estado de Pernambuco. Rev Árvore 31:427–435

Siqueira JO, Colozzi-Filho A, Oliveira E, Fernandes AB, Florence ML (1987) Micorrizas vesicular-arbusculares em mudas de cafeeiro produzidas no sul do estado de Minas Gerais. Pesq Agropec Bras 22(1):31–38

Siqueira JO, Colozzi Filho A, Oliveira E (1989) Ocorrência de micorrizas vesicular-arbusculares em agro e ecossistemas do estado de Minas Gerais. Pesq Agrop Bras 24:1499–1506

Smith SE, Read DJ (2008) Mycorrhizal symbiosis, 2nd ed. Academic Press Ltd, London

Soteras F, Becerra A, Cofré N, Bartoloni J, Cabello M (2012) Arbuscular mycorrhizal fungal species in saline environments of Central Argentina: seasonal variation and distribution of spores at different soil depths. Sydowia 64:301–311

Soteras F, Renison D, Becerra AG (2014) Restoration of high altitude forests in an area affected by a wildfire: *Polylepis australis* Bitt. seedlings performance after soil inoculation. Trees 28 (1):173–182

Soteras F, Grilli G, Cofré N, Marro N, Becerra A (2015) Arbuscular mycorrhizal fungal composition in high montane forests with different disturbance histories in Central Argentina. Appl Soil Ecol 85:30–37

Soteras, F, Moreira, BC, Grilli G, Pastor N, Mendes F C, Mendes DR, Reninson D, Megumi Kasuya NC, de Souza FA, Becerra A (2016) Arbuscular mycorrhizal fungal diversity in rhizosphere spores versus roots of an endangered endemic tree from Argentina: Is fungal diversity similar among forest disturbance types?. Appl Soil Ecol 98:272–277

Sousa CDS, Menezes RSC, Sampaio EVDSB, Lima FDS, Oehl F, Maia LC (2013) Arbuscular mycorrhizal fungi within agroforestry and traditional land use systems in semi-arid Northeast Brazil. Acta Scientiarum. Agronomy 35(3):307–314

Souza RG, Silva DKA, Mello CMA, Goto BT, Silva FSB, Sampaio EVSB, Maia, LC (2013) Arbuscular mycorrhizal fungi in revegetated mined dunes. Land Degrad Dev 24(2):147–155

Spain JL, Sieverding E, Oehl F (2006) *Appendicispora*: a new genus in the arbuscular mycorrhiza-forming Glomeromycetes, with a discussion of the genus *Archaeospora*. Mycotaxon 97:163–182

Stürmer SL, Bellei MM (1994) Composition and seasonal variation of spore populations of arbuscular mycorrhizal fungi in dune soils on the island of Santa Catarina. Brazil. Can J Botany 72(3):359–363

Stürmer SL, Siqueira JO (2011) Species richness and spore abundance of arbuscular mycorrhizal fungi across distinct land uses in western Brazilian Amazon. Mycorrhiza 21(4):255–67

Stürmer SL, Bever JD, Morton JB (2018) Biogeography of arbuscular mycorrhizal fungi (Glomeromycota): a phylogenetic perspective on species distribution patterns. Mycorrhiza 28:587–603

Tedersoo L, Bahram M, Põlme S, Kõljalg U, Yorou NS, Wijesundera R, Ruiz LV, Vasco-Palacios AM, Thu PQ, Suija A, Smith ME, Sharp C, Saluveer E, Saitta A, Rosas M, Riit T, Ratkowsky D, Pritsch K, Põldmaa K, Piepenbring M, Phosri C, Peterson M, Parts K, Pärtel K, Otsing E, Nouhra E, Njouonkou AL, Nilsson RH, Morgado LN, Mayor J, May TW, Majuakim L, Lodge DJ, Lee SS, Larsson KH, Kohout P, Hosaka K, Hiiesalu I, Henkel TW, Harend H, Guo LD, Greslebin A, Grelet G, Geml J, Gates G, Dunstan W, Dunk C, Drenkhan R, Dearnaley J, De Kesel A, Dang T, Chen X, Buegger F, Brearley FQ, Bonito G, Anslan S, Abell S, Abarenkov K (2014) Global diversity and geography of soil fungi. Science 346:1256688–1–10

Tedersoo L, Sánchez-Ramírez S, Kõljalg U, Bahram M, Döring M, Schigel D, May T, Ryberg M, Abarenkov K (2018) High-level classification of the Fungi and a tool for evolutionary ecological analyses. Fungal Divers 90(1):135–159

Urcelay C, Diaz S, Gurvich DE, Chapin FS III, Cuevas E, Domínguez LS (2009) Mycorrhizal community resilience in response to experimental plant functional type removals in a woody ecosystem. J Ecol 97:1291–1301

Vasconcellos RL, Bonfim JA, Baretta D, Cardoso EJ (2016) Arbuscular mycorrhizal fungi and glomalin-related soil protein as potential indicators of soil quality in a recuperation gradient of the Atlantic forest in Brazil. Land Degrad Dev 27(2):325–334

Veblen TT, Young KR, Orme AR (eds) (2015) The physical geography of South America. Oxford University Press

Velázquez S, Cabello M (2011) Occurrence and diversity of arbuscular mycorrhizal fungi in trap cultures from El Palmar National Park soils. Eur J Soil Biol 47:230–235

Velázquez M, Cabello M, Barrera M (2013) Composition and structure of arbuscular-mycorrhizal communities in El Palmar National Park, Argentina. Mycologia 105(3):509–520

Velázquez MS, Stürmer SL, Bruzone C, Fontenla S, Barrera M, Cabello M (2016) Occurrence of arbuscular mycorrhizal fungi in high altitude sites of the Patagonian Altoandina region in Nahuel Huapi National Park (Argentina). Acta Bot Bras 30(4):521–531

Vestberg M, Cardoso M, Mårtensson A (1999) Occurrence of arbuscular mycorrhizal fungi in different cropping systems at Cochabamba, Bolivia. Agric Food Sci Finl 8:309–318

Viglizzo EF, Frank FC, Carreno LV, Jobbagy EG, Pereyra H, Clatt J, Pincen D, Ricard MF (2011) Ecological and environmental footprint of 50 years of agricultural expansion in Argentina. Global Change Biol 17:959–973

Vilcatoma-Medina C, Kaschuk G, Zanette F (2018) Colonization and spore richness of arbuscular mycorrhizal fungi in Araucaria nursery seedlings in Curitiba, Brazil. Int J Agron Vol 2018, Article ID 5294295, 6 pages https://doi.org/10.1155/2018/5294295

Villagrán C, Hinojosa LF (1997) Historia de los bosques del sur de Sudamérica, II: análisis fitogeográfico. Rev Chil Hist Nat 70:241–267

Walker C, Cuenca G, Sánchez F (1998) *Scutellospora spinosissima* sp. nov., a newly described glomalean fungus from acidic, low nutrient plant communities in Venezuela. Ann Bot London 82(6):721–725

Walker C, Gollotte A, Redecker D (2018) A new genus, *Planticonsortium* (Mucoromycotina), and new combination (*P. tenue*), for the fine root endophyte, *Glomus tenue* (basionym *Rhizophagus tenuis*). Mycorrhiza 28:213–219

Wetzel K, Silva G, Matczinski U, Oehl F, Fester T (2014). Superior differentiation of arbuscular mycorrhizal fungal communities from till and no-till plots by morphological spore identification when compared to T-RFLP. Soil Biol Biochem 72:88–96

Young KR, Berry PE, Veblen TT (2007) Flora and Vegetation. In: Young KR, Berry PE, Veblen TT (eds)The Physical Geography of South America, Oxford University Press, UK

Zak DR, Holmes WE, White DC, Peacock AD, Tilman D (2003) Plant diversity, soil microbial communities, and ecosystem function: are there any links? Ecology 84(8):2042–2050

Zangaro W, Rostirola LV, de Souza PB, de Almeida Alves R, Lescano LEAM, Rondina ABL, Nogueira MA, Carrenho R (2013) Root colonization and spore abundance of arbuscular mycorrhizal fungi in distinct successional stages from an Atlantic rainforest biome in southern Brazil. Mycorrhiza 23(3):221–233

Chapter 4
Ectomycorrhizal Fungi in South America: Their Diversity in Past, Present and Future Research

Eduardo R. Nouhra, Götz Palfner, Francisco Kuhar, Nicolás Pastor, and Matthew E. Smith

4.1 Introduction. Primary Knowledge of Ectomycorrhizal Diversity at the Regional Level

Although the functional morphology and anatomy of the ectomycorrhizal (ECM) symbiosis was first explored in the 1880's (Frank 1885), pre-molecular monitoring of local or regional diversity of ectomycorrhizal fungi (EMF) required extensive time and effort. Determination of the mycorrhizal nature of thousands of potential fungal partners of ECM host trees worldwide was limited due to the available methodologies during most of the twentieth century. During that time the ECM habit was deduced primarily based on observations of growth patterns of fruiting bodies. Melin (1925) was the first to successfully establish *in vitro* synthesis of the ECM symbiosis, inoculating seedlings of European conifers and broad-leaved trees with axenic cultures of compatible EMF belonging to the genera *Amanita*, *Boletus*, *Cortinarius*, *Lactarius*, *Russula* and *Tricholoma*. However, only a few culturable species were proven to form mycorrhizas via this time-consuming and failure-prone method. Consequently, several EMF species from South American Nothofagaceae forests that were described in the late 19th and early 20th centuries were not known to form ectomycorrhizae at the time of their original description. Several emblematic taxa were assumed but not confirmed to be EMF, including *Boletus loyo* (Spegazzini 1912, Espinosa 1915), *Cortinarius magellanicus* (Spegazzini 1887)

E. R. Nouhra (✉) · F. Kuhar · N. Pastor
IMBIV/CONICET, (F.C.E.F.y N.) Universidad Nacional de Córdoba, Córdoba, Argentina

G. Palfner
Universidad de Concepción, Concepción, Chile

M. E. Smith
Department of Plant Pathology, University of Florida, Gainesville, FL, USA

M. C. Pagano, M. A. Lugo (eds.), *Mycorrhizal Fungi in South America*, Fungal Biology, https://doi.org/10.1007/978-3-030-15228-4_4

and *Descolea antarctica* (Singer 1950). Even Singer's classic phyto-mycogeographical studies of Nothofagaceae ectotrophic ecosystems relied on indirect methods to deduce which fungi were ECM and which were not. Singer and Moser (1965) used field observations of fruiting bodies presence beneath *Nothofagus* trees, and the absence of the same fungi from anectotrophic forest to determine the ECM status of a candidate fungi. Of course this only worked well with common taxa that regularly produced fruiting bodies.

During the second half of the twentieth century it became possible to directly observe the diversity of EMF on tree roots due to novel protocols for tracking mycelial connections between fruiting bodies and ectomycorrhizae. This worked well for comparative morphological-anatomical examination of cellular structures formed by the fungal mantle and other elements, which are often diagnostic and can allow identification of EMF to at least the genus level. These direct observational methods were initially applied to Northern Hemisphere ECM communities by Dominik (1969) and Zak (1973) and were later refined by Agerer and co-workers (Agerer 1991, 1996). Chilvers (1968) was the first to use this morpho-anatomical approach in the Southern Hemisphere, studying diversity of EMF on the roots of *Eucalyptus* from Australia.

Native ECM forests in South America (SA) are considered less extensive than those in the Northern Hemisphere, partly due to the fact that the ECM plants are almost all angiosperms (including members of the Betulaceae, Cistaceae, Fagaceae, Nothofagaceae, Nyctaginaceae and Fabaceae). Unlike the Northern Hemisphere Pinaceae, which are always ECM, the native South American conifers form arbuscular mycorrhizas and have not been documented to form ectomycorrhizae (Godoy et al. 1994; Fontenla et al. 1998). There is only one group of ECM gymnosperms known from SA; species of the enigmatic liana genus *Gnetum*, however the ECM status of this plant was determined for species occurring in Papua New Guinea (Tedersoo et al. 2012).

4.2 Current Knowledge and Biogeographical Considerations of Native Ectomycorrhizal Taxa in South America

Forest ecosystems cover extensive regions of the Earth's surface, and ectomycorrhizae are the most common and widespread mycorrhizal type in forests and woodlands of temperate and cold regions in both Hemispheres (Tedersoo et al. 2012). However, various studies carried out in tropical and subtropical regions have shown that ECM associations can be found in many terrestrial ecosystems around the world (Moyersoen et al. 1998a,b, 2001; Founoune et al. 2002; Onguene and Kuyper 2002). In addition, recent studies from tropical and subtropical habitats suggest that EMF can be diverse in these environments (Henkel et al. 2012; Kennedy et al. 2011; Riviere et al. 2007; Smith et al. 2013; Tedersoo et al. 2007; Vasco-Palacios et al. 2018, among others). The biogeographical patterns of South American mycorrhizal plant and fungal communities have been updated recently and are summarized in Tedersoo and Brundrett (2017). However, recent estimates suggest that the fungal diversity of tropical and subtropical habitats as well as locations throughout the Southern Hemisphere are highly understudied relative to Northern Hemisphere forests (Corrales et al. 2018; Hawksworth 1991; Tedersoo et al. 2007). This is evident from the large number of new taxa that are routinely being described and the many molecular inventories and molecular ecology studies being performed in SA. For example, recent studies have addressed fungal diversity in Andean Nothofagaceae forests (Truong et al. 2017), across elevation gradients in the Yungas forests (Geml et al. 2014), and in legume-dominated Neotropical forests (Henkel et al. 2012).

Considering the abundance of previous and contemporary scientific publications, and the inaccessibility of some articles, it is not possible to describe all of the studies of EMF and the ECM symbiosis from SA. Accordingly, this chapter provides an overview of the wide range of ECM habitats and EMF lineages from SA, with an emphasis on recent publications.

South America has a wide variety of unique biomes, but for the sake of simplicity we coarsely divide the ECM regions into three main areas:

1. The Northern Andes cordillera is a high altitude temperate region that harbors diverse Neotropical vegetation. At the time of the Great American interchange, Northern Hemisphere flora and funga crossed from North America into SA. The confluence of Central and SA and the uprising of the Northern Andes in the Miocene facilitated this migration (Mueller and Halling 1995; Halling 1996; Hoorn and Flantua 2015). As a consequence, Northern Hemisphere ECM trees such as *Alnus* (Betulaceae) and *Quercus* (Fagaceae), and probably *Salix* (Salicaceae), and their fungal associates migrated into South America (Kennedy et al. 2011). Native Fagaceae are restricted to Colombia (Kappelle et al. 1992) whereas naturally occurring populations of *Alnus* and *Salix* occur as far south as Argentina and Chile (Hauenstein et al. 2005; Nouhra et al. 2015). The ectotrophic forests of temperate Northern SA were previously divided into two regions by Singer and Morello (1960) based on the ECM host plants. The first region was montane forests of *Quercus* and *Colombobalanus* (Fagaceae) in Colombia and the second was the *Alnus acuminata* belt, which stretches as a narrow band along the Andean mountains, accompanied by alder-specific EMF. *Salix humboldtiana* is found across a wide area from Mexico to Chile but often occurs at low density along water courses (Ragonese 1987).
2. The sub Antarctic forests in far Southern SA are distributed along the Southern Andes and on the Pacific coast below 30° S latitude. These temperate to cold mixed forest are characterized by vegetation of Southern Gondwanan origin. This area was also discussed by Singer and Morello (1960). The region is dominated by ECM trees in the Nothofagaceae, which includes approximately 11 species distributed in the three genera *Fuscospora*, *Lophozonia* and *Nothofagus*. Trees in the Nothofagaceae are associated with a diverse community of EMF which are mostly of a Southern Gondwanan origin, with closest relatives in Australia and New Zealand (Nouhra et al. 2013; Truong et al. 2017). The ECM

relationships in Nothofagaceae forests have attracted the attention of numerous mycologists during the past 120 years and current studies are focused on fungal diversity, EMF community ecology, and the coevolution with host trees.

3. In the northeast corner of SA, numerous studies have explored the EMF of the Guiana Shield region and the coastal vegetation of the Atlantic rainforests of Brazil. In the heart of South America, the white sand forests in the Amazon basin have recently been studied and shown to house ECM plants and fungi (Vasco-Palacios et al. 2018). These forests range from sites with high plant diversity in lowland tropical rainforests (e.g. some forests in Amazonia, Caatinga and the Atlantic Rainforest) to sites almost totally dominated by one ECM plant species (e.g. some *Dicymbe*-dominated forests in the Guiana Shield). The ECM host plants in the tropical regions of SA have diverse phytogeographic origins and are highly variable in terms of growth habit and typical density (Corrales et al. 2018). Several unrelated plant genera have independently evolved the ability to form ECM symbioses with diverse EMF (Smith et al. 2013; Henkel et al. 2012). Among them *Dicymbe*, *Aldina* (Fabaceae), *Pseudomonotes* (Dipterocarpaceae) and *Pakaraimaea* (Cistaceae, previously considered in Dipterocarpaceae) are large forest trees that are often dominant or monodominant in their forests habitats. In contrast, there are also non-dominant taxa such as small trees, shrubs, and lianas within the genera *Coccoloba* (Polygonaceae), *Gnetum* (Gnetaceae) and *Guapira*, *Pisonia*, and *Neea* (Nyctaginaceae).

4.2.1 Ectomycorrhizal Fungi in the Tropical and Subtropical Andes

As early as the 1960's it was already known that some ECM plants and fungi were distributed south of the Panamanian peninsula in association with Northern Hemisphere hosts (Singer and Morello 1960; Singer 1963). The *Quercus* forests in Colombia were believed to be a continuation of the Central American oak distribution. The South American extension of the area of *Alnus acuminata* along the Andes was recognized during this period as was the discontinuous *Salix humboldtiana* populations along rivers and wetlands in the lowlands. At that time, the ECM fungal communities in the ectotrophic vegetation of the Guiana Shield and the Amazon Basin were virtually unknown.

Singer and Morello (1960) predicted that the absence of EMF in tropical and subtropical lowland forest was probably due to the adaptation of plants with vertical development of roots to the soft deep soils. These type of habitats also show little or no thermoperiodicity during the year and it was suggested that these two traits were perhaps incompatible with formation of ectomycorrhizae. Later Halling and Mueller (Halling 1996, 2001; Halling and Mueller 1999; Mueller and Halling 1995; Mueller and Singer 1988; Muller and Strack 1992), suggested the co-migration of the obligate ECM fungal communities from North America with their associated alder and oak symbionts, which have a North Temperate origin. They identified several ECM species that have a widespread distribution with oaks across multiple genera, including *Cortinarius*, *Lactarius*, *Laccaria* and *Strobilomyces*. They also identified several ECM species in the genera *Amanita*, *Laccaria*, *Lactarius*, *Leccinum* and *Boletus* that appear to be Neotropical endemics that are not found in temperate zones of North America. Within the Andean forest, approximately 65 ECM species were described among various fungal lineages associated with the montane Colombian *Quercus humboldtii* forests (Halling 1989; Mueller and Halling 1995; Vargas et al. 2017). Among them, the /boletus, /amanita, and /russula-lactarius were the most species-rich linages, followed by /laccaria, /inocybe, /paxillus-gyrodon, /cantharellus (including *Craterellus*), and /cortinarius (including *Rozites*), as shown in Fig. 4.1. Those studies defined on the basis of community composition that: (1) the ECM communities associated with oaks in Costa Rica, Panama and Colombia were more similar to those of North America than to any other Western Hemisphere communities and (2) the presence of endemic Neotropical EMF was an indicator of a well-diversified community (Mueller and Halling 1995).

Another interesting example of a Holarctic ECM host in SA is *Alnus acuminata* (Betulaceae). The genus *Alnus* is thought to have migrated multiple times from Eurasia to Western North America via Bering land bridge (Furlow 1979; Navarro et al. 2003; Chen and Li 2004). From there, some species apparently migrated to or speciated in Eastern North America whereas other species underwent a similar process towards the south. After the Isthmus of Panama was established, *A. acuminata* expanded along the Andes deep into SA (Furlow 1979; Chen and Li 2004). The current distribution of *A. acuminata* ranges from Mexico to northwestern Argentina (Weng et al. 2004; Ren et al. 2010), where it is usually found growing at high or moderate elevations in slopes, ravines, stream banks or roadsides (NAS 1980).

Alnus acuminata hosts a quadripartite symbiosis in which arbuscular mycorrhizal (AM) fungi, EMF and N-fixing actinomycete bacteria are involved (Carú et al. 2000; Becerra et al. 2005d). Through this highly effective symbiosis *Alnus* spp. are able to improve the fertility of mountainous land subject to erosive processes and to colonize poor substrates thereby initiating plant succession (Roy et al. 2007; Teklehaimanot and Mmolotsi 2007). It has been shown that the dominant symbionts in the South American alder roots are EMF (Becerra et al. 2005b, Nouhra et al. 2003). In general, *Alnus* species associate with more limited number of EMF compared to other ECM trees (Tedersoo et al. 2009; Põlme et al. 2013). This specialization is likely due to the high N and low pH that is created in *Alnus*-dominated forests as well as restricted receptivity by *A. acuminata* to only some EMF species (Molina et al. 1992).

Previous works on EMF associated with *A. acuminata* in SA focused on identifying the ECM morphotypes as well as a few sporocarps collections from the same sites. Those studies reported 15 taxa distributed in 7 genera, including *Alpova* (1), *Cortinarius* (2), *Gyrodon* (1), *Lactarius* (2), *Naucoria* (1), *Russula* (2) and *Tomentella* (3) as well as three unidentified "alnirhiza" morphotypes (Becerra et al. 2002, 2005a, b, c, d; Nouhra et al. 2005; Pritsch et al. 2010). Later, some of the same taxa were also detected based on ITS sequences matches to *Alnus* EMF communi-

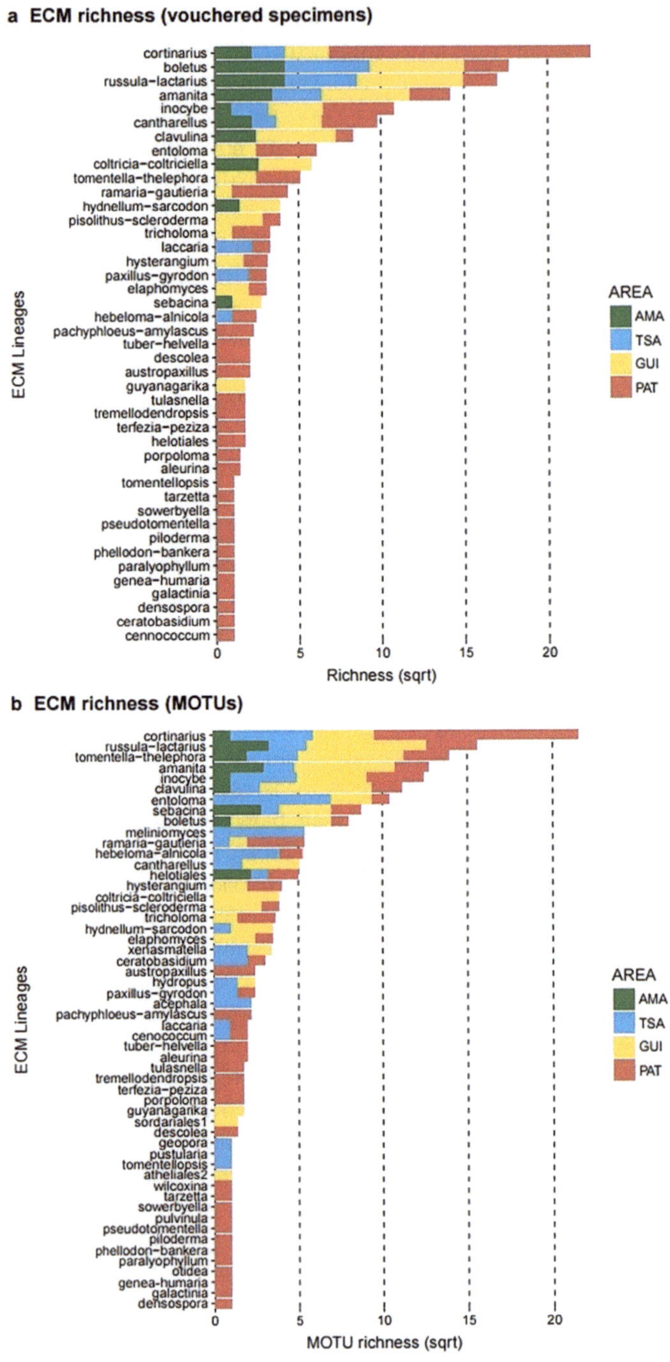

Fig. 4.1 ECM lineages of South American biogeographic areas. AMA = Amazon basin, GUI = Guiana shield, TSA = Tropical and subtropical Andes, PAT = Patagonian Nothofagaceae. Square root of the ECM richness based on vouchered specimens (**a**), and based on molecular operational taxonomic units (**b**)

ties in Mexico, thereby supporting the host EMF co-migration hypothesis (Kennedy et al. 2011). Another study based on sporocarp sampling identified additional taxa from the /cortinarius, /inocybe, and /russula-lactarius lineages (Niveiro 2012).

Most recently, Geml et al. (2014) used next-generation sequencing to detect a total of 181 OTUs belonging to ECM lineages as defined by Tedersoo et al. (2010b) and Tedersoo and Smith (2013), along an elevational gradient of the Yungas in Argentina. This study included sampling from montane *A. acuminata* forests at the highest elevation. The most frequent lineages were /meliniomyces (29 OTUs), /entoloma (48 OTUs), /cortinarius (24 OTUs), /hebeloma-alnicola (15 OTUs), /inocybe (15 OTUs) and /tomentella-thelephora (9 OTUs). Additional OTUs were also documented from 12 additional lineages (/amanita, /cantharellus, /ceratobasidium, /clavulina, /hydnellum-sarcodon, /hydropus, /laccaria, /paxillus-gyrodon, /ramaria-gautieria, /russula-lactarius, /sebacina and /tomentellopsis) (Fig. 4.1). Only a few of these OTUs from soil matched identified specimens, suggesting that most of these likely represent new species waiting to be described. This bias in the number of described species vs. the number of OTUs recovered from soil was also reported for species of Thelephoraceae, which exhibited a lack of specificity among *Alnus* species hosts (Nouhra et al. 2015).

Salix humboldtiana (Salicaceae) is of Laurasian origin and likely arrived in SA shortly after the formation of the Isthmus of Panama in the late Miocene or early Pliocene (Gentry 1982). *Salix humboldtiana* is distributed across South America but is most frequently found in riparian and flooded habitats. Thus far there are only a few limited studies of the native EMF with *Salix humboldtiana* in South America. Preliminary studies of *Salix humboldtiana* have documented several species of EMF as well as arbuscular mycorrhizas and dark septate mycorrhizal types (Becerra et al. 2009; Lugo et al. 2012). These studies documented seven ECM morphotypes that included members of the /tomentella-thelephora and /inocybe lineages. Ongoing studies have detected additional EMF species across the Argentinian distribution of *Salix humboldtiana* (Mujic et al. pers. comm.).

4.2.2 Ectomycorrhizal Fungi in the Neotropical Forests of Northeastern South America and the Amazon Basin

Although ECM associations were long hypothesized to be absent from Neotropical forests, Singer and Araujo (1979) documented typical EMF in white sand forests with *Aldina heterophylla* (Fabaceae subfamily Papilionoideae) in the Rio Negro region near Manaus, Brazil. This was an unexpected finding at the time and did not receive widespread attention or appreciation. Although *Aldina* was implicated as a ECM tree genus in 1979 (see also Singer et al. 1983; Moyersoen 1993), very little research on the ECM associations of South American tropical rainforests was initiated until the work of Henkel et al. (2002) and subsequent studies in the Guiana Shield region (see Henkel 2003; Henkel et al. 2012, among others).

Henkel et al. (2002) was the first study to show that species of *Dicymbe* (Fabaceae subfamily Caesalpinioideae) are symbiotically associated with EMF from many phylogenetic groups. Evidence suggests that the genus *Dicymbe* is related to other ectomycorrhizae-forming Caesalpinioideae species from West Africa (Smith et al. 2011). Henkel et al. (2002) also showed that a less frequently occurring taxon, *Aldina insignis*, was also ECM. Since the early 2000's Henkel and co-workers have sampled intensively for EMF sporocarps in Guiana Shield forests dominated by *Dicymbe* spp., but also in forests with other ECM hosts such as *Aldina* spp. and *Pakaraimaea dipterocarpacea* (Cistaceae). Henkel et al. (2012) documented the results of 7 years of sampling in *Dicymbe corymbosa* forests where 126 species from 25 genera of putative or confirmed EMF were recovered from three replicated 1-hectare plots. When sampling outside of the plots was considered, Henkel et al. (2012) documented 172 EMF species from a small area, indicating that high EMF biomass and diversity can be found in monodominant *Dicymbe* forests.

Henkel and co-authors also performed molecular sampling of ECM roots in the same Guiana Shield ecosystems (Smith et al. 2011, 2013, 2017). Taken together, these studies provide an overview of EMF communities in the Guiana Shield. Evidence from fruiting body studies and ECM root sampling suggest that several lineages are dominant and diverse both above and belowground (Fig. 4.1), specifically the /boletus, /russula-lactarius, /tomentella-thelephora and /clavulina lineages (Smith et al. 2017). Although the /boletus, /russula-lactarius, and /tomentella-thelephora lineages are considered diverse in other tropical regions, the high diversity of the /clavulina lineage is putatively unique to the Neotropics. It has been suggested that this region may be a historical center of origin or diversification for the /clavulina lineage (Smith et al. 2011; Kennedy et al. 2012). Based on sporocarp surveys, members of the /amanita and /coltricia lineages are also diverse in the Guiana Shield but these taxa were not regularly detected in ECM root studies, possibly due to problems with ITS rDNA amplification and sequencing (Smith and Henkel, unpublished). Studies of the ECM communities in the Guiana Shield also led to the discovery of the new genus *Guyanagarika* (Sánchez-García et al. 2016) which is currently the only known ECM lineage that is endemic to one specific biome.

Subsequent studies from other sites in tropical SA indicate that significant EMF diversity and additional unrecognized ECM plant species likely await future discovery. Recent studies such as those by Roy et al. (2016) and Sulzbacher et al. (2013) from Brazil and French Guiana highlight the fact that much more work is needed. For example, Roy et al. (2016) identified 62 morpho-species among 23 genera of EMF from typical Amazonian "white sand forests" with *Aldina heterophylla* as the dominant host. In addition, they documented putatively native EMF from Brazilian herbaria and found approximately 175 EMF species from 40 genera that were collected in native habitats. Although this was an impressive review of Brazilian herbarium data, some of the records require additional analysis to confirm their correct identification (i.e. *Octavianina*) or to examine their ECM status, (i.e. *Phaeocollybia* and *Phlebopus*). In the case of *Phaeocollybia*, one study has suggested a root parasitic lifestyle (Redhead and Malloch 1986) whereas recent work from *Abies* forests in Mexico suggest that *Phaeocollybia* is ECM (Argüelles-Moyao et al. 2017). In contrast, the non-ECM trophic mode of *Phlebopus* is now well established (Nouhra et al. 2008). Other genera included in the study, such as *Suillus* and *Rhizopogon*, are widely known as co-invasive species with *Pinus*, which are quite extensive in Brazil and other regions of South America.

Sulzbacher et al. (2013) studied the Atlantic Rainforest habitats in Rio Grande do Norte state of Brazil and found approximately 27 EMF species. As in Guiana, the /boletus, /russula-lactarius, /amanita and /coltricia lineages were among the most diverse lineages (Fig. 4.1). Although the host plants were not well known in all cases where these tropical EMF were collected, the suspected host plants are species of *Coccoloba* (Polygonaceae) as well as *Guapira* and *Neea* (Nyctaginaceae). These host genera are widespread in lowland tropical rainforests but they are mostly small subcanopy trees that occur at lower densities (Tedersoo et al. 2010a; Haug et al. 2005). This can make them difficult to locate and identify, particularly when they are growing in a highly diverse forest canopy of mostly non-ECM plants. Furthermore, this lower host density and lower host basal area may also partially explain the strong host preferences among the EMF found with these hosts (Tedersoo et al. 2010a). It is likely that other ECM host plants are present but have not yet been fully documented. Singer and Araujo (1979) reported the possibility of ECM associations in *Swartzia* (Fabaceae), *Glycoxylon* (Sapotaceae) and *Psychotria* (Rubiaceae) but work by other researchers has not substantiated these suggested associations (Tedersoo and Brundrett 2017). However, very few detailed studies on potentially new ECM plants have been conducted. In one recent study, Freire et al. (2018) synthesized ectomycorrhizae on the Neotropical native *Psidium cattleianum* (Myrtaceae), suggesting that this may be one of the hosts for EMF in the Atlantic Rainforests of Brazil.

In the Colombian Amazon one study focused on the ECM communities associated with *Dicymbe uaiparuensis* and *Aldina* sp. (Fabaceae) in white sand forests similar to those in Brazil (Vasco-Palacios et al. 2018). More recently, some studies have also examined the EMF associated with *Pseudomonotes tropenbosii* (Dipterocarpaceae), an endemic tree known only from a few patches of forests (Vasco-Palacios et al. 2014, 2018). The region is mostly structured by tropical wet floodplains, locally known as "varzea", which are subject to annual flooding with consequent soil enrichment (López-Quintero et al. 2012; Vasco-Palacios et al. 2018). Those studies identified at least 49 species of EMF from fruiting bodies and colonized root tips. Similar to many other tropical South American systems, the most diverse and common ECM lineages were /russula-lactarius, /amanita, /clavulina, and /coltricia. However, these sites were characterized by lower diversity of the /boletus lineage and higher diversity in the /sebacina lineage than in the Guiana Shield (Fig. 4.1).

One interesting finding is that several common EMF from Colombian Amazon sites (Vasco-Palacios et al. 2018) are shared with sites in the Guiana Shield approximately 1500 km away. This includes taxa from several different lineages, such as *Clavulina sprucei* and *Clavulina amazonensis* (/clavulina), *Craterellus atratoides* (/cantharellus), *Amanita xerocybe* (/amanita), *Singerocomus inundabilis* (/boletus), and *Lactifluus subiculatus* (/russula-lactarius). These findings suggest that at least some EMF taxa in the tropical regions of SA may be very widely distributed, with ranges across the Guiana Shield, Amazonia, and the Atlantic forests of Brazil.

4.2.3 *Ectomycorrhizal Fungi Associated with Nothofagaceae in the Andes of Patagonia*

In Patagonia, temperate forests dominated by Nothofagaceae species host a high species richness of EMF (Garrido 1988). There are 11 species of Nothofagaceae that occur across Southern SA: *Fuscospora alessandri, Lophozonia alpina, L. glauca L. leonii, L. macrocarpa, L. obliqua, Nothofagus antarctica, N. betuloides, N. dombeyi, N. nitida* and *N. pumilio*. The fungal communities associated with these forests have been studied since the early 1900's by Spegazzini. Later, Singer (1954), Singer and Smith (1958), Horak and Moser (1965, 1975), Horak (1979), Moser et al. (1975) and Garrido (1988) made extensive contributions to the taxonomy of agarics and gasteroid fungi of the region. Garrido (1986, 1988), also established the first checklists of EMF associated with Nothofagaceae in Chile. The most diverse genus of EMF in Argentinian and Chilean Patagonian forests is *Cortinarius* s.l. (Garnica et al. 2002, 2003; Valenzuela and Esteve-Raventos 1994; Truong et al. 2017), including morphologically defined genera such as *Dermocybe*, *Rozites*, *Stephanopus* and *Thaxterogaster*. Other important contributions are those of Gamundí (Gamundí 2010; Gamundí and de Halperín 1960; Gamundí et al. 2004) who described many of the ECM Pezizales of the region and Greslebin (2002) who studied the corticioid taxa.

Belowground approaches for studying the ECM communities associated with Nothofagaceae and other trees of the Andean-Patagonian forests started with studies by Garrido (1988) who examined roots of 38 native woody gymnosperm and angiosperm plant species in Southern Chile and performed inoculation experiments with seedlings. As expected, *Nothofagaceae* spp. were all ECM whereas neighbouring plants showed colonization by arbuscular or ericoid mycorrhizal fungi. Interestingly, *Ugni molinae* was colonized by an ECM morphotype attributed to *Cenococcum geophilum* whereas *Luma apiculata* (Myrtaceae) was colonized by an ECM morphotype attributed to an *Austropaxillus* species. Although no recent studies have verified these findings, further investigation of these claims should be attempted since *Nothofagaceae* have traditionally been considered the only native ECM trees in the Andean-Patagonian forests.

The work of Carrillo et al. (1992) and Godoy et al. (1994) supported the claim that *Nothofagaceae* are the main hosts of EMF in Patagonia with their extensive surveys of fine roots from 114 (Carrillo et al. 1992) and 83 (Godoy et al. 1994) herbaceous or woody gymnosperms, angiosperms and ferns from Southern Chile. Characterization and identification of ECM roots of *Nothofagaceae* spp. based on the macro and micro-morphological taxonomy was initiated by Agerer (1991) and was first applied by Godoy and Palfner (1997) and Flores et al. (1997) in greenhouse and nursery assays. Based on the same methodology, Palfner (2001) described 15 ECM morphotypes, associated with natural evergreen and deciduous *Nothofagaceae* forest in Southern Chile and Argentina. The corresponding mycobionts include genera from several agaricoid EMF groups such as *Austropaxillus*, *Boletus*, *Cortinarius*, *Descolea* and *Russula*, but also sequestrate and hypogeous genera like *Gautieria* and *Thaxterogaster* (currently *Cortinarius*). This study found no evidence for differences between the EMF communities on deciduous versus evergreen Nothofagaceae species. Recently some EMF have been shown to associate only with specific Nothofagaceae species, as in the case of *Cortinarius pyromyxa*, which can be found across a wide latitudinal range of Andean-Patagonian forests, but always in association with the evergreen *N. dombeyi* (Lam et al. in review). Further specific associations between individual taxa of ECM plants and EMF are likely waiting to be discovered as more molecular-based studies address the ECM communities of SA.

Although there have only been a few molecular studies of South American ECM communities (e.g. Nouhra et al. 2013; Fernández et al. 2013; Kuhar et al. 2016; Truong et al. 2017), these studies have confirmed the ECM trophic mode of many of the lineages that were assumed to form ECM (see above). Molecular studies have also revealed new ECM lineages for the Andean Patagonia. Ecological aspects related to the fructification and diversity correlated with environmental factors, showed that the ECM community structure is dependent on the altitude (Nouhra et al. 2012, 2013), precipitation patterns (Romano et al. 2017a) and fire (Longo et al. 2011). Special attention has focused on the phylogenetic arrangement of sequestrate ECM lineages and related taxa (Trierveiler-Pereira et al. 2015; Kuhar et al. 2017; Truong et al. 2017; Pastor et al. 2019; Salgado Salomon et al. 2018), indicating that many Patagonian EMF share a vicariant history via a Southern Gondwanan origin. The close biogeographic relationships of the Patagonian fungi with Oceania and Zealandia have recently been confirmed based on extensive molecular sampling (Truong et al. 2017), and the EMF show similar patterns to those seen in plants and animals (Sanmartin and Ronquist 2004). However, the fungal species distribution within Patagonia also seems to show a complex biogeographic history (Romano et al. 2017b). At the ECM lineage level (sensu Tedersoo et al. 2010b) there is extreme dominance by the /cortinarius lineage which accounts for a large fraction of the EMF diversity in Nothofagaceae forests. The /inocybe, /ramaria-gautieria, /tomentella-thelephora, and /tricholoma lineages are also notably diverse whereas the /russula-lactarius is remarkably species-poor compared with almost every other ECM region of the world. This biome is also home to a prominent set of exclusively Gondwanan taxa that includes at least 5 lineages, including /aleurina, /descolea, /austropaxillus, /porpoloma, /phaeohelotium as well as species of *Underwoodia* (an exclusively Southern Hemisphere branch of the /tuber-helvella lineage) (Truong et al. 2017; Tedersoo and Smith 2013, 2017). The ECM status of many putative EMF taxa still need to be examined carefully. For example, several taxa of Cantharellales, *Entoloma*, and *Rickenella* may form ECM but have not yet been recovered from ECM roots so their trophic mode remains in question (Nouhra et al. 2013). The diversity patterns in Southern SA suggest that the lineage diversification that followed the separation of austral land masses requires further study. Since ecophysiological functions and enzyme machinery are distinct among the different EMF lineages (Lindahl and Tunlid 2015), the unique taxonomic composition in Nothofagaceae forests of SA suggest that ecosystemic roles of the different groups need to be explored. Studies on ECM and the effects of EMF inoculation on the growth of Nothofagaceae seedlings are scarce (e.g. Salgado Salomón et al. 2013; Marín et al. 2018) and data on the role of EMF on the enzymatic processes of litter decomposition are needed in order to better understand the ecology of these forests.

4.3 Physiology and Cellular Biology of Nothofagaceae Ectomycorrhizae

General physiological benefits of the ECM symbiosis for both mycobionts and phytobionts are well known (van der Heijden et al. 2014) but more specific traits correlated to particular ECM fungal species remain relatively poorly studied, especially in the native Nothofagaceae forests of SA. However, Alvarez et al. (2004) performed pioneering research on this topic. Using confocal laser-scanning microscopy, Alvarez et al. (2004) demonstrated different cellular phosphorus-uptake strategies in axenic cultures of the native *Austropaxillus boletinoides* and *Descolea antarctica*, in comparison with the widespread *Paxillus involutus* and *Pisolithus tinctorius*. Hyphae of *A. boletinoides* and *D. antarctica* changed the number of surface-bound phosphomonoesterase (SBP) active centers depending on the pH and phosphorus concentration in the substrate. In contrast, *P. involutus* and *P. tinctorius* cells reacted by changing the intensity of activity but not number of SBP active centers. Later, Alberdi et al. (2007) showed that in vitro assays of synthetized ECM formed by *Nothofagus dombeyi* and *Descolea antarctica*, the colonized seedlings had higher relative water and leaf soluble carbon contents under drought conditions than seedlings inoculated with the widely distributed *Pisolithus tinctorius*. However, seedlings colonized with *D. antarctica* had less root soluble carbon than those with *P. tinctorius*. Following up on these findings, Alvarez et al. (2009) later showed that in pure culture *P. tinctorius* was more efficient in metabolizing reactive oxygen species than *D. antarctica* and thus suffered less cell damage, especially under drought conditions.

4.4 Chemistry and Chemotaxonomy of Nothofagaceae Ectomycorrhizal Fungi

Fungi are known for their enormous variety of secondary metabolites. Pigment chemistry has been a particularly valuable tool for detecting or verifying systematic relationships or separating lineages in higher fungi (Velíšek and Cejpek 2011). Within the South American EMF, Boletales and Cortinariaceae, have been the focus of pigment studies in order to reveal relationships with their counterparts in other continents.

Several studies have focused on the chemistry of the *Cortinarii* in subgenus *Dermocybe.* Gruber (1975) performed thin layer chromatography (TLC) of anthrachinone pigments extracted from 18 endemic *Dermocybe* species from Andean-Patagonian *Nothofagaceae* forest. She found endocrocin, dermolutein and endocrocinglycoside in most of the studied species. These yellow pigments are also known from Northern Hemisphere *Dermocybe* spp., whereas red pigments, quite common in Northern Hemisphere *Dermocybe* spp. are scarce in the South American taxa or possess different chemical properties. On the other hand, Gruber (1975) detected various unidentified pigments which did not match reference compounds available from Northern Hemisphere material. Keller et al. (1987) also used TLC in a study that compared pigment composition of South American and Australasian *Dermocybe* species. They found a higher diversity of anthrachinone compounds in Southern Hemisphere species than in their Northern counterparts. Interestingly, chemical analysis grouped species into several groups with similar pigment composition and these groups contained both Australasian and South American taxa. When viewed in combination with morphological attributes, it seems likely that these different groups reflect ancient Gondwanan lineages. Greff et al. (2017) separated and determined pigments of the Chilean endemic *Cortinarius nahuelbutensis* and identified flavomannin for the first time in a *Dermocybe* species. They also identified the novel compound emodinphyscion, suggesting a separate lineage at least for a part of South American *Dermocybe* spp.

Outside of the subgenus *Dermocybe* few other South American *Cortinarius* spp. have been subjected to chemical analysis. Recently, Lam et al. (2018) isolated novel diterpene pigments with a nor-guanacastane skeleton structure, called pyromyxons, from *Cortinarius pyromyxa. Cortinarius pyromyxa* is a mycobiont of *Nothofagus dombeyi* and is characterized by a conspicuous orange coloration of the pileus and stipe base. Arnold et al. (2012) analyzed the chemistry of *Cortinarius lebre*, which has a strong naphthalene-like odor, and identified indole as the principal compound of the characteristic aroma.

Pulvinic acid derivatives are long known to be key chemotaxonomic compounds within the Boletales (Winner et al. 2004). Garrido (1988) compared this pigment class between *Austropaxillus statuum*, a frequent native mycorrhizal partner of *Nothofagaceae* in Argentina and Chile, and the exotic introduced species *Paxillus involutus* from the Northern Hemisphere. Garrido (1988) found involutin, a characteristic compound of Northern Paxilli, to be absent in *A. statuum* and related taxa.

Instead *A. statuum* shared trihydroxyphenylglyoxylic acid with *Serpula lacrymans*. This chemistry-based finding is consistent with the phylogenetic separation of native Andino-Patagonian Paxilli in the genus *Austropaxillus* (Bresinsky et al. 1999) and also supports the placement of *Austropaxillus* in the family Serpulaceae (Skrede et al. 2011). Garrido also detected typical Boletales pigments such as atromentic, variegatic, and xerocomic acids, and variegatorubin in endemic boletes of Chile, such as *Butyriboletus loyo, Boletus loyita, B. putidus* and the poorly known species *B. araucarianus* and *B. bresinskyanus*.

4.5 Mycosociology of South American Nothofagaceae Forests

The highly specific associations between EMF and their phytobionts makes it possible to characterize an ECM forest type based on the tree composition, the associated fungal community, or by a combined "phyto-mycosociological approach" (Salazar 2016). In SA, Singer and Morello (1960) were the first to suggest this concept for the Andean-Patagonian and coastal forests of Southern Argentina and Chile. These forests are not just one large homogeneous area dominated by ECM *Nothofagaceae* but instead are a patchwork where *Nothofagaceae*-dominated stands coexist with non-ECM trees stands such as native conifers, Lauraceae, and sclerophyllus angiosperms. In some patches *Nothofagaceae*, the only native ECM trees, are naturally absent. According to the concepts of Singer (1971), this presence or absence of the native EMF community defines the forest as "ectotroph" or "anectotroph". On the species level, the ectotroph forest is defined by the presence of pairwise units of phytobiont and mycobiont (e.g. *Nothofagus dombeyi*/ and *Russula nothofaginea)* whose composition is variable, depending on local climate, soil conditions and topography. Although from a twenty-first century point of view, certain aspects of Singer's concept may appear self-evident or simple, one has to consider the conceptual and methodological limitations of the time when the only clues for presence of EMF in the field were spatial and temporal patterns of fruiting body formation of macrofungi that were assumed to be ECM. At that time only a few fungal species from the Northern Hemisphere, mostly associated with conifers, were definitely proven to form ectomycorrhizae by *in vitro* synthesis of mycorrhizal roots (Harley 1959). Singer (1971) also defined diagnostic parameters in order to classify a forest stand as ectotrophic or anectotrophic; the proportion of obligate EMF, the presence of host-specific fungal species and the percentage of endemic taxa. Singer and Morello (1960) also stated that forest plantations of introduced ECM trees like *Pinus* spp. in SA represent a separate ectotrophic influence, with mycobionts that were also introduced from the Northern Hemisphere.

A first mycosociological comparison between introduced and native ECM communities was published by Valenzuela et al. (1998). Their inventory of EMF fruiting bodies in *Nothofagaceae* forests and plantations of *Pinus radiata* in the Valdivian region of Southern Chile yielded different proportions of EMF and saprobic macrofungi in the different forest types. They also noted differences within each forest type across sites (coastal mountains, central valley, Andean slope). They also observed *Amanita rubescens*, originally introduced with *P. radiata* as an invader, in *Nothofagaceae* forest whereas no endemic EMF were found in pine plantations. Interestingly, we have made similar observations where the Northern Hemisphere symbiont *Amanita muscaria* (and its putative parasite *Chalciporus piperatus*) was observed in native *Nothofagaceae* forest in Southern Chile (M. Smith, personal observation).

Based on integration and further development of the approaches by the groups of Singer and Valenzuela, Palfner and Casanova-Katny (2018) completed a comparative study of the fungal communities of degraded and highly fragmented native *Nothofagaceae* forest and the surrounding forest plantations of introduced *Pinus radiata* and *Eucalyptus* spp. on Arauco Peninsula in central Southern Chile. They found that although total fungal species richness was relatively high in the dominant exotic plantations, species density and proportions of endemic taxa and specialists were highest in native forest patches. Palfner and Casanova-Katny (2018) determined that the fragmented *Nothofagaceae* forest represent important sanctuaries for native macrofungi, including many endemic EMF, and therefore should receive priority status in conservation programs.

4.6 Concluding Remarks

South America has a wide variety of unique biomes, landscapes, soil and climatic conditions that deeply influence the plant and fungal composition. Based on currently available data, the ectotrophic forest dominated by Nothofagaceae in Patagonia appears to host the highest ECM diversity. However, it is possible that this pattern is merely a reflection of the unbalanced distribution of available studies across SA. Our review of the literature suggests that significant EMF diversity remains to be discovered in all regions of SA.

As depicted in Fig. 4.1, Patagonia harbors the largest number of ECM lineages and also has the highest richness based either on vouchered specimens or molecular operational taxonomic units (MOTUs) recovered from ECM roots, if compared with any of the Neotropical sites considered. This is also consistent with global patterns of ECM distribution (Tedersoo et al. 2012). Patagonia clearly harbors a much higher diversity of ECM Ascomycota, particularly in the order Pezizales. This is expected since it appears that many ECM Ascomycota are absent, infrequent or species-poor in tropical regions (Corrales et al. 2018).

The Guiana Region also hosts a diverse EMF community and is home of at least one lineage that is unique at the global scale (/guyanagarika). To date, the tropical and subtropical Andes regions with their Northern Hemisphere-derived ECM hosts (e.g. Fagaceae and Salicaceae) and the Amazon basin with its widely dispersed Neotropical hosts (e.g. Nyctaginaceae, *Coccoloba* and *Gnetum*) are apparently the least diverse regions in terms of lineages and species richness. However, these two regions are also the least studied areas in SA. The ECM linages /cortinarius, /

russula-lactarius, /amanita and /clavulina are present in all of the areas treated in this chapter. The /cortinarius lineage is particularly rich in Patagonia and, at a lesser magnitude, the remaining three lineages are notably rich in the Guiana Shield. In the future it will be important to perform wider molecular sampling of ECM associations in plants, fungi, and soil across the South American subcontinent, including remote Andean and Neotropical areas such as the Amazon basin, Caatinga, Chaco, and other biomes to detect new and unique ECM taxa. Other aspects on the study of the EMF communities in SA, such as ecology, physiology, chemistry, cell biology, fungi and forest restoration, are mostly undeveloped but are promising for future research.

References

Agerer R (1991) Characterization of ectomycorrhiza. In: Norris JR, Read DJ, Varma AK (eds) Techniques for the study of mycorrhiza, Methods Microbiology 23, Academic Press, London, p 25–73

Agerer R (1996) Characterization of ectomycorrhizae: a historical overview. Descr Ectomyc 1: 1–22

Alberdi M, Alvarez M, Valenzuela E, Godoy R, Olivares E, Barrientos M (2007) Response to water deficit of *Nothofagus dombeyi* plants inoculated with a specific (*Descolea antarctica* Sing) and non-specific (*Pisolithus tinctorius* (Pers.) Coker & Couch) ectomycorrhizal fungi. Rev Chil Hist Nat 80: 479–491

Alvarez M, Godoy R, Heyser W, Härtel S (2004) Surface-bound phosphatase activity in living hyphae of ectomycorrhizal fungi of *Nothofagus obliqua*. Mycologia 96(3): 479–487

Alvarez M, Huygens D, Fernandez C, Gacitua Y, Olivares E, Saavedra I, Alberdi M, Valenzuela E (2009) Effect of ectomycorrhizal colonization and drought on reactive oxygen species metabolism of *Nothofagus dombeyi* roots. Tree Physiol 29: 1047–1057

Argüelles-Moyao A, Garibay-Orijel R, Márquez-Valdelamar LM, Arellano-Torres E (2017) *Clavulina-Membranomyces* is the most important lineage within the highly diverse ectomycorrhizal fungal community of *Abies religiosa*. Mycorrhiza 27(1): 53–65

Arnold N, Palfner G, Schmidt J, Kuhnt C, Becerra J (2012) Chemistry of the aroma bouquet of the edible mushroom "lebre" (*Cortinarius lebre*, Basdiomycota, Agaricales) from Chile. J Chil Chem Soc 58(3): 1333–1335

Becerra A, Daniele G, Domínguez L, Nouhra E, Horton T (2002) Ectomycorrhizae between *Alnus acuminata* H.B.K. and *Naucoria escharoides* (Fr.: Fr.) Kummer from Argentina. Mycorrhiza 12: 61–66

Becerra A, Nouhra E, Daniele G, Domínguez L, McKay D (2005a) Ectomycorrhizas of *Cortinarius helodes* and *Gyrodon monticola* with *Alnus acuminata* from Argentina. Mycorrhiza 15: 7–15

Becerra A, Pritsch K, Arrigo N, Palma M, Bartoloni N (2005b) Ectomycorrhizal colonization of *Alnus acuminata* Kunth in northwestern Argentina in relation to season and soil parameters. Ann For Sci 62: 325–332

Becerra A, Beenken L, Pritsch K, Daniele G, Schloter M, Agerer R (2005c) Anatomical and molecular characterization of *Lactarius* aff. *omphaliformis*, *Russula alnijorullensis* and *Cortinarius tucumanensis* ectomycorrhizae on *Alnus acuminata*. Mycologia 95: 1047–1057

Becerra A, Zak MR, Horton TR, Micolini J (2005d) Ectomycorrhizal and arbuscular mycorrhizal colonization of *Alnus acuminata* from Calilegua National Park (Argentina). Mycorrhiza 15: 525–531

Becerra A, Nouhra E, Silva M, McKay D (2009) Ectomycorrhizae, arbuscular mycorrhizae and dark septate fungi on *Salix humboldtiana* Willd. from Central Argentina: a first assessment in two riparian populations. Mycoscience 50: 343–352

Bresinsky A, Jarosch M, Fischer M, Schönberger I, Wittmann-Bresinsky B (1999) Phylogenetic relationships within *Paxillus* s. l. (Basidiomycetes, Boletales): Separation of a Southern Hemisphere genus. Plant Biol 1 (3): 327–333

Carrillo R, Godoy R, Peredo H (1992) Simbiosis micorrícica en comunidades boscosas del Valle Central en el sur de Chile. Bosque 13(2): 57–67

Carú M, Becerra A, Sepúlveda D, Cabello A (2000) Isolation of infective and effective *Frankia* strains from root nodules of *Alnus acuminata* (Betulaceae). World J Microbiol Biotechnol 16: 647–651

Chen ZD, Li JH (2004) Phylogenetics and biogeography of *Alnus* (Betulaceae) inferred from sequences of nuclear ribosomal DNA its region. Int J Plant Sci 165: 325–335

Chilvers GA (1968) Some distinctive types of eucalypt mycorrhiza. Aust J Bot 16: 49–70

Corrales A, Henkel TW, Smith ME (2018) Ectomycorrhizal associations in the tropics - biogeography, diversity patterns and ecosystem roles. New Phytol online, https://doi.org/10.1111/nph.15151

Dominik T (1969) Key to ectotrophic mycorrhizae. Folia For Pol. Seria A 15: 309–328

Espinosa MR (1915) Hongo nuevo chileno el loyo *Boletus loyus* Espinosa, n. sp. Bol Mus Nac Hist Nat Bol (Chile) 8: 5–1

Fernández NV, Marchelli P, Fontenla SB (2013) Ectomycorrhizas naturally established in *Nothofagus nervosa* seedlings under different cultivation practices in a forest nursery. Microb Ecol 66(3): 581–592

Flores R, Godoy R, Palfner G (1997) Morfo – anatomía de la ectomicorriza *Cenococcum geophilum* Fr. en *Nothofagus alessandrii* Esp. Gayana Bot 54(2): 157–167

Fontenla S, Godoy R, Rosso P, Havrylenko M (1998) Root associations in *Austrocedrus* forests and seasonal dynamics of arbuscular mycorrhizas. Mycorrhiza 8: 29–33

Founoune H, Duponnois R, Bâd AM, El Bouamib F (2002). Influence of the dual arbuscular endomycorrhizal / ectomycorrhizal symbiosis on the growth of *Acacia holosericea* (A. Cunn. ex G. Don) in glasshouse conditions. Ann For Scie 59: 93–98

Frank AB (1885) Ueber die Wurzelsymbiose beruhende Ernährung gewisser Bäume durch unterirdische Pilze. Berichte der Deutschen Botanischen Gesellschaft 3: 128–145

Freire CG, Giachini AJ, Gardin JP, Rodrigues AC, Vieira RL, Baratto CM, Werner SS, Abreu BH (2018) First record of in vitro formation of ectomycorrhizae in *Psidium cattleianum* Sabine, a native Myrtaceae of the Brazilian Atlantic Forest. PloS one, 13(5), p.e0196984

Furlow JJ (1979) The systematics of the American species of *Alnus* (Betulaceae). I. Rhodora 81: 1–121

Gamundí IJ (2010) Genera of Pezizales of Argentina 1. An updating of selected genera. Mycotaxon 113.1: 1–60

Gamundí IJ, de Halperín DR (1960) Flora criptogámica de Tierra del Fuego. Fundación para la Educación, la Ciencia y la Cultura, FECIC, Buenos Aires

Gamundí IJ, Minter DW, Romero AI, Barrera VA, Giaiotti AL, Messuti MI, Stecconi M (2004) Checklist of the Discomycetes (Fungi) of Patagonia, Tierra del Fuego and adjacent Antarctic areas. Darwiniana 42: 63–164

Garnica S, Weiss M, Oberwinkler F (2002) New *Cortinarius* species from *Nothofagus* Forests in South Chile. Mycologia 94 (1): 136–145

Garnica S, Weiss M, Oberwinkler F (2003) Morphological and molecular phylogenetic studies in South American *Cortinarius* species. Mycol Res 107 (10): 1143–1156

Garrido N (1986) Survey of ectomycorrhizal fungi associated with exotic forests trees in Chile. Nova Hedwigia 43(3–4): 423–442

Garrido N (1988) Agaricales sl und ihre Mykorrhizen in den *Nothofagus*-Wäldern Mittelchiles

Gentry AH (1982) Neotropical floristic diversity: Phytogeographical connections between Central and South America, Pleistocene climatic fluctuations, or an accident of the Andean orogeny? Ann Mo Bot Gard 69 (3):557–593

Geml J, Pastor N, Fernández L, Pacheco S, Semenova T, Becerra AG, Wicaksono CY, Nouhra ER (2014) Large-scale fungal diversity assessment in the Andean Yungas forests reveals strong community turnover among forest types along an altitudinal gradient. Molecular Ecology 23: 2452–2472

Godoy R, Palfner G (1997) Ectomicorrizas en *Nothofagus alpina* (Poepp. & Endl.) Oerst. y *N. dombeyi* (Mirb. Oerst.) del sur de Chile. Boletín Micológico 12(1–2):55–61

Godoy R, Romero R, Carrillo R (1994) Estatus micotrófico de la flora vascular en bosques de coníferas nativas del sur de Chile. Rev Chil Hist Nat 67: 209–220

Greff A, Porzel A, Schmidt J, Palfner G, Arnold N (2017) Pigment pattern of the Chilean mushroom *Dermocybe nahuelbutensis* Garrido & E. Horak. Records of Natural Products 11(6):547–551

Greslebin AG (2002) Flora Criptogámica de Tierra del Fuego. Fungi, Basidiomycota, Aphyllophorales: Coniophoraceae, Corticiaceae, Gomphaceae, Hymenochaetaceae, Lachnocladiaceae, Stereaceae, Thelephoraceae. Tulasnellales: Tulasnellaceae, FECIC, Buenos Aires

Gruber I (1975) Papierchromatographische Pigmentanalyse von südamerikanischen Dermocyben und Cortinarien. In: Moser M, Horak E (eds) *Cortinarius* Fr. und naheverwandte Gattungen in Südamerika, Beih. Nova Hewigia 52: 524–540

Halling RE (1989) A synopsis of Colombian boletes. Mycotaxon 34: 93–113

Halling RE (1996) Boletaceae (Agaricales): Latitudinal biodiversity and biological interactions in Costa Rica and Colombia. Revista de Biologia Tropical 44(4): 111–114

Halling RE (2001) Ectomycorrhizae: Co-Evolution, Significance, and Biogeography. Annals of the Missouri Botanical Garden 88: 5–13

Halling RE, Mueller G (1999) New boletes from Costa Rica. Mycologia 91: 893–899

Harley JL (1959) The Biology of Mycorrhiza. Plant Science Monographs, Leonard Hill, London, U.K., 233 p

Hauenstein E, González M, Peña-Cortés F, Muñoz-Pedreros A (2005) Diversidad vegetal en humedales costeros de la Región de La Araucanía. In: Smith-Ramírez C, Armesto JJ, Valdovinos C (eds) Historia, biodiversidad y ecología de los bosques costeros de Chile. Santiago: Editorial Universitaria, p 197–205.

Haug I, Weis M, Homeier J, Oberwinkler F, Kottke I (2005) Russulaceae and Thelephoraceae form ectomycorrhizas with members of the Nyctaginaceae (Caryophyllales) in the tropical mountain rainforest of southern Ecuador. New Phytol 165: 923–936

Hawksworth DL (1991) The fungal dimension of biodiversity: magnitude, significance, and conservation. Mycol Res 95: 641–655

Henkel TW, Terborgh J, Vilgalys RJ (2002) Ectomycorrhizal fungi and their leguminous hosts in the Pakaraima Mountains of Guyana. Mycol Res 106(5):515–531

Henkel TW (2003). Monodominance in the ectomycorrhizal *Dicymbe corymbosa* (Caesalpiniaceae) from Guyana. Journal of Tropical Ecology 19: 417–437

Henkel TW, Aime MC, Chin MML, Miller S, Vilgalys R, Smith M (2012) Ectomycorrhizal fungal sporocarp diversity and discovery of new taxa in *Dicymbe* monodominant forests of the Guiana Shield. Biodiversity and Conservation 21: 2195–2220

Hoorn C, Flantua S (2015) An early start for the Panama land bridge. Science 348 (6231): 186–187

Horak E (1979) Flora criptogámica de Tierra del Fuego. Fungi: Basidiomycetes agaricales y Gasteromycetes secotioides, 11, FECIC, Buenos Aires

Horak E, Moser M (1965) Fungi Austroamericani XII. Studien zur Gattung *Thaxterogaster* Sing. Nova Hedwigia 10: 211–241

Horak E, Moser M (1975) *Cortinarius* Fr. und nahe verwandte Gattungen in Sudamerika. Beihefte zur. Nova Hedwigia 52: 1–607

Kappelle M, Cleef AM, Chaverri A (1992) Phytogeography of Talamanca montane *Quercus* forest, Costa Rica. Journal of Biogeography 19: 299–315

Keller G, Moser M, Horak E, Steglich W (1987) Chemotaxonomic investigation of species of *Dermocybe* (Fr.) Wünsche (Agaricales) from New Zealand, Papua New Guinea and Argentina. Sydowia 40: 168–187

Kennedy P, Garibay-Orijel R, Higgins L, Angeles-Arguiz R (2011) Ectomycorrhizal fungi in Mexican *Alnus* forests support the host co-migration hypothesis and continental-scale patterns in phylogeography. Mycorrhiza 21(6):559–568

Kennedy PG, Matheny PB, Henkel TW, Uehling JK, Smith ME, Ryberg M (2012) Scaling up: examining the macroecology of ectomycorrhizal fungi. Mol Ecol 21: 4151–4154

Kuhar F, Barroetaveña C, Rajchenberg M (2016) New species of *Tomentella* (Thelephorales) from the Patagonian Andes forests. Mycologia 108(4): 780–790

Kuhar F, Smith ME, Mujic A, Truong C, Nouhra E (2017) A systematic overview of *Descolea* (Agaricales) in the Nothofagaceae forests of Patagonia. Fungal Biol 121 (10): 876–889

Lam HYT, Palfner G, Lima C, Porzel A, Brandt W, Frolov A, Wagner C, Merzweiler K, Wessjohann LA, Arnold N (2018) Nor-guanacastepene pigments from the Chilean mushroom *Cortinarius pyromyxa*. Journal of Natural Products (submitted)

Lindahl BD, Tunlid A (2015) Ectomycorrhizal fungi–potential organic matter decomposers, yet not saprotrophs. New Phytol 205(4): 1443–1447

Longo MS, Urcelay C, Nouhra E (2011) Long term effects of fire on ectomycorrhizas and soil properties in *Nothofagus pumilio* forests in Argentina. Forest ecology and management 262(3): 348–354

López-Quintero CA, Straatsma G, Franco-Molano AE, Boekhout T (2012) Macrofungal diversity in Colombian Amazon forests varies with regions and regimes of disturbance. Biodivers Conserv 21: 2221–2243

Lugo MA, Becerra AG, Nouhra ER, Ochoa AC (2012) Mycorrhizal diversity in native and exotic willows (*Salix humboldtiana* and *S. alba*) in Argentina. In: Pagano M (ed) Mycorrhiza: Occurrence in Natural and Restored Environments. Nova Science Publishers, New York, p 201–222

Marín C, Valenzuela E, Godoy R, Palfner G (2018) Diversity and growth-effects of ectomycorrhizal fungi of a *Nothofagus pumilio* forest in the Andes of Southern Chile. Boletín Micológico 33(1): 9–20

Melin E (1925) Untersuchungen über die Bedeutung der Baummykorrhiza. Eine ökologisch-physiologische Studie. Gustav Fischer, Jena, 152 p

Molina R, Massicotte H, Trappe JM (1992) Specificity phenomena in mycorrhizal symbioses: community-ecological consequences and practical implications. In: Allen MF (ed) Mycorrhizal functioning: an integrative plant-fungal process. Chapman and Hall, New York, pp 357–423

Moser M, Horak E, Gruber I (1975) *Cortinarius* Fr. und nahe verwandte Gattungen in Südamerika. Beihefte zur Nova Hedwigia 52: 1–628

Moyersoen B (1993) Ectomicorrizas y micorrizas vesículo-arbusculares en Caatinga Amazónica del Sur de Venezuela. Sci Guaianae (Caracas) 3: 1–82

Moyersoen B, Fitter A, Alexander I (1998a) Spatial distribution of ectomycorrhizas and arbuscular mycorrhizas in Korup National Park rainforest, Cameroon, in relation to edaphic parameters. New Phytol 139(2): 311–320

Moyersoen B, Alexander I, Fitter A (1998b) Phosphorus nutrition of ectomycorrhizal and arbuscular mycorrhizal tree seedlings from a lowland tropical rain forest in Korup National Park, Cameroon. J Trop Ecol 14(1): 47–61

Moyersoen B, Becker P, Alexander I (2001). Are ectomycorrhizas more abundant than arbuscular mycorrhizas in tropical heath forests? New Phytol (2001) 150: 591–599

Mueller GM, Halling R (1995) Evidence for high biodiversity of Agaricales (Fungi) in Neotropical montane *Quercus* forests. In: Churchill S, Balslev H, Forero E, Luteyn J (eds) Biodiversity and Conservation of Neotropical Montane Forests, New York Botanical Garden, Bronx, p 303–312

Muller GM, Singer R (1988) *Laccaria gomezii*, a new agaric species from the querceta of Colombia and Costa Rica. Mycotaxon 33: 223–227

Muller G, Strack BA (1992) Evidence for a mycorrhizal host shift during migration of *Laccaria trichodermophora* and other agarics into Neotropical oak forests. Mycotaxon 45: 249–256

NAS (National Academy of Sciences) (1980) Firewood Crops: Shrub and Tree Species for Energy Production. Report of an ad hoc panel of the Advisory Committee on Technology Innovation, Board on Science and Technology for International Development, Commission on International

Relations National Research Council (U.S.). Advisory Committee on Technology Innovation, National Academy Press, Washington DC

Navarro E, Bousquet J, Moiroud A, Munive A, Piou D, Normand P (2003) Molecular phylogeny of *Alnus* (Betulaceae), inferred from nuclear ribosomal DNA ITS sequences. Plant Soil 254: 207–217

Niveiro N (2012) Agaricales sensu lato (Agaricomycetes) de las Selvas del Dominio Amazónico de la Argentina. Diversidad, Distribución y Abundancia. Tesis Doctoral. Universidad Nacional de Córdoba

Nouhra ER, Domínguez L, Becerra A, Mangeaud A (2003) Colonización micorrícica y actinorrícica en plantines de *Alnus acuminata* (Betulaceae), cultivados en suelos nativos de *Alnus rubra*. Bol Soc Argent Bot 38: 199–206

Nouhra ER, Domínguez L, Becerra AG, Trappe JM (2005) Morphological, molecular and ecological aspects of the South American hypogeous fungus *Alpova austroalnicola* sp. nov. Mycologia 97: 598–604

Nouhra ER, Domínguez L, Urcelay C, Becerra AG (2008) Mycorrhizal status of *Fagara coco* (Rutaceae): does it form ectomycorrhizas with *Phlebopus bruchii* (Boletaceae)? Symbiosis 46: 113–120

Nouhra ER, Urcelay C, Longo MS, Fontenla S (2012) Differential hypogeous sporocarp production from *Nothofagus dombeyi* and *N. pumilio* forests in southern Argentina. Mycologia 104(1): 45–52

Nouhra ER, Urcelay C, Longo S, Tedersoo L (2013) Ectomycorrhizal fungal communities associated to *Nothofagus* species in Northern Patagonia. Mycorrhiza 23(6): 487–496

Nouhra ER, Pastor N, Becerra A, Sarrionandia Areitio E, Geml J (2015) Greenhouse grown seedlings of *Alnus* species showed lack of specificity and a strong preference for *Tomentella* ectomycorrhizal associates. Microb Ecol 69: 813–825

Onguene N, Kuyper T (2002) Importance of the ectomycorrhizal network for seedling survival and ectomycorrhiza formation in rain forests of south Cameroon. Mycorrhiza 12: 13–17

Palfner G (2001) Taxonomische Studien an Ektomykorrhizen aus den-*Nothofagus*-Wäldern Mittelsüdchiles

Palfner G, Casanova-Katny A (2018) Comparación de las micocenosis de remanentes de bosque nativo y plantaciones forestales en la península de Arauco, Región del Biobío, con énfasis en aspectos funcionales y de conservación. In: Smith-Ramirez C, Armesto JJ, Valdovinos C (eds) Historia, biodiversidad y ecología de los bosques costeros de Chile. 2nd edition, Editorial Universitaria, Santiago, Chile (in press)

Pastor N, Chiapella J, Kuhar F, Mujic A, Crespo E, Nouhra R. (2019) Unveiling new sequestrate *Cortinarius* species from northern Patagonian Nothofagaceae forests based on molecular and morphological data. Mycologia 111(1): 103–117

Põlme S, Bahram M, Yamanaka T, Nara K, Dai YC, Grebenc T, Kraigher H, Toivonen M, Wang PH, Matsuda Y, Naadel T, Kennedy PG, Koljalg U, Tedersoo L (2013) Biogeography of ectomycorrhizal fungi associated with alders (*Alnus* spp.) in relation to biotic and abiotic variables at the global scale. New Phytol 198: 1239–1249

Pritsch K, Becerra A, Polme S, Tedersoo L (2010) Description and identification of *Alnus acuminata* ectomycorrhizae from Argentinean alder stands. Mycologia 102: 1263–1273

Redhead SA, Malloch DW (1986) The genus *Phaeocollybia* (Agaricales) in eastern Canada and its biological status. Can J Bot 64(6): 1249–1254

Ragonese A. (1987). Familia Salicaceae. Flora Ilustrada de Entre Ríos, III. Colección Científica del INTA. Tomo VI, Buenos Aires, p 6–14

Ren BQ, Xiang XG, Chen ZD (2010) Species identification of *Alnus* (Betulaceae) using nrDNA and cpDNA genetic markers. Mol Ecol Resour 10: 594–605

Riviere T, Diedhiou A, Diabate M, Senthilarasu G, Natarajan K, Verbeken A, Buyck B, Dreyfus B, Bena G, Ba A (2007) Genetic diversity of ectomycorrhizal Basidiomycetes from African and Indian tropical rain forests. Mycorrhiza 17: 415–428

Romano GM, Greslebin AG, Lechner BE. (2017a) Modelling agaricoid fungi distribution in Andean forests of Patagonia. Nova Hedwigia 105(1–2): 95–120

Romano GM, Ruiz EV, Lechner BE, Greslebin AG, Morrone JJ (2017b) Track analysis of agaricoid fungi of the Patagonian forests. Aust Syst Bot 29(6): 440–446

Roy S, Khasa DP, Greer CW (2007) Combining alders, frankiae, and mycorrhizae for the revegetation and remediation of contaminated ecosystems. Can J Bot 85: 237–251

Roy M, Schimann H, Braga Neto R, Da Silva RA, Duque J, Frame D, Wartchow F, Neves MA (2016) Diversity and distribution of ectomycorrhizal fungi from Amazonian lowland white sand forests in Brazil and French Guiana. Biotropica 48(1): 90–100

Salazar VE (2016) Micosociología: antecedentes históricos, evolución y proyecciones. Boletín Micológico 31(2): 23–35

Salgado Salomón ME, Barroetaveña C, Rajchenberg M (2013) Occurrence of dark septate endophytes in *Nothofagus* seedlings from Patagonia, Argentina. Southern Forests: a Journal of Forest Science 75(2): 97–101

Salgado Salomon ME, Dreschdm P, Horak E, Galleguillos F, Barroetaveña C, Peintner U (2018) The enigmatic *Cortinarius magellanicus* complex occurring in Nothofagaceae forests of the Southern Hemisphere. Fungal Biol 122: 1077–1097

Sánchez-García M, Henkel TW, Aime MC, Smith ME, Matheny PB (2016) *Guyanagarika*, a new ectomycorrhizal genus of Agaricales from the Neotropics. Fungal Biol 120(12): 1540–1553

Sanmartin I, Ronquist F (2004) Southern hemisphere biogeography inferred by event-based models: plant versus animal patterns. Systematic biology 53(2): 216–243

Singer R (1950) *Descolea antarctica*, género y especie nuevos de Tierra del Fuego. Lilloa 23: 255–258

Singer R (1954) Agaricales von Nahuel Huapi. Sydowia 8: 100–157

Singer R (1963) Oak mycorrhiza fungi in Colombia. Mycopathol Mycol Appl 20 (3–4): 239–252

Singer R (1971) Forest mycology and communities in South America II. Mycorrhiza sociology and fungus succession in the *Nothofagus-Austrocedrus chilensis* Woods of Patagonia. In: Hacskaylor R (ed) Proceedings of the First North American Conference on Mycorrhizae – USDA Forest Service Publication 1189: 204–215

Singer R, Araujo I (1979) Litter decomposition and Ectomycorrhiza in Amazonian forests. A comparison of litter decomposing and ectomycorrhizal Basidiomycetes in latosol-terra-firme rain forest and white podzol campinarana. Acta Amazonica 9: 25–41

Singer R, Morello JH (1960) Ectotrophic forest tree mycorrhizae and forest communities. Ecology 41:549–551

Singer R, Moser M (1965) Forest mycology and forest communities in South America I. The early fall aspect of the mycoflora of the Cordillera Pelada (Chile), with a mycogeographic analysis and conclusions regarding the heterogeneity of the Valdivian floral district. Mycopathologia et Mycologia Applicata 26 (2–3): 129–191

Singer R, Smith AH (1958) Studies on secotiaceous fungi—I: A monograph of the genus *Thaxterogaster*. Brittonia 10(4):201–216

Singer R, Araujo I, Ivory MH (1983) The ectotrophically mycorrhizal fungi of the neotropical lowlands, especially Central Amazonia (Litter decomposition and ectomycorrhiza). Beihefte zur Nova Hedwigia 77: 1–339

Skrede I, Engh IB, Binder M, Carlsen T, Kauserud H, Bendiksby M (2011) Evolutionary history of Serpulaceae (Basidiomycota): molecular phylogeny, historical biogeography and evidence for a single transition of nutritional mode. BMC Evol Biol 11: 230

Smith ME, Henkel TW, Aime MC, Fremier AK, Vilgalys R (2011) Ectomycorrhizal fungal diversity and community structure on three co-occurring leguminous canopy tree species in a Neotropical rainforest. New Phytol 192: 699–712

Smith ME, Henkel TW, Uehling JK, Fremier AK, Clarke HD, Vilgalys R. (2013) The Ectomycorrhizal Fungal Community in a Neotropical Forest Dominated by the Endemic Dipterocarp *Pakaraimaea dipterocarpacea*. PLoS ONE 8(1): e55160

Smith ME, Henkel TW, Williams GC, Aime MC, Fremier AK, Vilgalys R (2017) Investigating niche partitioning of ectomycorrhizal fungi in specialized rooting zones of the monodominant leguminous tree *Dicymbe corymbosa*. New Phytol 215(1): 443–453

Spegazzini (1887) Fungi Patagonici. Boletin Academia Nacional de Ciencias Córdoba 11(1): 15

Spegazzini C (1912) Mycetes Argentinenses. Anales Museo Nacional de Historia Natural Buenos Aires 23: 9

Sulzbacher MA, Giachini AJ, Grebenc T, Silva BDB, Gurgel FE, Loiola MIB, Neves MA, Baseia IG (2013) A survey of an ectotrophic sand dune forest in the northeast Brazil. Mycosphere 4(6): 1106–1116

Tedersoo L, Brundrett MC (2017) Evolution of ectomycorrhizal symbiosis in plants. In: Tedersoo L (ed) Biogeography of Mycorrhizal Symbiosis. Ecological Studies (Analysis and Synthesis), vol 230. Springer, Cham

Tedersoo L, Smith ME (2013) Lineages of ectomycorrhizal fungi revisited: foraging strategies and novel lineages revealed by sequences from belowground. Fungal Biol Rev 27: 83–99

Tedersoo L, Smith ME (2017) Ectomycorrhizal fungal lineages: Detection of four new groups and notes on consistent recognition of ectomycorrhizal taxa in high-throughput sequencing studies. In: Tedersoo L (eds) Biogeography of Mycorrhizal Symbiosis. Ecological Studies (Analysis and Synthesis), vol 230. Springer, Cham

Tedersoo L, Suvi T, Beaver K, Kõljalg U (2007) Ectomycorrhizal fungi of the Seychelles: diversity patterns and host shifts from the native *Vateriopsis seychellarum* (Dipterocarpaceae) and *Intsia bijuga* (Caesalpiniaceae) to the introduced *Eucalyptus robusta* (Myrtaceae), but not *Pinus caribea* (Pinaceae). New Phytol 175: 321–333

Tedersoo L, Suvi T, Jairus T, Ostonen I, Polme S (2009) Revisiting ectomycorrhizal fungi of the genus *Alnus*: differential host specificity, diversity and determinants of the fungal community. New Phytol 182: 727–735

Tedersoo L, Sadam A, Zambrano M, Valencia R, Bahram M (2010a) Low diversity and high host preference of ectomycorrhizal fungi in Western Amazonia, a neotropical biodiversity hotspot. ISME J 4(4): 465–471

Tedersoo L, May TW, Smith ME (2010b) Ectomycorrhizal lifestyle in fungi: global diversity, distribution, and evolution of phylogenetic lineages. Mycorrhiza 20: 217–263

Tedersoo L, Põlme S (2012) Infrageneric variation in partner specificity: multiple ectomycorrhizal symbionts associate with *Gnetum gnemon* (Gnetophyta) in Papua New Guinea. Mycorrhiza 22: 663–668

Tedersoo L, Bahram M, Toots M, Diédhiou A, Henkel t, Kjøller R, Morris MH, Nara K, Nouhra E, Peay K, Põlme, Ryberg M, Smith ME, Kõljalg U (2012) Towards global patterns in the diversity and community structure of ectomycorrhizal fungi. Mol Ecol 21: 4160–4170

Teklehaimanot Z, Mmolotsi RM (2007) Contribution of red alder to soil nitrogen input in a silvopastoral system. Biol Fertil Soils 43: 843–848

Trierveiler-Pereira L, Smith M, Trappe J, Nouhra ER (2015) Sequestrate fungi from Patagonian *Nothofagus* forests: *Cystangium* (Russulaceae, Basidiomycota). Mycologia 107(1): 90–103

Truong C, Mujic A, Healy R, Kuhar F, Furci G, Torres D, Niskanen T, Sandoval-Leiva P, Fernández N, Escobar J, Moretto A, Palfner G, Pfister D, Nouhra E, Swenie R, Sánchez-García M, Matheny B, Smith M (2017). How to know the fungi: combining field inventories and DNA-barcoding to document fungal diversity. New Phytol 214: 913–919

Valenzuela E, Esteve-Raventós F (1994) *Cortinarius horakii*, a new species from Chile. Mycol Res 98(8): 937–938

Valenzuela E, Moreno G, Garnica S, Ramírez C (1998) Micosociología en bosques nativos de *Nothofagus* y plantaciones de *Pinus* radiata en la X Región de Chile: diversidad y rol ecológico. Rev Chil Hist Nat 71: 133–146

van der Heijden MGA, Martin FM, Selosse MA, Sanders IR (2014) Mycorrhizal ecology and evolution: the past, the present, and the future. New Phytol 205: 1406–1423

Vargas N, Pardo-de La Hoz C J, Danies G, Franco-Molano A E, Jiménez P, Restrepo S, Grajales A (2017) Defining the phylogenetic position of *Amanita* species from Andean Colombia. Mycologia 109(2):261–276

Vasco-Palacios AM, López-Quintero CA, Franco-Molano AE, Boekhout T (2014) *Austroboletus amazonicus* sp. nov. and *Fistulinella campinaranae* var. *scrobiculata*, two commonly occurring boletes from a forest dominated by *Pseudomonotes tropenbosii* (Dipterocarpaceae), in Colombian Amazonia. Mycologia 106: 1004–1014

Vasco-Palacios AM, Hernandez J, Peñuela-Mora MC, Franco-Molano AE, Boekhout T (2018) Ectomycorrhizal fungi diversity in a white sand forest in western Amazonia. Fungal Ecol 31: 9–18

Velíšek J, Cejpek K (2011) Pigments of Higher Fungi: A Review. Czech J Food Sci 29 (2): 87–102

Weng C, Bush MB, Chepstow-Lusty AJ (2004) Holocene changes of Andean alder (*Alnus acuminata*) in highland Ecuador and Perú. J Quat Sci 19: 685–691

Winner M, Giménez A, Schmidt H, Sontag B, Steffan B, Steglich W (2004) Unusual pulvinic acid dimers from the common fungi *Scleroderma citrinum* (ommon earthball) and *Chalciporus piperatus* (peppery bolete). Angewandte Chemie 43(14): 1883–1886

Zak B (1973) Classification of ectomycorrhizae. In: Marks GC, Kozlowski TT (eds) Ectomycorrhizae: Their Ecology and Physiology, Academic Press, New York, London, p 43–78

Chapter 5
A Systematic Review of South American and European Mycorrhizal Research: Is there a Need for Scientific Symbiosis?

César Marín and C. Guillermo Bueno

5.1 Introduction

The study of global biodiversity presents spatially biased data distribution with large differences in the sampling efforts and data resolution between the Northern and Southern hemispheres (Meyer et al. 2015; Wetzel et al. 2018). These differences are exacerbated by monetary, linguistic, geographic, and political barriers, more prevalent in the Southern hemisphere (Amano and Sutherland 2013). Besides regional differences, our knowledge regarding biodiversity information differs among different organisms, being belowground organisms relatively unknown, even though they are fundamental to terrestrial ecosystem functioning and aboveground biodiversity dynamics (Bardgett and van der Putten 2014; Carey 2016). Despite the appearance of new and more efficient molecular and macroecological biodiversity approaches in the last decades (Wiens 2007), which have boosted regional and global biodiversity studies, geographical data gaps are still large on general soil biodiversity (Cameron et al. 2018), and on soil and mycorrhizal fungi in particular (Tedersoo et al. 2014; Davison et al. 2015; Bueno et al. 2017a). This has led to a biodiversity-knowledge paradox: while areas in the Southern hemisphere, such as South America (SA), host the most diverse biodiversity hotspots, they have been largely understudied, particularly on belowground fundamental organisms and associations such as the mycorrhizal symbiosis.

Research efforts via specific scientific networks are efficient strategies to overcome local limitations in resources and to extent the research aims in ecological time or space, which is ideal for answering largely unknown exploratory or

C. Marín (✉)
Universidad de O'Higgins, Rancagua, Chile
e-mail: cesar.marin@uoh.cl

C. G. Bueno
University of Tartu, Tartu, Estonia

M. C. Pagano, M. A. Lugo (eds.), *Mycorrhizal Fungi in South America*,
Fungal Biology, https://doi.org/10.1007/978-3-030-15228-4_5

mechanistic questions (Richter et al. 2018). There are increasing networking efforts to monitor changes in global aboveground biodiversity, functions, and ecosystem services (Scholes et al. 2008, 2012; Pereira et al. 2010, 2013; Tallis et al. 2012), while belowground and fungal biodiversity, including mycorrhizal fungi, have barely been covered by any scientific network (Wetzel et al. 2015, 2018). In this context, the integration of Southern needs and perspectives of mycorrhizal research into operative networking efforts in collaboration with Northern researchers, can strengthen local and global research, creating successful and mutualistic collaborative efforts. One recent example of these collaborative efforts is the South American Mycorrhizal Research Network (SAMRN) (Bueno et al. 2017a; Godoy et al. 2017), which is an horizontal scientific community directed towards the progress of mycorrhizal research and knowledge, along with applications and public outreach in SA. The SAMRN is constructed on the basis of collaborative efforts, to overcome the lack of funding or collaboration between South American and European or North American funding agencies (Amano and Sutherland 2013). Despite these local constraints, collaborative networking initiatives are effective and promising tools. For instance, in over two years, this network has strengthened scientific collaboration between and within local and foreign researchers and students through the organization of conferences, symposia, and technical specific workshops (https://southmycorrhizas.org/). These activities have in turn led to several scientific outreach activities and publications, providing a solid ground for the announcement and development of the present book.

Overall, in a context of unbalanced geographical resources and needs, it is important to understand the research efforts done in different regions to ultimately enhance future strategies that will focus on the research needs and knowledge gaps of local and global biodiversity. In this context, the following review focuses on the mycorrhizal symbiosis as a key player of the main terrestrial processes and ecosystem functions (Bardgett and van der Putten 2014), present in most terrestrial plant species (Brundrett and Tedersoo 2018). The objective of this systematic review was to compile, characterize, classify, and compare the scientific literature on mycorrhizal research in South America and Europe from 1975 to 2018. This study represents the first effort to understand South American and European differences in aims and perspectives, which can enable the integration of South American mycorrhizal information and research initiatives into global initiatives and models.

5.2 Systematic Review of Mycorrhizal Literature

In order to develop our review, we followed the PRISMA protocol ("Preferred Reporting Items for Systematic Reviews and Meta-Analyses"; Liberati et al. 2009) which consisted of several steps. First, we conducted a literature search (on 27th of November 2018) using Web of Science with the terms "mycorrhiza*" AND terms for the geographical region. We used the term "mycorrhiza*" to include all the variants of the word mycorrhiza (i.e. "mycorrhizae", "mycorrhizal", etc.). For the geographical terms we used "South America" and "Europe", in adition to all countries within South America (i.e. "Argentina", "Bolivia", "Brazil", "Chile", "Colombia", "Ecuador", "French Guiana", "Guyana", "Paraguay", "Peru", "South Georgia", "Uruguay", and "Venezuela") and Europe (i.e. "Armenia", "Austria", "Azerbaijan", "Belarus", "Belgium", "Bulgaria", "Czech Republic", "Denmark", "Estonia", "Finland", "France", "Georgia", "Germany", "Greece", "Hungary", "Iceland", "Ireland", "Italy", "Kazakhstan", "Latvia", "Lithuania", "Luxembourg", "Macedonia", "Malta", "Moldova", "Montenegro", "Netherlands", "Norway", "Poland", "Portugal", "Romania", "Russia", "Serbia", "Slovakia", "Slovenia", "Spain", "Sweden", "Switzerland", "Turkey", "Ukraine", and "United Kingdom"). Second, we compiled the list of articles for each continent after removing duplicates, using the EndNote Web software for all the European and South American mycorrhizal published scientific articles between 1975 and 2018 in any language (>99% of articles were in English). All research articles as well as reviews and meta-analyses were included. Third, the articles not directly related to the mycorrhizal symbiosis (less than 5% of all articles) were manually excluded. Then, an article dataset was compiled for each continent with information about the country or countries where the studies were conducted, year of publication, and number of citations (2724 articles in total). Fourth, we selected the most influential articles (the ones cited 70 or more times until the 27th of November, 2018) and carefully checked and assigned them to one of the following nine general mycorrhizal topics: "rhizosphere interactions", "plant invasions", "phylo/biogeography", "morphology", "molecular methods", "ecosystem remediation", "community structure", "biogeochemistry", and "anthropogenic effects". These topics referred to the role of the mycorrhizal symbiosis or mycorrhizal fungi in relation to each specific topic. For instance, "rhizosphere interactions" comprised those articles regarding the interaction of mycorrhizal plant roots and their immediate surroundings with other organisms in the soil. When an article included more than one topic, the assigned topic was the most prevailing in the article.

5.3 South American and European Mycorrhizal Research Trends

The literature search yielded a total of 1927 scientific articles for Europe and 797 for SA (Fig. 5.1), showing that even though Europe has a territory roughly four times smaller than SA, its mycorrhizal research was more than double. From the beginnings of the 1990's, there has been a steady increase in the yearly production of scientific articles in both continents, with an outstanding research increase over the last 10 years (Fig. 5.1). At the end of the 1980's and up to the 1990's, several factors could have boosted mycorrhizal research, i.e. the availability of molecular methods that allowed for more efficient and detailed taxonomic and biochemical studies of the mycorrhizal symbiosis (White et al. 1990; Gardes and Bruns 1993;

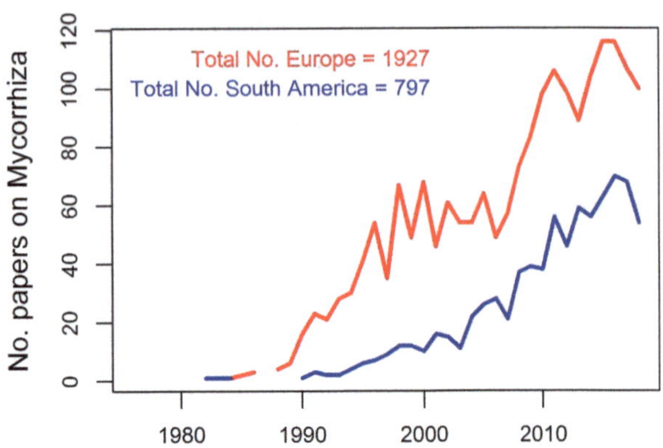

Fig. 5.1 Number of scientific articles on mycorrhizal research yearly published for South America (blue lines) and Europe (red lines) for the period 1975–2018

Harrison and van Buuren 1995; Bianciotto et al. 1996). As a consequence of the new information available, new conceptual developments on mycorrhizal networks (Toju et al. 2014; van der Heijden et al. 2015) and biogeography (Read 1991) were extended. A second research wave on mycorrhiza occurred at the end of the 2000's (Fig. 5.1), which could be explained by the arrival of the 'omics' era (Bonfante 2018). This technological revolution boosted the identification of mycorrhizal genes and their expression in relation to plant physiological processes in fungal colonization and subsequent mycorrhizal activities. Derived from the arrival of this technology, one specific topic that attracted great attention was the identification and classification of fungal mycorrhizal species (Schüßler et al. 2001; Öpik et al. 2010; Nilsson et al. 2018; Tedersoo et al. 2018; Wijayawardene et al. 2018).

Regarding the mycorrhizal scientific production of SA, Brazil was the country with the highest number of scientific articles published, followed by Argentina, Chile, Venezuela, and Ecuador (Fig. 5.2). Brazil is by far the most populated of these countries, and hosts the largest number of scientists in SA. Together with Brazil, Argentina and Chile have a long historical tradition in natural history studies, including mycological studies, which goes back to Darwin (Berkeley 1841). It is worth mentioning the studies carried out in the Patagonian region, which were developed by important European naturalists and mycologists such as Claudio Gay (Montagne 1850) and Rodolfo and Federico Philippi (Philippi 1893; Castro et al. 2006). The interest of European mycologists in Argentinian and Chilean fungi continued over the Twentieth century, resulting in detailed descriptions of fungi and fungal communities (Spegazzini 1921; Singer and Morello 1960; Singer et al. 1965; Singer 1969, 1970).

In Europe, Norway, Spain, Sweden, Germany, and Poland were respectively the five top countries with the highest number of published scientific articles on mycorrhizal research (Fig. 5.3). These five countries are historically well known in the mycorrhizal research field. For instance, the ectomycorrhizal associations in the Swedish and Norwegian coniferous forests have been studied for almost 200 years (Bonfante 2018). These two Nordic countries have also increasingly focused their

Fig. 5.2 Country relevance regarding mycorrhizal scientific production in South America (indicated by blue circle size). The number of articles is only indicated for the top five most productive countries from 1975 to 2018; 59 of the articles were developed in more than one country

research on mycorrhizal signaling and metabolic pathways, and more recently, on soil microbiome interactions with the plant rhizosphere (Bonfante 2018; Sterkenburg et al. 2018). In contrast, the mycorrhizal research in Spain and Poland has been more focused on arbuscular mycorrhizal fungi, particularly on either the anthropogenic effects on fungi or its role in ecosystem remediation (in Spain), and on morphology and taxonomy (in Poland). On the other hand, German mycorrhizal research comprises a wider range of topics that vary from taxonomy and morphology to phylogeography and biogeography of mycorrhizal fungi.

We found that the mycorrhizal topics covered in the most cited articles (70 or more citations) were notably different in SA and Europe. In South America, studies describing the mycorrhizal fungi community structure predominated (7 out of 20 most influential articles; Fig. 5.4), while in Europe the articles were more devoted to the study of the anthropogenic impact on the mycorrhizal symbiosis (41 out of

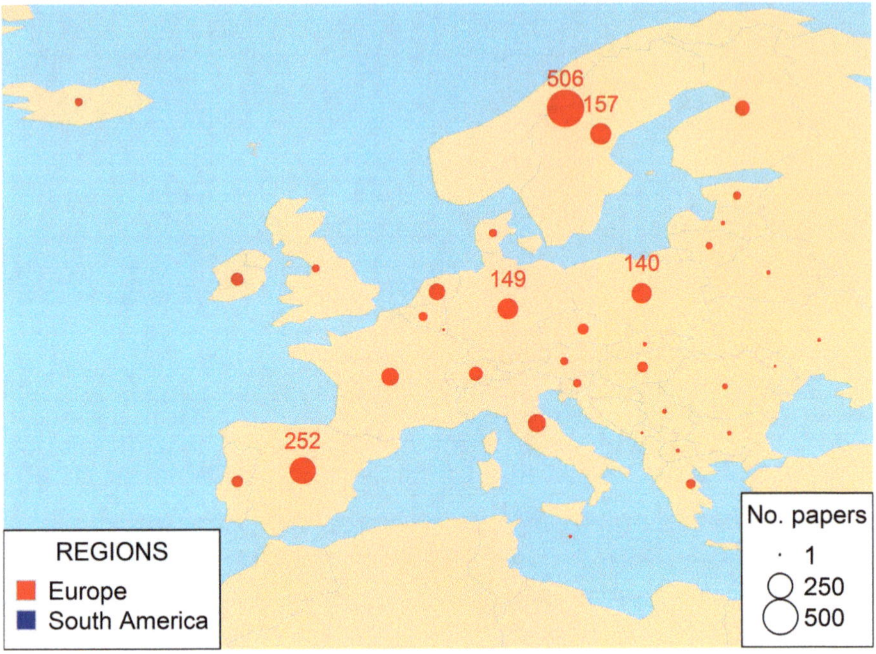

Fig. 5.3 Relative relevance per country following the mycorrhizal scientific production in Europe (indicated by different circle sizes in red). The number of articles is only indicated for the top five most productive countries from 1975 to 2018; 258 mycorrhizal European articles were developed in more than one country

Fig. 5.4 Percentages of the most influential articles for each mycorrhizal topic (20 articles) from 1975 to 2018 in South America

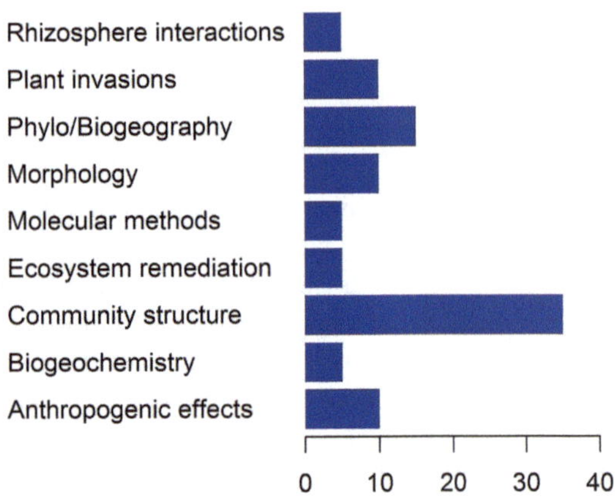

141 most influential articles; Fig. 5.5). This difference indicates that more 'basic science' research was needed in South America, whereas a more applied and specific research was developed in Europe, such as the study of mycorrhizal roles on rhizosphere interactions and on biogeochemistry (Figs. 5.4, 5.5). In any case, this

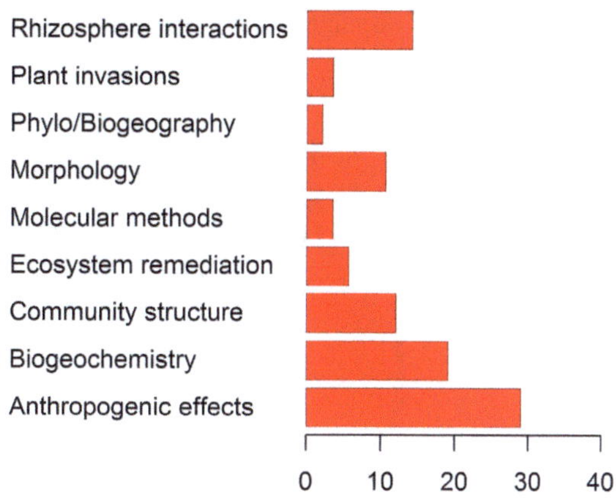

Fig. 5.5 Percentages of the most influential articles for each mycorrhizal topic (140 articles) from 1975 to 2018 in Europe

can also indicate that both territories may have different mycorrhizal research needs. In Europe, the smaller geographical extent and diversity, as well as the higher population densities and larger historical research (Bonfante 2018), could have contributed to be better explored and known in terms of mycorrhizal ecology, where the main concerns are the mycorrhizal roles for nature conservation under the current scenarios of global changes. In contrast, South American highest research interest was on its relatively unknown biodiversity and local mycorrhizal knowledge, which may have enhanced research on more descriptive and fundamental questions (Figs. 5.4, 5.5). It seems logical that the knowledge development has followed some clear steps: after studying the biodiversity patterns of mycorrhizal fungi communities, their effects on the rhizosphere and on biogeochemical cycles will follow.

Regarding the historical impact (number of citations) of each mycorrhizal research topic per continent, SA showed a very different pattern from Europe (Fig. 5.6). South American studies focused on the use and description of molecular methods, highly cited from the middle 1980's to the late 1990's (Fig. 5.6). This was followed by a very influential paper on ecosystem remediation issued during the late 1990's (Franco and de Faria 1997). These trends were temporally replaced by a more diverse group of topics related to mycorrhizal fungal morphology, phylogeography/biogeography (or phylo/biogeography), and mycorrhizal-related plant invasion research (Fig. 5.6). In addition, four influential papers set rhizosphere interactions (Rubiales et al. 2009), biogeochemistry (Ryan et al. 2010), and anthropogenic effects (Cornejo et al. 2008; Stürmer and Siqueira 2011) as very popular topics of the South American literature in the late 2000's and early 2010's (Fig. 5.6). Europe had seven times more influential papers and thus the impact of single papers was less pronounced. Overall, the main influential topics during this period were phylo/biogeography from the middle 1980's to the early 1990's, ecosystem remediation during the 2000's, and plant invasions and molecular methods afterwhile (Fig. 5.6).

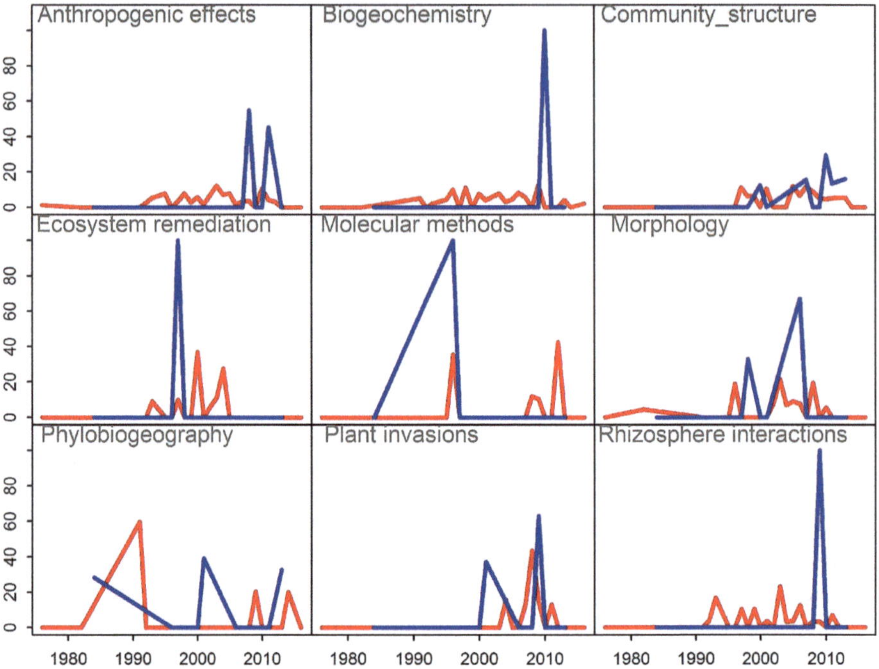

Fig. 5.6 Percentage of citations of highly cited (70 or more citations) scientific articles in South America (blue lines) and Europe (red lines) from 1975 to 2018

5.4 Geographical and Thematic Gaps on South American Mycorrhizal Research

Global studies on soil fungal and mycorrhizal diversity (Tedersoo et al. 2014; Davison et al. 2015) have so far excluded large portions of South American ecosystems and even some countries (Bueno et al. 2017a). These global studies have focused on the southern or northern regions of SA, leaving out much of the most important biodiversity hotspots, such as large areas of the Amazon basin, continental savannas, most of the Andes, and the Chocó biogeographic region, which hosts one of the rainiest and most diverse forests in the world (Galeano et al. 1998). This has already been shown in a review on hypogeous sequestrate fungi (Sulzbacher et al. 2017), indicating potential limitation for research in those areas. This pattern also is more or less consistent with our own findings (Fig. 5.2) since countries as Perú, Bolivia, Paraguay, and Uruguay have been barely studied. The main exception to this pattern among those global studies (Tedersoo et al. 2014; Davison et al. 2015) and our findings (Fig. 5.2) is Brazil, which is the most productive South American country regarding mycorrhizal literature (Fig. 5.2). But even though there is much Brazilian mycorrhizal research conducted on its southern part, closer to their most important universities, it is still scarce in the northern ecosystems, including the Amazon basin or the savanna.

Our review suggests that there are still large and relevant South American areas and ecosystems with a lack of basic mycorrhizal knowledge (Figs. 5.2–5.5), being urgent the development of descriptive science such as the analysis of these areas' fungal mycorrhizal biodiversity and community composition, to ultimately enhance our local and global understanding of biodiversity. Contrarily, in better known areas such as the south of Brazil or the Patagonian region (Fig. 5.2; Bueno et al. 2017a), we suggest to develop more specific and applied research, such as the study of the role of the mycorrhizal symbiosis in relation to rhizosphere interactions, biogeochemistry, or in relation to anthropogenic impacts on ecosystems. This mycorrhizal research would enhance the development of more applied science in terms of sustainable development, environmental conservation, or functional and mechanistic aspects. Needless to say, all these research topics are not mutually exclusive and the research aims need to be aligned with social and environmental local needs, as such is the case of some understudied areas being strongly affected by poverty, anthropogenic pressure, and ecosystem degradation.

Overall, and accounting for the current mycorrhizal research trends presented here and promising lines of mycorrhizal research (Bonfante 2018; Waller et al. 2018), we strongly believe that two main data types might need to be urgently collected in SA: (1) molecular data on mycorrhizal fungi, especially data obtained from environmental samples and which leads to DNA-based classification systems ('species hypothesis', Nilsson et al. 2018; 'virtual taxa', Öpik et al. 2010), and (2) plant roots to define plant mycorrhizal traits (Moora 2014; Bueno et al. 2018). There is a large gap of fungal mycorrhizal molecular data missing on databases, as it was illustrated by a quick search (on 19th of December, 2018) on the database MaarjAM (Öpik et al. 2010), which contains arbuscular mycorrhizal fungi DNA sequences from environmental or cultured samples. In MaarjAM, 373 sampling locations were situated in Europe, while only 97 were in SA. Moreover, in terms of plant mycorrhizal traits (Moora 2014) there are still many areas and entire countries in SA where the distribution of plant species and plant communities is not completely known, in contrast with a relatively well documented distribution of the European flora and vegetation (Kalwij et al. 2014; Soudzilovskaia et al. 2015, 2017). In fact, the plant mycorrhizal traits of most of the South American flora remain unknown. For instance, a recent study conducted in Chile dealing with the latitudinal distribution of plant mycorrhizal traits obtained after a systematic and thorough literature search, showed a coverage of about 13% of continental Chile plant species with geographical information (Silva-Flores pers. com.). Thus, considering Chile as one of the countries where mycorrhizal research has been further developed (Fig. 5.2), this percentage is relatively low in relation to a recent European study which covered around 45% of the European species with available geographic information (Bueno et al. 2017b). Thus, plant mycorrhizal trait collection is clearly needed in SA to estimate the prevalence of the mycorrhizal symbiosis in plant communities and ecosystems, and assign their response to biotic and abiotic conditions. This will ultimately lead to the understanding of the ecological roles of the mycorrhizal symbiosis in SA's unique ecosystems and its future responses to current global changes and local anthropogenic activities.

5.5 Suggested Directions

We believe that there are three steps that can be followed in order to boost South American mycorrhizal research and facilitate its global integration. First, it is necessary to strengthen the communication among South American mycorrhizal researchers and to channel and optimize research efforts through collaborative networking initiatives, such as the South American Mycorrhizal Research Network (SAMRN) (https://southmycorrhizas.org/; Bueno et al. 2017a). This aim could be efficiently achieved through the coordination of international funding public agencies, coordination which has not yet been implemented in SA. For instance, an ideal funding possibility in SA will be a call for large consortiums of international researchers with shared aims. This, in turn, will lead to homogenize discrepancies among methodological and study designs, ultimately enhancing the consolidation of strong international scientific groups in SA. For example, the European Union and their scientific funding agencies provide large international funding calls, which has led to large and significant studies on mycorrhizal diversity (Davison et al. 2015; van der Linde et al. 2018). Second, we suggest to facilitate the integration of intercontinental and global projects. European and North American global projects have not systematically integrated South American researchers, which could enlarge and optimize their sampling schemes as well as our scientific global knowledge. Furthermore, South American research network initiatives such as the SAMRN could enhance the communication among researchers from different continents and the optimization of potential collaborations. For instance, in terms of effective information and resources exchange, South American researchers could contribute with local knowledge, reduce local bureaucracy, offer access to unknown localities, and provide southern research perspectives, research needs and conceptual gaps. In turn, northern researchers could provide a more theoretical approach to global research questions, linked to a higher availability of technological and funding resources. Some integrating initiatives have been started, such as the analysis of worldwide leaf microbiomes (FunLeaf project; https://sisu.ut.ee/funleaf); however, as regards of the mycorrhizal symbiosis, further integration between descriptive and applied research on mycorrhizal diversity, ecosystem functions, and the effects of global changes still need to be promoted. Finally, the scientific interaction between the two continents could be considerably improved by the exchange of graduate students and postdocs. This exchange will facilitate the flow of ideas and research opportunities, and a starting point of the much needed scientific symbiosis between SA and Europe.

Acknowledgments C.M. was funded by the Universidad de O'Higgins postdoctoral research fund and by the Fondecyt project No. 1190642 (Chilean Goverment). C.G.B. was funded by the

Estonian Research Council (IUT 20–28) and the European Regional Development Fund (center of excellence: EcolChange).

References

Amano T, Sutherland WJ (2013) Four barriers to the global understanding of biodiversity conservation: wealth, language, geographical location and security. Proc R Soc B 280(1756):20122649

Bardgett RD, van der Putten WH (2014) Belowground biodiversity and ecosystem functioning. Nature 515(7528):505–511

Berkeley MJ (1841) On an edible fungus from Tierra del Fuego, and an allied Chilean species. Trans Linn Soc Lond 19(1):37–44. https://doi.org/10.1111/j.1096-3642.1842.tb00073.x

Bianciotto V, Bandi C, Minerdi D, Sironi M, Tichy HV, Bonfante P (1996) An obligately endosymbiotic fungus itself harbors obligately intracellular bacteria. Appl Environ Microbiol 62(8):3005–3010

Bonfante P (2018) The future has roots in the past: the ideas and scientists that shaped mycorrhizal research. New Phytol 220(4):982–995

Brundrett MC, Tedersoo L (2018) Evolutionary history of mycorrhizal symbioses and global host plant diversity. New Phytol 220(4):1108–1115

Bueno CG, Marín C, Silva-Flores P, Aguilera P, Godoy R (2017a) Think globally, research locally: emerging opportunities for mycorrhizal research in South America. New Phytol 215(4):1306–1309

Bueno CG, Moora M, Gerz M, Davison J, Öpik M, Pärtel M, Helm A, Ronk A, Kühn I, Zobel M (2017b) Plant mycorrhizal status, but not type, shifts with latitude and elevation in Europe. Global Ecol Biogeogr 26(6):690–699

Bueno CG, Gerz M, Zobel M, Moora M (2018) Conceptual differences lead to divergent trait estimates in empirical and taxonomic approaches to plant mycorrhizal trait assignment. Mycorrhiza 29(1):1–11. https://doi.org/10.1007/s00572-018-0869-1

Cameron EK, Martins IS, Lavelle P, Mathieu J, Tedersoo L, Gottschall F, Guerra CA, Hines J, Patoine G, Siebert J, Winter M, Cesarz S, Delgado-Baquerizo M, Ferlian O, Fierer N, Kreft H, Lovejoy TE, Montanarella L, Orgiazzi A, Pereira HM, Phillips HRP, Settele J, Wall DH, Eisenhauer N (2018) Global gaps in soil biodiversity data. Nature Ecol Evol 2:1042–1043

Carey J (2016) News feature: Crucial role of belowground biodiversity. P Natl A Sci USA 113(28): 7682–7685

Castro SA, Camousseight A, Muñoz-Schick M, Jaksic FM (2006). Rodulfo Amando Philippi, el naturalista de mayor aporte al conocimiento taxonómico de la diversidad biológica de Chile. Rev Chil Hist Nat 79: 133–143. https://doi.org/10.4067/S0716078X2006000100011

Cornejo P, Meier S, Borie G, Rillig MC, Borie F (2008) Glomalin-related soil protein in a Mediterranean ecosystem affected by a copper smelter and its contribution to Cu and Zn sequestration. Sci Total Environ 406(1–2):154–160

Davison J, Moora M, Öpik M, Adholeya A, Ainsaar L, Bâ A, Burla S, Diedhiou AG, Hiiesalu I, Jairus T, Johnson NC, Kane A, Koorem K, Kochar M, Ndiaye C, Pärtel M, Reier Ü, Saks Ü, Singh R, Vasar M, Zobel M (2015) Global assessment of arbuscular mycorrhizal fungus diversity reveals very low endemism. Science 127(6251):970–973

Franco AA, de Faria SM (1997) The contribution of N2-fixing tree legumes to land reclamation and sustainability in the tropics. Soil Biol Biochem 29(5–6): 897–903

Galeano G, Suárez S, Balslev H (1998) Vascular plant species count in a wet forest in the Chocó area on the Pacific coast of Colombia. Biodivers Conserv 7(12):1563–1575

Gardes M, Bruns TD (1993) ITS primers with enhanced specificity for basidiomycetes – application to the identification of mycorrhizae and rusts. Mol Ecol 2:113–118

Godoy R, Silva-Flores P, Aguilera P, Marín C (2017) Microbial Interactions in the plant-soil continuum: Research results presented at the Workshop Mycorrhizal Symbiosis in the Southern Cone of South America. J Soil Sci Plant Nutr 17(4):1–3

Harrison MJ, van Buuren ML (1995) A phosphate transporter from the mycorrhizal fungus *Glomus versiforme*. Nature 378: 626–629

Kalwij JM, Robertson MP, Ronk A, Zobel M, Pärtel M (2014) Spatially-explicit estimation of geographical representation in large-scale species distribution datasets. PloS One 9(1): e85306

Liberati A, Altman DG, Tetzlaff J, Mulrow C, Gøtzsche PC, Ioannidis JP, Clarke M, Devereaux PJ, Kleijnen J, Moher D (2009) The PRISMA statement for reporting systematic reviews and meta-analyses of studies that evaluate health care interventions: explanation and elaboration. PLoS Med 6(7): e1000100

Meyer C, Kreft H, Guralnick R, Jetz W (2015) Global priorities for an effective information basis of biodiversity distributions. Nature Commun 6: 8221

Montagne C (1850) Plantas Celulares. In: Gay C (ed) Historia Física y Política de Chile – Flora Chilena. Museo de Historia Natural de Santiago, Santiago, v7, p 1–515

Moora M (2014) Mycorrhizal traits and plant communities: perspectives for integration. J Veg Sci 25(5):1126–1132

Nilsson RH, Larsson K-H, Taylor AFS, Bengtsson-Palme J, Jeppesen TS, Schigel D, Kennedy P, Picard K, Glöckner FO, Tedersoo L, Saar I, Kõljalg U, Abarenkov K (2018) The UNITE database for molecular identification of fungi: handling dark taxa and parallel taxonomic classifications. Nucleic Acids Res 47(D1):D259-D264. https://doi.org/10.1093/nar/gky1022

Öpik M, Vanatoa A, Vanatoa E, Moora M, Davison J, Kalwij JM, Reier U, Zobel M (2010) The online database MaarjAM reveals global and ecosystemic distribution patterns in arbuscular mycorrhizal fungi (Glomeromycota). New Phytol 188(1):223–241

Pereira HM, Belnap J, Brummitt N, Collen B, Ding H, Gonzalez-Espinosa M, Gregory RD, Honrado J, Jongman RHG, Julliard R, McRae L, Proença V, Rodrigues P, Opige M, Rodriguez JP, Schmeller DS, van Swaay C, Vieira C (2010) Global biodiversity monitoring. Front Ecol Environ 8(9): 459–460

Pereira HM, Ferrier S, Walters M, Geller GN, Jongman RHG, Scholes RJ, Bruford MW, Brummitt N, Butchart SHM, Cardoso AC, Coops NC, Dulloo E, Faith DP, Freyhof J, Gregory RD, Heip C, Höft R, Hurtt G, Jetz W, Karp D, McGeoch MA, Obura D, Onoda Y, Pettorelli N, Reyers B, Sayre R, Scharlemann JPW, Stuart SN, Turak E, Walpole M, Wegmann M (2013) Essential biodiversity variables. Science 339(6117):277–278

Philippi F (1893) Die Pilze Chiles. Hedwigia 32:115–118

Read DJ (1991) Mycorrhizas in ecosystems. Experientia 47(4):376–391

Richter DD, Billings SA, Groffman PM, Kelly EF, Lohse KA, McDowell WH, White TS, Anderson S, Baldocchi DD, Banwart S, Brantley S, Braun JJ, Brecheisen ZS, Cook CW, Hartnett HE, Hobbie SE, Gaillardet J, Jobbagy E, Jungkunst HF, Kazanski CE, Krishnaswamy J, Markewitz D, O'Neill K, Riebe CS, Schroeder P, Siebe C, Silver WL, Thompson A, Verhoef A, Zhang G (2018) Ideas and perspectives: Strengthening the biogeosciences in environmental research networks. Biogeosciences 15(15): 4815–4832

Rubiales D, Fernández Aparicio M, Wegmann K, Joel DM (2009) Revisiting strategies for reducing the seedbank of *Orobanche* and *Phelipanche* spp. Weed Res 49: 23–33

Ryan MG, Stape JL, Binkley D, Fonseca S, Loos RA, Takahashi EN, Silva CR, Silva SR, Hakamada RE, Ferreira JM, Lima AMN, Gava JL, Leite FP, Andrade HB, Alves JM, Silva GGC (2010) Factors controlling *Eucalyptus* productivity: how water availability and stand structure alter production and carbon allocation. Forest Ecol Manag 259(9):1695–1703

Scholes RJ, Mace GM, Turner W, Geller GN, Jürgens N, Larigauderie A, Muchoney D, Walther BA, Mooney HA (2008) Toward a global biodiversity observing system. Science 321(5892):1044–1045

Scholes RJ, Walters M, Turak E, Saarenmaa H, Heip CHR, Tuama ÉÓ, Faith DP, Mooney HA, Ferrier S, Jongman RHG, Harrison IJ, Yahara T, Pereira HM, Larigauderie A, Geller G (2012) Building a global observing system for biodiversity. Curr Opin Env Sust 4(1):139–146

Schüßler A, Schwarzott D, Walker C (2001) A new fungal phylum, the Glomeromycota: phylogeny and evolution. Mycol Res 105: 1413–1421

Singer R (1969) Mycoflora Australis. Beihefte Nova Hedwigia 29, Lehre: J. Cramer, Lehre

Singer R (1970) *Phaeocollybia* (Cortinariaceae, Basidiomycetes). Flora Neotropica 4: 1–13

Singer R, Morello JH (1960) Ectotrophic forest tree mycorrhizae and forest communities. Ecology 41(3):549–551

Singer R, Moser M, Gamundí I, Ellas R, Sarmiento G (1965) Forest mycology and forest communities in South America. Mycopath Mycol Appl 26(2–3):129–191

Soudzilovskaia NA, Douma JC, Akhmetzhanova AA, van Bodegom PM, Cornwell WK, Moens EJ, Treseder KK, Tibbett M, Wang Y-P, Cornelissen JHC (2015) Global patterns of plant root colonization intensity by mycorrhizal fungi explained by climate and soil chemistry. Global Ecol Biogeogr 24(3):371–382

Soudzilovskaia NA, Vaessen S, van't Zelfde M, Raes N (2017) Global patterns of mycorrhizal distribution and their environmental drivers. In: Tedersoo L (ed) Biogeography of mycorrhizal symbiosis. Springer, Cham, p 223–235

Spegazzini C (1921) Mycetes Chilenos. Bol Acad Nac Cienc Córdoba 25: 1–124

Sterkenburg E, Clemmensen KE, Ekblad A, Finlay RD, Lindahl BD (2018) Contrasting effects of ectomycorrhizal fungi on early and late stage decomposition in a boreal forest. ISME J 12(9):2187–2197

Stürmer SL, Siqueira JO (2011) Species richness and spore abundance of arbuscular mycorrhizal fungi across distinct land uses in Western Brazilian Amazon. Mycorrhiza 21(4):255–267

Sulzbacher MA, Grebenc T, Giachini AJ, Baseia IG, Nouhra ER (2017) Hypogeous sequestrate fungi in South America–how well do we know them?. Symbiosis 71(1):9–17

Tallis H, Mooney H, Andelman S, Balvanera P, Cramer W, Karp D, Balvanera P, Cramer W, Karp D, Polasky S, Reyers B, Ricketts T, Running S, Thonicke K, Tietjen B, Walz A (2012) A global system for monitoring ecosystem service change. BioScience Mag 63(11):977–986

Tedersoo L, Bahram M, Põlme S, Kõljalg U, Yorou NS, Wijesundera R, Ruiz L. V, Vasco-Palacios AM, Thu PQ, Suija A, Smith ME, Sharp C, Saluveer E, Saitta A, Rosas M, Riit T, Ratkowsky D, Pritsch K, Põldmaa K, Piepenbring M, Phosri C, Peterson M, Parts K, Pärtel K, Otsing E, Nouhra E, Njouonkou AL, Nilsson RH, Morgado LN, Mayor J, May TW, Majuakim L, Lodge DJ, Lee SS, Larsson K-H, Kohout P, Hosaka K, Hiiesalu I, Henkel TW, Harend H, Guo L, Greslebin A, Grelet G, Geml J, Gates G, Dunstan W, Dunk, Drenkhan R, Dearnaley J, De Kesel A, Dang T, Chen X, Buegger F, Brearley FQ, Bonito G, Anslan S, Abell S, Abarenkov K (2014) Global diversity and geography of soil fungi. Science 346(6213): 1256688–1–10

Tedersoo L, Sánchez-Ramírez S, Kõljalg U, Bahram M, Döring M, Schigel D, May T, Ryberg M, Abarenkov K (2018) High-level classification of the Fungi and a tool for evolutionary ecological analyses. Fungal Divers 90(1):135–159

Toju H, Guimarães PR, Olesen JM, Thompson JN (2014) Assembly of complex plant–fungus networks. Nature Commun 5: e5273

van der Heijden MG, Martin FM, Selosse M, Sanders IR (2015) Mycorrhizal ecology and evolution: the past, the present, and the future. New Phytol 205(4):1406–1423

van der Linde S, Suz LM, Orme CDL, Cox F, Andreae H, Asi E, Atkinson B, Benham S, Carroll C, Cools N, De Vos B, Dietrich H-P, Eichhorn J, Gehrmann J, Grebenc T, Gweon HS, Hansen K, Jacob F, Kristöfel F, Lech P, Manninger M, Martin J, Meesenburg H, Merilä P, Nicolas M, Pavlenda P, Rautio P, Schaub M, Schröck H-W, Seidling W, Šrámek V, Thimonier A, Thomsen IM, Titeux H, Vanguelova E, Verstraeten A, Vesterdal L, Waldner P, Wijk S, Zhang Y, Žlindra D, Bidartondo MI (2018) Environment and host as large-scale controls of ectomycorrhizal fungi. Nature 558(7709):243–248

Waller LP, Felten J, Hiiesalu I, Vogt-Schilb H (2018) Sharing resources for mutual benefit: crosstalk between disciplines deepens the understanding of mycorrhizal symbioses across scales. New Phytol 217(1):29–32

Wetzel FT, Saarenmaa H, Regan E, Martin CS, Mergen P, Smirnova L, Tuama ÉÓ, García Camacho FA, Hoffmann A, Katrin Vohland K, Häuser CL (2015) The roles and contributions of Biodiversity Observation Networks (BONs) in better tracking progress to 2020 biodiversity targets: a European case study. Biodiversity 16(2–3):137–149

Wetzel FT, Bingham HC, Groom Q, Haase P, Kõljalg U, Kuhlmann M, Martin CS, Penev L, Robertson T, Saarenmaa H, Schmeller DS, Stoll S, Tonkin JD, Häuser CL (2018) Unlocking biodiversity data: Prioritization and filling the gaps in biodiversity observation data in Europe. Biol Conserv 221: 78–85

White TJ, Bruns TD, Lee SB, Taylor JW (1990) Amplification and direct sequencing of fungal ribosomal RNA genes for phylogenetics. In: Innis M, Gelfand DH, Sninsky JL, White TJ (eds) PCR protocols: a guide to methods and applications. San Diego, Academic Press, p 315–322

Wiens JJ (2007) Species delimitation: new approaches for discovering diversity. Syst Biol 56(6): 875–878

Wijayawardene NN, Pawłowska J, Letcher PM, Kirk PM, Humber RA, Schüßler A, Wrzosek M, Muszewska A, Okrasińska A, Istel L, Gęsiorska A, Mungai P, Azeez Lateef A, C. Rajeshkumar KC, Singh RV, Radek R, Walther G, Wagner L, Walker C, Wijesundara DSA, Papizadeh M, Dolatabadi S, Shenoy BD, Tokarev YS, Lumyong S, Hyde KD (2018) Notes for genera: basal clades of Fungi (including Aphelidiomycota, Basidiobolomycota, Blastocladiomycota, Calcarisporiellomycota, Caulochytriomycota, Chytridiomycota, Entomophthoromycota, Glomeromycota, Kickxellomycota, Monoblepharomycota, Mortierellomycota, Mucoromycota, Neocallimastigomycota, Olpidiomycota, Rozellomycota and Zoopagomycota). Fungal Div 92(1): 43–129

Chapter 6
Endo- and Ectomycorrhizas in Tropical Ecosystems of Colombia

Clara P. Peña-Venegas and Aída M. Vasco-Palacios

6.1 Introduction

It is affirmed that fungi kingdom is more diverse in tropical ecosystems than in temperate ones. However, not all fungi in nature follow that pattern. A good example occurs with symbiotic fungi that form mycorrhiza. Arbuscular (endo-) mycorrhizal (AM) fungi are more diverse in tropical ecosystems but ectomycorrhizal (EM) fungi are more diverse in temperate and boreal ecosystems (Tedersoo et al. 2014). Most important, endo- and ecto-mycorrhizal fungi coexist, even in tropical ecosystems, if host plants with high affinity for mycorrhization are present (Neuenkamp et al. 2018).

Colombia is the fourth world's most biodiverse country, and the most megadiverse per square kilometer (IUCN 2009). So far, 35,000 flowering plant species have been listed for the country (Bernal et al. 2016). It is suggested that plant and mycorrhizal fungal communities are correlated and both determine important features on natural and anthropic ecosystems. Due to the close ecological relationship that exist between plants and soil fungi, it is expected that it harbors a high mycological diversity of fungi as well, with near of 100,000 species of fungi in general and 11,000 of macrofungi. Unfortunately, the fungal inventory of Colombia is still far from complete; and symbiotic fungi and the ecological roles of fungi in the ecosystem functioning is notably unknown. Mycorrhizal symposiois ocurr about 90% of vascular and non vascular plants (Smith and Read 2008; Van der Heijden et al.

C. P. Peña-Venegas (✉)
Instituto Amazónico de Investigaciomnes Científicas Sinchi, Leticia, Amazonas, Colombia
e-mail: cpena@sinchi.org.co

A. M. Vasco-Palacios
Biomicro - Grupo de Microbiología Ambiental, Escuela de Microbiología, Universidad de Antioquia, Medellín, Colombia

Fundación Biodiversa Colombia, Bogotá, Colombia

M. C. Pagano, M. A. Lugo (eds.), *Mycorrhizal Fungi in South America*, Fungal Biology, https://doi.org/10.1007/978-3-030-15228-4_6

2015). Mycorrhizae played a key role on soil carbon and nutrient cycling. In Colombia, studies on both endo and ectomycorrhizal fungi has been developed but that have been focused mainly on AM associated with crops and EM associated with *Quercus humboldtii*. In this review we compiled all the studies that have been developed in both AM and EM in Colombia, one of the pioneer countries in studying mycorrhizal associations in South America. We highlight the main topics studied, main results, applications, shortcomings and future challenges of this study area. Through a case study, we show how endo- and ecto-mycorrhizal fungi co-exist in the soil of a tropical rainforest. We pretent to stablish which issues are still unexplored and which aspects need to be addressed by new studies to advance in the understanding of these associations in Colombia and in neighboring South American countries.

6.2 The Study of Arbuscular Mycorrhizal Associations in Colombian Ecosystems

Arbuscular mycorrhization is one of the most studied plant-fungi associations. However, studies are unequally distributed around the world. Tropical areas are some of the less studied ones in when compared to temperate areas. One example of this inequality was presented by Alexander and Selosse (2009), who reviewed the number of publications on arbuscular mycorrhizal (AM) association in natural ecosystems. The authors estimated that between 2000 and 2009 approximately 5600 papers about AM association in forests were published. From them, only 170 papers were about AM association in tropical forests.

Colombia was one of the first countries in South America in studying AM fungi. An important number of documents related with this plant-fungi association had been produced. For this chapter, a total of 64 documents were reviewed: 25 of them were written between 1981 and 1995, and 39 of them were written after 1995 until 2017 (Fig. 6.1). Publications include works in 21 of the 23 states of the continental Colombia (Fig. 6.2).

However, few of those studies were published in English in indexed journals (30%) although in the last years the number of papers published in English had been increasing. Around 70% of the studies about AM association have not been published, or were published in Spanish in thesis, in non-indexed journals and in books, with limited access for academics (Fig. 6.3). The International Center for Tropical Agriculture – CIAT, placed in Palmira-Valle, Colombia, and the academy are the main contributors on AM fungal bibliography.

During 70's and 80's Colombia experienced a boom in the study of AM association and AM fungal diversity with the arrival of American and European researchers to the CIAT, who taught Colombian researchers about this association. From that time, 12 new species of AM fungi from Colombia were described: *Acaulospora appendiculata*, *A. longula*, *A. mellea*, *A. morrowae*, *Glomus manihotis*, and

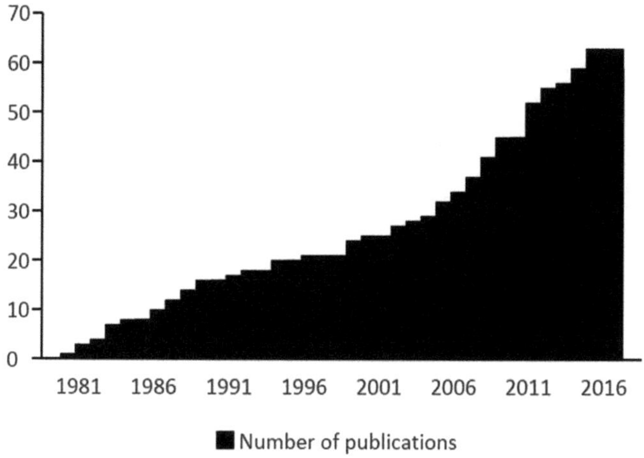

Fig. 6.1 Number of publications of arbuscular mycorrhizal association from Colombia in a range of 36 years (from 1981 until 2017)

Entrophospora colombiana (Schenck et al. 1984), *Acaulospora myrocarpa* (Schenck et al. 1986), *Entrophospora schenckii* (Sieverding and Toro 1987), *Glomus glomerulatum* (Sieverding 1987), *Acaulospora denticulata* and *A. rehmii* (Sieverding and Toro 1987), and *Scutellospora biornata* (Spain et al. 1989). With these species, the CIAT started one of the most important arbuscular mycorrhizal fungal collection of the world which conserve 44 species, and where 86% of the registers included in its database are from Colombia (CIAT 2000). Mostly of the samples were collected by Dr. Ewald Sieverding who is considered the father of the AM studies in Colombia (Sieverding 1984, 1989a, 1989b; Sieverding and Howeler 1985).

The taxonomy of AM fungi changed in 2001 when molecular approaches were used to place AM fungi in a new monophyletic phylum (Glomeromycota). New molecular taxonomy of AM fungi suggested that the traditional taxonomy of AM fungi based on morphological data should be re-evaluated according to their natural phylogenetic organization (Schüßler et al. 2001). Since that time, different authors made efforts to harmonize molecular and morphological information to produce a new taxonomy and suggest taxonomic keys to classify correctly spore AM morphotypes (Oehl et al. 2011a; Oehl et al. 2011b; Sieverding et al. 2014; Oehl et al. 2008; Walker et al. 2007; Sieverding and Oehl 2006; Spain et al. 2006). It looks that since 2011, all AM fungal experts agree in a new taxonomic classification of AM fungi, even when it is known that many AM fungal will be described only by molecular approaches, without knowing a corresponding spore morphotype. For the development of this new AM fungal taxonomic classification, the AM strains deposited at the CIAT had been used as reference cultures.

Although the report of new species of AM fungi from Colombia and the creation of a collection were important increasing the inventories of these organisms, it has not been the focus of most studies regarding the study of the AM association in

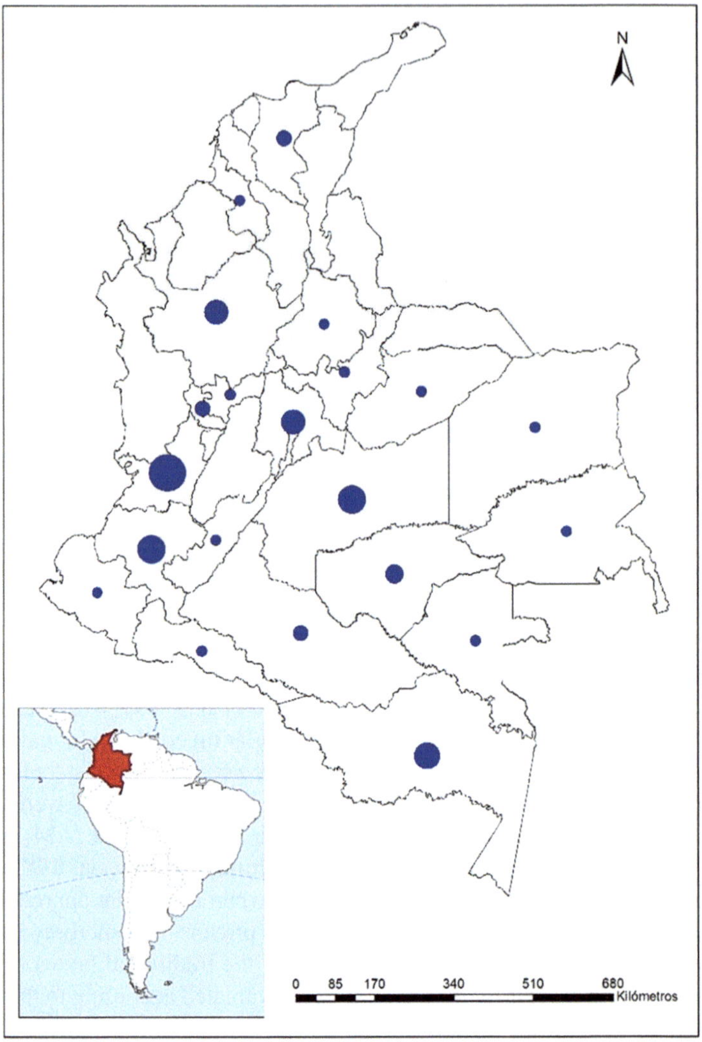

Fig. 6.2 Map with the Colombian states where works on arbuscular mycorrhizal associations had been done. The circle size represents the number of publications reported in each state (Map author: Delio Mendoza – Sinchi Institute)

Colombia. Arbuscular mycorrhization is recognized as an important association related with plant nutrition (Sánchez de Prager 2004). As in Colombia, only 2.6% of soils are suitable for agriculture, most of the papers studied AM association in relation with crops or commercial plants. AM association has been described in no less than 29 commercial species including: algarroba (*Prosopis juliflora*), avocado (*Persea Americana*) (Prada 2009), azalea flower (*Rhododendron* sp.), beans (*Phaseolus vulgaris*) (Barrios et al. 2006), blueberry (*Vaccinium meridionale*) (Ávila Díaz-Granados et al. 2009), borojo (*Borojoa patinoi*) (Possú et al. 2004),

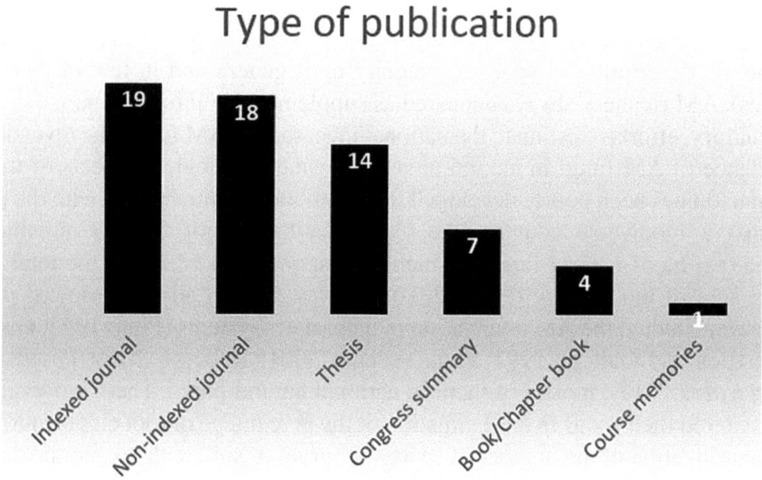

Fig. 6.3 Type of publications about arbuscular mycorrhizal association produced from Colombia

cassava (*Manihot esculenta*) (Howeler and Cadavid 1990; Howeler et al. 1982; Howeler and Sieverding 1982; Howeler and Sieverding 1983; Howeler et al. 1987; Peña-Venegas 2015; Rodríguez and Sanders 2015; Sieverding and Leihner 1984a; 1984b), cocoa (*Theobroma cacao*) (Possú et al. 2004), coffee (*Coffea Arabica*) (Bolaños and Rivillas-Osorio 2000; Estrada and Sánchez de Prager 1995; Rivillas 1995), gooseberry (*Physalis peruviana*) (Ramírez et al. 2000), guadua bamboo (*Guadua angustiflia*), green pea (*Pisum sativum*) (Triana 2015), inchi (*Caryodendron orinocence*) (Pinto 1988; Pinto and Pedraza 1988), indian couch grass (*Bothriochloa pertusa*) (Pérez, Botero and Usma 2012), leucaena (*Leucaena leucocephala*) (Osorio and Habte 2013), Mexican sunflower (*Tithonia diversiflora*) (Phiri et al. 2003), oil palm (*Elaeis guineensis*) (Phosri et al. 2010), onion (*Allium fistulosum*) (Montenegro-Gómez et al. 2017), maize (*Zea mays*) (Gómez and Sánchez de Prager 2012; Zabala 2012; Zabala and Sánchez de Prager 2014), molasses grass (*Melinis minutiflora*) (Quiroga et al. 2009), passion fruit (*Paciflora edulis*) (Sánchez de Prager 2003), peach palm (*Bactris gasipaes*) (Possú et al. 2004), pitch pine (*Pinus caribea*) (Pedraza 1981), plantain (*Musa paradisiaca*) (Osorio et al. 2008a; 2008b), signalgrass (*Brachiaria decumbens*) (Posada-Almanza et al. 2006), sorghum (*Sorghum* sp.), sugarcane (*Saccharum officinarum*), sweet pepper (*Capsicum* sp.) (Peña-Venegas 2010; Sánchez de Prager et al. 2010) and tomato (*Solanum lycopersicum*) (Guzmán et al. 2013). Studies about the AM association in these species focused mainly in the estimation of the percentage of root colonization, the quantification of the number of spores per gram of soil, and the relation of root colonization and spore production with edaphic conditions, chemical or organic fertilization, and herbicide application. Most of these studies emphasised the role of AM fungi as promotor of plant growth and yield. However, some papers also include other functions of AM as its role in controlling nematodes and plant pathogens, AM fungal synergims with other beneficial microorganisms, and the effect of AM fungal external mycelia on soil aggregation. Although some of these studies include the taxonomic description of spores (generally until genera and in few of them until species), AM richness always appeared as supplementary information.

Contrary, efforts to estimate the national inventory of AM fungi, the diversity and abundance of AM fungi in natural areas or to study ecological aspects of the AM associacion has been poorly developed in Colombia in contradiction with the potential that a biodiverse country like Colombia has. Until 2015, Colombia had 59,558,000 ha of natural forests, which corresponded to 52.2% of the total continental area of the country (SIAC 2016). However, from all the reviewed papers, only seven studied the AM association in natural ecosystems (Pinto 1992; Guerrero 1993; Ríos and Gallego 1997; Peña-Venegas 2001; Restrepo 2006; Posada et al. 2012; Ardila 2017), mostly of them in national natural parks. These works in general centered their focus in the estimation of the percentage of root colonization, and the quantification of the number of spores per gram of soil, with taxonomic description of spores in some cases.

From the reviewed documents it is possible to conclude that AM association continue being a relevant topic in Colombia, as there is an important number of papers recently published. All domesticated plants studied until now form AM associations, independently the geographic location or environmental conditions in which plants were cultivated. Although soil AM fungal composition are commonly studied (mainly from AM fungal spores), soil AM fungal communities differ from those colonizing plant roots (Saks et al. 2014). More emphasis on root mycorrhization is needed in order to advance in the selection and evaluation of AM fungal strains as possible inoculants for specific plant cultures. The use of molecular tools in this case is relevant. However, from all documents reviewed, only two thesis included taxonomic determination of AM fungi from molecular data (Peña-Venegas 2015; León 2015), being a novel tool poorly explored in Colombia to study AM fungal communities. Colombia is a diverse country with different ecosystems, climatic conditions and soils, but few works on AM fungi had been done in natural ecosystems. We need to encourage students and researchers to study more these organisms in natural ecosystems to understand more AM fungal communities, their ecology and diversity.

6.3 The Study of Ectomycorrhizal Associations in Colombian Ecosystems

Tropical forests were supposed to be dominated by AM fungi while EM fungi were restricted to temperate regions, except on tropical montane forests where Holarctic element such *Juglandaceae, Betulaceae* and *Fagaceae* occur (Singer and Morello 1960; Halling 1996; Mueller 1996; González et al. 2006; Corrales et al. 2018). A recent interest in the tropical EM association revealed that the number of plant families associated with EM fungi is higher than expected (Tedersoo and Brundrett 2017, Corrales et al. 2018). Families of tropical plants such as Dipterocarpaceae or the subfamilies Caesalpinioideae and Papilionoideae of the family Fabaceae are specifically EM in lownland forests (Moyersoen 2012; Vasco-Palacios et al. 2014; Corrales et al. 2018). Other families of plants that present this association but a different ecological pattern are Gnetaceae, Nyctaginaceae, and Polygonaceae (Singer et al. 1983; Tedersoo et al. 2010; Moyersoen 2012; Moyersoen and Weiss 2014; Corrales et al. 2018; Vasco-Palacios et al. 2018). In tropical ecosystems, the functional role of EM fungi is poorly understood. Fungal communities drive tree population dynamics and can alter decomposition rates. EM association may help the hosts plant to establish in poor soils. Ectomycorrhizal-forests seems to storage higher levels of soil carbon than AM-forests (Averill et al. 2014), but this important feature has not been studied yet in tropical ecosystems.

An exhaustive review of literature yielded a total of 97 publications related to EM fungi in Colombia, which include 27 theses, 18 books or book chapters, 48 papers from indexed journals and four from non-indexed journals. Reviewed literature showed that a total of 172 species of EM fungi have been reported for this country: 116 species from *Quercus*-forests *Pinus*-forests, two species from *Colombobalanus*-forests, 27 species from white sand forests with *Dicymbe* and *Aldina*, and five species from terra-firme forests with *Pseudomonotes tropenbosii* (Table 6.1). *Amanita crebresulcata* was reported from dry-forests in the state of Cesar, but not data about the putative host was included. EM species reported from Colombia belonged to typical lineages of Basidiomycota such as Amanitaceae, Cantharellaceae, and genera of the families Russulaceae and Boletaceae (Table 6.1). As singular EM taxa we can mention *Polyporoletus sublividus* (Albatrellaceae), *Tremellogaster surinamensis* (Diplocystidiaceae) and four endemic species of *Sarcodon* (Bankeraceae). The genus *Sarcodon* was considered to be distributed in the Northern Hemisphere, but now 10 species are known from lowland areas in tropics, expanding the range of distribution and the knowledge about plants host-associated to this genera. The EM lineages *Tomentella* and *Sebacina* are commonly detected on root analysis (Vasco-Palacios 2016). However, those have not been reported in Colombia yet, probably because the basidiomata of these genera can be easily overlooked as they occur erratic and are resupinate and/or cryptic (Moyersoen and Weiss 2014).

In Colombia, studies about EM association have been carried-out mainly in *Quercus*-dominated montane forests. The first mention on EM fungi in Colombia is a list of EM fungi from this oak forests was published by Singer (1963) which included descriptions of new species. The number of publications increased considerably between 80's and 90's, time in which several specialists such as Dr. Dumont, Dr. G. Guzman, Dr. R. Halling and Dr. G. Mueller visited the country (Fig. 6.3a). From all documents reviewed, 58% studied the EM association with *Q. humboldtii*, mostly from Antioquia and Boyacá states (Fig. 6.4). The states with most records are Antioquia, Boyacá and Tolima, manly with records from *Quercus*-forests, and Amazonas from forests with *Dicymbe, Aldina* and *Pseudomonotes tropenbosii* host trees (Table 6.1).

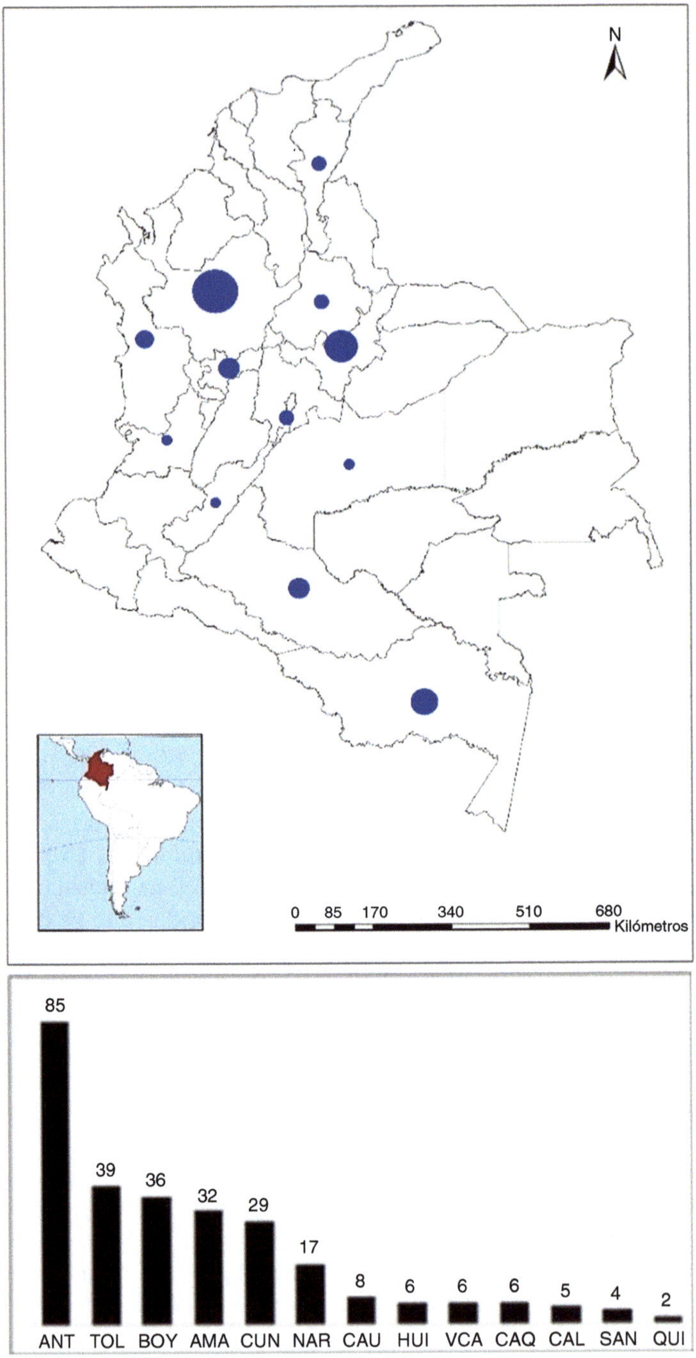

Fig. 6.4 Records of EM fungal species in Colombian states (states with only one record were not included). Amazonas (AMA), Antioquia (ANT), Boyacá (BOY), Caldas (CAL), Cauca (CAU), Caquetá (CAQ), Cundinamarca (CUN), Huila (HUI), Nariño (NAR), Quindío (QUIN), Santander (SAN), Tolima (TOL), Valle del Cauca (VCA). The circle size represents the number of records reported in each state (Map author: Delio Mendoza – Sinchi Institute)

Table 6.1 List of reported ectomycorrhizal (EM) species between 1963 and 2018, based on literature review and specimens from the Herbarium of the Universidad de Antioquia (HUA)

TAXA	Distribution	Hosts	Type	References
Ascomycota				
Pezizales				
Helvellaceae				
Helvella lacunosa	TOL	*Q. humboldtii*		Tobón (1991), HUA (Gómez 7)
Helvella macropus	ANT, TOL	*Q. humboldtii*		Tobón (1991), HUA (Vasco-Palacios 1070)
Helvella sulcata	ANT	*Q. humboldtii*		HUA (Halling 5065)
Leotiales				
Leotiaceae				
Leotia lubrica	ANT, BOY	*Quercus humboldtii*		López-Q et al. (2007), Mecanismo de Facilitación (2001)
Leotia viscosa	ANT	*Quercus humboldtii*		Tobón (1991)
Basidiomycota				
Agaricales				
Amanitaceae				
Amanita advena	ANT	*Q. humboldtii*	Type	Tulloss et al. (1992)
Amanita arocheae	ANT	*Q. humboldtii, C. excelsa*	Type	Halling and Mueller (2005), Parra-Aldana et al. (2011), Tulloss et al. (1992), Tulloss (2005)
Amanita aureomonile	VCA	*Q. humboldtii, C. excelsa*	Type	Parra-Aldana et al. (2011), Tulloss et al. (1992)
Amanita brunneolocularis	ANT, BOY, SAN, TOL	*Q. humboldtii*	Type	AMVA-Área Metropolitana del Valle de Aburrá (2000), Franco-Molano and Uribe-Calle (2000), Halling and Mueller (2005), Saldarriaga et al. (1988a), Tulloss et al. (1992)
Amanita brunneolocularis var. *pallida*	BOY	*Q. humboldtii*	Type	Vargas-Estupiñán et al. (2017)
Amanita campinarae	AMA	*Dicymbe sp.*		Vasco-Palacios et al. (2018)
Amanita ceciliae	BOY	*Q. humboldtii*		Singer (1963), Tulloss et al. (1992)
Amanita citrina	BOY, TOL	*Q. humboldtii*		Vargas-Estupiñán et al. (2017)
Amanita colombiana	ANT, BOY, TOL	*Q. humboldtii*	Type	Franco-Molano and Uribe-Calle (2000), Tulloss et al. (1992), Vargas-Estupiñán et al. (2017)

(continued)

Table 6.1 (continued)

TAXA	Distribution	Hosts	Type	References
Amanita crebresulcata	AMA, CES	*Dicymbe sp.*		Palacio M. et al. (2015)
Amanita flavoconia	ANT, BOY, CUN, NAR, TOL	*Q. humboldtii, Pinus sp.*		AMVA-Área Metropolitana del Valle de Aburrá (2000), Halling and Mueller (2005), Palacio et al. (2015), Saldarriaga et al. (1988a), Tulloss et al. (1992), Vargas-Estupiñán et al. (2017)
Amanita fuligineodisca	ANT, BOY, CUN, NAR, SAN, TOL	*Q. humboldtii*	Type	Halling and Mueller (2005), Mecanismo de Facilitación (2001), Pulido (1983), Saldarriaga et al. (1988a), Tulloss et al. (1992), Vargas-Estupiñán et al. (2017)
Amanita gemmata	CUN	*Q. humboldtii*		Guzmán and Varela (1978), Pulido (1983), Tulloss et al. (1992)
Amanita humboldtii	CUN, NAR, TOL	*Q. humboldtii*	Type	Dennis (1970), Guzmán and Varela (1978), Mueller and Wu (1997), Pulido (1983), Singer (1963), Tulloss et al. (1992)
Amanita lanivolva	AMA	*Dicymbe sp.*		Vasco-Palacios. et al. (2018)
Amanita muscaria	ANT, BOY, CAL, CUN, SAN	*Pinus patula*		AMVA-Área Metropolitana del Valle de Aburrá (2000), Henao-M and Ruiz (2006), Mecanismo de Facilitación (2001), Montoya et al. (2005), Tulloss et al. (1992)
Amanita picea	BOY	*Q. humboldtii*	Type	Tulloss et al. (1992)
Amanita rubescens	ANT, CUN, BOY, TOL	*Q. humboldtii, Pinus sp.*		AMVA-Área Metropolitana del Valle de Aburrá (2000), Pulido (1983), Saldarriaga et al. (1988a), Vargas-Estupiñán et al. (2017)
Amanita savannae	MET	Non data	Type	Tulloss and Franco-Molano (2008)
Amanita sororcula	ANT, BOY	*Q. humboldtii*	Type	Franco-Molano and Uribe-Calle (2000), Tulloss et al. (1992), Vargas-Estupiñán et al. (2017)
Amanita virosa	ANT, BOY, VCA	*Q. humboldtii*		Vargas-Estupiñán et al. (2017)
Amanita xerocybe	AMA	*Dicymbe sp.*		Vasco-Palacios et al. (2018)
Amanita xylinivolva	ANT, BOY, CAU, CUN, NAR, SAN, TOL	*Q. humboldtii*	Type	Franco-Molano and Uribe-Calle (2000), Saldarriaga et al. (1988a), Tulloss et al. (1992)
Cortinariaceae				
Cortinarius boyacensis	ANT, BOY, TOL	*Q. humboldtii*	Type	Dennis (1970), Franco-Molano and Uribe-Calle (2000), Mueller and Wu (1997), Singer (1963)

(continued)

Table 6.1 (continued)

TAXA	Distribution	Hosts	Type	References
Cortinarius iodes	ANT, BOY, NAR, TOL	*Q. humboldtii*		Franco-Molano et al. (2000, 2010), López-Q et al. (2007),
Cortinarius violaceus	ANT, TOL	*Q. humboldtii*		Franco-Molano et al. (2000, 2010)
Rozites colombiana	ANT	*Q. humboldtii*		Halling and Mueller (2005), Halling and Obrevo (1987), López-Q et al. (2007), Saldarriaga et al. (1988a)
Hydnangiaceae				
Laccaria amethystina	ANT, CUN TOL	*Q. humboldtii*		Franco-Molano and Uribe-Calle (2000), Guzmán and Varela (1978), Halling and Mueller (2005), Mueller (1996)
Laccaria gomezii	ANT, HUI	*Q. humboldtii*	Type	Franco-Molano and Uribe-Calle (2000), Halling and Mueller (2005), Mueller (1996), Mueller and Singer (1988)
Laccaria laccata	ANT, BOY, CAL, CUN, MAG, QUI, TOL	*Pinus sp.*		Betancur et al. (2007), Franco-Molano (2002), Guzmán and Varela (1978), Halling and Mueller (2005), López-Q et al. (2007), Mecanismo de Facilitación (2001), Montoya et al. (2005), Mueller (1996), Nieves-Rivera et al. (1997), Pulido (1983), Saldarriaga et al. (1988b)
Laccaria ohiensis		*Non data*		Mueller (1996)
Laccaria proxima		*Non data*		Mueller (1996)
Hygrophoraceae				
Hygrophorus cossus	BOY	*Q. humboldtii*		Franco-Molano and Uribe-Calle (2000)
Hygrophorus obconicus	PNN	*Non data*		Boekhout and Pulido (1989), Pulido and Boekhout (1989)
Hygrophorus quercuum	BOY	*Q. humboldtii*	Type	Mueller and Wu (1997)
Hymenogastraceae				
Phaeocollybia ambigua	ANT, NAR	*Quercus humboldtii*	Type	Franco-Molano and Uribe-Calle (2000), Halling and Mueller (2005), Horak and Halling (2018)
Phaeocollybia caudata	ANT	*Quercus humboldtii*	Type	Franco-Molano and Uribe-Calle (2000), Halling and Mueller (2005), Horak and Halling (2018)
Phaeocollybia Columbiana	VCA	*Quercus humboldtii*	Type	Franco-Molano and Uribe-Calle (2000), Mueller and Wu (1997), Singer (1970)

(continued)

Table 6.1 (continued)

TAXA	Distribution	Hosts	Type	References
Phaeocollybia oligoporpa	ANT, NAR	*Quercus humboldtii*		Franco-Molano and Uribe-Calle (2000), Halling and Mueller (2005), Horak and Halling (2018)
Phaeocollybia quercetorum	ANT, CAL	*Quercus humboldtii*		Betancur et al. (2007), Franco-Molano and Uribe-Calle (2000), Halling, Mueller (2005), Horak and Halling (2018)
Phaeocollybia singularis	NAR	*Quercus humboldtii*	Type	Halling and Mueller (2005), Horak and Halling (2018)
Inocybaceae				
Inocybe calamistrata	ANT	*Q. humboldtii*		Franco-Molano et al. (2010)
Inocybe hystrix	ANT	*Q. humboldtii*		López-Q et al. (2007)
Inocybe jalopensis	CUN	*Q. humboldtii*		Dennis (1970), Guzmán and Varela (1978), Pulido (1983), Singer (1963)
Inocybe rimosa	CUN	*Non data*		Guzmán and Varela (1978), Pulido (1983)
Inocybe tequendamae	CUN	*Non data*	Type	Dennis (1970), Guzmán and Varela (1978), Mueller and Wu (1997), Singer (1963)
Tricholomataceae				
Tricholoma caligatum	BOY	*Q. humboldtii*		HUA (Echeverry 36)
Boletales				
Boletaceae				
Aureoboletus auriporus	ANT	*Q. humboldtii*		Franco-Molano and Uribe-Calle (2000)
Aureoboletus russellii	CAU	*Q. humboldtii*		Halling (1989)
Austroboletus amazonicus	AMA	*P. tropenbosii*	Type	Vasco-Palacios et al. (2014)
Austroboletus subflavidus	NAR	*Q. humboldtii*		Vasco-Palacios et al. (2014)
Austroboletus subvirens	ANT, HUI	*Q. humboldtii*		Franco-Molano and Uribe-Calle (2000), Halling (1989)
Boletellus ananas	ANT, VCA	*Q. humboldtii*		Franco-Molano and Uribe-Calle (2000), Halling (1989, 1996), Singer (1970)
Boletus fuligineotomentosus	VCA	*Q. humboldtii*	Type	Franco-Molano and Uribe-Calle (2000), Halling (1989), Mueller and Wu (1997)
Boletus neoregius	ANT, CUN	*Q. humboldtii*		Franco-Molano et al. (2000, 2010), Halling and Mueller (2005)

(continued)

Table 6.1 (continued)

TAXA	Distribution	Hosts	Type	References
Boletus pavonius	SAN	*Q. humboldtii*		
Boletus pyrrhosceles	ANT	*Q. humboldtii*		Franco-Molano and Uribe-Calle (2000), Halling (1992)
Boletus subtomentosus	ANT, BOY	*Q. humboldtii*		Franco-Molano and Uribe-Calle (2000), Halling (1989), Mecanismo de Facilitación (2001)
Chalciporus caribaeus	ANT	*Q. humboldtii*		López-Q et al. (2007)
Chalciporus piperatus	ANT, TOL	*Q. humboldtii, Pinus sp.*		Franco-Molano et al. (2010)
Chalciporus pseudorubinellus	ANT, CAU, TOL	*Q. humboldtii*		Franco-Molano and Uribe-Calle (2000), Halling (1989)
Cyanoboletus pulverulentus	CUN	*Q. humboldtii*		Franco-Molano and Uribe-Calle (2000), Halling (1989)
Fistulinella campinaranae var. *scrobiculata*	AMA	*P. tropenbosii*		Vasco-Palacios et al. (2014)
Leccinum andinum	ANT, TOL	*Q. humboldtii*	Type	Franco-Molano and Uribe-Calle (2000), Halling (1989, 1996), Halling and Mueller (2005)
Leccinum rugosiceps	ANT, CAU, TOL	*Q. humboldtii*		Franco-Molano et al. (2010), Halling (1996), Halling and Mueller (2005)
Leccinum talamancae	ANT	*Q. humboldtii*		Halling and Mueller (2005), López-Q et al. (2007)
Phylloporus centroamericanus	ANT	*Q. humboldtii*		Halling and Mueller (2005), Franco-Molano et al. (2010)
Phylloporus fibulatus	ANT, NAR, TOL	*Q. humboldtii*	Type	Franco-Molano and Uribe-Calle (2000), Halling et al. (1999), Mueller and Wu (1997), Singer et al. (1990)
Phylloporus phaeoxanthus	ANT, BOY, TOL	*Q. humboldtii*		Halling and Mueller (2005), Franco-Molano et al. (2010)
Phylloporus purpurellus	CAU	*Q. humboldtii*	Type	Franco-Molano and Uribe-Calle (2000), Halling et al. (1999), Mueller and Wu (1997)
Porphyrellus indecisus	BOY	*Q. humboldtii*		
Pulveroboletus atkinsonianus	ANT	*Q. humboldtii*		Halling (1989, 1996)
Pulveroboletus ravenelii	ANT	*Q. humboldtii*		Boekhout and Pulido (1989), Franco-Molano et al. (2010)
Singerocomus inundabilis	AMA	*Dicymbe sp.*		Vasco-Palacios et al. (2018)

(continued)

Table 6.1 (continued)

TAXA	Distribution	Hosts	Type	References
Strobilomyces confusus	ANT, HUI	*Q. humboldtii*		Franco-Molano and Uribe-Calle (2000), Halling (1989), Halling and Mueller (2005), HUA (Vasco-P. 2332)
Suillus luteus	ANT, CAL, CUN, TOL	*Pinus sp.*,		AMVA-Área Metropolitana del Valle de Aburrá (2000), Dennis (1970), Franco-Molano and Uribe-Calle (2000), Guzmán and Varela (1978), Montoya et al. (2005), Saldarriaga et al. (1988a)
Tylopilus obscurus	ANT, TOL	*Q. humboldtii*	Type	Franco-Molano et al. (2000), Halling (1989), Halling and Mueller (2005)
Tylopilus umbrosus	NAR	*Q. humboldtii*		Franco-Molano et al. (2010)
Xanthoconium separans	ANT, NAR	*Q. humboldtii*		Franco-Molano et al. (2000, 2010)
Xerocomellus chrysenteron	ANT, NAR, TOL	*Q. humboldtii*		Franco-Molano et al. (2010)
Xerocomellus truncatus	ANT	*Q. humboldtii*		Franco-Molano and Uribe-Calle (2000), Halling (1989)
Xerocomus orquidianus	ANT	*Q. humboldtii*	Type	Franco-Molano and Uribe-Calle (2000), Halling (1989)
Xerocomus tenax	ANT	*Q. humboldtii*		HUA (Halling s.n.)
Calostomataceae				
Calostoma cinnabarinum	ANT, CAL, HUI, TOL	*Q. humboldtii, Pinus sp.*		AMVA-Área Metropolitana del Valle de Aburrá (2000), Betancur et al. (2007), Dumont and Umaña (1978), López-Q et al. (2007), Saldarriaga et al. (1988a)
Diplocystidiaceae				
Tremellogaster surinamensis	AMA, CAQ	*P. tropenbosii*		Vasco-Palacios et al. (2005)
Gyroporaceae				
Gyroporus castaneus	CAU	*Q. humboldtii*		Franco-Molano et al. (2010)
Sclerodermataceae				
Scleroderma albidum	CUN	*Non data*		Guzmán and Varela (1978)
Scleroderma areolatum	ANT, BOY	*Q. humboldtii*		AMVA-Área Metropolitana del Valle de Aburrá (2000), Mecanismo de Facilitación (2001)
Scleroderma citrinum	QUI	*Q. humboldtii*		Franco-Molano (2002)
Cantharellales				
Cantharellaceae				

(continued)

Table 6.1 (continued)

TAXA	Distribution	Hosts	Type	References
Cantharellus cibarius	ANT, CAQ, CUN	*Q. humboldtii*		Franco-Molano and Uribe-Calle (2000), Guzmán and Varela (1978), López-Q et al. (2007). Vasco-P. et al. (2005)
Cantharellus cinereus	CUN	*Q. humboldtii*		Guzmán and Varela (1978)
Cantharellus cinnabarinus	CUN	*Q. humboldtii*		Franco-Molano and Uribe-Calle (2000)
Cantharellus guyanensis	AMA, CAQ	*Dicymbe sp.*		Franco-Molano et al. (2005, 2010)
Cantharellus lateritius var. *colombianus*	NAR, TOL	*Q. humboldtii*		Franco-Molano and Uribe-Calle (2000), Petersen and Mueller (1992)
Cantharellus rhodophyllus	CHO	*Non data*		Guzmán et al. (2004)
Craterellus atratoides	AMA	*Dicymbe sp.*		Vasco-Palacios et al. (2018)
Craterellus atratus	AMA	*Dicymbe sp.*		Vasco-Palacios et al. (2018)
Craterellus boyacensis	ANT, BOY, HUI, TOL	*Q. humboldtii*	Type	Dennis (1970), Franco-Molano and Uribe-Calle (2000), Halling and Mueller (2005), Mueller and Wu (1997), Singer (1963), Wu and Mueller (1995)
Craterellus cinereofimbriatus	AMA	*Dicymbe sp.*		Vasco-Palacios et al. (2018)
Craterellus fallax	CUN, HUI	*Q. humboldtii*		Franco-Molano and Uribe-Calle (2000), Wu and Mueller (1995)
Craterellus strigosus	AMA	*Dicymbe sp.*		Vasco-Palacios et al. (2018)
Pseudocraterellus undulatus	ANT	*Q. humboldtii*		Franco-Molano and Uribe-Calle (2000), Wu and Mueller (1995)
Hydnaceae				
Hydnum albidum	ANT	*Pinus sp.*		Herbarium specimens
Hydnum repandum	ANT, BOY, TOL	*Q. humboldtii, Pinus sp.*		Herbarium specimens
Clavulinaceae				
Clavulinaceae				
Clavulina craterelloides	AMA	*P. tropenbosii*		Franco-Molano et al. (2005), Vasco-P. et al. (2005)
Clavulina amazonensis	AMA	*Dicymbe sp.*		Vasco-Palacios et al. (2018)
Clavulina connata	AMA	*Dicymbe sp.*		Vasco-Palacios et al. (2018)
Clavulina effusa	AMA	*Dicymbe sp.*		Vasco-Palacios et al. (2018)
Clavulina kunmudlutsa	AMA	*Dicymbe sp.*		Vasco-Palacios et al. (2018)
Clavulina sprucei	AMA	*Dicymbe sp.*		Vasco-Palacios et al. (2018)

(continued)

Table 6.1 (continued)

TAXA	Distribution	Hosts	Type	References
Gomphales				
Gomphaceae				
Gloeocantharellus uitotanus	AMA	*Non data*	Type	Vasco-Palacios and Franco-Molano. 2005
Ramaria botrytis	BOY	*Q. humboldtii*		
Ramaria chocoënsis	CHO	*Non data*	Type	Hahn and Christan (2002)
Ramaria cyaneigranosa	BOY	*Q. humboldtii*		
Ramaria flava	BOY, NAR	*Q. humboldtii*		
Ramaria formosa	CAL	*Q. humboldtii*		Betancur et al. (2007)
Ramaria stricta	CAL	*Q. humboldtii*		Betancur et al. (2007)
Hymenochaetales				
Hymenochaetaceae				
Coltricia cinnamomea	ANT, AMA, TOL	*Q. humboldtii, Dicymbe sp.*		Vasco-Palacios et al. (2018)
Coltricia focicola	CUN	*Non data*		Guzmán and Varela (1978)
Coltricia hamata	AMA	*Dicymbe sp.*		Vasco-Palacios et al. (2018)
Coltricia perennis	ANT	*Pinus sp.*		AMVA-Área Metropolitana del Valle de Aburrá (2000))
Coltricia verrucata	AMA	*Dicymbe sp.*		Vasco-Palacios et al. (2018)
Coltriciella dependens	AMA	*Dicymbe sp.*		Vasco-Palacios et al. (2018)
Russulales				
Albatrellaceae				
Polyporoletus sublividus	CAQ	*Non data*		Vasco-Palacios et al. (2005)
Russulaceae				
Lactarius atroviridis	ANT, TOL	*Q. humboldtii*		Franco-Molano et al. (2000, 2010), Halling and Mueller (2005)
Lactarius brasiliensis	AMA	*Dicymbe sp.*		Vasco-Palacios et al. (2018)
Lactarius caucae	CAU	*Q. humboldtii*	Type	Franco-Molano and Uribe-Calle (2000), Mueller and Wu (1997)
Lactarius chrysorrheus	ANT, CUN, TOL	*Q. humboldtii*		Franco-Molano and Uribe-Calle (2000), Guzmán and Varela (1978)
Lactarius costaricensis	NAR	*Q. humboldtii*		Franco-Molano et al. 2010, Halling and Mueller (2005)
Lactarius deceptivus	ANT, BOY, TOL	*Q. humboldtii*		Franco-Molano et al. (2000, 2010), Halling and Mueller (2005), Mecanismo de Facilitación (2001)

(continued)

Table 6.1 (continued)

TAXA	Distribution	Hosts	Type	References
Lactarius fragilis	ANT, TOL	*Q. humboldtii*		Franco-Molano et al. (2000, 2010)
Lactarius gerardii	ANT, BOY	*Q. humboldtii*		Franco-Molano et al. (2000), HUA (Vasco-P. 1071)
Lactarius indigo	ANT, BOY, CUN, NAR, TOL	*Q. humboldtii*		AMVA-Área Metropolitana del Valle de Aburrá (2000), Franco-Molano et al. (2010), Halling and Mueller (2005), Mecanismo de Facilitación (2001)
Lactarius quercuum	BOY	*Q. humboldtii*	Type	Dennis (1970), Mueller and Wu (1997), Singer (1963)
Lactarius rimosellus	ANT	*Q. humboldtii*		Franco-Molano et al. (2000, 2010)
Lactarius subumbrinus	ANT	*Q. humboldtii*		HUA (López 7)
Lactifluus annulifer	AMA	*Dicymbe sp.*		Vasco-Palacios et al. (2018)
Lactifluus subiculatus	AMA	*Dicymbe sp.*		Vasco-Palacios et al. (2018)
Russula boyacensis	BOY	*Q. humboldtii*	Type	Dennis (1970), Franco-Molano and Uribe-Calle (2000), Mueller and Wu (1997), Singer (1963)
Russula brevipes	CAL, CUN	*Q. humboldtii*		Guzmán and Varela (1978), Montoya et al. (2005)
Russula caucaensis	CAU	*Q. humboldtii*	Type	Franco-Molano and Uribe-Calle (2000), Mueller and Wu (1997)
Russula columbiana	CUN	*Q. humboldtii*	Type	Dennis (1970), Franco-Molano and Uribe-Calle (2000), Mueller and Wu (1997), Singer (1963)
Russula compacta	ANT	*Q. humboldtii*		Franco-Molano et al. (2010), Halling and Mueller (2005)
Russula cyanoxantha	ANT, CUN, TOL	*Q. humboldtii*		Franco-Molano and Uribe-Calle (2000), Guzmán and Varela (1978)
Russula emetica	ANT, BOY, TOL, VCA	*Q. humboldtii*		Franco-Molano and Uribe-Calle (2000), Saldarriaga et al. (1988a), Sierra et al. (2011), García and Rojas (2010)
Russula emetica var. *lacustris*	BOY	*Q. humboldtii*	Type	Dennis (1970), Singer (1963)
Russula humboldtii	CUN	*Q. humboldtii*	Type	Dennis (1970), Mueller and Wu (1997)
Russula hygrophytica	AMA	*Dicymbe sp.*		Vasco-Palacios et al. (2018)
Russula idroboi	CUN	*Q. humboldtii*	Type	Mueller and Wu (1997), Singer (1963)
Russula peckii	ANT	*Q. humboldtii*		Franco-Molano et al. (2010)

(continued)

Table 6.1 (continued)

TAXA	Distribution	Hosts	Type	References
Russula puiggarii	AMA, ANT	*P. tropenbosii, Dicymbe sp.*		López-Q et al. (2007), Vasco-Palacios et al. (2018)
Russula rosea	AMA	*Dicymbe sp.*		Vasco-Palacios et al. (2018)
Russula semililacea	CUN	*Q. humboldtii*	Type	Franco-Molano and Uribe-Calle (2000), Mueller and Wu (1997)
Russula silvestris	ANT	*Q. humboldtii*		López-Q et al. (2007)
Russula virescens	ANT	*Q. humboldtii*		Franco-Molano et al. (2000, 2010)
Thelephorales				
Bankeraceae				
Sarcodon bairdii	CAQ	*Dicymbe sp.*	Type	Grupe et al. (2016)
Sarcodon colombiensis	AMA	*P. tropenbosii*	Type	Grupe et al. (2016)
Sarcodon pallidogriseus	CAQ	*Dicymbe sp.*	Type	Grupe et al. (2016)
Sarcodon rufobrunneus	AMA	*Dicymbe sp.*	Type	Grupe et al. (2016)
Thelephoraceae				
Thelephora cervicornis	QUI	*Q. humboldtii*		Franco-Molano (2002)
Thelephora palmata	ANT	*Q. humboldtii*		

The largest collections of EM fungi are hosted in the Herbarium of the University of Antioquia (HUA), the Colombian National Herbarium (COL), and the Herbarium of the University of Los Andes (ANDES). In addition, some small collections from thesis are in the Herbarium of the University of Caldas (FAUC) and the Herbarium of the Pedagogical and Technological University of Colombia (UPTC).

Despite the high number of studies in *Quercus*-forests, there are no works that included detailed studies of EM in roots. Around 50% of the studies focus on the characterization of fungal diversity based on fruiting bodies and from those, only two publications included EM inventories in roots (Vasco-Palacios 2016, Vasco-Palacios et al. 2018). Twenty-three percent of all works are about systematics and taxonomy of specific EM taxa (Fig. 6.5) and a significant number of new species have been published from these investigations (Table 6.1). The symbiosis *Quercus*-EM fungi is important for the establishment and development of seedlings, and therefore, for the management of this species. *Quercus humboldtii* is the only oak species that grows in Colombia in the southern boundary of the geographic distribution of this important Holarctic lineage (Avella and Rangel, 2014). The diversity of fungi associated with *Q. humboldtii* is only known from the fungal fruiting bodies collected around trees, but the EM in *Q. humboldtii* roots had been never observed (ca. Singer 1963; Halling 1996; Mueller 1996; Mueller and Wu 1997;

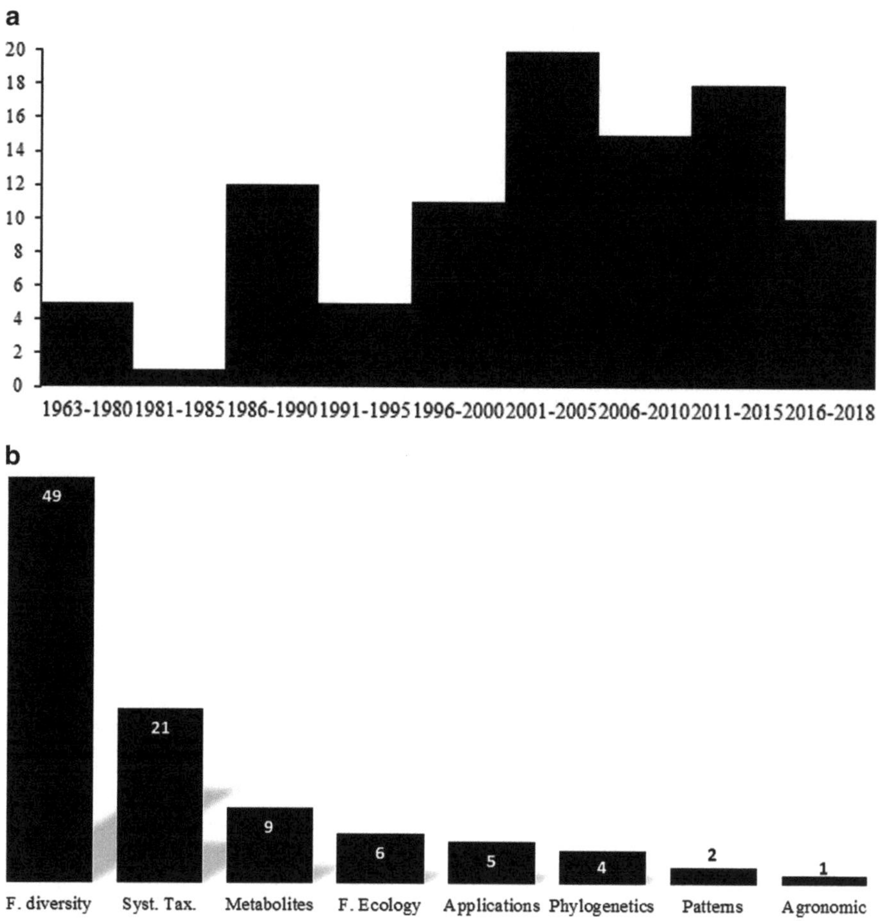

Fig. 6.5 (**a**) Number of works about ectomycorrhizal fungi, first reference is Singer 1963. References considered in this review including thesis, books, books chapter, indexed and not indexed journals; (**b**) Topics about ectomycorrhizal fungi association that include data from Colombia

Tulloss et al. 1992; Franco-Molano et al. 2000; Tulloss and Franco-Molano 2008). Thirty-four new fungal EM species has been described from *Quercus*-forests in Colombia from fungal fruiting bodies (Table 6.1). DNA sequences of most of these EM species are unknown. Sequences are important for identification of EM fungi at root level, for geographical comparisons and for meta-barcoding studies of soil samples. Nowadays, *Quercus humboldtii* is cataloged as a "vulnerable" species in Colombia (UICN red lists) due to its timber exploitation and the transformation of forests into agricultural fields (Cárdenas and Salinas 2006). Fungi play an integral role in shaping and maintaining *Quercus*-forests, as they are intimately involved with processes such as nutrient cycling, nutrient uptake, and decomposition of organic matter. In spite of, researchers in *Quercus*-forests have ignored the existence and role of fungi in establishing and maintaining oak populations. Just few plans of management and conservation include or mention ectomycorrhizal fungi as part of the management of this species (CAR-Corporación Autónoma Regional de Cundinamarca 2016). Mushrooms are not important only for the host plants there are food sources for small mammals and insects (Pyare and Longland 2001, Amat-García et al. 2004). In addition, some species such as *Tylopilus, Russula* and *Ramaria* are important for local communities whom harvested and commercialized those in Boyacá and Santander states (Piragauta and Pérez 2006).

The black-oak *Colombobalanus excelsa* is an endemic Fagaceae species that also presents EM associations (Tedersoo and Brundrett 2017). Its distribution is restricted to four distant populations in montane forests and has been cataloged as a "vulnerable" (Cárdenas and Salinas 2006; Parra-Aldana et al. 2011). Like other species of Fagaceas, black oak forms stand with high dominance (Parra-Aldana et al. 2011). Two species of *Amanita*, *A. arocheae* and *A. aureomonile* are the only species registered as fungal symbionts of this tree (Tulloss 2005; Parra-Aldana et al. 2011). There is no more information available so far about EM of black oak. In mountain areas of Colombia, species of the genus *Pinus* (Pinaceae) and *Eucalyptus* (Myrtaceae) occur. Those plant genera were introduced and fungal symbionts were introduced together with these trees, and as a result of this process, species such as *Amanita muscaria* and *Suillus luteus* are part of the fungal diversity of Colombia (Table 6.1). Recent studies in the Amazonian region described the presence of EM fungi in tropical rainforests associated with endemic trees such as *Pseudomonotes tropenbosii* (Dipterocarpaceae), *Dicymbe uaiparuensis* and *Aldina* sp. (Fabaceae) (Vasco-Palacios et al. 2014, 2018). These are the only studies available in public databases that include morphological and molecular information (ITS sequences) of fungal fruiting bodies (Vasco-Palacios 2016, Vasco-Palacios et al. 2018).

It is possible presume that there is a high number of undescribed EM fungal taxa in Colombia based on the results obtained from recently studies. From the total EM diversity reported for Colombia, 39 records (23% of total diversity) were new species described from Colombian specimens. An important number of them are endemic and only known from the specimen's type. Thirty-five of those are associated with *Quercus*-forests, and five with *Dicymbe, Aldina* or *Pseudomonotes tropenbosii* as EM hosts from tropical lowland forests. Our studies on tropical lowland forests revealed that about 25% of all fungal specimens collected were new species (Grupe et al. 2016; Vasco-Palacios et al. 2018).

Despite the interest in studying EM fungi, there is not enough information to fully know the EM fungal communities, EM species richness, EM fungal host plants, EM species distribution and even less our understanding of the ecological role that EM symbiosis plays in tropical ecosystems and its relation with carbon sequestration. This knowledge can contribute to the conservation of vulnerable ecosystems in Colombia such as tropical mountain forests that are under heavy pressure and from which only 4% still remains. Efforts to document the EM communities of unstudied plant hosts present in Colombia are needed. *Colombobalanus excelsa* (Fagaceae), *Juglans neotropica* (Juglandaceae), *Salix humboldtiana* (Salicaceae) and *Alnus acuminata* (Betulaceae) are EM plants which occur also in mountain areas and same than to *Quercus*, *Colombobalanus* and *Juglans* are categorized as endangered or vulnerable according to IUCN criteria (Cárdenas and Salinas 2006). In tropical lowland forests, it is necessary to explore deeply ecosystems such as white sand forests with *Aldina* and *Dicymbe* (Fabaceae) and terra-firme forests with *Coccoloba* (Polygonaceae), *Guapira, Neea* and *Pisonia* (Nyctaginaceae). Tropical dry forests are another endangered ecosystem in the country, from which only 8% remains today (Pizano et al. 2014). This forest possesses high degrees of endemism and speciation (Dexter et al. 2018) but studies on its EM fungal community have not been conducted so far. A particular and endemic EM fungal community, with undescribed species can be associated with hosts such as *Achatocarpus* (Achatocarpaceae), *Coccoloba* (Polygonaceae), *Neea*, *Pisonia* (Nyctaginaceae), and *Acacia* (Fabaceae) (Pizano et al. 2014). Expanding the study of EM in Colombia might greatly increase the number of EM fungal species and their host plants. Studies on EM fungal diversity should combine fruiting bodies surveys with a detailed analysis of root morphology and metagenomic sequencing. Furthermore, an enrichment of EM sequence reference databases from vouchered specimens of tropical species can contribute to identify new EM at root level. Additionally, root-based research might allow to identify new plant hosts and confirm EM symbioses, later can be particularly interesting in lowland tropical rainforests and tropical dry forests.

States: Amazonas (AMA), Antioquia (ANT), Boyacá (BOY), Caldas (CAL), Cauca (CAU), Caquetá (CAQ), Cesar (CES), Chocó (CHO), Cundinamarca (CUN), Huila (HUI), Nariño (NAR), Magdalena (MAG), Meta (MET), Quindío (QUIN), Risaralda (RIS), Santander (SAN), Tolima (TOL), Valle del Cauca (VCA). *Types: Species described from specimens collected in Colombia.

6.4 Case Study: Co-Existence of Endo- and Ecto-Mycorrhizas in a Tropical Rain Forest of the Colombian Amazon

The aim of this study was to evaluate (endo- and ecto-) mycorrhizal fungal communities in a terra-firme forest of the Colombian Amazon. Ecto-mycorrhization is a specific relation between plants and fungi, while arbuscular mycorrhization not, as plants have different dependence for the mycorrhizal association. Comparison of ecto- and endo-mycorrhization in a certain place might be dissimilar as we could be comparing specific and dependent plant-fungus EM associations with facultative endomycorrhizal associations. To deal with this issue, we compare EM plants with the AM-dependent manioc, assuming both associations have the same nutritional relevance and, therefore, mycorrhizal community composition might be comparable at root level.

The study was performed in the Middle Caquetá region of Colombia between 00°22′14.9″ S and 00°55′11″ S, and 72°06′36.3″ W and 71°26′18.3″ W. Elevation ranges between 200 and 300 m, with slopes between 7–25%. Annual rainfall is unimodal with 3000 mm in average (Duivenvoorden and Lips 1993). This region is located at the intersection of sedimentary plains of Tertiary origin (dissected terraces and hills), with rocky outcrops of Paleozoic origin creating elevated plateaus. The area is dominated by Oxisols with low pH and low-fertility, and Podzols as inclusions. The dominant vegetation is a mosaic of mature and secondary tropical forest of different ages combined with indigenous shifting agricultural plots. There, four major forest units have been recognized: floodplain forests, white sand forests, terra-firme forests and secondary forests (Parrado-Rosselli 2005). Terra-firme forests present a high species richness with members of the plan families Mimosaceae, Fabaceae, Lecythidaceae, Arecaceae and it is considered an AM-dominant forests. However, in some areas the presence of the family Dipterocarpaceae occur. The EM symbiosis is an ecological feature of all members of the family Dipterocarpaceae. *Pseudomonotes tropenbosii* is an endemic dipterocarp tree that forms EM symbiosis and accounts around 19% of canopy trees (Parrado-Rosselli 2005; López-Q et al. 2012; Vasco-Palacios 2016). In AM-dominant forests the EM plant hosts *Coccoloba polystachya* (Polygonaceae) occur in low densities. For EM fungi searching, soil samples were collected from terra-firme forests with and without *Pseudomonotes tropenbosii,* following Tedersoo et al. (2014) methodology. DNA was extracted from 2.0 g of soil using the PowerSoil DNA Isolation Kit (MoBio, Carlsbad, CA). The internal transcribed spacer (ITS) regions 2 was amplified by polymerase chain reaction (PCR) using a mixture of six forward primers as described by Tedersoo et al. (2014). Sequences were obtained by 454 pyrosequencing (Roche GS FLX+, Beckman Coulter Genomics, Danvers, MA). Bioinformatics and statistical analyses were performed as described by Vasco et al. (Submitted).

Natural forest is disturbed by indigenous shifting plots, which are established after logging and burning mature or secondary forest (Peña-Venegas et al. 2017). Manioc (*Manihot esculenta* Crantz), a crop with high dependence for AM association, is the dominant crop species cultivated. As manioc is cultivated from asexual pieces of harvested plant branches, AM inocula come almost exclusively from soils and surrounded forest plants. For AM fungi associated with manioc, two swiddens nearby the forest were visited from sample collection. Fine roots of manioc were collected from plants that farmers were harvesting at the time of our visit. A total of 11 root samples were collected and two soil samples, one of each swidden, from 5 sub-samples of around 100 g. AM fungal DNA was isolated from 5 g of soil or 70 mg of dry fine manioc roots using the PowerSoil® DNA Isolation kit (MoBio laboratories, Inc.). Glomeromycota sequences were amplified using the nuclear SSU rRNA gene primers NS31 and AML2 (Simon et al. 1992; Lee et al. 2008), as described by Öpik et al. (2013), using 454-sequencing. Chimeric sequences were detected and removed using UCHIME v7.0.1090 (Edgar et al. 2011), and the MaarjAM database (status February 2015, 5264 sequences, 348 VT) as reference. The MaarjAM database contains representative sequences covering the NS31/AML2 amplicon of Glomeromycotina sequences classified as virtual taxa (VT) (Öpik et al. 2014; Öpik et al. 2009). A VT is a group of closely related SSU rRNA gene sequences phylogenetically grouped with sequence identity ≥97% (Öpik et al.

2014). Reads were identified against Glomeromycotina in the MaarjAM database with BLAST+ v2.5.0 (Camacho et al. 2009) using an open reference operational taxonomic unit picking approach (Bik et al. 2012). Sequences that did not match any VT in the MaarjAM database were compared against the International Nucleotide Sequence Database Collaboration (INSDC) using the same criteria except a similarity threshold of 90%, an alignment length at least 90% of the shorter of the query, and an alignment length not differing from the shorter of the query and subject sequences by more than 10%. To identify sequences as new VT, sequences receiving no match against MaarjAM but a match against Glomeromycotina in the INSDC were clustered at 99% similarity level using BLASTclust (BLAST v2.2.26) (Altschul et al. 1990). Clustered sequences were aligned with all sequences available in the MaarjAM database using the MAFFT multiple sequence alignment web service in JALVIEW version 2.8 (Waterhouse et al. 2009) and subjected to a neighbor-joining phylogenetic analysis in TOPALi v2.5 (Milne et al. 2004). Novel VT were identified on the basis of sequence similarity and tree topology with AM fungal genus and species on the phylogeny of all Glomeromycotina VT (Öpik et al. 2013).

6.4.1 Fungal Community Composition of the Studied Terra-Firme Forest

A total of 1127 OTUs were recovered from soil samples. Close to 20% of all OTUs are "unknown" species supporting the gap of knowledge on fungal diversity in tropics that exist (Hawksworth and Lücking 2017; Corrales et al. 2018). Sixty percent of fungi were saprotrophs, 10% plant pathogens and 5% EM (Table 6.2). Some arbuscular mycorrhizal were detected (0.4%), in a very low number this due ITS sequencing is not the recommended barcode marker for this fungal group (Simon *et al.* 1992; Lee et al. 2008).

6.4.2 Ectomycorrhizal Fungi in Terra-Firme Forest

The EM fungal diversity account 57 different OUTs. In general, a relatively large number of EM taxa were present in terra-firme forests associated to *P. tropenbosii* and in AM-dominant areas associated to *Coccoloba polystachya*. The EM fungal taxa included *Thelephora-Tomentella* (Thelephoraceae, 17 OTUs), *Lactarius-Russula* (Russulaceae, 8 OTUs), *Clavulina* (Clavulinaceae, 6 OTUs) and *Scleroderma* (Sclerodermataceae, 6 OTUs) (Fig. 6.6). Clavulinaceae is highly diverse in the neotropics (Henkel et al. 2012; Uehling et al. 2012). Species of Scleroderma dominates plant roots of species such as *Gnetum* and *Coccoloba* (Tedersoo and Põlme 2012; Sene et al. 2015). A continuum between AM-forests and EM-forests occurs in Amazonian ecosystems, which favor the distribution of

Table 6.2 Number of soil fungal OTUs per specific trophic groups in a terra-firme forests, of the Colombian Amazonian region, based on ITS searching

Trophic categories	# OTUS
AM	**64**[a]
Mycoparasite	7
Biotrophic	12
Animal parasite	14
EM	**57**
Plant pathogen	111
Unknown	239
Saprotroph	679

[a]Corrected AM richness as the number of AM OUTs (VT) obtained in shifting plots from sequences amplified by SSU rRNA, using NS31 and AML2 as primers

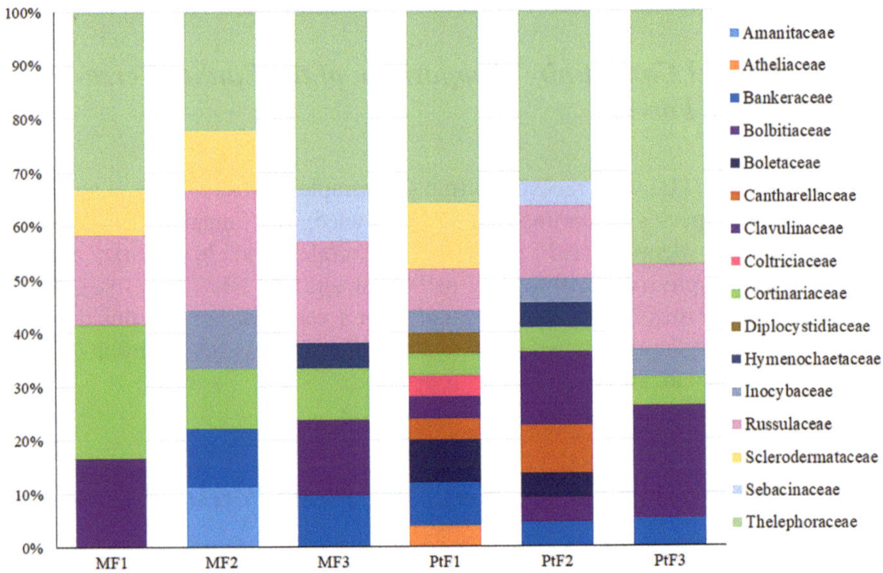

Fig. 6.6 Relative abundance of EM fungi per soil sample

EM fungal species in Amazonian forests (Vasco-Palacios et al. 2018; Corrales et al. 2018). A deeply study is need to fully understand how EM fungi may drive plant hosts distribution in Amazonian forests and its relation with changes in edaphic conditions. In the case of *P. tropenbosii*, the species only had been reported in small patches within terra-firme forests of Colombia. A community of *P. tropenbosii* might exist in a terra-firme forests and few meters later disapperared.

6.4.3 *Arbuscular Mycorrhizal Fungi Associated to Manioc in Shifting Terra-Firme Plots*

A total of 64 AM fungal VT were recovered from the study area. From them only 11% are AM fungal species reported before in the country. As it has been observed in previous works (Saks et al. 2014; Varela-Cervero et al. 2015), root AM fungal richness was higher than soil AM fungal richness: 24 VT from soil samples and 64 from root samples (Table 6.3) (Öpik et al. 2006). From the total AM fungal richness recovered, five VT were not colonizing manioc roots (only recovered from soil samples) (Fig. 6.7).

It is suggested that forested areas conserve a higher number of Glomeraceae species, which is considered a species indicator of non-disturbed areas in contrast to Archaeosporaceae, Claroideoglomaceae and Diversisporaceae as indicators of disturbed areas (Moora et al. 2014). AM fungi associated with manioc roots cultivated in shifting plots conserved a high number of Glomeraceae species, but also, *Claroideoglomus* was the most abundant genera associated with manioc roots. Results suggest that shifting cultivation, plots established in mature or old secondary forests used intensively for 2–3 years and abandoned later, can be considered a transitory disturb as it conserve both AM fungal species indicators of disturbed and undisturbed areas (García de León et al. 2018).

In terms of OUTs, the studied Amazonian forest had similar EM and AM fungal richness, although there are less EM plants than AM plants in that forest. It suggests that other EM host plants might be in Amazonian forests that had not been identified. As presented before, one of the reason is that some EM are not evident in plant roots and only researchers with some experience on it can identify clearly those root structures. Addinionally, the searching of EM in forest are based on previous reported EM species, underlooking the real mycorrhizal status of plant species. We can conclude that indeed ecto- and endo-mycorrhizal fungi co-exist in tropical forests.

It is presumed that EM might be more sensitive to disturbs, being elimitated from previous forested areas in which shifting plots are placed for food cultivation. As a consequence of it, forest in which EM fungi were present will be transformed in secondary forest dominated mainly by plants associated with AM fungi. It is important to indicate that the study area has been inhabited continuously for more than 2000 years as there it is possible to find Amazonian Dark Earths, an indicator of ancient human occupation (Eden et al. 1984). Even though, human interventions in the area for thousands of years, EM plant host species and EM fungi still exist. As explained before, shifting cultivation can be considered a transitory disturb as a place used for agriculture remain for a longer period of time under forest than under cropping conditions. The ecological implication human activities had in the plant composition of secondary forests and in the abundance of EM host species in successional Amazon forest had never been addressed, and might help to clarify how human interventions might condition the presence or not of EM fungal species in tropical areas.

Table 6.3 Arbuscular mycorrhizal fungi associated to manioc roots cultivated in swiddens

Virtual taxa	AM fungus		Type of sample	
(VT)	Genus	Species	Root	Soil
VTX00280	*Claroideoglomus*	sp.	6156	0
VTX00090	*Rhizoglomus*	*manihotis*	4718	6
VTX00024	*Acaulospora*	sp.	4244	0
VTX00126	*Glomus*	sp.	3933	371
VTX00093	*Glomus*	sp.	2087	30
VTX00082	*Glomus*	sp.	1803	1
VTX00418	*Glomus*	sp.	1170	0
VTX00028	*Acaulospora*	sp.	780	42
VTX00248	*Glomus*	sp.	577	0
VTX00178	*Glomus*	sp.	502	0
VTX00227	*Acaulospora*	sp.	401	0
VTX00113	*Glomus*	sp.	230	0
VTX00115	*Glomus*	sp.	240	0
VTX00264	*Rhizoglomus*	*clarum*	177	0
VTX00312	*Glomus*	sp.	162	2
VTX00070	*Glomus*	sp.	127	74
VTX00403	*Glomus*	sp.	127	0
VTX00087	*Glomus*	sp.	123	0
VTX00269	*Glomus*	sp.	110	0
VTX00359	*Glomus*	sp.	107	0
VTX00292	*Glomus*	sp.	99	0
VTX00153	*Glomus*	sp.	82	0
VTX00270	*Glomus*	sp.	75	1
VTX00080	*Glomus*	sp.	57	88
VTX00238	*Paraglomus*	*occultum*	56	68
VTX00163	*Glomus*	sp.	33	0
VTX00039	*Gigaspora*	*decipiens*	31	16
VTX00109	*Glomus*	sp.	31	0
VTX00129	*Glomus*	sp.	25	0
VTX00084	*Glomus*	sp.	18	0
VTX00026	*Acaulospora*	sp.	15	87
VTX00255	*Dentistucata*	*heterogama*	14	0
VTX00370	*Glomus*	sp.	11	0
VTX00089	*Glomus*	sp.	9	35
VTX00030	*Acaulospora*	sp.	8	1
VTX00368	*Glomus*	sp.	7	0
VTX00108	*Glomus*	sp.	4	0
Mo-P3	*Paraglomus*	sp.	4	0
VTX00399	*Glomus*	sp.	3	219
VTX00398	*Glomus*	sp.	3	3
VTX00105	*Rhizoglomus*	*intraradices*	3	0

(continued)

Table 6.3 (continued)

Virtual taxa	AM fungus		Type of sample	
(VT)	Genus	Species	Root	Soil
VTX00092	*Glomus*	sp.	2	1
VTX00091	*Glomus*	sp.	2	0
VTX00112	*Glomus*	sp.	2	0
VTX00375	*Glomus*	sp.	1	2
VTX00253	*Glomus*	sp.	1	1
VTX00069	*Glomus*	sp.	1	0
VTX00079	*Glomus*	sp.	1	0
VTX00166	*Glomus*	sp.	1	0
VTX00364	*Glomus*	sp.	1	0
VTX00318	*Scutellospora*	sp.	1	0
VTX00199	*Glomus*	sp.	1	0
VTX00419	*Glomus*	sp.	1	0
VTX00397	*Glomus*	sp.	1	0
VTX00041	*Racocetra*	*castanea*	1	0
VTX00072	*Glomus*	sp.	1	0
VTX00327	*Glomus*	sp.	1	0
VTX00268	*Glomus*	sp.	1	0
VTX00096	*Glomus*	sp.	1	0
VTX00167	*Glomus*	sp.	0	48
VTX00004	*Archaeospora*	sp.	0	3
VTX00057	*Claroideoglomus*	sp.	0	1
VTX00117	*Glomus*	sp.	0	1
VTX00064	*Glomus*	sp.	0	1

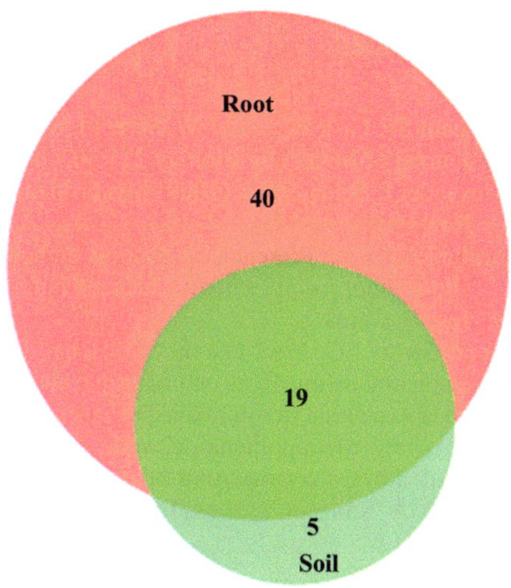

Fig. 6.7 Arbuscular mycorrhizal fungi richness (as virtual taxa) associated to manioc in swidden fields of an Amazonian forest

6.5 Conclusions and Highlights for Future Research

Even in tropical countries like Colombia in which endo- and ecto-mycorrhizal fungi had been studied for almost 50 years, there are still many work to do. More mycologist in Colombia and South American countries are required to fill all gaps on the knowledge that we still have. EM and AM fungal diversity inventories are far from complete and studies on natural environments are limited as well as the results obtenied from those. The ecological rol that AM and EM fungi plays in natural enviroments are still ununderstood, even more when is an issue that EM and AM co-exist in tropical forests. The use of molecular approaches to study EM and AM fungal communities in soil and root samples are still incipient in Colombia, and had limited the advance on the understanding of mycorrhizal symbiosis. A next step is necessary to take. Although it is important the report of mycorrhization of different plant species, what it is relevant is to understand where, when and how it occurs and which variables might be important for the occurrence of the plant-fungi association. Even more, there are relevant areas in which we can take advantage of EM and AM associations, further than agriculture or timber production. One of them is soil bioremediation. It is already known that through mycorrhizal associations it is possible to mobilize toxic compounds that later can be immobilized in host plants. In the case of myorrhizal perennial species such as many tropical trees, the immobilization of toxic substances can be done *in situ* thorugh this alternative with less cost and risk than mechanical methods to remove and dispose contaminated soil in a different place. We need to encourage students and researchers to study more these organisms in natural ecosystems to understand more about EM and AM fungal communities, their ecology and diversity. Another important aspect is to understand how plant- and fungal-symbionts can be affected by the climate change and how this may affect their role of them in ecosystems and food production.

References

Altschul SF, Gish W, Miller W, Myers EW, Lipman DJ (1990) Basic local alignment search tool Journal of Molecular Biology 215:403–410.

Alexander I, Selosse MA (2009) Mycorrhizas in tropical forests: a neglected research imperative New Phytologist 182: 14–16.

Ardila PA (2017) Presencia de esporas de hongos micorrícico arbusculares en suelos del bosque alto andino, parque natural Chicaque Forestry engineer thesis. Universidad Distrital Francisco José de Caldas.

Amat-García EC, Amat-García GD, Henao FG (2004) Diversidad taxonómica y ecológica de la entomofauna micófaga en un bosque altoandino de la cordillera oriental de Colombia. Revista de la Academia Colombiana de Ciencias Exactas, Físicas y Naturales 107: 223–231

AMVA-Área Metropolitana del Valle de Aburrá (2000) Hongos y musgos del Valle de Aburrá. Medellín (Colombia). Área Metropolitana del Valle de Aburrá, 153 pp.

Avella MA, Rangel CJ (2014) Oak forest: conservation and sustainability. Colombia Forestal 17: 100–117

Averill C, Turner BL, Finzi AC (2014) Mycorrhiza-mediated competition between plants and decomposers drives soil carbon storage. Nature 505: 543–545

Ávila Díaz-Granados RA, Orozco OJ, Moreno GL, Magnitskiy S, Rodríguez A (2009) Influence of mycorrhizal fungi on the rooting of stem and stolon cuttings of the Colombian Blueberry (*Vaccinium meridionale* Swartz) International Journal of Fruit Science, 9: 372–384.

Barrios E, Mahuku G, Cortés L, Asakawa N, Jara C, Quintero J (2006) Green manure impacts in nematodes, arbuscular mycorrhizal and pathogenic fungi in tropical soils planted to common beans. 18th Congress of Soil Science. Philadelphia, Pennsylvania, USA.

Bernal R, Gradstein R, Celis M (2016) Catálogo de plantas y líquenes de Colombia Volumen I. First Edition.

Betancur MA, Calderón MH, Betancourt OG, Sucerquia AG (2007) Hongos macromycetes en dos relictos de bosque húmedo tropical montano bajo de la vereda la cuchilla, Marmato, Caldas. Boletín Científico del Museo de Historia Natural Universidad de Caldas 11: 19–31

Bik HM, Porazinska DL, Creer S, Caporaso JG, Knight R, Thomas WK (2012) Sequencing our way towards understanding global eukariotic biodiversity Trends in Ecology and Evolution 27: 233–243

Boekhout T, Pulido MM (1989) The occurrence of macrofungi and their habitats in vegetations along the Parque Los Nevados transect. En: van der Hammen T, Díaz-P. S, Álvarez VJ (eds.) La cordillera central colombiana transecto Parque Los Nevados (Segunda. Parte). Berlin: J. Cramer; Estudios de Ecosistemas Tropandinos: 3.600 pp

Bolaños MM, Rivillas-Osorio CA, Suárez-Vásquez S (2000) Identificación de micorrizas arbusculares en suelos de la zona cafetera de Colombia Cenicafé 51: 245–262.

Camacho C, Coulouris G, Avagyan V, Ma N, Papadopoulos J, Bealer K, Madden TL (2009) BLAST+: architecture and applications BMC bioinformatics 10: 421

CAR-Corporación Autónoma Regional de Cundinamarca (2016) Plan de manejo y conservación del Roble (*Quercus humboldtii* Bonpl.) en la jurisdicción Car Cundinamarca. Bogotá D.C., Colombia, 41 pp.

Cárdenas D, Salinas N (2006) Libro Rojo de plantas de Colombia. Especies Maderables Amenazadas. SINCHI, Ministerio de Ambiente, Vivienda y Desarrollo territorial. Bogotá.

Corrales A, Henkel TW, Smith ME (2018) Ectomycorrhizal associations in the tropics–biogeography, diversity patterns and ecosystem roles. New Phytologist https://doi.org/10.1111/nph.15151

Dennis RW (1970) Fungus flora of Venezuela and adjacent countries. Kew Bulletin Additional Series III. J. Cramer. Vaduz

Dexter K, Pennington T, Oliveira-Filho A, Bueno M, Silva de Miranda P, Neves D (2018) Inserting Tropical Dry Forests Into the Discussion on Biome Transitions in the Tropics. Frontiers in Ecology and Evolution 6

Duivenvoorden JF, Lips JM (1993) Landscape ecology of the Middle Caquetá basin. Tropenbos Colombia, Bogotá

Dumont KP, Umaña MI (1978) Los hongos de Colombia – V. *Laterna triscapa* y *Calostoma cinnabarina* en Colombia. Caldasia 12: 349–352

Eden MJ, Bray W, Herrera L, McEwan C (1984) Terra Preta Soils and Their Archaeological Context in the Caqueta Basin of Southeast Colombia American Antiquity 49: 125–140.

Edgar RC, Haas BJ, Clemente JC, Quince C, Knight R (2011) UCHRIME improves sensitivity and speed of chimera detection Bioinformatics 27: 2194–2200

Estrada G, Sánchez de Prager M (1995) Dependencia del café *Coffea arabica* L var. *colombia* por micorriza vesiculo-arbuscular Acta agronómica 45: 85–88.

Franco-Molano AE (2002) Hongos. En: Mecanismo de Facilitación. IAvH. Resultados del Segundo Día de la Biodiversidad. Departamento del Quindío. Resultados. IAvH. 52 pp.

Franco-Molano AE, Aldana R, Halling R (2000) Setas de Colombia (Agaricales, Boletales y otros hongos). Guía de Campo. Colciencias Universidad de Antioquia, Medellín, 156 pp

Franco-Molano AE, Corrales A, Vasco-Palacios AM (2010) Macrohongos de Colombia ii. Listado de especies de los ordenes Agaricales, Boletales, Cantharellales y Russulales (Agaricomycetes, Basidiomycota). Actualidades biológicas 32: 89–113
Franco-Molano AE, Uribe-Calle E (2000) Hongos Agaricales y Boletales de Colombia. Biota Colombiana 1: 25–43
Franco-Molano AE, Vasco-Palacios AM, López-Quintero C, Boekhout T (2005) Macrohongos de la región del Medio Caquetá. Guía de campo. Multimpresos. Medellín. 219 pp.
García AG, Rojas AC (2010) Macrohongos presentes en el bosque seco tropical de la región del Valle del Cauca, Colombia. Revista de Ciencias 14: 45–54
García de León D, Neuenkampa L, Moora M, Öpik M, Davison J, Peña-Venegas CP, Vasar M, Jairus T, Zobel (2018) Arbuscular mycorrhizal fungal communities in tropical rain forest are resilient to slash-and-burn agriculture Journal of Tropical Ecology.
Gómez AF, Sánchez de Prager M (2012) Actividad biológica de hongos formadores de micorriza arbuscular en un suelo Humic dystrudepts cultivado con maíz y diferentes fuentes de fertilización Acta Agronómica Special Issue.
González C, Jarvis A, Palacio JD (2006) Biogeography of the Colombian oak, *Quercus humboldtii* Bonpl.: geographical distribution and their climatic adaptation. International Centre for Tropical Agriculture (CIAT)/Museo de Historia Natural, Universidad del Cauca, Popayán.
Grupe AC, Vasco-Palacios AM, Boekhout T, Smith M, Henkel T (2016) *Sarcodon* in the Neotropics: four new species from Colombia and a key to selected species. Mycologia 108: 791–805, https://doi.org/10.3852/15-254
Guerrero E (1993) Evaluación de la micorriza en ecosistemas andinos del parque Nacional Natural Chingaza. Primera etapa: Paramo. Pontificia Universidad Javeriana.
Guzmán G, Torres M, Ramirez-Guillén F, Ríos-Hurtado A (2004) Introducción al conocimiento de los Macromicetos del Chocó, Colombia. Revista Mexicana de Micología 19: 33–43
Guzmán G, Varela (1978) Los Hongos de colombia III. Observaciones sobre los hongos, líquenes y mixomicetos de colombia. Caldasia 12: 309–437
Guzmán OA, Castaño J, Sánchez de Prager M (2013) Estudio preliminar del efecto de microorganismos benéficos sobre el tomate (*Solanum lycopersicum* L.) y el nematodo del nudo radical (*Meloidogyne* spp.) Revista Agronomía 21: 51–64.
Hahn C, Christan J (2002) *Ramaria chocoënsis* sp. nov., a gomphoid member of *Ramaria* sect. dendrocladium from Colombia, El Chocó, with special regards to rhizomorph anatomy. Mycological Progress 1: 383–398. https://doi.org/10.1007/s11557-006-0035-6
Halling R (1989) A Synopsis of Colombian Boletes. Mycotaxon 44: 93–113
Halling R (1992) A new species of *Boletus* section Luridi from Colombia. Brittonia 43: 322–325 https://doi.org/10.2307/2806931
Halling R (1996) Boletaceae (Agaricales): Latitudinal biodiversity and biological interactions in Costa Rica and Colombia. Revista de Biología Tropical 44 (Suppl 4): 111–114
Halling R, Mueller G (2005) Common mushrooms of the Talamanca mountains. Costa Rica. The New York Botanical Garden Press, N.Y. U.S.A. 198 pp.
Halling R, Mueller G, Dallwitz M (1999) A new *Phylloporus* (Basidiomycetes, Boletaceae) with a key to species in Colombia and Costa Rica. Mycotaxon 73: 63–67
Halling R, Obrevo L (1987) A new species of *Rozites* from oak forests of Colombia, with notes on biogeography. Mycologia 79: 674–678 https://doi.org/10.2307/3807818
Hawksworth DL, Lücking R (2017) Fungal Diversity Revisited : 2. 2 to 3. 8 Million Species, 1–17. Microbiol Spectr. 5(4) https://doi.org/10.1128/microbiolspec.FUNK-0052-2016.
Henao-M LG, Ruiz A (2006) Investigación y gestión local de robledales alrededor del uso tradicional de macromicetes en la cordillera Oriental colombiana. pp 215–224. En: Solano-Vargas CN. I Simposio Internacional de Robles y Ecosistemas Asociados, Memorias. Fundación Natura, 292pp.
Henkel TW, Aime MC, Chin M, Miller SL, Vilgalys R, Smith ME (2012) Ectomycorrhizal fungal sporocarp diversity and discovery of new taxa in Dicymbemonodominant forests of

the Guiana Shield. Biodiversity and Conservation 21: 2195–2220. https://doi.org/10.1007/s10531-011-0166-1

Horak E, Halling RE (2018) New records of Phaeocollybia from Colombia. Mycologia 83(4): 464–472.

Howeler RH, Cadavid LF (1990) Short-and long-term fertility trials in Colombia to determine the nutrient requirements of cassava Fertilizer Research 26: 61–80.

Howeler RH, Cadavid LF, Burckhardt E (1982) Response of cassava to VA mycorrhizal inoculation and phosphorus application in greenhouse and field experiments Plant and Soil 69: 327–339.

Howeler RH, Sieverding E (1982) La importancia de las micorrizas en la absorción de fósforo por yuca Suelos Ecuatoriales 12: 182–192.

Howeler RH, Sieverding E (1983) Potentials and limitations of mycorrhizal inoculation illustrated by experiments with field-grown cassava Plant and Soil 75: 245–26.

Howeler RH, Sieverding E, Saif S (1987) Practical aspects of mycorrhizal technology in some tropical crops and pastures Plant and Soil 100: 249–283.

International Center of Tropical Agriculture – CIAT (2000) Catálogo de cepas de micorrizas arbusculares.

International Union for Conservation of Nature- IUCN (2009). IUCN Red List of Threatened Species Version 2009.2. On line: http://www.iucnredlist.org [03/12/2009].

Lee J, Lee S, Young P (2008) Improved PCR primers for the detection and identification of arbuscular mycorrhizal fungi. FEMS Microbiology Ecology 65: 339–349.

León D (2015) Comunidades nativas de HFMA asociadas a yuca silvestre en la Amazonia colombiana en época seca y lluviosa bajo dos tipos de paisaje MSc thesis Pontificia Universidad Javeriana

López-Q C, Vasco-Palacios AM, Franco-Molano AE (2007) Macrohongos de un bosque de roble, *Quercus humboldtii* Bonpl., en la vereda Contrafuerte, municipio de Andes (Colombia) pp. 21–34. In: Reserva natural regional Cuchilla Jardín Támesis Antioquia, Una mirada a su biodiversidad. CORANTIOQUIA, RedBio, Gobernación de Antioquia, Corporación Ambiental, Medellín.156 pp.

López-Q C, Straatsma G, Franco-Molano AE, Boekhout T (2012) Macrofungal diversity in Colombian Amazon forests varies with regions and regimes of disturbance. Biodiversity and Conservation 21: 2221–2243

Mecanismo de Facilitación (2001) Día de la biodiversidad. Resultados. http://www.humboldt.org.co/diadeladiversidad/resultados/hongos.html (Uploaded on Abril 22, 2008)

Milne I, Wright F, Rowe G, Marshal DF, Husmeier D, McGuire G (2004) TOPALi: software for automatic identification of recombinant sequences within DNA multiple alignments Bioinformatics 20:1806–1807

Montenegro-Gómez SP, Gómez-Posada S, Barrera-Berdugo SE (2017) Efecto de la gallinaza sobre *Azotobacter* sp., *Azospirillum* sp., y hongos micorrízicos arbusculares en un cultivo de cebolla (*Allium fitulosum*) Revista Entramado 26: 250–257.

Montoya F, Arias D, Betancur-Agudelo M (2005) Contribución al conocimiento de los hongos Macromicetos del resguardo indígena Nuestra Señora de la Candelaria de la Montaña Riosucio, Caldas. Boletín Científico, Museo de Historia Natural 9: 19–30

Moora M, Davison J, Öpik M, Metsis M, Saks Ü, Jairus T, Vasar M, Zobel M (2014) Anthropogenic land use shapes the composition and phylogenetic structure of soil arbuscular mycorrhizal fungal communities. FEMS Microbiology Ecology 90(3): 609–621

Moyersoen B (2012) Dispersion, an important radiation mechanism for ectomycorrhizal fungi in Neotropical lowland forests? In: Sudarshana, P., (Ed.), Tropical Forest. InTech, pp. 93–116

Moyersoen B, Weiss M (2014) New neotropical sebacinales species from a *Pakaraimaea dipterocarpacea* forest in the guayana region, southern Venezuela: structural diversity and phylogeography. Plos One 9. e103076 https://doi.org/10.1371/journal.pone.0103076

Mueller G (1996) Distribution and species composition of *Laccaria* in tropical and subtropical America. Revista de Biología Tropical 44 (Supl. 4): 131–135

Mueller G, Singer R (1988) *Laccaria gomezii*, a new agaric species from the querceta of Colombia and Costa Rica. Mycotaxon 23: 223–227
Mueller G, Wu Q (1997) Mycological contributions of Rolf Singer: Field itinerary, index to new taxa, and list of publications. Fieldiana Botany. New Series 38. 124 pp
Neuenkamp L, Moora M, Öpik M, Davison J, Gerz M, Männistö M, Jairus T, Vasar M, Zobel M (2018) The role of plant mycorrhizal type and status in modulating the relationship between plant and arbuscular mycorrhizal fungal communities New Phytologist https://doi.org/10.1111/nph.14995.
Nieves-Rivera AM., Santos-Flores CJ, Betancourt C (1997) Notas sobre los Agaricales del páramo de Guasca, departamento de Cundinamarca, Colombia. Caldasia 19: 349–351
Oehl F, Adriano de Souza F, Sieverding E (2008) Revision of *Scutellospora* and description of five new genera and three new families in the arbuscular mycorrhiza-forming *Glomeromycetes* Mycotaxon 106: 311–360.
Oehl F, Alves da Silva G, Goto BT, Sieverding E (2011a) Glomeromycota: three new genera and glomoid species reorganized Mycotaxon 116: 75–120.
Oehl F, Sieverding E, Palenzuela J, Ineichen K, Alves da Silva G (2011b) Advances in Glomeromycota taxonomy and classification IMA Fungus 2: 191–199.
Öpik M, Davison J, Moora M, Zobel M (2014) DNA-based detection and identification of Glomeromycota: the virtual taxonomy of environmental sequences Botany 92: 135–147
Öpik M, Metsis M, Daniell TJ, Zobel M, Moora M (2009) Large-scale parallel 454 sequencing reveals host ecological group specificity of arbuscular mycorrhizal fungi in a Boreonemoral forest New Phytologist 184: 424–437
Öpik M, Moora M, Liira J, Zobel M (2006) Composition of root-colonizing arbuscular mycorrhizal fungal communities in different ecosystems around the globe Journal of Ecology 94: 778–790.
Öpik M, Zobel M, Cantero JC, Davison J, Facelli JM, Hiiesalu I, Jairus T, Kalwij JM, Koorem K, Leal ME, Lüra J, Metsis M, Neshataeva V, Paal J, Phosri C, Pölme S, Reier Ü, Saks Ü, Schimann H, Thiéry O, Vassar M, Moora M (2013) Global sampling of plant roots expands the described molecular diversity of arbuscular mycorrhizal fungi. Mycorrhiza 23: 411–430.
Osorio CEU, Sánchez DAC, Molano AEF (2008a) Multiplicación de hongos micorriza arbuscular (HMA) y efecto de la micorrización en plantas micropropagadas de banano (*Musa* AAA cv Gran Enano) (Musaceae) Revista de la Facultad de Agricultura de Medellín 61: 4279–4290.
Osorio NW, Díez MC, Sierra J, Paternina L (2008b) Ecological considerations on the mycorrhizal association in soils of Andean-highlands. Workshop UTPL Mycorrhizas in Tropical Forests.
Osorio NW, Habte M (2013) Synergistic effect of a phosphate-solubilizing fungus and an arbuscular mycorrhizal fungus on Leucaena seedlings in an Oxisol fertilized with rock phosphate Botany 91: 274–281.
Palacio M, Gutiérrez Y, Franco-Molano AE, Callejas-Posada (2015) Nuevos registros de macrohongos (Basidiomycota) para Colombia procedentes de un bosque seco tropical. Actualidades Biológicas 37: 319–339
Parra-Aldana A, Botero V, Díez MC (2011) El roble negro, patrimonio natural del Huila ¿Qué hacer para su conservación?. Centro de Publicaciones, Universidad Nacional de Colombia. ISBN: 978–958–719–747–1
Parrado-Rosselli A (2005) Fruit availability and seed dispersal in terra firme forest of Colombian Amazonia. (Doctoral dissertation). University of Amsterdam. Amsterdam, The Netherlands.
Pedraza JE (1981) Respuesta de plántulas de *Pinus caribaea* Morelet var. hondurensis Barr. et Golf. a micorrización y fertilización fosfatada en un oxisol de los Llanos Orientales de Colombia 1er Congreso Nacional de la Ciencia del Suelo Villavicencia, Colombia.
Peña-Venegas CP (2001) Dinámica de la comunidad micorriza arbuscular en bosques de la Amazonia sur de Colombia Revista Suelos Ecuatoriales 31: 103–107.
Peña-Venegas CP (2010) Arbuscular mycorrhizal fungi in the Amazon región. In: Mycorrhiza: occurrence in natural and restored environments. M. Pagano eds. Nova Publishers, New York: 75–86.

Peña-Venegas CP (2015) People, soil and manioc interactions in the upper Amazon region. PhD Thesis. Wageningen University ISBN: 978–94–6257–322–2.
Peña-Venegas CP, Verschoor G, Stomph TJ, Struik P C (2017) Challenging current knowledge on Amazonian Dark Earths: Indigenous manioc cultivation on different soils of the Colombian Amazon. Culture, Agriculture, Food and Environment: 1–11.
Pérez Botero JC, Usma J. (2012) Caracterización de hongos comestibles en bosques de roble (Quercus humboldtii Bonpl.). Tesis pregrado, Universidad de Antioquia
Petersen RH, Mueller G (1992) New South American taxa of *Cantharellus*, *C. nothofagorum*, *C. xanthoscyphus* and *C. lateritius* var. *colombianus*. Boletín de la Sociedad Argentina de Botánica 28: 195–200
Phiri S, Rao IM, Barrios E, Singh BR (2003) Plant growth, mycorrhizal association, nutrient uptake and phosphorus dynamics in a volcanic-ash soil in Colombia as affected by the establishment of *Tithonia diversiflora* Journal of Sustainable Agriculture 21: 43–61.
Phosri C, Rodriguez A, Sanders IR, Jeffries P (2010) The role of mycorrhizas in more sustainable oil palm cultivation Agriculture, Ecosystems and Environment 135: 187–193
Pyare S, Longland WS (2001) Patterns of ectomycorrhizal-fungi consumption by small mammals in remnant old-growth forests of the Sierra Nevada. Journal of Mammalogy 82: 681–689
Pinto JB (1992) Evaluación de poblaciones micorrizales en suelos degradados y de bosque maduro en Araracuara-Amazonas. Corporación Araracuara-COA.
Pinto VJ (1988) Evaluación de la ocurrencia de la simbiosis micorrizal en plantas de inchi (*Caryodendron orinocense* Kast) de diferentes edades en San José del Guaviare. Thesis Universidad Distrital Francisco José de Caldas.
Pinto VJ, Pedraza J (1988) Cambios estacionales de la población micorrizal relacionada con las características climáticas y edáficas en suelos donde crece el inchi (*Caryodendron orinocense* Kast) en San José del Guaviare Suelos Ecuatoriales 18: 226–239.
Piragauta M, Pérez E (2006) Uso de hongos por campesinos de los Municipios de Moniquirá y Arcabuco, departamento de Boyacá. Tesis pregrado, Universidad Pedagógica y Tecnológica de Colombia
Pizano C, González R, González MF, Castro-Lima F, López R, Rodríguez N, Idárraga-Piedrahíta A, Vargas W, Vergara-Varela, H, Castaño-Naranjo A, Devia W, Rojas A, Cuadros H, Toro JL (2014) Las plantas de los bosques secos de Colombia. In: Pizano C, García H (eds.). El Bosque Seco Tropical en Colombia. Bogotá, DC, Colombia: Instituto de Investigación de Recursos Biológicos Alexander von Humboldt (IAvH).
Posada RH, Madriñán S, Rivera EL (2012) Relations between the litter colonization by saprophytic and arbuscular mycorrhizal fungi with depth in a tropical forest Fungal Biology 116: 747–755.
Posada-Almanza RH, Franco-Castro LA, Medina-Girón E (2006) El tiempo de establecimiento de pasturas y su relación con la micorriza arbuscular en paisajes de loma y vega Acta biológica Colombiana 11: 55–64.
Possú WB, Unigarro A, Rosero SC, Solarte AF (2004) Determinación de hongos formadores de micorrizas (HMA) en *Theobroma cacao* L, *Musa* sp., Simmonds, *Borojoa patinoi*. Cuatr y *Bactris gasipaes* HBK en el municipio de Tumaco, Nariño Revista de Ciencias Agrícolas 21.
Prada PH (2009) Identificación de hongos formadores de micorriza arbuscular (HMA) en agroecosistemas de aguacate (*Persea americana*) y su relación con algunas características edáficas, en tres regiones de Antioquia. Biology thesis Universidad de Antioquia.
Pulido M (1983) Estudios en Agaricales Colombianos – los hongos de Colombia IX- Biblioteca José Jerónimo Triana No 7 Instituto de Ciencias Naturales, Universidad Nacional de Colombia, Bogotá, 143 pp.
Pulido M, Boekhout T (1989) Distribution of macrofungi along The Parque Los Nevados transect pp. 485–505 In: Van der Hammen T, Díaz-Piedrahita S, Alvarez V (eds.) La cordillera Central Colombiana transecto Parque Los Nevados (segunda parte). Studies on Tropical Andean Ecosystems, Volume 3

Quiroga LMZ, Sánchez de Prager M, Barrios E (2009) Response of *Melinis minutiflora* to inoculation with arbuscular mycorrhizal fungi in an Inceptisol of Colombia Acta Agronómica 58: 225–233.
Ramírez MM, Oehl F, Pérez A, Rodríguez A (2000) Arbuscular mycorrhizal fungi (AMF) diversity obtained from gooseberry plantations (*Physalis peruviana*) in a Colombian Andean region Silva Fennica 752: 1.
Restrepo JF (2006) Comparación de la diversidad y abundancia de micorrizas arbusculares en diferentes ecosistemas naturales en el parque nacional natural Amacayacu, Amazonas, Colombia. Biology thesis. Universidad de Antioquia.
Ríos RG, Gallego RR (1997) Hongos del orden Glomales (Zygomycetes) presentes en una porción de bosque andino localizado en el parque nacional natural Las Orquídeas. Biology degree. Universidad de Antioquia.
Rivillas OCA (1995) Efecto y detección bioquímica de hongos-micorrícico-arbusculares en las variedades de café caturra y Colombia 16th congreso ASCOLFI January 1995.
Rodríguez A, Sanders IR (2015) The role of community and population ecology in applying mycorrhizal fungi for improved food security The ISME journal 9: 1053–1061.
Saks Ü, Davison J, Öpik M, Vasar M, Moora M, Zobel M (2014) Root-colonizing and soil-borne communities of arbuscular mycorrhizal fungi in a temperate forest understorey Botany 92: 277–285.
Saldarriaga Y, Pineda F, García G, Velásquez L, Guzmán G (1988a) Nuevos registros de Agaricales en Colombia. Revista Mexicana de Micología 4: 333–342
Saldarriaga Y, Velásquez L, Pineda F, García G (1988b) Hongos de Antioquia. Universidad de Antioquia. 206pp
Sánchez de Prager M (2003) Actividad biológica en la rizosfera del maracuya-*Paciflora edulis*, f flavicarpa Degener en diferentes sistemas de manejo, estados de desarrollo y condiciones fitosanitarias. PhD Thesis. Universidad Pontificia de Madrid.
Sánchez de Prager M (2004) Las endomicorrizas: expresión bioedáfica de importancia en el trópico Universidad Nacional de Colombia Palmira, Valle 352 p.
Sánchez de Prager M, Almanza RP, Velásquez D, Narváez M (2010) Metodologías básicas para el trabajo con micorriza arbuscular y hongos formadores de micorriza arbuscular Universidad Nacional de Colombia Sede Palmira. 139 p.
Schenck NC, Spain JL, Sieverding E (1986) A new sporocarpic species of *Acaulospora* (Endogonaceae) Mycotaxon 25: 111–117.
Schenck NC, Spain JL, Sieverding E, Howeler RH (1984) Several new and unreported vesicular-arbuscular mycorrhizal fungi (Endogonaceae) from Colombia Mycologia 76: 685–699.
Schüßler A, Schwarzotti D, Walker C (2001) A new fungal phylum, the Glomeromycota: phylogeny and evolution Mycological Resources 105: 1413–1421.
Sene S, Avril R, Chaintreuil C, Geoffroy A, Ndiaye C, Diedhiou AG, Sadio O, Courtecuisse R, Sylla SN, Selosse MA et al. (2015) Ectomycorrhizal fungal communities of *Coccoloba uvifera* (L.) L. mature trees and seedlings in the neotropical coastal forests of Guadeloupe (Lesser Antilles). Mycorrhiza 25: 547–559.
SIAC (2016) Sistema de Información Ambiental de Colombia – SIAC. Ministerio de Medio Ambiente.
Sierra JD, González JA, Sáenz M (2011) Registro preliminar de macrohongos (ascomycetes y basidiomycetes) en el bosque húmedo montano del alto el romeral (municipio de Angelópolis, departamento de Antioquia Colombia). Revista Facultad Nacional de Agronomía 64: 6159–6174
Sieverding E (1984) Aspectos básicos de la investigación de MVA In: Investigaciones sobre micorrizas en Colombia Universidad Nacional de Colombia.
Sieverding E (1987) A VA-mycorrhizal fungus, *Glomus glomerulatum* sp. nov. with two hyphal attachments and spores formed only in sporocarps Mycotaxon 29: 73–79.
Sieverding E (1989a) La micorriza: Un componente biotecnológico en la producción vegetal Revista Colombia Ciencia y Tecnología 7: 9–11.

Sieverding E (1989b) Ecology of VAM fungi in tropical ecosystems Agriculture, Ecosystems and Environment 29: 369–390.
Sieverding E, Alves da Silva G, Berndt R, Oehl F (2014) *Rhizoglomus*, a new genus of the Glomeraceae Mycotaxon 129: 373–386.
Sieverding E, Howeler RH (1985) Influence of spores of VA mycorrhizal fungi on cassava yield response to phosphorus fertilization Plant and Soil 88: 213–221.
Sieverding E, Leihner DE (1984a) Effect of herbicides on population dynamics of VA-mycorrhiza with cassava Angew Botanik 58: 283–294.
Sieverding E, Leihner DE (1984b) Influence of crop rotation and intercropping of cassava with legumes on VA mycorrhizal symbiosis of cassava Plant and Soil 80:143–146.
Sieverding E, Oehl F (2006) Revision of *Entrophospora* and description of *Kuklospora* and *Intraspora*, two new genera in the arbuscular mycorrhizal Glomeromycetes Journal of Applied Botany and Food Quality 80: 69–81.
Sieverding E, Toro T (1987) *Acaulospora denticulata* sp. nov. and *Acaulospora rehmii* sp. nov. (Endogonaceae) with ornamented spore walls Angew Botanik 61: 217–223.
Simon L, Lalonde M, Bruns TD (1992) Specific amplification of 18S fungal ribosomal genes from VA endomycorrhizal fungi colonizing roots. Appl Environ Microbiol 58: 291–295.
Singer R (1963) Oak mycorrhiza fungi in Colombia. Mycopathologia 20: 239–252 doi:https://doi.org/10.1007/BF02089212
Singer R (1970) Strobilomycetaceae (Basidiomycetes). Flora Neotropica, Monograph 5. 34 pp.
Singer R, Araujo I, Ivory MH (1983) The ectotrophically mycorrhizal fungi of the neotropical lowlands, especially Central Amazonia. Nova Hedwigia Beihefte. 77: 1–352
Singer R, Garcia J, Gomez LD (1990) The Boletineae of Mexico and Central America I & II. Nova Hedwigia, Beiheft, Beiheft 98. 70 pp.
Singer R, Morello JH (1960) Ectotrophic forest tree mycorrhizae and forest communities. Ecology 41: 549–551
Smith SE, Read DJ (2008) Mycorrhizal symbiosis. 3rd. Academic Press New York, ISBN, 440026354, 605.
Spain JL, Sieverding E, Oehl F (2006) *Appendicispora*: a new genus in the arbuscular mycorrhiza-forming Glomeromycetes, with a discussion of the genus *Archaeospora* Mycotaxon 97: 163–182.
Spain JL, Sieverding E, Toro T (1989) *Scutellospora biornata*: a new species in the Endogonaceae from the Llanos Orientales of Colombia Mycotaxon 35: 219–227.
Tedersoo L, Bahram M, Põlme S, Kõljalg U, Yorou NS, Wijesundera R, Ruiz RV, Vasco-Palacios AM, Thu PQ, Suija A, Smith ME, Sharp C, Saluveer E, Saitta A, Rosas M, Riit T, Ratkowsky D, Pritsch K, Põldmaa K, Piepenbring M, Phosri C, Peterson M, Parts K, Pärtel K, Otsing E, Nouhra E, Njouonkou AL, Nilsson RH, Morgado LN, Mayor J, May TW, Majuakim L, Lodge DJ, Lee SS, Larsson KH, Kohout P, Hosaka K, Hiiesalu I, Henkel TW, Harend H, Guo L, Greslebin A, Grelet G, Geml J, Gates G, Dunstan W, Dunk C, Drenkhan R, Dearnaley J, De Kesel A, Dang T, Chen X, Buegger F, Brearley FQ, Bonito G, Anslan S, Abell S, Abarenkov K (2014) Global diversity and geography of soil fungi Science. 346, 1256688.
Tedersoo L, Brundrett (2017) Evolution of ectomycorrhizal symbiosis in plants. In Tedersoo (edt) Biogeography of Mycorrhizal Symbiosis. Springer, Cham. 407–467
Tedersoo L, May TW, Smith ME (2010) Ectomycorrhizal lifestyle in fungi: global diversity, distribution, and evolution of phylogenetic lineages. Mycorrhiza 20: 217e263 doi:https://doi.org/10.1007/s00572-009-0274-x.
Tedersoo L, Põlme S (2012) Infrageneric variation in partner specificity: multiple ectomycorrhizal symbionts associate with *Gnetum gnemon* (Gnetophyta) in Papua New Guinea. Mycorrhiza 22: 663–668.
Tobón LE (1991) Ascomicetos de Colombia: Discomicetos del Departamento de Antioquia. Caldasia: 327–335.

Triana CV (2015) Evaluación de la especificidad entre plantas e inóculos comerciales de micorrizas para el desarrollo y producción de arveja (*Pisum sativum* L) Thesis Agroecological Engineer Corporación Universitaria Minuto de Dios.

Tulloss RE (2005) *Amanita*-distribution in the Americas, with comparison to eastern and southern Asia and notes on spore character variation with latitude and ecology. Mycotaxon, 93(189), e231.

Tulloss RE, Franco-Molano AE (2008) Studies in *Amanita* subsection Vittadiniae II a new species from Colombia savanna. Mycotaxon 105: 317–323

Tulloss RE, Ovrebo C, Halling R (1992) Studies on *Amanita* (Amanitaceae) from Andean Colombia, Memories of the New York Botanical Garden. 66: 27–30

Uehling JK, Henkel TW, Aime MC, Vilgalys R, Smith ME (2012) New species and distribution records for *Clavulina* (Cantharellales, Basidiomycota) from the Guiana Shield, with a key to the lowland Neotropical taxa. Fungal Biology 116: 1263–1274

Van Der Heijden MG, Martin FM, Selosse MA, Sanders IR (2015) Mycorrhizal ecology and evolution: the past, the present, and the future New Phytologist 205: 1406–1423

Varela-Cervero S, Vassar M, Davison J, Bares JM, Öpik M, Azcón-Aguilar C (2015) The composition of arbuscular mycorrhizal fungal communities differs among the roots, spores and extraradical mycelia associated with five Mediterranean plant species Environmental Microbiology

Vargas-Estupiñán N, Pardo-de La Hoz CJ, Danies G, Franco-Molano AE, Jiménez P, Restrepo S, Grajales A (2017) Defining the phylogenetic position of *Amanita* species from Andean Colombia. Mycologia 1–16.

Vasco-Palacios AM (2016) Ectomycorrhizal fungi in Amazonian tropical forests in Colombia. Doctoral Thesis, Utrecht University. Editorial Panamericana, Bogotá. 201 pp.

Vasco-Palacios AM, Franco-Molano AE (2005) A new species of *Gloeocantharellus* (Fungi-Basidiomycetes) from Colombian Amazonia. Mycotaxon 91: 87–92

Vasco-Palacios AM, Franco-Molano AE, López-Quintero C, Boekhout T (2005) Macromicetes (Ascomycota, Basidiomycota) de la región del Medio Caquetá, departamentos de Caquetá y Amazonas (Colombia). Biota Colombiana 6: 127–159

Vasco-Palacios AM, Hernández J, Peñuela-Mora MC, Franco-Molano AE, Boekhout T (2018) Ectomycorrhizal fungi diversity in a white sand forest in western Amazonia. Fungal Ecology 31: 9–18. doi:https://doi.org/10.1016/j.funeco.2017.10.003

Vasco-Palacios AM, López-Quintero C, Franco-Molano AE, Boekhou, T (2014) *Austroboletus amazonicus* sp. nov. and *Fistulinella campinaranae* var. *scrobiculata*, two commonly occurring boletes from a forest dominated by *Pseudomonotes tropenbosii* (Dipterocarpaceae) in Colombian Amazonia. Mycologia 106: 1004–1014.

Walker C, Vestberg M, Demircik F, Stockinger H, Saito M, Sawaki H, Nishmura I, Schüßler A (2007) Molecular phylogeny and new taxa in the Archaeosporales (Glomeromycota): *Ambispora fennica* gen. sp. nov., Ambisporaceae fam. nov., and emendation of *Archaeospora* and Archaeosporaceae Mycological Research 3: 137–153.

Waterhouse AM, Procter JB, Martin DMA, Clamp M, Barton GJ (2009) Jalview Version 2 - a multiple sequence alignment editor and analysis workbench Bioinformatics 25: 1189–1191

Wu Q, Mueller GM (1995) The genus *Craterellus* (Basidiomycetes, Aphyllophorales) in Costa Rica and Colombia. Documents Mycologiques XXV: 487–496

Zabala FJV (2012) Efecto de abonos verdes en la agregación y micorrización en el cultivo de maíz (*Zea mays* L.) en un suelo de ladera de Palmira (Colombia) MSc thesis Universidad Nacional de Colombia.

Zabala FJV, Sánchez de Prager M (2014) Dinámica de los hongos micorriza arbuscular (MA) en un Humic Dystrudepts sembrado con maíz *Zea mays* L. y abonos verdes (AV) Revista de Investigación Agraria y Ambiental (5): 69–79

Chapter 7
How Does the Use of Non-Host Plants Affect Arbuscular Mycorrhizal Communities and Levels and Nature of Glomalin in Crop Rotation Systems Established in Acid Andisols?

Paula Aguilera, Fernando Borie, Alex Seguel, and Pablo Cornejo

7.1 Introduction

Agriculture intensification including soil disturbance, monoculture and increased fertilization affect soil biota communities (Wardle et al. 2004) reducing its abundance and the overall diversity of soil organisms and, consequently, affecting ecosystem functionality like plant nutrient acquisition and cycling of resources between communities (van der Heijden et al. 2008; Wagg et al. 2014). One of the main threats is the loss of organic matter, which is the support to microbial life. It has been reported that agricultural intensification affects more negatively the abundances of taxonomic groups with larger body size compared with smaller ones like protozoa, bacteria and fungi (Postma-Blaauw et al. 2010). Some other studies also found reduced bacterial biomass but not fungal biomass contrasting with those reports showing a reduction of fungal-bacterial biomass ratio, especially under arable conditions. However, intensification of agriculture includes an adequate use of the main management practices like tillage, fertilization and crop rotation, all of them modifying microbial diversity and community structure. Crop rotation and tillage are the most studied practices affecting soil microbial communities. Adoption

P. Aguilera (✉) · A. Seguel · P. Cornejo
Scientific and Technological Bioresources Nucleus (BIOREN-UFRO), Centro de Investigación en Micorrizas y Sustentabilidad Agroambiental (CIMYSA-UFRO), Universidad la Frontera, Temuco, Chile
e-mail: paula.aguilera@ufrontera.cl

F. Borie
Scientific and Technological Bioresources Nucleus (BIOREN-UFRO), Centro de Investigación en Micorrizas y Sustentabilidad Agroambiental (CIMYSA-UFRO), Universidad la Frontera, Temuco, Chile

Facultad de Recursos Naturales, Universidad Católica de Temuco, Temuco, Chile

M. C. Pagano, M. A. Lugo (eds.), *Mycorrhizal Fungi in South America*,
Fungal Biology, https://doi.org/10.1007/978-3-030-15228-4_7

of crop rotation have been largely motivated by the associated crop yield increase (Bullock 1992) mainly due to an enhanced soil fertility (particularly when legumes are used in rotation), maintenance of soil structure, disruption of pest cycles and weed suppression (Smith et al. 2008). On the other hand, by increasing the quantity, quality and chemical diversity of residues together high diversity rotations can sustain soil biological communities, with positive effects on soil organic matter and soil fertility (Tiemann et al. 2015). The relationship between above and below-ground biodiversity in agroecosystems is still controversial. A number of studies have shown that microbial diversity increase, not change, and decrease under crop rotations. For example, Lupwayi et al. (1998) reported that microbial diversity was higher under a rotation of wheat and clover or field peas than under continuous wheat. On the contrary, Navarro-Noya et al. (2013) found that continuous maize versus maize-wheat rotation had no effect on soil microbial diversity. However, Yin et al. (2011) reported that in soybean in rotation with wheat decreased diversity indices compared to continuous wheat. In spite of soil microbiota is crucial for several ecosystem processes, such as nutrient acquisition (Smith and Read 2008), N and C cycling and soil formation and stabilization (Rillig and Mummey 2006) their impact on plant productivity and how soil management are affecting their functionality is still unclear. It has been estimated that in one gram of soil exist around 10 to 200 m fungal hyphae (Leake et al. 2004). However, the main groups of such important soil microbes regulating plant productivity are N-fixing bacteria and mycorrhizal fungi, the first responsible of 5–20% and the last one for up to 75% of all N and P acquired by plants annually. Higher positive effects of this two symbiosis on plant productivity are found in nutrient poor ecosystems where they enhance up to 90% (van der Heijden et al. 2008). Therefore, this chapter will be focused in the behavior of arbuscular mycorrhizal fungi (AMF) in the productivity of acid volcanic soils.

Arbuscular mycorrhizal (AM) symbiosis is a mutualistic association established between some soil fungi and plant roots being extremely abundant in the plant kingdom. Therefore, it has been estimated that about 75% of all plants species form symbiosis with fungi of phylum Glomeromycota (Smith and Read 2008) including almost all the most important agricultural species. Although AMF are not host-specific symbionts some reports have shown that there exists some host preference and host selectivity (Torrecillas et al. 2012; Aguilera et al. 2014, 2017). Richness and composition of AMF communities are associated to host plant, climate and soil conditions (Öpik et al. 2006) and consequently soil intensive management have leaded to a decreased diversity (Verbruggen et al. 2010). Globally, AMF play a key role in ecosystems influencing several important functions like plant productivity, plant P uptake, N acquisition and reduction of N-leaching, regulation of plant diversity (van der Heijden et al. 1998), soil formation and soil aggregation (Smith and Read 2008). Mutual benefits for both partners are based on a bidirectional interchange of nutrients, particularly P provided by the fungus and photosynthetic carbonaceous compounds provided by the plant (Smith and Read 2008; Smith and Smith 2013; Borie et al. 2010). Moreover, AM provides a series of additional benefits to plants like protection against abiotic (Seguel et al. 2013) and biotic stress

(Azcón-Aguilar and Barea 1996) enhancing aggregation and contribution to a better soil structure (Rillig and Mummey 2006). Main abiotic stresses include drought, salinity, heavy metals (Miransari 2011; Meier et al. 2012; Cornejo et al. 2013; Lenoir et al. 2016; among others) including Al phytoxicity (Seguel et al. 2013), a plant stressor habitually found in acidic soils. It has been suggested that one of the main mechanisms developed by AM symbiosis against such stressors is glomalin, a glycoprotein copiously produced by AMF (Wright and Upadhyaya 1996, 1998); which is present in the fungal wall hyphae in a high percent (80%, Driver et al. 2005). Role, benefits and problems associated to its extraction from soils and chemical nature of such protein will be discussed on a next paragraph. Summarizing, all beneficial effects produced by AM symbiosis suggest that plants have survived along the time due to their roots have been colonized by fungi supplying nutrients and providing a protected niche against diverse environmental stresses. It has even been proposed that mycorrhizal plants extend their influence to nearby plant roots in such way that many of them would not be able to coexist with other plants without AMF participation. Agroecosystems are managed biological systems that may involve the use of several practices, some of them detrimental to AM symbiosis. Among agricultural management systems soil tillage and crop rotation are the two main events, which significantly affect the behavior of AM symbiosis on plant productivity. Whereas soil disturbance and fragmentation of net fungal mycelia produced by plough in conventional tillage with beneficial or detrimental effects to crops and soil quality affecting AMF community have been well documented (Jansa et al. 2002, 2003; Oehl et al. 2010; Kabir 2005 for references), the incidence of crop rotations has been much less studied in terms of plant nutrient acquisition as well as providing other ecosystem services.

7.2 Effect of Crop Rotation on AMF Functionality

Although a wide range of plant species forming AM have been described there are relatively few of them which either form no mycorrhizal or sparse infections with AMF (see Javaid 2007 for references). This is the case of plant species belonging to Brassicaceae, Chenopodiaceae, Caryophyllaceae and Cyperaceae families (Brundrett 2009). A notable exception in the largely mycotrophic Fabaceae family is the non-mycotrophic genus *Lupinus*. Therefore, the use of such crops in rotations tend to lead to a reduction in mycorrhizal propagules affecting subsequent cultivation of AM host crops which just increase AMF populations for maintaining mycorrhizal activity and functionality in soil. The lack of root AM colonization in that species is due to the presence of some allellochemical compounds exuded by plant roots such as thioglucosides and isothiocianates, compounds containing S in their chemical structure (Schreiner and Koide 1993). However, the mechanisms involved in the compatibility roots/fungi are still not well understood. On the other hand, *Chenopodium albus* contains saponins in their roots suggesting to be responsible for preventing fungal root colonization (Lavaud et al. 2000). In addition, Arihara and

Karasawa (2000) have reported the effects of fallow and five mycotrophic crops (sunflower, maize, soybean, potato and wheat) and two non-mycotrophic ones (rape and sugar beet) on root colonization and growth under field conditions from 1990 to 1992 as pre-culture of maize. They found depressed root colonization, grain yield and P uptake when using rape, sugar beet and fallow although available P was almost not altered. In summary, the effects of having a host or non-host as proceeding crop in the rotation system conditions the differences in AM fungal inocula density (Karasawa et al. 2002).

For instance, in Argentina is common the rotation of rapeseed (*Brassica napus* –Myc) with soybean (*Glicine max* L + Myc) and Valetti et al. (2016) in a recent study showed that the inclusion of rapeseed in the soybean-based system decreased by a 30% AMF soybean root colonization. Similar trends reported Koide and Peoples (2013) in a system rapeseed-maize showing a decrease in mycorrhizal colonization, yield and shoot P concentration in maize at the first year, but such negative effects were temporary and they did not occur in the second year of maize. The same behavior had been reported by Karasawa et al. (2001) when using mustard, radish, sugar beet and backwheat (all-Myc) in maize growth compared with other four +Myc crops in greenhouse using an Andosol. This type of −Myc crops generates thioglucosides with fungicidal characteristics thereby affecting fungal sporulation and AM colonization. Other −Myc crops like Proteaceae do not appear to exudate harmful substances, but Lambers and Teste (2013) reported some evidences of mycorrhizal sporulation inhibition near cluster roots of *Branksia prionotes*. However, in the case of *Lupinus albus* (−Myc specie) which excretes high amounts of citric acid by its roots under P deficiency, this exudation may limit AM fungal growth. In summary, the effect of crop rotation or the use of crop sequences on AMF functionality have been focused in +Myc crops in terms of fungal diversity or the effect on plant growth, P acquisition and spore germination. Quite different is for −Myc crops where studies in connection with AMF diversity are practically absent. Surprisingly, no references related to glomalin status left in the soil after cropping such plant species have been found. This lack of information highlights the importance, novelty and relevance of this chapter.

7.3 Volcanic Soils and AMF

Acids soils (pH<5.5) constrain productivity on about 1.6 billion hectares worldwide, which represent about 50% of arable land especially in tropical and subtropical zones most concentrated in under developed countries. These soils are characterized by its high Al activity and P deficiency, which limit crop productivity and sustainability. A significant area of Southern Chile is covered by volcanic ash derived soils being the younger soils (Andisols) more acidic than older ones (Ultisols). In such soils, due to mineral matrix including elevated Fe and Al oxides, P bioavailability is typically low as a consequence of low P diffusion rates. This high P fixation together free Al which is phytotoxic to plant roots (Kochian et al.

2005), the two most important factors affecting plant productivity representing an economic challenge for local farmers which must apply lime/gypsum and P fertilizers for supplying crop nutritional needs. Lime application produce a decrease in free Al ions, maximize P applied (avoiding precipitation of $AlPO_4$) and increase root elongation growth (Kochian et al. 2005). For enhancing P acquisition from soils, plant roots and their associated microbiota have developed some strategies being the main: a) adaptations of root geometry and architecture for a better soil exploring (Lambers et al. 2006); b) root exudations of protons (H^+) (Hinsinger 2001), chelant organic acid anions (Ryan et al. 2001) or phosphatases enzymes which hydrolyses soil organic P; c) root association with free-living or symbiotic microorganisms including AMF (Richardson et al. 2011; Smith and Smith 2013). Some of this strategies that give to the plants efficiency in P acquisition are common with that reported for plant Al tolerance (Kochian et al. 2005; Seguel et al. 2013). Bearing in mind that Al-toxicity and P deficiency coexist in acidic soils and the mechanisms/root traits could be similar it is expected that in general Al-tolerant plants may have a greater efficiency in P acquisition. However, discrepant results have been reported from different studies relating Al-tolerance and P efficiency when comparing performance of the same genotypes growing at laboratory and at field conditions (Ferrufino et al. 2000; Villagarcia et al. 2001). More recently, studies have demonstrated that citrate exudation does not totally explain the greater P-uptake efficiency observed in an old recognized Al-tolerant and P-efficient wheat cultivar, suggesting that this mechanism could be complementary to other root traits. In this context, several studies have shown that AMF help to biological adaptations of cereals growing under stressed conditions like P bioavailability and Al-phytotoxocity (Cumming and Ning 2003; Seguel et al. 2013, 2016) which is suggesting that Al-tolerance/P efficiency is highly influenced by the AM symbiosis performance. However, the study of the strategies involving the decrease of Al-toxicity and the increase of P availability (Al-P interactions) have been scarcely studied under field conditions in spite of the extended area where plants habitually grow worldwide. In this context, Aguilera et al. (2011) using confocal microscopy showed the fluorescence emitted by glomalin-Al complex in AMF spores extracted from a soil high in extractable Al suggesting the key role played by fungal structures in decreasing Al phytotoxicity (Aguilera et al. 2011). Additionally, in a study of Al-P interactions on wheat genotypes contrasting in Al-tolerance when growing at field in a Chilean acid Andisol (pH 5.0; Al sat 32%), Seguel et al. (2017) found that grain production, P in shoots and roots, root P/Al ratio, root AM colonization, and AM spore density were higher in Al-tolerant plants than in Al-sensitive ones. In contrast, Al concentration in shoots and roots was higher in the sensitive genotype with a concomitant decrease in P concentration. All these findings suggest that plant traits such as Al tolerance, P efficiency, and mycorrhizal activity/functionality are co-operating in overcoming adverse acid soil conditions.

Some efforts have been made in relation to the study of AMF diversity in soils from Southern Chile. For instance, Aguilera et al. (2014) identified 24 fungal species belonging to 12 genera present in the rhizosphere of six winter wheat cultivars cropped in an Andisol showing no significantly differences in AM diversity associated to cultivars but species richness was different among cultivars. On the other hand, Castillo et al. (2010, 2016) have identified 29 fungal species and 8 genera being *Acaulospora* and *Glomus* the most representative genera. In a bibliographic review of some records from croplands, grassland and forests soils from Southern-Central Chile generated by collections made during the period 2004–2014, Castillo et al. (2016) recorded 21 genera and 66 species which represents 24% of AMF species known so far. For us it is interesting that in such work they found the smaller AMF spore number in systems grassland-rape, potato-rape, grassland-lupine and potato-lupine associations when growing in pots, all of them including −Myc crops.

It is recognized that glomalin is present in large amounts in soils being a distinct component of soil organic matter with particular characteristics. It is a thermostable glycoprotein discovered by Wright and Upadhyaya (1996), which is present originally in AM wall hyphae in the 80% (Driver et al. 2005) and released to the environment in a minor proportion (Rillig and Mummey 2006). At hyphal senescence glomalin is stabilized and accumulated in the soil through interaction with soil matrix constituents, which gives to this protein a high recalcitrance. Glomalin represents an important C reservoir ranging from 2 to 5% in agricultural soils (Rillig et al. 2003), even accounting for about 8% in tropical rainforests Andisols from Martinique in soil with high allophane content (Woignier et al. 2014), more than 10% in soils from highland Chilean Andisols forests (Seguel et al. 2008), and even near to 80–90% in soil highly contaminated with Cu in central Chile (Cornejo et al. 2008). Its original function was related with soil aggregation and C storage and also numerous studies have reported relationships between soil glomalin levels and soil aggregates stability (Wright and Upadhyaya 1996; Rillig 2004; Borie et al. 2008; Fokom et al. 2012; Qiang-Sheng et al. 2014) and glomalin and total SOC (Lovelock et al. 2004; Wang et al. 2015; Zhang et al. 2017). Recently, it has been postulated that glomalin has a dual functionality, primarily playing a key role in fungal physiology and secondarily the effects observed in soil aggregation and soil decontamination (Purin and Rillig 2007). Like SOM levels, glomalin stocks are governed by its production and decomposition rates, but environmental conditions could affect both fluxes independently (Rillig 2004). Accordingly, whereas glomalin production and accumulation is dependent of abundance and diversity of AMF, plant community composition and land use systems (Treseder and Turner 2007) its decomposition in turn is determined by mesofauna/microbial activity, but it has been shown that AM fungal hyphae is scarcely palatable for microarthropods (Klironomos and Kendrick 1996). This low palatability and the strong linkages with soil matrix are associated to its high recalcitrance being stabilized and accumulated in some soils.

Glomalin is quantified either by the non-specific Bradford assay for proteins or by ELISA test using an antibody against fresh fungal hyphae or spores (Wright and Upadhyaya 1996). The use of this antibody to identify pure fungal proteins has been questioned because it reacts positively to non-fungal proteins (Rosier et al. 2006). As many laboratories are not equipped to apply ELISA assay almost all reports have been carried out using the non-specific Bradford total protein assay. However, it has been demonstrated that other compounds extracted by so drastic conditions as soil autoclaving at 121 °C at pH 8.0 by 1 h, according the ad-hoc protocol, also react with Bradford reactive producing an overestimation of soil glomalin stocks (Araújo et al. 2015; Schindler et al. 2007; Whiffen et al. 2007; Purin and Rillig 2007).

The extracted contaminants complicates the characterization of the structure and composition of glomalin. As material extracted by citrate buffer is not exclusively glomalin (pure protein from AMF origin) it has been coined a new conceptual term for defining the whole similar compounds namely Glomalin Related Soil Protein (GRSP). However, here we will indistinctly use the name glomalin when referring to protein extracted from soils as well as the protein obtained from soilless systems. The basic extraction method originally reported involves the use of sodium citrate as extractant at different concentrations (0.2 M, 0.5 M), pHs (7.0, 8.0) and time of autoclaving (30 or 60 min) (Wright and Upadhyaya 1996, 1998). At mild extraction conditions easily extractable glomalin is obtained (EE-GRSP) suggesting a type of glomalin more labile or more recently produced whereas at stronger extraction conditions gives total glomalin levels (T-GRSP). Recent reports have shown that GRSP, in terms of chemical structure, constitutes a mixture of many compounds, some of them not related to AMF (Araújo et al. 2015).

Therefore, Schindler et al. (2007) applying NMR techniques to GRSP extracted from soils with different organic matter contents and purified by precipitation and dialysis concluded that spectra obtained resemble that of humic acids which are co-precipitated with proteinaceous material. In addition, Guillespie et al. (2011) using X-Ray absorption spectroscopy, pyrolysis-mass spectrometry and proteomics found that GRSP obtained directly from soils is a mixture of proteinaceous, humic, lipid and inorganic substances.

They conclude that in the soils analyzed and following the protocol of glomalin extraction detailed in Wright and Upadhyaya (1998) the protein as a product of AMF is poorly represented in the GRSP mixture. More recently, Wang et al. (2014, 2015) with the application of XR and SIR Spectroscopy reported that GRSP from plantations and forest from Eastern Chine consists of 49 fluorescent substances, 7 measures of functional groups and several elements including Si, Fe and Al.

Summarizing we agree with the authors reporting that is necessary to deep in obtaining more purified proteinaceous materials from soils (Guillespie et al. 2011; Kanerva et al. 2013) especially in those with high organic matter/humic acids contents as Andisols. However, we believe that still is possible to minimize extra co-extracted compounds by improvements in the extraction step by including: (1) ultrasonic stirring to break some SOM-clay interactions, (2) shortening the times for separation of centrifugate from solution, and (3) to compare the use of membrane dialysis tubing of 3.5 KD and 8.0 KD. Guillespie et al. (2011) propose an interesting area for future GRSP, which is the use of monoxenic carrot root hyphae of several AMF species for extensive proteomics fingerprinting. In the same context, we have extracted glomalin produced in the hyphosphere of plants grown in compartmentalized rhizoboxes using soilless as substrate (Aguilera et al. 2018) and inoculated with pure AMF species coming from our collection. Advanced instrumental methods including proteomic fingerprinting analysis like MALDI-TOF-MS (Kanerva et al. 2013) need to be applied to GRSP obtained from soils and glomalin obtained from pure AM fungal species.

Finally, all glomalin analyses obtained from soils and from axenic conditions must be contrasted to obtain more information about production and nature of this protein in acid soils and how it is affected by agronomical management, especially considering the use of usual rotation of host and non-host agricultural plant species.

7.4 General Statements

Plant growth in acid soils as Andisols is generally depressed by a complex of severe conditions where Al phytotoxicity and P-defficiency are the two more prominent stress agents. For overcoming such limitations farmers must use lime, Al-tolerant genotypes and P fertilizers for profitable yields. Therefore, in wheat (+Myc) production it is normally that farmers use wheat/lupine/wheat, wheat/rapeseed/wheat or wheat/barley/wheat as crop sequence in the rotation. But, lupine and rape are non-mycotrophic crops (−Myc) thereby affecting mycorrhizal fungal propagules left in the soil for subsequent crop and perhaps AMF diversity which is not the case for barley or oats, two +Myc crops. Moreover, we do not know how transient those negative events are. Our research group, using wheat as a plant model, have reported that mycorrhizal symbiosis not only gives higher tolerance to Al-phytotoxicity but also increases P acquisition by root plants. Those beneficial effects are much higher in Al-tolerant genotypes than in Al-sensitive ones. We have postulated that one of the main mechanisms operating in such stressed conditions is the production of glomalin, a glycoprotein produced by arbuscular mycorrhizal fungi, highly recalcitrant, which complex free Al decreasing their activity concomitantly with an increase in P availability. Glomalin also produces higher soil aggregation and higher water holding capacity. However, chemical structure of glomalin is still not well understood being a controversial topic due to the complexity of its extraction from soils without interferences but it is clear that is found in higher amounts in organic soils contributing to C storage and C stability, being a reliable indicator of soil health/fertility and functioning maintenance of the agroecosystems. Effects on P-use and mycorrhizal behavior on crop productivity when using −Myc crops in rotation sequences have been scarcely studied.

Acknowledgements Financial support of FONDECYT 11170641 (P. Aguilera), FONDECYT 1170264 (P. Cornejo), FONDECYT 11160385 (A. Seguel) and FONDECYT 1191551 (F. Borie) Grants from Comisión Nacional Científica y Tecnológica de Chile.

References

Aguilera P, Borie F, Seguel A, Cornejo P (2011) Fluorescence detection of aluminum in arbuscular mycorrhizal fungal structures and glomalin using confocal laser scanning microscopy. Soil Biol Biochem 43: 2427–2431

Aguilera P, Larsen J, Borie F et al (2018) New evidences on the contribution of arbuscular mycorrhizal fungi inducing Al tolerance in wheat. Rhizosphere 5: 43–50

Aguilera P, Marín C, Oehl F, Godoy R, Borie F, Cornejo P (2017) Selection of aluminum tolerant cereal genotypes strongly influences the arbuscular mycorrhizal fungal communities in an acidic Andosol. Agric Ecosyst Environm 246: 86–93

Aguilera P (2014) Diversity of arbuscular mycorrhizal fungi and their incidence in aluminum tolerance of *Triticum aestivum* L. growing in acidic soils with phytotoxic aluminum levels Doctoral Thesis, Universidad de La Frontera, 150 p

Aguilera P, Cornejo P, Borie F, Barea JM, von Baer E, Oehl F (2014) Diversity of arbuscular mycorrhizal fungi associated with *Triticum aestivum* L. plants growing in an Andosol with high aluminum level. Agr Ecosyst Environ 186: 178–184

Araújo PJ, Quiquampoix H, Matumoto-Pintro PT, Staunton S (2015) Glomalin-related soil protein in French temperate forest soils: interference in the Bradford assay caused by co-extracted humic substances. Eur J Soil Sci 66: 311–319

Arihara J, Karasawa T (2000) Effect of previous crops in arbuscular mycorrhizal formation on growth of succeeding maize. Soil Sci Plant Nut 46: 43–51

Azcón-Aguilar C, Barea JM (1996) Arbuscular mycorrhizas and biological control of soil-borne plant pathogens. An overview. Mycorrhiza 6: 457–464

Borie F, Rubio R, Morales A (2008) Arbuscular mycorrhizal fungi and soil aggregation. J Soil Sci Plant Nut 8: 9–18

Borie F, Rubio R, Curaqueo G, Cornejo P (2010) Arbuscular mycorrhizae in agricultural and forests ecosystems in Chile. J Soil Sci Plant Nut 10: 185–206

Bullock D (1992) Crop rotation. Crit Rev Plant Sci 11: 309–326

Brundrett M (2009) Mycorrhizal associations and other means of nutrition of vascular plants: understanding the global diversity of host plants by resolving conflicting information and developing reliable means of diagnosis. Plant Soil 320: 37–77

Castillo C, Borie F, Oehl F, Sieverding E (2016) Arbuscular mycorrhizal fungal biodiversity: prospecting in Southern- Central Zone of Chile. A review. J Soil Sci Plant Nut 16: 400–422

Castillo C, Rubio R, Borie F, Sieverding E (2010) Diversity of arbuscular fungi in horticultural production systems in Souhtern Chile. J Soil Sci Plant Nut 10:407–413

Cornejo P, Meier S, Borie G, Rillig M, Borie F (2008) Glomalin-related soil protein in a Mediterranean ecosystem affected by a copper smelter and its contribution to Cu and Zn sequestration. Sci Total Environ 406: 154–160

Cornejo P, Pérez-Tienda J, Meier S, Valderas A, Borie F, Azcón-Aguilar C, Ferrol N (2013) Copper compartmentalization in spores as survival strategy of arbuscular mycorrhizal fungi in copper-polluted environments. Soil Biol Biochem 57: 925–928

Cumming J, Ning J (2003) Arbuscular mycorrhizal fungi enhance aluminum resistance of broomsedge (*Andropogon virginicus* L). J Exp Bot 54: 1447–1459

Driver J, Holben W, Rillig M (2005) Characterization of glomalin as a hyphal wall component of arbuscular mycorrhizal fungi. Soil Biol Biochem 37: 101–106

Ferrufino A, Smyth T, Israel D, Carter E (2000) Root elongation of soybean genotypes in response to acidity by constraints in a subsurface solution compartment. Crop Sci 40:413–421

Fokom R, Adamou S, Teugwa MC, Begoude Boyoguenob AD, Nanaa WL, Ngonkeu MEL, Tchameni NS, Nwaga D, Tsala Ndzomo G, Amvam Zollo PH (2012). Glomalin-related soil protein, carbon, nitrogen, and soil aggregate stability as affected by land use variation in the humid forest zone of South Cameroon. Soil Till Res126: 69–75

Guillespie A, Farrel L, Walley F, Ross A, Leinweber P, Eckhardt K, Regier T, Blyth R (2011) Glomalin-related soil protein contains non-mycorrhizal related heat stable proteins, lipids and humic materials. Soil Biol Biochem 43: 766–777

Hinsinger P (2001) Bioavailability of soil inorganic P in the rhizosphere as affected by root-induced chemical changes: a review. Plant Soil 237: 173–195

Jansa J, Mozafar A, Anken T, Ruh R, Sanders I R, Frossard E (2002) Diversity and structure of AMF communities as affected by tillage in a temperate soil. Mycorrhiza 12: 225–234

Jansa J, Mozafar A, Kuhn G, Anken T, Ruh R, Sanders R, Frossard E (2003) Soil tillage affects the community structure of mycorrhizal fungi in maize roots. Ecol Appl 13(4): 1164–1176

Javaid A (2007) Allellophatic interactions in mycorrhizal associations. Allellophaty J, 20: 29–42.

Kabir Z (2005) Tillage or no-tillage: impact on mycorrhizae. Can J Plant Sci 85: 23–29

Kanerva S, Smolander A, Kitunen V, Ketola RA, Kotiaho T (2013) Comparison of extractants and applicability of MALDI-TOF- MS in the analysis of soil proteinaceous materials from different types of soil. Organic Geochem 56: 1–9

Karasawa T, Kasahara Y, Takebe M (2001) Variable response of growth and arbuscular mycorrhizal causing by fluctuation in the populations of indigenous arbuscular mycorrizal fungi. Soil Biol and Biochem 34: 851–857

Karasawa T, Kasahara Y, Takebe M (2002) Differences in growth responses of maize to preceding cropping caused by fluctuation in the population of indigenous arbuscular mycorrhizal fungi. Soil Biol Biochem 34(6): 851–857

Kochian L, Piñeros M, Hoekenga O (2005) How do crop plants tolerate acid soils?. Mechanisms of aluminum tolerance and phosphorus efficiency. Ann Rev Plant Biol 55: 459–493

Klironomos J, Kendrick W (1996) Palability of microfungi to soil arthropods in relation to the functioning of arbuscular mycorrhizae. Biol Fert Soils 21: 43–52

Koide R, Peoples M (2013) Behavior of Bradford-reactive substances is consistent with predictions for glomalin. Appl Soil Ecol 63: 8–14

Lambers H, Shane W, Cramer M, Pearse S, Veneklaas E (2006) Root structure and functioning for efficient acquisition of phosphorus: Matching morphological and physiological traits. Ann Bot 98: 693–213

Lambers H, Teste F (2013) Interactions between arbuscular mycorrhizal and non-mycorrhizal plants: do mycorrhizal species at both extremes of plant availability play the same play. Plant Cell Environ 36: 1911–1919

Lavaud C, Voutquenne L, Bal P, Pouny I (2000) Saponins from Chenopodium album. Fitoterapia 71: 338–340

Leake J, Jhonson D, Donnelly D, Muckle G, Boddy L, Read D (2004) Netwoks of power and influence: the role of mycorrhizal mycelium in controlling plant communities and agrosystem functioning. Can J Bot 82: 1016–1045

Lenoir I, Fontaine J, Sahraoui A (2016) Arbuscular mycorrhizal fungal responses to abiotic stresses: a review. Phytochemistry 123: 4–15

Lovelock C, Right S, Clark D, Ruess R (2004) Soil stocks of glomalin produced by arbuscular mycorrhizal fungi across a tropical rain forest landscape. J Ecol 92: 278–287

Lupwayi N, Rice W, Clayton G (1998) Soil microbial diversity and community structure under wheat as influenced by tillage and crop rotation. Soil Biol Biochem 30: 1733–1741

Meier S, Borie F, Curaqueo G, Bolan N, Cornejo P (2012) Effects of arbuscular mycorrhizal inoculation on methallophyte and agricultural plants growing at increasing copper levels. Appl Soil Ecol 61: 280–287

Miransari M (2011) Hiperaccumulators, arbuscular mycorrhizal fungal and stress of heavy metals. Biothec Adv 29: 645–653

Navarro-Noya YE, Gómez-Acata S, Montoya-Ciriaco N, Rojas-Valdez A, Suárez-Arriaga MC, Valenzuela-Encinas C, Jiménez-Bueno N, Verhulst N, Govaerts B, Dendooven L (2013) Relative impacts of tillage, residue management and crop-rotation on soil bacterial communities in a semi-arid agroecosystem. Soil Biol Biochem 65:86–95

Oehl F, Laczko E, Bogenrieder A, Stahr K, Bösch R, van der Heijden M, Sieverding E (2010) Soil type and land use intensity determine the composition of arbuscular mycorrhizal fungal communities. Soil Biol Biochem 42: 724–738

Öpik M, Moora M, Liira J, Zobel M (2006) Composition of root-colonizing arbuscular mycorrhizal fungal communities in different ecosystems around the globe. J Ecol 94: 778–790

Potsma-Blaauw M, de Goede R, Bloem J, Faber J, Brussaard L (2010) Soil biota community structure and abundance under agricultural intensification and extensification. Ecology 91: 460–463

Purin S, Rillig M (2007) The arbuscular mycorrhizal fungal protein glomalin: limitations, progress, and a new hypothesis for its function. Pedobiologia 51: 123–130

Qiang-Sheng W, Ming-Qin C, Ying-Ning Z, Xin Hua H (2014) Direct and indirect effects of glomalin, mycorrhizal hyphae, and roots on aggregate stability in rhizosphere of trifoliate orange. Nature DOI:https://doi.org/10.1038/srep05823.

Richardson A, Lynch E, Jonathan P, Delhaize E, Smith F, Smith S, Harvey P, Ryan M, Veneklaas E, Lambers H, Oberson A, Culbernor R, Simpon R (2011) Plant and microbial strategies to improve the phosphorus efficiency of agriculture. Plant Soil 349: 121–156

Rillig M, Mummey D (2006) Mycorrhizas and soil structure. New Phytol 171: 41–53

Rillig M (2004) Arbuscular mycorrhizae, glomalin and soil aggregation. Can J Soil Sci 84: 355–363

Rillig M, Ramsey P, Morris S, Paul E (2003). Glomalin, an arbuscular mycorrhizal fungal soil protein, responds to land-use change. Plant Soil 253: 293–299

Rosier C, Hoye A, Rillig M (2006) Glomalin-related soil protein: assessment of current detection and quantification tools. Soil Biol Biochem 38: 2205–2211

Ryan P, Delhaize E, Jones L (2001) Function and mechanisms of organic anion exudation from plant roots. Annual Review of Plant Physiology Plant Molecular Biology 52: 527–560

Schindler F, Mercer E, Rice J (2007) Chemical characteristics of glomalin-related soil protein (GRSP) extracted from soils of varying organic matter content. Soil Biol Biochem 39: 320–329

Schreiner R, Koide R (1993) Antifungal compounds from the root of mycothrophic and non-mycothrophic plant species. New Phytol 123: 99–105

Seguel A, Rubio R, Carrillo R, Borie F (2008) Levels of glomalin and their relation with soil chemical and soil and biological soil characteristics in a relict of native forest of Southern Chile. Bosque 29: 11–22

Seguel A, Cumming J, Klug-Stewart K, Cornejo P, Borie F (2013) The role of arbuscular mycorrhizas in decreasing aluminium phytotoxicity in acidic soils: a review. Mycorrhiza 23: 167–183

Seguel A, Cumming J, Cornejo P, Borie F (2016) Aluminum tolerance of wheat cultivars in a non-limed and limed Andisol. Appl Soil Ecol 108: 228–237

Seguel A, Cornejo, Ramos A, von Baer E, Cumming J, Borie F (2017) Phosphorus acquisition by three wheat cultivars contrasting in aluminum tolerance growing in an aluminum-rich Andisol. Crop Pasture Sci 68: 315–316

Smith S, Read D (2008) Mycorrhizal symbiosis. Cambridge, UK: Academic Press.

Smith S, Smith F (2013) How useful is the mutualism-parasitism continuum of arbuscular mycorrhizas functioning?. Plant Soil 349: 121–156

Smith R, Gross K, Robertson G (2008) Effects on crop diversity on agroecosystem function. Ecosystems 11: 355–366

Tiemann L, Grandy A, Atkinson, Marin E, McDaniel M (2015) Crop rotational diversity enhances belowground communities and function in an agroecosystem. Ecol Lett 18: 761–771

Torrecillas E, Alguacil M, Roldan A (2012) Host preferences of arbuscular mycorrhizal fungi colonizing annual herbaceous plant species in semiarid Mediterranean prairies. Appl Environ Microbiol 78: 6180–6186

Treseder K, Turner K (2007) Glomalin in ecosystems. Soil Science Society of America Journal 71: 1257–1266

Valetti L, Iriarte L, Fabra A (2016) Effect of previous cropping of rapeseed (Brassica napus L) on soybean (Glycine max) root mycorrhization, nodulation and plant growth. Eur J Soil Biol 76: 103–106 van der Heijden M, Bardgett R, van Straalen N (2008) The unseen majority: Soil microbes as drivers of plant diversity and productivity in terrestrial ecosystems. Ecol Lett 11: 296–310

van der Heijden M, Klironomos J, Ursic M, Moutoglis P, Streitwolf-Engel R, Boller T, Wiemken A, Sanders IR (1998) Mycorrhizal fungal diversity determines plant biodiversity, ecosystem variability and productivity. Nature 396: 69–72

Verbruggen E, Röling W, Gamper H, Kowalchuk G, Verhoef H, van der Heijden M (2010) Positive effects of organic farming on below- groundmutualists: large-scale comparison of mycorrhizal fungal communities in agricultural soils. New Phytol 186: 968–979

Villagarcia M, Thomas E, Carter J (2001) Genotypic rankings for aluminum tolarnace of soybean roots grown in hydroponics and sand cultura. Crop Sci 41: 1499–1507

Wang Q, Wu Y, Wang W, Zhong Z, Pei Z, Ren J, Wang H, Zu Y (2014) Spacial variations in concentration, compositions of glomalin related soil protein in poplar plantations in Northeastern China, and possible relations with soil physicochemical properties. The Scientific World Journal Volume 2014, Article ID 160403, 13 pages

Wang Q, Wang W, He X (2015) Role and variation of the amount and composition of glomalin in soil properties in farmland and adjacent plantations with reference to a primary forest in North-Eastern China. PLoS One DOI: https://doi.org/10.1371/journal.pone.01 39623. October 2, 2015 (19 pages)

Wardle D, Bardgett R, Klironomos J (2004) Ecological linkages between aboveground and belowground biota. Science, 304 (5677): 1629–1633

Wagg C, Franz Bender S, Widmer F (2014) Soil biodiversity and soil community composition determines ecosystem multifunctionality. Pro. Natl Acad. Si USA 111: 5266–5670.

Whiffen L, Midley D, Mc Gee P (2007) Polyphenolic compounds interfere with quantification of protein extracts using the Bradford method. Soil Biol Biochem 39: 691–694

Wright S, Upadhyaya A (1996) Extraction of an abundant and unusual protein from soil and comparison with hyphal protein of arbuscular mycorrhizal fungi. Soil Sci 161: 575–586

Wright S, Upadhyaya A (1998) A survey of soils fro aggregate stability and glomalin, a glycoprotein produced by hyphae of arbuscular mycorrhizal fungi. Plant Soil 198: 97–107

Woignier T, Etcheverria P, Borie F (2014) Role of allophanes in the accumulation of glomalin-related soil protein in tropical soils (Martinique, French West Indes) Eur J Soil Sci 65: 531–538

Yin C, Jones L, Peterson D et al (2011) Members of soil bacterial communities sensitive to tillage and crop rotation. Soil Biol Biochem 42: 2111–2118

Zhang J, Tang X, Zhong S (2017) Recalcitrant carbon components in glomalin-related soil protein facilitate soil organic carbon preservation in tropical forests. Nature https://doi.org/10.1038/s41598-017-02486-6

Chapter 8
Ecology and Biogeography of Arbuscular Mycorrhizal Fungi Belonging to the Family Gigasporaceae in La Gran Sabana Region (Guayana Shield), Venezuela

Milagros Lovera, Gisela Cuenca, Pablo Lau, and Jesús Mavárez

8.1 Introduction

La Gran Sabana (LGS) is a large plateau in southeast Venezuela. It is part of the Canaima National Park, a protected area characterized by a high diversity and endemism of both plants and animals (Huber 1994; Berry et al. 1995). La Gran Sabana is located within 4° 30′ – 6° 45′ N and 60° 34′ – 62° 50′ W with altitudes from 1440 meters above sea level (m.a.s.l) in the North to 800 m.a.s.l. in the South (Fig. 8.1). With a complex mosaic of ecosystems, it is dominated by open savannas intermixed with other vegetation formations like forests, shrublands, meadows and palm swamps. The presence of sandstone mountains with a flat summit and vertical walls that can reach up to 3000 m.a.s.l. (*tepuis*), are a distinctive feature of the area. The parental substrate of the Guayana Shield, to which LGS belongs, are Precambrian quartzite and sandstone, one of the oldest materials on Earth (Briceño and Schubert 1990). High precipitation levels (2500 mm/year) and a very ancient parental substrate results in very acidic soils, highly weathered and low in nutrients (Fölster and Dezzeo 1994). In this harsh environment, the associations with soil microorganisms are crucial for the survival of plants.

M. Lovera (✉) · G. Cuenca
Centro de Ecología, Instituto Venezolano de Investigaciones Científicas (IVIC), Caracas, Venezuela

P. Lau
Centro de Agroecología Tropical, Universidad Nacional Experimental Simón Rodríguez (UNESR), Caracas, Venezuela

J. Mavárez
Laboratoire d'Ecologie Alpine, UMR 5553 CNRS-Université Grenoble Alpes, Grenoble, France

Departamento de Ciencias Biológicas y Ambientales, Universidad Jorge Tadeo Lozano, Bogotá, Colombia

M. C. Pagano, M. A. Lugo (eds.), *Mycorrhizal Fungi in South America*, Fungal Biology, https://doi.org/10.1007/978-3-030-15228-4_8

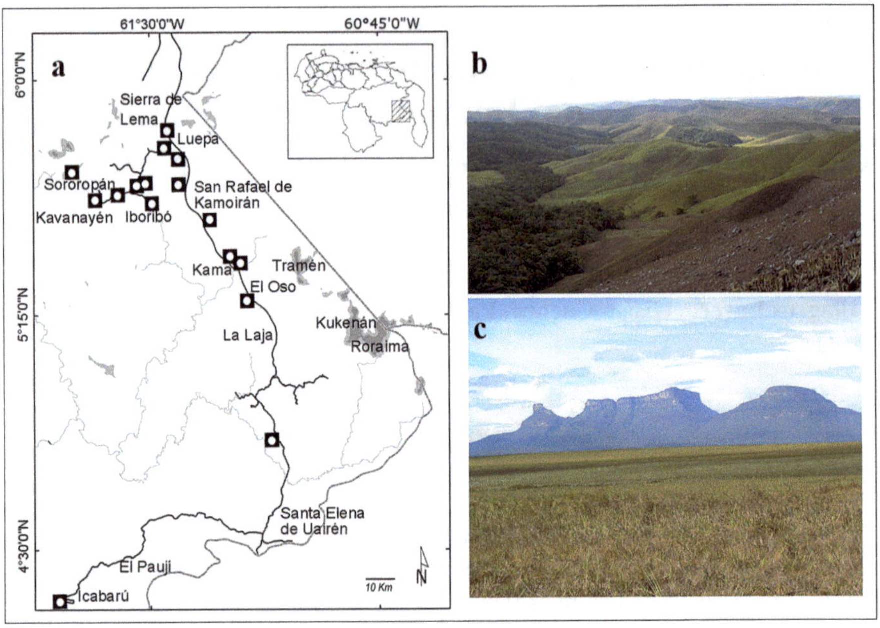

Fig. 8.1 (**a**) Location of La Gran Sabana (LGS). Squares indicate sampling site locations, (**b**) View of LGS landscape, and (**c**) Eastern *tepui* chain with Ilú-tepui at South and Tramen-tepui at North

Arbuscular mycorrhizas (AM) are ancient relations between soil fungi belonging to Glomeromycota phylum and the majority of land plants (Schüβler et al. 2001). In this association, the fungi depend on carbon supplied by host plants while the plants receive many benefits which include among others, a better P, N, Zn and Cu nutrition, increased tolerance to water stress, protection against root pathogens and enhanced soil structure (Kiers et al. 2011). In LGS oligotrophic soils, many indigenous plants are expected to depend on mycorrhizas for survival. Indeed, most of the plants evaluated so far in this region, present high levels of mycorrhizal colonization (Cuenca and Lovera 1992; Lovera and Cuenca 1996; Rosales et al. 1997). Savannas have been present on LGS for at least 10000 years (Rull et al. 2016). Its area has increased at forests expense due to recurrent burning and other anthropic disturbances, including soil removal and mining that cause serious problems as soil erosion, water contamination and damage to plant communities.

Since the 1990s, our research group has conducted several studies in LGS aimed at assessing the diversity of arbuscular mycorrhizal fungi (AMF) present in both natural and disturbed areas. Findings indicated an important loss of AMF diversity produced by anthropic disturbances (Cuenca and Lovera 1992; Lovera and Cuenca 1996; Cuenca et al. 1998; Lovera and Cuenca 2007). The inventories carried out in the region indicate the existence of a high diversity of AMF and have allowed the detection of a large number of new species, many of which belong to the

Gigasporaceae family (Cuenca et al. 2003). This family has been associated to sandy soils and it is particularly efficient transferring P to the plant (Lekberg et al. 2007; Chagnon et al. 2013; Veresoglou et al. 2013), being very frequent in the oligotrophic soils characteristics of LGS. So far, 4 new species belonging to the family Gigasporaceae have been described in LGS: *Scutellospora spinosissima* (Walker et al. 1998), *S. crenulata* (Herrera-Peraza et al. 2001), *S. striata* (Cuenca and Herrera-Peraza 2008) and *S. tepuiensis* (De Andrade et al. 2017). Additionally, other AMF morphotypes present in the region have unique characteristics, which are consistent with the idea that a significant proportion of the Glomeromycota diversity is expected to be found in the tropics (Cuenca and Lovera 2010; Chaudhary et al. 2017). Most of the new species of Gigasporaceae described for LGS have not been detected until now outside this region. Some of them are restricted to specific environments on a local scale and others are found coexisting in different ecosystems present throughout the study area. In the case of *S. spinosissima*, the restriction in its range of distribution is considered to be regional since this species has also been found in the Colombian Amazonian area (Peña-Venegas et al. 2006) and the Northeast of Brazil (Pereira et al. 2018). It is generally considered that microorganisms do not have dispersion restrictions, and therefore their taxa tend to be globally distributed (Finlay 2002). However, studies on the distribution of AMF on a global scale have found the existence of some patterns in their distribution among continents, climatic zones and ecosystems (Öpik et al. 2010; Kivlin et al. 2011; Stürmer et al. 2018a). Other studies such as Davison et al. (2015), find that AMFs have a predominantly cosmopolitan distribution, with a low proportion of endemism.

In this framework, the aim of this work is to evaluate the presence of the Gigasporaceae family in LGS in order to identify biotic and abiotic factors that could be involved in the presence of many endemism of this group of AMF in this region of Venezuelan Guayana Shield. Additionally, information on the phylogenetic relationships of some species described in LGS region (*S. spinosissima*, *S. crenulata* and *S. striata*) and two putative new species of *Scutellospora*, is included in order to explore the influence of its evolutionary history on the restricted nature of their distribution.

8.2 Methods

8.2.1 Study Area and Characteristics of the Biomes Evaluated

The presence or absence of Gigasporaceae was evaluated in different localities of LGS visited by our research group in previous field work. These localities covering the main types of natural ecosystems present in the region (savannas, shrublands, forests, herbaceous meadows and palm swamps). Most sampling sites were located in the northern sector of the plateau (Fig. 8.1), between 1000–1400 m.a.s.l. Ecosystems are sub-mesothermal mountainous with mean annual temperature between 18–24 °C. The most outstanding features of the natural ecosystems evaluated are the following:

Savannas These are a mixture of grasses and sedges without woody elements, with a continuous herbaceous cover. Principal plant species are *Trachypogon spicatus, Axonopus anceps, Paspalum carinatum, Leptocoryphium lanatum* (Poaceae) and *Rhynchospora barbata, Hypolytrum pulchrum, Scleria cyperina, Bulbostylis paradoxa, Lagenocarpus rigidus* (Cyperaceae), among others. Soils are oxisols or entisols, very low in nutrients and strongly leached (Berry et al. 1995; Huber 2006).

Herbaceous meadows These are herbaceous-type savannas where the dominant species are not grasses but broadleaf herbs. They are restricted in their extension and are always associated with humic soils (peat) saturated with water during a large part of the year. This type of ecosystem is dominated by species of Rapateaceae family as *Stegolepis ptaritepuiensis* and *S. guianensis*, rossete plants of Xyridaceae and Bromeliaceae families and subshrubs belonging to Rubiaceae and Ochnaceae. Its flora is related to the highland-tepuian flora and has a high proportion of plant endemism (Huber 1995).

Sclerophyllous shrubs This type of ecosystem grows on rocky sandstone substrates, or on deep sand areas of alluvial origin (Huber 1994). It presents a woody component of low height (2–3 m). Its flora is largely autochthonous, with many endemic plant species. It consists mainly of slow growth sclerophyllous leaf shrubs adapted to edaphic environments with severe water and nutritional deficiencies. Plant species that occur frequently are: *Bonnetia sessilis* (Bonnetiaceae), *Clusia pusilla* (Clusiaceae), *Gongylolepis benthamiana* (Asteraceae), *Euphronia guianensis* (Vochysiaceae) (Huber 1995).

Forests Growing generally associated to diabase outcrops, adjacent to lowland bottoms of *tepuis* and areas bordering water bodies. They are mostly in the form of islands of variable size, medium height (15–25 m) and form a closed canopy. The dominant families are usually Legumes, Lauraceae, Vochysiaceae, Rubiaceae, Annonaceae and Burseraceae. In the understorey, shrubs, palms and herbs (heliconia type) occur (Huber 1995).

Palm swamps This type of ecosystem is formed by large patches of the palm *Mauritia flexuosa* (Arecaceae) that grow on alluvial plains covered by seasonally flooded savannas. They develop at lower altitude (750–1000 m.a.s.l.). In the herbaceous stratum, occur frequently *Hypogynum virgatum*, *Andropogon* spp. and *Panicum* spp. (Poaceae), *Rhynchospora* and *Bulbostylis* spp. (Cyperaceae), and other herbs such as *Waltheria* spp. (Sterculiaceae). There are also low shrubs of the families Melastomataceae, Clusiaceae and Piperaceae (Huber 1995).

8.2.2 Collection of Presence Data of Gigasporaceae in LGS

The information about the presence of the family Gigasporaceae in LGS was obtained from the Glomeromycotan Herbarium of Venezuela (HGV), which harbors the reference specimens and the ecological information of the localities visited by our research group along two decades in this region. Many of the specimens of the HGV herbarium were re-evaluated in order to verify their identification according to the current status of the AMF taxonomy. The classification of Redecker et al. (2013) was followed together with other specialized literature: original published descriptions of the species, INVAM at the West Virginia University, USA (http://invam.wvu.edu), Department of Plant Pathology, University of Agriculture in Szczecin, Poland (http://www.agro.ar.szczecin.pl/~jblaszkowski/) and Blaszkowski (2012).

The Gigasporaceae species present in LGS were divided into two groups according to its geographical distribution range for comparative purposes: (1) **Endemic distribution (ED):** species restricted to South America (include local and regional endemism present in LGS), and (2) **Wide distribution (WD):** species which have been documented in South America and also in other continents or in the North American subcontinent. Gigasporaceae species were classified as ED or WD on the basis of information available in the public databases MaarjAM (Öpik et al. 2010) and BD (Stürmer et al. 2018a) and scientific literature including AMF inventories, however this classification should be considered as provisional, subject to changes as new geographic regions are evaluated.

8.2.3 Statistical Analyses

The proportion of species with endemic distribution for different biomes was compared using a Pearson chi-square test for categorical data. Relationships between soil properties and species composition of Gigasporaceae were explored using a canonical correspondence analysis with Montecarlo permutation test (*cca function*, vegan package 2.5–2, R, Oksanen et al. 2018).

8.3 Gigasporaceae in Different Biomes of LGS

Through the revision of the HGV herbarium, 18 species of Gigasporaceae are present in LGS, representing 34% of family diversity. The records of 5 scutellosporoid morphotypes that are considered putative new species were included in this analysis, adding a total of 23 species for the region. Sclerophyllous shrublands harbored the greatest diversity of Gigasporaceae, with 82% of the total species present in LGS (Table 8.1).

All endemic species of Gigasporaceae (ED group) was found in sclerophyllous shrublands being *S. spinosissima*, *S. crenulata* and *Scutellospora* sp.$_{1}$ the most frequent species in this biome (Table 8.1). This result is coincident with the proposal of *S. spinosissima* and *S. crenulata* as indicator species for shrublands of LGS (Chaudhary et al. 2017).

The predominance of ED group species in the herbaceous meadows, suggests ecological associations of the ED Gigasporaceae with this particular biome. However, scarce information available in the database for herbaceous meadows, and especially for palm swamps, represented a limitation for the analysis of these ecosystems. Some species such as *S. tepuiensis*, *Scutellospora* sp.$_{3}$, *Scutellospora* sp.$_{5}$ were restricted to one type of biome (shrubland) and in the case of *Scutellospora* sp.$_{5}$ and *S. tepuiensis* were restricted to a single locality. *S. tepuiensis* has been found only at the top of the Sororopan *tepui*, it would be very interesting to carry out samplings in other *tepuis* of LGS to establish if its distribution is more widespread in highland *tepuian* ecosystems. Other Gigasporaceae species present in the Sororopan *tepui, S. spinosissima* and *R. tropicana*, are also found in other ecosystems of LGS. In the case of *R. tropicana*, it is a species belonging to the WD group that is also found in Africa.

Table 8.1 Relative frequency of Gigasporaceae species with endemic or wide distribution present in different biomes of LGS and mean soil properties of the sites where Gigasporaceae were collected. The number of localities evaluated is shown in parentheses below each biome

			Savanna (6)	Meadow (3)	Shrubland (9)	Forest (5)	Palm swamp (1)
		HMA	**Relative frequency**				
Endemic distribution (ED)	*Scre*	*Scutellospora crenulata*		0.33	**0.66**		1.00
	Sspi	*Scutellospora spinosissima*	0.33	**0.66**	**1.00**	0.40	
	Sstr	*Scutellospora striata*	0.17		0.22		
	Step	*Scutellospora tepuiensis*			0.11		
	Ssp1	*Scutellospora* sp.$_{1}$	0.33		**0.66**	0.20	1.00
	Ssp2	*Scutellospora* sp.$_{2}$	0.17	**0.66**	0.44	**0.60**	
	Ssp3	*Scutellospora* sp.$_{3}$			0.22		
	Ssp4	*Scutellospora* sp.$_{4}$		0.33	0.22	0.20	
	Ssp5	*Scutellospora* sp.$_{5}$			0.11		

(continued)

Table 8.1 (continued)

			Savanna (6)	Meadow (3)	Shrubland (9)	Forest (5)	Palm swamp (1)
Wide distribution (WD)	*Cgil*	*Cetraspora gilmorei*	0.17			0.20	
	Dbio	*Dentiscutata biornata*	**0.66**[a]		0.22		
	Dcer	*Dentiscutata cerradensis*			0.11		
	Dret	*Dentiscutata reticulata*	0.17		0.33		
	Dsav	*Dentiscutata savannicola*	0.33		0.33	0.20	
	Gdec	*Gigaspora decipiens*	0.33			0.20	
	Ggig	*Gigaspora gigantea*			0.44	0.40	
	Gmar	*Gigaspora margarita*			0.11	0.20	
	Gram	*Gigaspora ramisporophora*			0.11	0.40	
	Rtro	*Racocetra tropicana*	0.17	0.33	0.22	0.40	
	Sare	*Scutellospora arenicola*	0.17		0.22		
	Scal	*Scutellospora calospora*	0.17		**0.66**	0.40	1.00
	Sdip	*Scutellospora dipapillosa*					1.00
	Sper	*Scutellospora pernambucana*				**0.80**	
		Species richness	12	5	19	13	4
		Soil properties	**Mean values (Standar deviation)**				
		pH	5.3 (0.4)	5.1 (0.6)	5.2 (0.5)	4.8 (0.4)	
		OM (%)	3.9 (1.6)	17.1 (5.1)	10.4 (7.2)	6.3 (3.1)	
		N (%)	0.2 (0.1)	0.5 (1.0)	0.2 (0.6)	0.2 (1.5)	
		P (μg/g)	1.2 (1.0)	1.1 (1.0)	1.3 (0.6)	4.4 (1.5)	
		Sand (%)	65.2 (11.7)	80.0 (0.0)	88.6 (7.0)	83.3 (5.1)	

[a]Frequency values greater than 0.50 are indicated in bold, with the exception of the palm swamp ecosystem in which only one location was evaluated

Biogeographical data obtained by Stürmer et al. (2018a) points to *Scutellospora* as one of the genera with the highest proportion of endemic species (>50%). This is a trend strongly supported by our findings, given that the most of the ED Gigasporacea found in LGS belongs to the genus *Scutellospora*. Additionally, the remarkable richness of Gigasporaceae species in LGS is according with the overrepresentation of this family in South America found by Stürmer et al. (2018a). In shrublands and forests of LGS, the Gigasporaceae family represents 26–32% of the species richness within each location (Cuenca and Lovera 2010; Chaudhary et al. 2017), while even greater dominance of Gigasporaceae (50–87%) has been found in the savannas of Roraima (Brazil) (Stürmer et al. 2018b). Interestingly, both regions are part of the Guayana Shield, which suggests that could be an association of Gigasporaceae to the particular ecological conditions of this geological basement.

8.4 Ecological Factors Associated with the Presence of Endemic Gigasporaceae in LGS

8.4.1 Vegetation

The presence of ED Gigasporaceae in the different biomes evaluated could be associated with their levels of plant endemism. Both shrublands and herbaceous meadows have a higher proportion of endemic plants than other biomes in LGS (Berry et al. 1995). Then, it was interesting to perform a Contingency Table Analysis (CTA) to evaluate if the proportion of ED Gigasporaceae species is greater in these biomes in comparison with the one found in the other types of vegetation of the region (savannas, forests and palm swamps). This comparison evidenced that ED species are present in shrublands and meadows in a significantly higher proportion than in other vegetation types pooled together (Pearson Chi-Square = 8.090, df = 1, P = 0.004) (Fig. 8.2). This result supports earlier findings obtained by Öpik et al. (2009) in which endemic putative AM fungi were associated with host plant species with narrower ecological and geographical ranges suggesting that preferential associations with endemic plants present in meadows and shrublands could have played a significant role in restricting the distribution of some of the ED Gigasporaceae species of LGS. This hypothesis could be tested in the future using molecular tools to characterize the AMF community associated with the group of endemic plants of these ecosystems.

Differences in the structure of AMF communities found in ecosystems with different plant communities composition are commonly associated with variations in plant-fungus function compatibility (Öpik et al. 2010; Kiers et al. 2011). However, the existence of specificity by the host in arbuscular mycorrhizal association is still a subject of debate. It has been proposed that the specificity of the mycorrhizal association in relation to the host plant can be better understood at a broader level, in ecological groups of plants, implying that plants with similar traits select symbi-

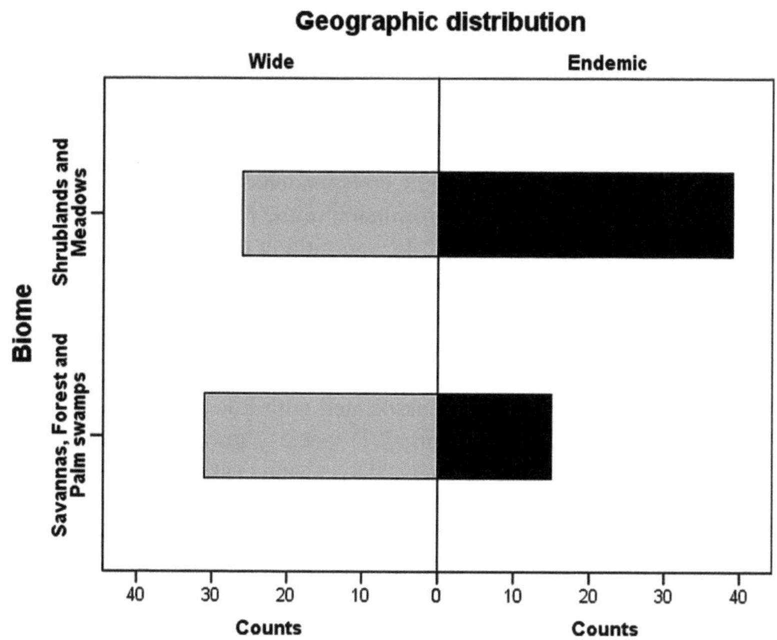

Fig. 8.2 Presence records of the group of species of Gigasporaceae with wide and restricted range of distribution present in different ecosystems of LGS

onts that complement these traits (Öpik et al. 2009; Chagnon et al. 2015). Most of the ED Gigasporaceae species in this study, could be kept in trap pots only when they were found associated with native plants from herbaceous meadows (*Stegolepis* sp.) and shrubs (*Clusia pusilla*, *Eucerea nitida*, *Gongylolepis benthamiana*, *Calliandra resupina*, *Pagameopsis garryoides*, *Macairea parviflora* and *Humiria balsamifera*), plants that share adaptation strategies to edaphic environments with strong nutritional deficiencies.

8.4.2 Soils

The presence of AMF can be strongly linked with soil properties such as pH gradients, nutrient content and soil texture. In particular, oligothrophic soils and sandy texture are conditions favourable for the Gigasporaceae family (Landis et al. 2004; Lekberg et al. 2007; Cuenca and Lovera 2010). The species of this family develop a greater volume of extra-radical mycelium and, therefore, have a greater ability to capture water and nutrients, which makes them successful in environments where these resources are scarce (Hart and Reader 2002).

Soils in LGS are generally acidic, sandy and particularly low in available phosphorus, nevertheless sclerophyllous shrublands and herbaceous meadows grow on specific soil patches with singular characteristics: the former are found in deep sand areas and rock outcrops, while the latter grows on peat bogs with high levels of organic matter and nitrogen. These observations allow suggesting that the higher proportion of ED Gigasporaceae species in shrublands and meadows in LGS (Fig. 8.2) could also be related to the particular edaphic conditions of these ecosystems. To explore this idea, a Canonical Correspondence Analysis (CCA) was made on the set of records with edaphic information in the HGV database. It was found that the variables that explain most of the variation in the ordination of the species were the sand content, phosphorus and nitrogen levels (Fig. 8.3). Organic matter was not included in the analysis because it had low significance and a high correlation with the percentage of nitrogen in previous analyses.

Edaphic variables allowed segregation in the ordination of the ED and WD groups. In general, ED species were associated with lower phosphorus levels and higher sand and nitrogen contents, while WD species showed an opposite tendency. The ED species that were mostly influenced by edaphic variables were *S. tepuiensis*, *S. crenulata*, *S. striata*, *Scutellospora* sp.$_4$ and *Scutellospora* sp.$_5$. The restricted geographic distribution in these species could be linked to their adaptations to the extreme limit of the range of soil textures and low availability of phosphorus that constitutes the niche of the family. The association between some ED species and high levels of nitrogen is interesting; however, N and sand vectors have similar directions in the analysis, so the interpretation of this result must be made with caution. Gigasporaceae are predominantly found in environments with low levels of nitrogen and organic matter (Landis et al. 2004; Stürmer et al. 2018b), nevertheless, N: P stoichiometry can also influence AM fungal communities in according to Johnson (2010). In the case of shrublands and meadows in LGS, prevalent condition of low availability of P plus high N seems to favor the presence of several species of the ED group of Gigasporaceae.

Scutellospora spinosissima, *Scutellospora* sp.$_1$ y *Scutellospora* sp.$_2$, show a lower influence of the edaphic variables (Fig. 8.3). In the case of *S. spinosissima*, its greater adaptability to different edaphic conditions in LGS soils could be related with its more widespread distribution. So far it is the only species of the ED group of Gigasporaceae that has been found outside of LGS.

The others two localities with presence records for *S. spinosissima* in South America are: Leticia (an adjacent region to the Guayana Shield in the Colombian Amazon) and the Atlantic Forest in North-eastern Brazil, a region with very poor and lixiviated soils that belongs to the Atlantic Shield (Peña-Venegas et al. 2006; Pereira et al. 2018). The finding of *S. spinosissima* in another geological shield of South America could have important implications from a biogeographical point of view. Considering that ecologically similar areas around LGS and Northeast of Brazil have been little explored regarding the presence of AMF, it is feasible that *S. spinosissima* and other species of ED Gigasporaceae group could indeed have a more widespread distribution in this region.

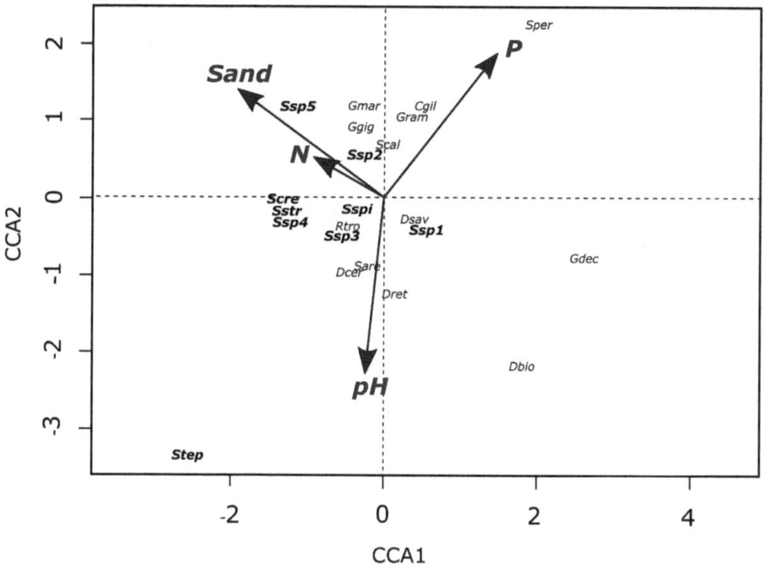

Fig. 8.3 Canonical Correspondence Analysis (CCA) ordination plot of species of Gigasporaceae in La Gran Sabana (LGS). Species with restricted distribution (ED) are highlighted in bold. The AMF species are abbreviated (see full names in Table 8.1). The constrained model is significant ($P = 0.002$) and explains 33.4% of inertia. Explanatory variables: nitrogen ($P = 0.034$), phosphorus ($P = 0.033$) and sand ($P = 0.004$) are significant after a Permutation test by anova.cca function of vegan package in R. The range of variation of edaphic parameters analysed were: pH (4.3–5.8), sand (60–95%), P (0.01–6.00 μg. g^{-1}), organic matter (1.9–20.8%) and N (0.08–0.55%)

8.5 Biogeography and Phylogeny of Endemic Gigasporaceae in LGS

In addition to the ecological issues mentioned above, historical or evolutionary factors such as speciation-extinction processes and the rise and fall of barriers against dispersion could have modelled AMF communities and have impacted the observed geographic distribution patterns. Indeed, the presence of endemic species in a particular locality could tentatively be attributed to two such historical scenarios: (1) these species originated in their present locality and did not get dispersed or (2) they persist today only in a small area of their previous geographical distribution, which was larger originally (Brown and Lomolino 1998).

The use of phylogenetic data to make inferences about the history of speciation and biotic assembly of AM communities in different geographic regions could be important for a better understanding of Glomeromycotan biogeography (Chaudhary et al. 2008). In this framework, the molecular characterization of the SSU-RNA_r gene and phylogenetic analyses of several Gigasporaceae species belonging to ED group (*S. spinosissima*, *S. crenulata*, *S. striata*, *Scutellospora* sp.$_1$ and *Scutellospora* sp.$_2$) made by Lovera (2012) resulted particularly useful to understand the influence of past processes related to the evolutionary history of these species on their current geographic distributions. The molecular phylogeny inferred showed that *S. spinosissima*, *S. crenulata*, *S. striata* and *Scutellospora* sp.$_1$, constitute a monophyletic group within the genus *Scutellospora* (Fig. 8.4). The monophyletic origin of these species suggests that the most plausible explanation for their regional endemism in LGS or in a bigger area in the Northeast of South America is the result of a process of *in situ* diversification combined with limitations for dispersal and/or establishment in other regions (scenario 1).

The *in situ* diversification scenario implies that speciation processes that occurred in sympatry, lead to the coexistence of phylogenetically close species. Congeneric coexistence is expected to be enhanced by divergence in functional traits (*i.e.* character displacement), yet among AMF, the existence of similar functional traits in nearby species (*i.e.* phylogenetic conservatism) has been well established at the family level (Hart and Reader 2002; Maherali and Klironomos 2012; Chagnon et al. 2013). In the case studied here, all the species that make up the clade *S. spinosissima*, *S. crenulata*, *S. striata* and *Scutellospora* sp.$_1$ coexist sympatrically in many of the ecosystems evaluated, particularly in the sclerophyllous shrubland (Table 8.1). The adaptation to particular niches, such as different soil types within a habitat edaphically heterogeneous, or the development of different functional compatibilities with host plants, would diminish inter-specific competition among this set of sympatric AMF. If this turns to be correct, niches in the species of this group would be segregated by their relatively high specialization to particular biotic or abiotic conditions (Drumbell et al. 2009). Such specialisation could also explain, at least partially, the endemism of these taxa and their inability to establish themselves in environments with ecological conditions different from those that prevail in this region.

Species in the family Gigasporaceae are generally considered to be inefficient dispersers due to their characteristically large spores produced in low numbers, root colonization only from spores and spore dormancy (Chagnon et al. 2013; Kivlin et al. 2014). However, the existence of cosmopolitan species within the family implies that in a large span of time, they can reach ranges of global distribution (Davison et al. 2015). The low genetic divergence found among *S. spinosissima*, *S. crenulata*, *S. striata* and *Scutellospora* sp.$_1$ (Fig. 8.4), suggests that this clade diversified relatively recently, providing little time for dispersal events out of its hypothetical center of origin in LGS. The finding of *S. spinosissima* in the region of the Atlantic Shield in Brazil could be due to a recent event of dispersion and establishment in another environment with similar ecological conditions, or alternatively, could imply the existence of a larger previous distribution in the Northeast of South America that is now reduced only to the areas of the geological shields. The joint use of genetic markers more variable than the SSU gene and better analytical tools (*e.g.* coalescence-based models, approximate bayesian computation, etc.) will be necessary to understand appropriately the patterns, tempo and mode of diversification in this clade of AMF. The recent diversification of this group of ED Gigasporaceae

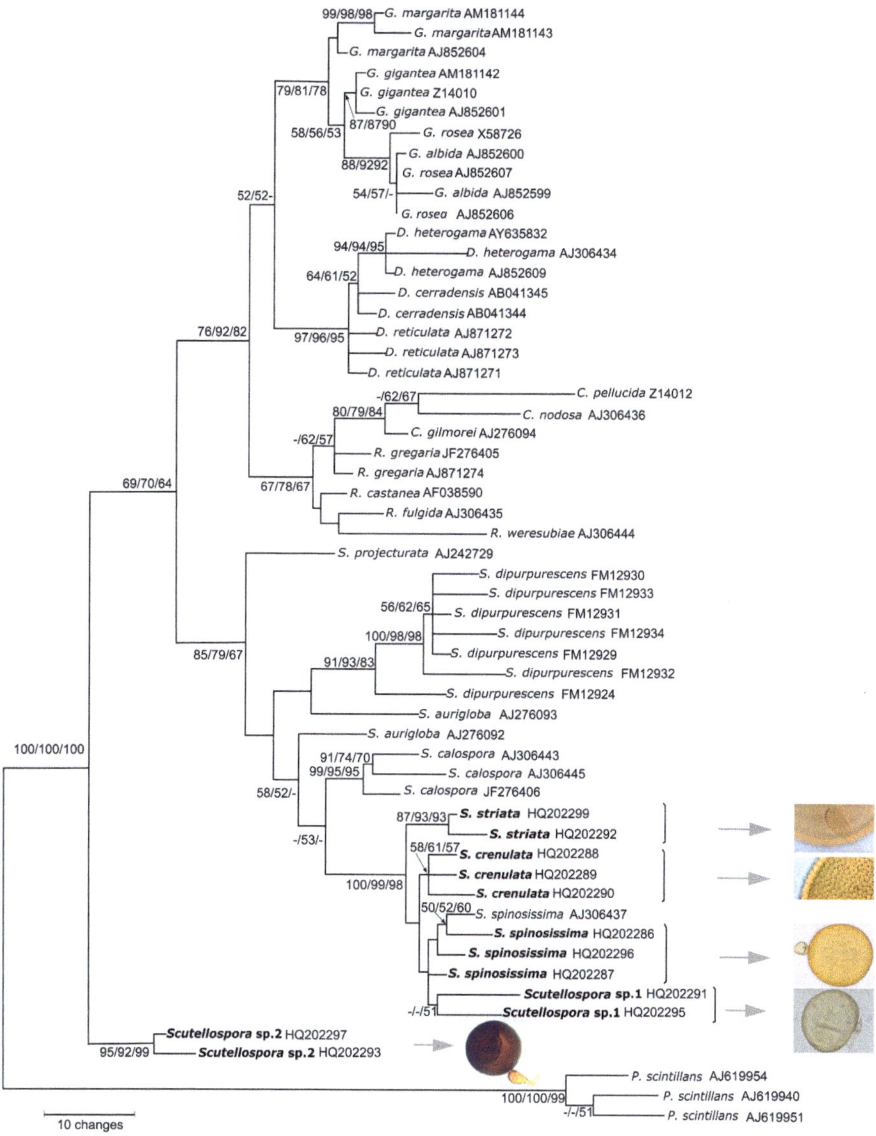

Fig. 8.4 Phylogenetic inference of the Gigasporaceae family based on ribosomal SSU gene sequences, including *Pacispora scintillans* as external group. The tree shown is the consensus of 36 most parsimonious trees obtained with the Maximum Parsimony method with 100 replicates (length: 476, Consistency index, CI: 0.67437). Node support values (>50%) obtained according to Neighbor-Joining, Maximum Likelihood and Maximum Parsimony methods (NJ/ML/MP) are indicated (only values >50 are shown). Sequences of species obtained in Lovera (2012) are highlighted in bold. Genera abbreviation: G: *Gigaspora*, D: *Dentiscutata*, C: *Cetraspora*, R: *Racocetra* and S: *Scutellospora*. Images of endemic species are shown on the right of its respective clade

is according to Davison et al. (2015) who suggested that the diversification of the majority of current AM fungal virtual taxa occurred within the period of 4–30 million yr. ago.

Finally, the phylogenetic analysis carried out in Lovera (2012) also showed that *Scutellospora* sp.$_2$ is phylogenetically distant from the rest of the species evaluated, forming part of a new basal lineage of Gigasporaceae (Fig. 8.4). Morphologically, *Scutellospora* sp.$_2$ has a germination shield with the shape of an orb, which is a feature that has been observed as well in *Bulbospora minima* and *S. pernambucana*, two species recently discovered and described from Brazil, that also constitute ancestral lineages of the Gigasporaceae family (Silva et al. 2008; Marinho et al. 2014). *Scutellospora* sp.$_2$ and *S. pernambucana* coexist in forest habitats in LGS, unfortunately, the phylogenetic relationships among these species are yet unknown due to the use of different molecular markers. Likewise, is important to obtain molecular data of *S. tepuiensis* and the rest of the putative new species found in this region, to improve understanding of the diversification processes of Gigasporaceae present in LGS.

8.6 Conclusions

The existence of abundant undescribed species, high rates of endemism probably associated with an *in situ* diversification processes, and the presence of a basal lineage to the family, make LGS a very interesting region for the study of the biogeography and evolution of Gigasporaceae. Furthermore, these results indicate that the region could represent a center of diversification for the genus *Scutellospora* or even for the whole family Gigasporaceae. Similarly, during the last decade, ten new species of Gigasporaceae have been described from Northeast Brazil, and about 60% of the species diversity of the family has been recorded in this region leading to propose that Brazil could constitute a center of diversification for this family (Marinho et al. 2014; de Souza et al. 2016). The ensemble of these results suggest that LGS in Venezuela, and probably the entire area of the Guayana Shield, in addition to the Atlantic Shield in the Northeast Brazil, must be considered as hotspots of diversification for Gigasporaceae. Based on this striking result we propose that the ecological conditions associated with these ancient geological shields (acidic, sandy and oligotrophic soils with presence of numerous endemism of plants) promote diversification processes in Gigasporaceae. Additionally, the biogeographic history of these tectonically stable areas, which have never been covered by the sea, could represent a refuge area for AMF species, as has been proposed by Stürmer et al. (2018a). Our results pointed out to shrublands as the biome that hosts the greatest diversity and endemism of Gigasporaceae in LGS, a type of vegetation that is included in the category *tropical grasslands, savannas, and shrublands* which is considered an evolutionary hotspot for AMF by Pärtel et al. (2017). The high endemism of plants and the particular edaphic conditions present in LGS shrublands looks like the main drivers of the important diversification processes within Gigasporaceae detected in these ecosystems.

The interesting patterns discovered in this study, together with the exciting questions that arise from them, highlight the importance of integrating phylogenetic information in studies about the ecology and biogeography of AMF. In this regard, it is necessary to increase sampling effort and enhance molecular data on the AMF species in this region in order to increase the understanding of the biogeographical patterns glimpsed so far.

Acknowledgments We are grateful to Grisel Velázquez, who prepared the map with the sampling locations and Pedro Borges, who kindly read the manuscript and offered helpful comments and English reviews. We also thank Parupa Scientific Station (PSS) for their logistic support in La Gran Sabana during the previous field work in which the information analyzed in this study was collected.

References

Berry PE, Huber O, Holst BK (1995) Floristic analysis and phytogeography. In: Steyermark JA, Berry PE, Holst BK (eds) Flora of the Venezuelan Guayana, Vol 1: Introduction. Missouri Botanical Garden, St Louis & Timber Press, Portland, OR

Blaszkowski J (2012) Glomeromycota. W. Szafer Institute of Botany Polish Academy of Sciences, Kraków, p 304

Briceño HO, Schubert C (1990) Geomorphology of the Gran Sabana, Guayana Shield, southeastern Venezuela. Geomorphology 3:125–141

Brown JH, Lomolino MV (1998) Biogeography. 2nd ed. Sinauer Associates, Inc., Sunderland, Massachusetts, xii + 691 p

Chagnon P, Bradlley RL, Klironomos JN (2015) Trait-based partner selection drives mycorrhizal network assembly. Oikos 124:1609–1616

Chagnon PL, Bradley RL, Maherali H, Klironomos JN (2013) A trait-based framework to understand life history of mycorrhizal fungi. Trends Plant Science 18:484–491

Chaudhary VB, Cuenca G, Johnson NC (2017) Tropical-temperate comparison of landscape-scale arbuscular fungal species distributions. Diversity and Distributions 2017:1–13

Chaudhary VB, Lau MK, Johnson NC (2008) Macroecology of microbes. Biogeography of the Glomeromycota. In: Varma A (ed) Mycorrhiza. Springer-Verlag, Berlin Heidelberg

Cuenca G, De Andrade Z, Escalante G (1998) Diversity of Glomalean spores from natural disturbed and revegetated communities growing on nutrient-poor tropical soils. Soil Biology and Biochemistry 30:711–719

Cuenca G, Herrera-Peraza RA (2008) *Scutellospora striata* sp nov, a newly described glomeromycotan fungus from La Gran Sabana, Venezuela. Mycotaxon 105:79–87

Cuenca G, Lovera M (1992) Vesicular-arbuscular mycorrhizae in disturbed and revegetated sites from La Gran Sabana, Venezuela. Canadian Journal of Botany 70:73–79

Cuenca G, Lovera M (2010) Seasonal variation and distribution at different soil depths of arbuscular mycorrhizal fungi spores in a tropical sclerophyllous shrubland. Botany 88:54–64

Cuenca G, Lovera M, De Andrade Z, Meneses E (2003) Why are so diverse the *Scutellospora* at La Gran Sabana Venezuela? Proceedings of the 4th International Conference on Mycorrhizae ICOM 4 Montreal, Quebec, Canada p 593

Davison J, Moora M, Öpik M, Adholeya A, Ainsaar L, Bâ A, Burla S, Diedhiou AG, Hiiesalu I, Jairus T, Johnson NC, Kane A, Koorem K, Kochar M, Ndiaye C, Pärtel M, Reier Ü, Saks Ü, Singh R, Vasar M, Zobel M (2015) Global assessment of arbuscular mycorrhizal fungus diversity reveals very low endemism. Science 349:970–973

De Andrade Z, Furrazola E, Cuenca G (2017) *Scutellospora tepuiensis* sp. nov. from the highland tepuis of Venezuela. Mycotaxon 132:9–18

de Souza TAF, Rodriguez-Echeverría S, de Andrade LA, Freitas H (2016) Arbuscular mycorrhizal fungi in *Mimosa tenuiflora* (Willd.) Poir from Brazilian semi-arid. Brazilian J Microbiol 47 (2):359–366

Drumbell AJ, Nelson M, Helgason T, Dytham C, Fitter AH (2009) Relative roles of niche and neutral processes in structuring a soil microbial community. ISME Journal 1–9

Finlay BJ (2002) Global dispersal of free-living microbial eukaryote species. Science 296:1061–1063

Fölster H, Dezzeo N (1994) La degradación de la vegetación. In: Ecología de la altiplanicie de la Gran Sabana (Guayana Venezolana) I. Investigaciones sobre la dinámica bosque-sabana en el sector S-E: Subcuencas de los ríos Yuruaní, Arabapó y Alto Kukenán. Dezzeo N (ed) Scientia Guaianae N°4. Caracas, Venezuela

Hart M, Reader RJ (2002) Taxonomic basis for variation in the colonization strategy of arbuscular mycorrhizal fungi. New Phytologist 153:335–344

Herrera-Peraza RA, Cuenca G, Walker C (2001) *Scutellospora crenulata* a new species of Glomales from La Gran Sabana Venezuela. Canadian Journal of Botany 79: 674–678

Huber O (1994) Arbustales. In: Ecología de la altiplanicie de la Gran Sabana (Guayana Venezolana) I. Investigaciones sobre la dinámica bosque-sabana en el sector S-E: Subcuencas de los ríos Yuruaní, Arabapó y Alto Kukenán. Dezzeo N (ed) Scientia Guaianae N°4. Caracas, Venezuela

Huber O (1995) Vegetation. In: Steyermark JA, Berry PE, Holst BK (eds) Flora of the Venezuelan Guayana, Vol 1: Introduction. Missouri Botanical Garden, St Louis & Timber Press, Portland, OR

Huber O (2006) Herbaceous ecosystems on the Guayana Shield, a regional overview. Journal of Biogeography 33:464–475

Johnson NC (2010) Resource stoichiometry elucidates the structure and function of arbuscular mycorrhizas across scales. New Phytologist 185(3):631–647

Kiers ET, Duhamel M, Beesetty Y, Mensah JA, Franken O, Verbruggen E, Fellbaum CR, Kowalchuk GA, Hart M, Bago A, Palmer TM, West SA, Vandenkoornhuyse P, Jansa J, Bücking H (2011) Reciprocal rewards stabilize cooperation in the mycorrhizal symbiosis. Science 333:880–882

Kivlin SN, Hawkes CV, Treseder K (2011) Global diversity and distribution of arbuscular mycorrhizal fungi. Soil Biology & Biochemistry 43:2294–2303

Kivlin SN, Winston GC, Goulden ML, Treseder KK (2014) Environmental filtering affects soil fungal community composition more than dispersal limitation at regional scales. Fungal Ecology 12:14–25

Landis FC, Gargas A, Givnish TJ (2004) Relationships among arbuscular mycorrhizal fungi vascular plants and environmental conditions in oak savannas. New Phytologist 164 (3):493–504

Lekberg Y, Koide RT, Rohr JR, Aldrich-Wolfe L, Morton JB (2007) Role of niche restrictions and dispersal in the composition of arbuscular mycorrhizal fungal communities. Journal of Ecology 95:95–105

Lovera M (2012) Filogenia molecular de especies de hongos micorrízicos arbusculares del género *Scutellospora* presentes en La Gran Sabana (Edo. Bolívar, Venezuela). MSc Thesis, Center for Advanced Studies, IVIC, Caracas, 91 p

Lovera M, Cuenca G (1996) Arbuscular mycorrhizal infection in Cyperaceae and Gramineae from natural, disturbed and restored savannas in La Gran Sabana, Venezuela. Mycorrhiza 6:111–118

Lovera M, Cuenca G (2007) Diversidad de hongos micorrrízicos arbusculares (HMA) y potencial micorrízico del suelo de una sabana natural y una sabana perturbada de La Gran Sabana, Venezuela. Interciencia 32:108–114

Maherali H, Klironomos JN (2012) Phylogenetic and Trait-Based Assembly of Arbuscular Mycorrhizal Fungal Communities. PLoS ONE 7 (5):e36695

Marinho F, Silva GA, Ferreira ACA, Veras JSN, Sousa NMF, Goto BT, Maia L, Oehl F (2014) *Bulbospora minima*, a new genus and a new species in the Glomeromycetes from semi-arid Northeast Brazil. Sydowia 66:313–323

Oksanen J, Blanchet FG, Friendly M, Kindt R, Legendre P, McGlinn D, Minchin PR, O'Hara RB, Simpson GL, Solymos P, Stevens MHH, Szoecs E, Wagner H (2018) vegan: Community Ecology Package version 2.5–2

Öpik M, Metsis M, Daniell TJ, Zobel M, Moora M (2009) Large-scale parallel 454 sequencing reveals host ecological group specificity of arbuscular mycorrhizal fungi in a boreonemoral forest. New Phytologist 184:424–437

Öpik M, Vanatoa A, Vanatoa E, Moora M, Davison J, Kalwij JM, Reier Ü, Zobel M (2010) The online database MaarjAM reveals global and ecosystemic distribution patterns in arbuscular mycorrhizal fungi (Glomeromycota). New Phytologist 188:223–241

Pärtel M, Öpik M, Moora M, Tedersoo L, Szava-Kovats R, Rosendahl S, Rillig MC, Lekberg Y, Kreft H, Helgason T, Eriksson O, Davison J, de Bello F, Caruso T, Zobel M (2017) Historical biome distribution and recent human disturbance shape the diversity of arbuscular mycorrhizal fungi. New Phytologist 216 (1):227–238

Peña-Venegas CP, Cardona Vanegas GI, Mazorra Valderrama A (2006) Micorrizas arbusculares de la Amazonia colombiana. Catálogo ilustrado. Publicaciñn del Instituto Amazónico de Investigaciones Científicas, SINCHI., Editorial Scripto Ltda., Colombia, 90 p

Pereira CMR, Silva DKA, Goto BT, Rosendahl S, Maia L (2018) Management practices may led to loss of arbuscular mycorrhizal fungal diversity in protected areas of the Brazilian Atlantic Forest. Fungal Ecology 34:50–58

Redecker D, Schüssler A, Stockinger H, Stürmer SL, Morton JB, Walker C (2013) An evidence-based consensus for the classification of arbuscular mycorrhizal fungi (Glomeromycota). Mycorrhiza 23 (7):515–31

Rosales J, Cuenca G, Ramírez N, De Andrade Z (1997) Native colonizing species and degraded land restoration in La Gran Sabana, Venezuela. Restoration Ecology 3:147–155

Rull V, Vegas-Villarrubia T, Montoya E (2016) The neotropical Gran Sabana region: palaeoecology and conservation. The Holocene 26 (7):1162–1167

Schüßler A, Schwarzott D, Walker C (2001) A new fungal phylum the Glomeromycota: phylogeny and evolution. Mycological Research 105:1413–1421

Silva DKA, Freitas NO, Cuenca G, Maia LC, Oehl F (2008) *Scutellospora pernambucana*, a new fungal species in the Glomeromycetes with a characteristic germination orb. Mycotaxon 106:361–370

Stürmer SL, Bever JD, Morton JB (2018a) Biogeography of arbuscular mycorrhizal fungi (*Glomeromycota*): a phylogenetic perspective on species distribution patterns. Mycorrhiza 28 (7):587–603

Stürmer SL, Kemmelmeier K, Moreira BC, Kasuya MCM, Pereira GMD, Silva K. (2018b) Arbuscular mycorrhizal fungi (Glomeromycota) communities in tropical savannas of Roraima, Brazil. Mycological Progress 17 (10):1149–1159

Veresoglou SD, Caruso T, Rillig MC (2013) Modelling the environmental and soil factors that shape the niches of two common arbuscular mycorrhizal fungal families. Plant and Soil 368:507–518

Walker C, Cuenca G, Sánchez F (1998) *Scutellospora spinosissima* sp. nov. a newly described Glomalean fungus from low nutrient communities in Venezuela. Annals Botany 82:721–725

Chapter 9
Tropical Dry Forest Compared to Rainforest and Associated Ecosystems in Brazil

Marcela C. Pagano, Danielle K. da Silva, Gladstone A. da Silva, and Leonor C. Maia

9.1 Introduction

The arbuscular mycorrhizal fungi (AMF) links the plants and geochemical components of the ecosystems providing important ecosystem services. Research on Mycorrhizas has gone through different stages (Pagano 2016), but their importance in natural ecosystems is nowadays increasingly recognized. Commom AMF species composition differs in each paleocontinent, while endemic species are usually rare (Davison et al. 2015). For example, *Funneliformis mosseae* is considered a widespread generalist (Öpik et al. 2006), frequent in grasslands and arable lands and less associated to forests (Bouffaud et al. 2016). AMF density and distribution vary both spatially and temporally within and between species, being influenced by soil types and host plant species diversity. Some AM fungi, for example, are only found in specific soil nutrient conditions (Valyi et al. 2016).

Natural forest ecosystems and their associated vegetation types were not fully investigated, and the deep soil layers should also be included in studies to get a complete picture of AMF diversity, as they can show different composition than the topsoil (Oehl et al. 2005). In the same way, trap cultures are a useful tool to better understand the AMF ecology of native plant communities and are useful to confirm the results of a higher or lower species richness obtained from different vegetation types.

M. C. Pagano (✉)
Federal University of Minas Gerais, Belo Horizonte, Brazil

D. K. da Silva
Programa de Pós-Graduação em Ecologia e Monitoramento Ambiental, Centro de Ciências Aplicadas e Educação, Universidade Federal da Paraíba, João Pessoa, PB, Brazil

G. A. da Silva · L. C. Maia
Laboratório de Micorrizas, Departamento de Micologia, Universidade Federal de Pernambuco, Recife, PE, Brazil

M. C. Pagano, M. A. Lugo (eds.), *Mycorrhizal Fungi in South America*, Fungal Biology, https://doi.org/10.1007/978-3-030-15228-4_9

A recent review about AMF in tropical forests worldwide (Marinho et al. 2018) showed that 228 species belonging to 14 families and 35 genera of Glomeromycotina were registered in tropical forests. This number of species represents 75% of the known richness of this group of symbiotic fungi, and the authors pointed out that the largest numbers of these AMF species are from Dry forests.

In Brazil, the tropical dry forests are most located in the Northeast semiarid region and in the Midwest and Southeast, being characterized by high temperature and low humidity due to severe periods of drought. In the semiarid Northeast the characteristic vegetation is known as 'Caatinga', and the plants are adapted to tolerate the dry season through different mechanisms including deciduousness, presence of spines, bodies for water storage, small leaf area and deep roots (Giulietti et al. 2006). The dry forests of the other regions are savannas, most known as 'Cerrado' and 'Cerradão', that extends from the margin of the Amazonian forest to the Midwest and Southeast region, presenting, diverse physiognomies ranging from dense grassland, in general sparsely covered by shrubs and small trees, to an almost closed woodland with a canopy that reaches 12–15 m height (Ratter et al. 1997). All native plants are adapted to live in its variable arid climate and many support fire. Highlands and rupestrian fields and grasslands at altitudes between 800 m and 2000 m are sub-physiognomies of Cerrado (Rizzini 1997).

The Brazilian rainforest is mainly represented by the Amazon forest and the Atlantic forest, that includes the Araucaria forest in the South of Brazil, and encompasses, among other associated ecosystems: sand dunes, restingas and mangroves along the Atlantic coast of the country. An open forest, characteristic of the Pantanal biome and that differs from the Atlantic and Amazon forests occurs as a transition between the Amazonia forest and the Cerrado, in the Midwest.

In all these dry and humid forest environments AMF associations have been registered. This chapter discusses advances on diversity of arbuscular mycorrhizal fungi in natural forest ecosystems drawing on results of research in Brazil.

9.2 The Arbuscular Mycorrhizal Symbioses in Tropical Dry Forest

9.2.1 Caatinga

Arbuscular Mycorrhizal Fungi are well represented in semiarid lands, which are characterized by diverse vegetation types due to its soil, topography and climatic variation. Most thorny dry woody vegetation (caatinga vegetation); non-thorny dry forest and closed, non-thorny dry tall-shrubby vegetation (carrasco vegetation) presented AMF association and high AMF diversity (Pagano et al. 2013). As described by these authors, the *Arum*-type AM morphology was prevalent in roots and 32 AM fungal taxa (spore-based taxonomy) were isolated from rhizospheric soil samples collected in Caatinga areas, with *Glomus*, *Gigaspora* and *Cetraspora* being commonly found (Pagano et al. 2013). In general, compared to dry forest, woody caatinga presented higher sporulation and AMF diversity; however, carrasco vegetation was more similar to dry forest in their AMF species composition. Total AMF spore numbers were consistently similar in all sites, but the AMF spore ontogeny, varied between vegetational types: Deciduous Forest and Carrasco presented higher gigasporoid spore numbers, followed by glomeroid and acaulosporoid spores type (Fig. 9.1); while, the woody caatinga showed lower Gigasporales representants (Pagano et al. 2013).

Spore populations of AMF communities in dry forest are generally low in numbers and can vary between one to 2.8 spores g^{-1} soil (Mello et al. 2012; Pagano et al. 2013; Da Silva et al. 2014); however, it also depends on soil type and vegetation, as range of 0–10 glomerospores/g^{-1} soil was registered in different areas of tropical dry forest (Maia et al. 2015a). Studies in a Caatinga environmental gradient (a dry

Fig. 9.1 Trees associated with mycorrhizas in native forests of Brazil and AM spores recovered from rhizospheric soils. Clockwise, from upper: dry forest and rainforest vegetation, spores (Gigasporales and *Acaulospora*) recovered from their rhizospheric soils (photos by M. Pagano and E. A. Correa)

forest, a transitional zone and a moist forest) showed the presence of AM association and high AMF diversity, trap cultures being of great importance as more species can be identified, such as those of *Glomus* (Da Silva et al. 2014).

Rarely more than 42 AMF species were reported in field studies performed in dry forests (Table 9.1). In general, species of Acaulosporaceae, Gigasporales, and Glomerales can be commonly found in dry forest; however, *Glomus* can predominate in transitional zone and moist forest (above 670 m.a.s.l.) (Da Silva et al. 2014). Near 60% of AMF species identified in dry forest ecosystems were retrieved from trap cultures, and among them 17% belonged to Glomeraceae, a few species such as *Glomus ambisporum* were obtained only from trap cultures (Da Silva et al. 2014). This is due to the fact that species of Glomeraceae display more extensive root colonization than other families and lower soil colonization by extraradical hyphae in addition to rapid colonization of new plant hosts from colonized roots fragments (Hart and Reader 2002).

New species have being described from Caatinga areas such as *Bulbospora minima* (Marinho et al. 2014), *Paraglomus pernambucanum* (Mello et al. 2013), *Racocetra intraornata* (Goto et al. 2009), *Septoglomus furcatum* (Błaszkowski et al. 2013) and *Septoglomus titan* (Goto et al. 2013). These new records show the need to better study the great estimated AMF species richness in Brazil, and, therefore, to the discovery of new species (de Souza et al. 2010).

9.2.2 Cerrado

Earlier studies that described AMF communities in Cerrado (e.g. Cordeiro et al., 2005; Ferreira et al. 2012) registered~11 AMF species. Plants in areas of a Cerrado *stricto sensu* presented low root colonization (30%) and AMF density varied from 7 to 8 spores g^{-1} soil (Cordeiro et al. 2005).

More recent studies analyzed the Murundu fields (characterized by termite mounds varying from 2 to 10 m in diameter and 2 m in height) that occur in some parts of the Cerrado biome in Goiás state, which were considered hotpots for AMF diversity (Assis et al. 2014). In that study, AM fungal community was represented by 27 species from eight genera and five families; *Acaulospora mellea, A. cavernata, A. colombiana, Oehlia diaphana* and *Dentiscutata reticulata* were commonly found.

The tropical wetland (Pantanal biome) was only recently investigated and in vegetation areas with different flooding regimes (flood-free, occasional flooding and seasonal flooding), 37 AMF species were registered (Gomide et al. 2014). The authors observed increasing spore numbers in "Cerradão", the tallest Cerrado vegetation with a continuous and moderately closed canopy according to Andrade et al. (2002), and grassland soils and higher richness in Cerrado > areas exposed at low water/lowlands > "Cerradão", corroborating the observations that AMF diversity is related to heterogeneity of vegetation (Gomide et al. 2014).

Table 9.1 Total number of identified AM fungal species in some natural Brazilian ecosystems

Ecosystem	Biome	Vegetation type/State/Region	AMF species	Reference
Forest	Amazonia	Terra firme forest, Central Amazonia	39	Freitas et al. (2014)
Forest	Atlantic forest	Mature forest, Paraná state, Southern	47	Zangaro et al. (2013)
Forest	Atlantic forest	Mature forest, Pernambuco state/Northeast	17	Pereira et al. (2014)
Forest	Atlantic forest/Cerrado	Riparian vegetation	27	Pagano and Cabello (2012)
Forest	Atlantic forest	Araucaria forest *(Araucaria angustifolia)*	18	Moreira et al. (2016)
Tropical wetland	Pantanal	Semi-deciduous forest, Cerrado, Cerradão, grasslands	21–25	Gomide et al. (2014)
Savanna forests	Cerrado	Natural Cerrado forest, Midwest	29–33	Pontes et al. (2017)
Savanna forests	Cerrado	Murundu fields[a], Goiás, Midwest	27	Assis et al. (2014)
Savanna forests	Cerrado	Highland fields[a], Minas Gerais, Southeast	8	Pagano and Scotti (2009)
			51	Oki et al. (2016)
			20	Costa et al. (2016)
Savanna forests	Cerrado	Ferruginous fields, Iron mining areas[a], Minas Gerais, Southeast	6	Pagano and Scotti (2010)
			59	Teixeira et al. (2017)
Savanna forests	Cerrado	High altitude Cerrado savanas, Bahia, Northeast	49	Vieira et al. (2019)
Dry forest	Caatinga	Montane forest, Pernambuco Montane forest, southern Ceará, Northeast	47	da Silva et al. (2014)
			52	Assis et al. (2018)
Dry forest	Caatinga	Dry forest, moist forest	27–42	da Silva et al. (2014), Assis et al. (2018)
Dry forest	Caatinga	Caatinga, Pernambuco, Northeast	16	Mello et al. (2012)
Dry forest	Caatinga	Deciduous Forest, Northeast	13–15	Pagano et al. (2013)
Dry tall-shrubby vegetation	Caatinga	Carrasco, Northeast	16–18	Pagano et al. (2013)

(continued)

Table 9.1 (continued)

Ecosystem	Biome	Vegetation type/State/ Region	AMF species	Reference
Dry woody savanna vegetation	Caatinga	Woody caatinga, Northeast	9–23	Pagano et al. (2013)
Sand dunes and Coastal ecosystems	Atlantic forest	Sand dunes and Restinga forest, South	10–53	da Silva et al. (2012), Stürmer et al. (2013), Souza et al. (2013), Silva et al. (2015a), Silva et al. (2015b), Silva et al. (2017)
Sand dunes and Coastal ecosystems	Atlantic Forest	Mangrove forest, restinga forest, Northeast	17–22	Silva et al. (2017)

aReported as AMF hotspots sites

The southeast Brazilian Highlands or Rupestrian fields were also investigated regarding AM fungal communities. These areas have shrubby, tortuous and sclerophyllous vegetation or grasses that grow in stones, in sandy soils and present varied adaptations (Rizzini 1997).

For a recent list of AMF species in Cerrado Rupestrian grasslands see Oki et al. (2016). The specialized vegetation types in such environment present high herbaceous species richness, high endemism (species of Asteraceae, Euphorbiaceae, Melastomataceae and Velloziaceae), and unique plant and fungal species compositions, among other organisms, resulting in megadiverse ecosystems (Fernandes 2016). Thus, these areas have been pointed out as a hotspot of diversity for AMF and endophytic fungi, whose relations need to be more understood (Oki et al. 2016). In this rupestrian ecosystem Coutinho et al. (2015) reported the presence of 22% of the known world diversity of AMF.

Recent studies in tropical mountain ecosystems, such as in the Chapada Diamantina in Bahia (NE Brazil) where predominates rupestrian fields showed that the AMF communities were related to the heterogeneity of habitats, including silt and coarse sand contents as the main factors. Among the 49 identified AMF species, members of Glomeraceae and Acaulosporaceae were the most representative. The AMF communities did not follow the shifts in plant communities. The high altitude savannas (Cerrado) and natural rocky rupestrian fields (shrublands), differed in the composition of the AMF communities (Vieira et al. 2019).

In other Cerrado areas of Minas Gerais, earlier studies have shown AM colonization and glomerospores occurrence in the root zone of *Paepalanthus bromelioides* and *Bulbostylis* sp. (Pagano and Scotti 2009) and a high root colonization in native *Centrosema coriaceum* (Matias et al. 2009). Those rhizospheric sandy soils (sand >78%) presented low organic matter content (2.72%), low base saturation and P content and moderate acidity (pH 5.3) (Pagano and Scotti 2009). Eight AMF species were identified in the rhizosphere of the studied plants: *Acaulospora spinosa*, *A. elegans*, *A. foveata*, *Gigaspora margarita*, *Dentiscutata biornata*, *D. cerradensis*, *D. heterogama* and *Racocetra verrucosa*. The low number of AMF species reported

Fig. 9.2 Highland fields of Brazil. Clockwise, from upper left: native vegetation, colonized root by AM, and spores recovered from rhizospheric soils (photos by M. Pagano)

could be related to the small sample effort or the few plant species evaluated (Pagano and Scotti 2009).

Rupestrian ferruginous fields are characterized by soils of the iron mines region called Quadrilátero Ferrífero, in Minas Gerais State, and present small woody plants which support environmental stress (Rizzini 1997). In that areas a high percentage (90%) of root colonization was observed in preserved vegetation dominated by *Eremanthus incanus* (Pagano et al. 2010) (Fig. 9.2).

Teixeira et al. (2017) found 59 AMF species in an iron mining area in Minas Gerais State representing 15% of the 289 known species. These authors retrieved 57% of the AMF species by trap cultures and proposed this region as a hotspot of AMF diversity. Hotspots are priority places for AMF conservation, and besides the high diversity, they are threatened in the highest degree according to Myers et al. (2000).

Also in Minas Gerais, studies at "Parque Nacional das Sempre-Vivas", in Diamantina, revealed endangered plant species (*Syngonanthus elegans*) associated to 26 AMF species (Costa et al. 2016), which demonstrate the importance of mutualistic partners for plant establishment and survival.

New reports investigated the occurrence and density of AMF spores across different vegetation types in Bahia state, in the northeast region, showing that Glomeraceae and Acaulosporaceae were the most representative families in species number. The AMF community composition differ between habitats types, and the soil physical characteristics (silt and coarse sand) were the main factors related to the AMF community. A checklist of AMF in the Brazilian Cerrado was provided by Jobim et al. (2016) and on that a total of 92 species were reported; two of these species (*Ambispora brasiliensis* and *Cetraspora auronigra*) were exclusively found in rupestrian fields. These data indicate that more efforts should be made to investigate the diversity of AMF in areas of Cerrado, which is still little known regarding the occurrence of these soil fungi that probably also contribute for the establishment of vegetation and the ecosystem balance.

9.3 The Arbuscular Mycorrhizal Symbioses in Tropical Forest and Associated Ecosystems

Several studies were performed in areas of rainforest in Brazil, but most of them included sites of Atlantic rainforest, with less attention to the Amazonian rainforest, which probably also has a high diversity of AMF. Below we comment these studies, showing the need for further investigations.

9.3.1 The Arbuscular Mycorrhizal Symbioses in the Amazonian Rainforest

As mentioned, little consideration has been given to investigate mycorrhizal associations in the Amazonian rainforest (Stürmer and Siqueira 2006). However, as shown by Freitas et al. (2014), common AMF are widely dispersed in plant communities of this biome, where the spore density can attain nine spores per gram of dry soil and, interesting, circa 80% can be identified. In that study, 39 species were registered, with taxa of Glomeraceae dominating the AMF community. The highest number of species belonged to *Glomus* followed by *Acaulospora, Claroideoglomus* and *Scutellospora* (Freitas et al. 2014). Other research on Amazonian forest pointed to no alteration of species richness and abundance distribution across the conversion of pristine tropical forest to pastures (Leal et al. 2013). Other recent reports showed the decrease of AMF with the reduction of secondary forest cover in eastern Amazonia (Maia et al. 2015b) and AMF spore communities in the terra firme forest, a vegetation type from Central Amazonia (Freitas et al. 2014).

9.3.2 The Arbuscular Mycorrhizal Symbioses in the Atlantic Rainforest

In an updated review provided by Jobim et al. (2018), a total of 128 AMF species were registered in the Brazilian Atlantic Forest, and 18 of them, as well as one family and three genera, were first described from this biome, which is a hotspot of biodiversity.

In areas of Atlantic rainforest, 40 AMF species were associated with different vegetation types (Table 9.1), and these AMF communities were dominated by species belonging to the families Glomeraceae > Acaulosporaceae > Gigasporaceae.

Acaulospora and *Glomus* can be commonly found, accounting for 70–80% of the total spores recovered (Bonfim et al. 2013). In some rainforests in the South and Southeast of Brazil, AMF richness is lower at initial stages of the succession compared to mature forests although the number of spores can be greater at initial stages (Aidar et al. 2004; Stürmer et al. 2006; Zangaro et al. 2008). Recent studies conducted in a fragment of a seasonal semidecidual mountain forest in Vitória da Conquista, Bahia State, showed that the edge effect can modify the AMF communities with some exclusive species in the first 0–10 m from the edge (Santos et al. 2018).

In some areas of Atlantic rainforest, the spore density can attain ~20 spores per gram of dry soil, and ~72% can be identified by morphological methods; among 57 species identified from 79 spore types isolated from soil under different vegetational stages, Glomeraceae dominated the AMF community, and the highest number of species in mature forest belonged to *Glomus* and *Acaulospora* followed by *Claroideoglomus*, *Scutellospora* and *Gigaspora* (Zangaro et al. 2013). In that study, AM fungal community structure differed in 11 species along plant succession under Grass (43), Scrub (52), Secondary (41) and Mature forest (47 species). *Glomus* and *Acaulospora* predominated.

In other sites in Brazilian Atlantic forests, earlier studies also showed mean number of glomerospores from <1 to >10 spores g^{-1} soil (Aidar et al. 2004) and AMF richness of 13 (Silva et al. 2006) to 25 species (Aidar et al. 2004) in secondary forest and 14–27 species in riparian forests (Fernandes et al. 2016; Pagano and Cabello 2012).

Comparing areas of Atlantic Forest with different land uses, In Pernambuco State, Pereira et al. (2014) found 50 AMF species distributed in 15 genera; 52% of them belonged to *Acaulospora* and *Glomus*. These authors found that AMF community composition was more influenced by land use than by the physical and chemical soil characteristics and that "diversity, evenness, and richness indices tended to be lower in communities established in climax environments of Atlantic Forest, rather than in the ones established in cultivated areas".

The AMF in natural *Araucaria* forests, an ecosystem of Atlantic Forest occurring in the South and also in areas of the Southeast region of Brazil, was largely studied by the group of Elke Cardoso (University of São Paulo). In pioneer research by Moreira et al. (2003, 2006, 2007a, b, 2009) in the southeast of Brazil 18 AMF species of the genera *Glomus, Funneliformis, Rhizoglomus, Gigaspora, Acaulospora,* and *Archaeospora* were observed in the root zone of *A. angustifolia* (Moreira et al. 2016). They also found a rate of root colonization varying from 30 to 50%. *Acaulospora* and *Glomus* are very common in these forests, as previously reported (Moreira et al. 2007a, b, 2009). However, auxiliary cells (typical of Gigasporales) were also observed. The authors confirmed this plant species (*A. angustifolia*) as very AMF dependent for developing and survival, as trees that grow in relatively poor soil can obtain the nutrients necessary for growth and formation from AMF networks and, probably, by legumes that also fix nitrogen in their roots that grow in the forests (Cardoso and Vasconcellos 2015).

9.3.3 *The Mycorrhizal Symbioses in Sand Dunes, Restingas, and Mangroves*

In South America, studies on the AMF symbioses in Sand dunes and Restingas are concentrated in Brazil, with most reports from the States of Santa Catarina (Stürmer and Bellei 1994, 2011; Córdoba et al. 2001; Silva et al. 2017) and Rio Grande do Sul (Cordazzo and Stürmer 2007), in the South region; in São Paulo (Trufem 1995; Trufem et al. 1994), and Rio de Janeiro (Silva et al. 2017), in the Southeast region; in Paraíba (da Silva et al. 2012; Silva et al. 2015a, 2015b, 2017; Goto et al. 2009, 2010, 2012b; de Souza et al. 2013), Bahia (Santos et al. 1995; Goto et al. 2012b, Assis et al. 2016), and Rio Grande do Norte (Goto et al. 2012a, b; Błaszkowski et al. 2014, 2015; Silva et al. 2017) in the Northeast region.

In São Paulo State, a high number of glomerospores was observed in the rhizosphere of plants from the restinga da Ilha do Cardoso and this number increased with increasing temperature, precipitation and insolation (Trufem et al. 1994; Trufem 1995). The authors pointed out the dominance of *Acaulospora, Gigaspora* and *Scutellospora* over *Glomus* and *Sclerocystis.*

Stürmer's work in sand dunes improved the knowledge regarding native AMF communities in coastal ecosystems of South Brazil, being useful to be applied in conservation programs. His results confirmed the beneficial effects of AMF on dune stabilization (Stürmer and Siqueira 2006). In the coast of Santa Catarina, reports of Stürmer and Bellei (1994) showed the seasonal variation of AMF populations associated with *Spartina ciliata* in dunes, with four species being commonly retrieved: *Gigaspora albida*, *Racocetra weresubiae*, *Acaulospora scrobiculata* and *Scutellospora* sp. Twelve AMF species were recorded in Praia da Joaquina, and Gigasporales dominated in the fixed dunes whereas Acaulosporaceae dominated in the frontal dunes; the number of spores and species richness increased with dune stabilization (Córdoba et al. 2001). In Rio Grande do Sul, in contrast, Gigasporales dominated the fledgling dunes, while Glomeraceae dominated the fixed dunes with *Panicum racemosum* (Cordazzo and Stürmer 2007). *Racocetra weresubiae*, *Dentiscutata cerradensis* and *Racocetra gregaria* predominated in three areas of coastal dunes with similar AMF community structure (Stürmer et al. 2013).

The pioneer work on sand dunes from Northeastern Brazil (Bahia) was conducted by Santos et al. (1995), who reported that the majority of coastal dunes plants investigated were associated with arbuscular mycorrhiza, and *Glomus microcarpum* was the dominant species. Later, the research group led by LC Maia performed some studies on sand dunes and Restinga areas from Northeastern Brazil, mostly in Paraíba State. In these works, glomoid and gigasporoid spores were predominant both in natural and revegetated dunes (da Silva et al. 2012; Souza et al. 2013). Silva et al. (2015a) recorded 50 AMF species belonging to 18 genera in a vegetational gradient in dunes; higher AMF diversity was observed in herbaceous than in shrubby and arboreal dunes. These authors also found that pH, cation exchange capacity, Fe and fine sand were the main structuring factors of the AMF communities.

In studies along a transect crossing a fluvial-marine island, in short distances sites showing different edaphic characteristics and vegetation physiognomies (preserved mangrove forest, degraded mangrove forest, natural Restinga forest, and two regeneration Restinga forests) 22 AMF species were identified without differences in species richness. Conversely, spore abundance per family, diversity and composition of AMF communities and rate of mycorrhizal colonization differed statistically among the sites (Silva et al. 2017). These authors also mention that soil characteristics, especially the sum of exchangeable bases were strongly related to composition of AMF communities and that even within short distances the different habitat types host diverse AMF communities.

In Rio Grande do Norte state, Jobim and Goto (2016) recorded 46 AMF species distributed in 15 genera in maritime sand dunes of Brazilian northeast, while 24 AMF species were recorded by Silva et al. (2017) in dune areas from Rio Grande do Norte.

Several new species were described from sand dunes, such as *Racocetra tropicana* (Goto et al. 2011), *Paradentiscutata maritima* (Goto et al. 2012b), *Glomus trufemii* (Goto et al. 2012a), *Rhizoglomus natalensis* (Błaszkowski et al. 2014) and *Acaulospora ignota* (Błaszkowski et al. 2015), what indicates that these ecosystems are also of great interest for studies on diversity of AMF.

9.4 Conclusion

In this chapter, the occurrence of arbuscular mycorrhizas in different forest ecosystems has been compared. The diversity of AM fungi in different vegetation types was listed compiling recent results in dry forests, rainy forests, Cerrado (savanah) and rupestrian fields showing particular species composition. Three hotspots for AMF diversity were proposed: the Murundu fields, the Highland fields and the Rupestrian ferruginous fields. Besides, there are more challenges related to the need of improving the methodology of collection. Whitcomb and Stutz (2007) stressed the need of a higher sample effort (eg. from 30 to 75 samples) to deal with underestimated AMF diversity. Usually, to detect 70–80% of AMF species present in a community is difficult due to problems in quantifying the diversity and the complexity of AMF communities. These authors showed randomly distributed AMF species with no detectable belowground hotspots (associated to the location of plants). Thus, it is necessary to increase the sample efforts for detecting more species of AMF in natural communities. Morphological identification of AMF continues to be important, as well as molecular identification, considering that 11 to 40% of commonly present morphotypes continue to be unidentifiable.

Finally, this chapter shows that several natural ecosystems present high AMF diversity; with forests and Restingas (sand dunes) presenting high glomerospore richness. Further research is necessary, especially regarding the occurrence, identification and ecology of mycorrhizas for a better understanding of the Brazilian unique ecosystems.

References

Aidar MPM, Carrenho R, Joly CA (2004) Aspects of arbuscular mycorrhizal fungi in an Atlantic Forest chronosequence. Biota Neotropica 4: 1–15

Andrade LAZ, Felfili JM, Violatti L (2002) Fitossociologia de uma área de cerrado denso na RECORIBGE, Brasília, DF. Acta Botanica Brasilica 16:225–240

Assis DMA, Melo MAC, da Silva DKA, Oehl F, da Silva GA (2018) Assemblages of arbuscular mycorrhizal fungi in tropical humid and dry forests in the Northeast of Brazil. Botany 96: 859–871. https://doi.org/10.1139/cjb-2018-0068

Assis DMA, Oehl F, Gonçalves, CM, Silva, DKA; Silva, GA (2016) Community structure of arbuscular mycorrhizal fungi in fluvial and maritime dunes of Brazilian Northeast. Applied Soil Ecology 108: 136–146

Assis PCR, Saggin Júnior OJ, Paulino HB, Stürmer SL, Siqueira JO, Carneiro MAC (2014) Arbuscular mycorrhizal fungi in "murundu fields" after conversion to farm systems in the Cerrado. R. Bras. Ci. Solo 38: 1703–1711

Błaszkowski J, Chwat G, Kovács GM, Gáspár BK, Ryszka P, Orłowska E, Pagano MC, Araújo FS, Wubet T, Buscot F (2013) *Septoglomus fuscum* and *S. furcatum,* two new species of arbuscular mycorrhizal fungi (Glomeromycota). Mycologia 105(3):670–80. doi: https://doi.org/10.3852/12-127

Błaszkowski J, Chwat G, Góralska A, Goto BT (2014) *Rhizophagus natalensis*, a new species in the Glomeromycota. Mycotaxon 129: 97–108

Błaszkowski J, Chwat G, Góralska A (2015) *Acaulospora ignota* and *Claroideoglomus hanlinii*, two new species of arbuscular mycorrhizal fungi (Glomeromycota) from Brazil and Cuba. Mycological progress 14:18

Bonfim JA, Vasconcellos RLF, Stürmer SL, Cardoso EJBN (2013) Soil Arbuscular mycorrhizal fungi in the Brazilian Atlantic forest: Agradient of environmental restoration Applied Soil Ecology 71: 7–14

Bouffaud ML, Creamer RE, Stone D, Plassart P, van Tuinen D, Lemanceau P, Wipf D, Redecker D (2016) Indicator species and co-occurrence in communities of arbuscular mycorrhizal fungi at the European scale. Soil Biology & Biochemistry 103: 464–470

Cardoso EJBN, Vasconcellos RLF (eds.) (2015) Floresta com araucária: composição florística e biota do solo. FEALQ, Piracicaba, São Paulo.

Cordazzo CV, Stürmer SL (2007) Ocorrência de fungos micorrízicos arbusculares em Panicum racemosum (P. Beauv.) Spreng (Poaceae) em dunas costeiras do extremo sul do Brasil. Atlântica, Rio Grande, 29: 65–68

Cordeiro MAS, Carneiro MAC, Paulino HB, Saggin Júnior OJ (2005) Colonização e densidade de esporos de fungos micorrízicos em dois solos do cerrado sob diferentes sistemas de manejo. Pesquisa Agropecuária Tropical 35: 147–153

Córdoba AS, Mendonça MM, Stürmer SL, Rygiewicz PT (2001) Diversity of arbuscular mycorrhizal fungi along a sand dune stabilization gradient: A case study at Praia da Joaquina, Ilha de Santa Catarina, South Brazil. Mycoscience 42: 379–387

Costa HAO, Stürmer SL, Ragonezi C, Grazziotti PH, Grazziotti DCFS, de Barros Silva E (2016) Species richness and root colonization of arbuscular mycorrhizal fungi in *Syngonanthus elegans*, an endemic and threatened species from the Cerrado domain in Brazil. Ciência e Agrotecnologia 40(3):326–336.

Coutinho ES, Fernandes GW, Berbara RLL, Valério HM, Goto BT (2015) Variation of arbuscular mycorrhizal fungal communities along an altitudinal gradient in rupestrian grasslands in Brazil. Mycorrhiza 25: 627–638

Davison J, Moora M, Öpik M, Adholeya A, Ainsaar L, Bâ A et al. (2015) Global assessment of arbuscular mycorrhizal fungus diversity reveals very low endemism. Science 349: 970–973

da Silva DKA, Pereira CMR, Souza RG, Silva GA, Oehl F, Maia LC (2012) Diversity of arbuscular mycorrhizal fungi in restinga and dunes area in Brazilian Northeast. Biodivers. Conserv. 21: 2361–2373 da Silva I R, Mello CMA, Ferreira Neto RA, da Silva DKA, de Melo AL, Oehl F, Maia LC (2014) Diversity of arbuscular mycorrhizal fungi along an environmental gradient in the Brazilian semiarid. Applied Soil Ecology 84: 166–175

de Souza FA, Stürmer SL, Carrenho R, Trufem SFB (2010) Classificação e taxonomia de fungos micorrízicos arbusculares e sua diversidade e ocorrência no Brasil. In: Siqueira, J.O., de Souza, F.A., Cardoso, E.J.B.N., Tsai, S.M. (Eds.), Micorrizas: 30 anos de Pesquisas no Brasil. UFLA, Lavras, pp. 15–73

de Souza RG, Silva DKA, Mello CMA, Goto BT, Silva FSB, Sampaio EVSB, Maia LC (2013) Arbuscular mycorrhizal fungi in revegetated mined dunes. Land Degradation & Development 24: 147–155

Fernandes GW (2016) The Megadiverse Rupestrian Grassland pp 3–14, In: Ecology and Conservation of Mountaintop grasslands in Brazil. Springer, *Switzerland*

Fernandes RA, Ferreira DA, Saggin-Junior OJ et al. (2016) Occurrence and species richness of mycorrhizal fungi in soil under different land use, Can. J. Soil Sci. 96: 271–280

Ferreira DA, Carneiro MAC, Saggin Junior OJ (2012) Arbuscular Mycorrhizal Fungi in an Oxisol under Managements and Uses in Cerrado. R. Bras. Ci. Solo 36: 51–61

Freitas RO, Buscardo E, Nagy L, Maciel ABS, Carrenho R, Luizão RCC (2014) Arbuscular mycorrhizal fungal communities along a pedo-hydrological gradient in a Central Amazonian terra firme forest. Mycorrhiza 24:21–32

Giulietti, A. M.; Conceição, A.; de Queiroz, L. P. 2006. Diversidade e caracterização dos fungos do semiárido brasileiro. Recife: APNE - Associação Plantas do Nordeste, v. II. 219 p

Gomide PHO, Silva MLN, Soares CRFS, Cardoso EL, de Carvalho F, Leal PL, Marques RM, Stürmer SL (2014) Arbuscular mycorrhizal fungi in vegetation types in the pantanal of Nhecolândia, Mato Grosso do Sul, Brazil, R. Bras. Ci. Solo, 38:1114–1127

Goto BT, Maia LC, Silva GA, Maia LC, Oehl F (2010) *Dentiscutata colliculosa*, a new species in the Glomeromycetes from Northeastern Brazil with colliculate spore ornamentation. Nova Hedwigia 90: 383–393

Goto BT, Maia LC, Silva GA, Oehl F (2009) *Racocetra intraornata*, a new species in the *Glomeromycetes* with a unique spore wall structure. Mycotaxon 109: 483–491

Goto BT, Silva GA, Maia LC, Souza RG, Coyne D, Tchabi A, Lawouin L, Hountondjo F, Oehl F (2011) *Racocetra tropicana*, a new species in the *Glomeromycetes* from tropical áreas. Nova Hedwigia 92: 69–82

Goto BT, Jardim JG, Silva GA, Furrazola E, Torres-Arias Y, Oehl F (2012a) *Glomus trufemii* (*Glomeromycetes*), a new sporocarpic species from Brazilian sand dunes. Mycotaxon 120: 1–9

Goto BT, Silva GA, Assis DMA, Silva DKA, Souza RG, Ferreira ACA, Jobim K, Mello CMA, Vieira HEE, Maia LC, Oehl F (2012b) *Intraornatosporaceae* (*Gigasporales*), a new family with two new genera and two new species. Mycotaxon 119: 117–132

Goto BT, Araújo AF, Soares ACF, Ferreira AC, Maia LC, Souza CS, Silva GA (2013) Septoglomus titan, a new fungus in the Glomeraceae (Glomeromycetes) from Bahia, Brazil. Mycotaxon 124:101–109

Hart MM, Reader RJ (2002) Taxonomic basis for variation in the colonization strategy of arbuscular mycorrhizal fungi. New Phytol. 153, 335–344

Jobim K, Goto BT (2016) Diversity of arbuscular mycorrhizal fungi (Glomeromycota) in maritime sand dunes of Brazilian northeast. Studies in Fungi 1 (1): 43–55

Jobim K, Oliveira BIS, Goto BT (2016) Checklist of the Glomeromycota in the Brazilian Savanna. Mycotaxon. 131: 1–13

Jobim K, Vista XM, Goto BT (2018) Updates on the knowledge of arbuscular mycorrhizal fungi (Glomeromycotina) in the Atlantic Forest biome –an example of very high species richness in Brazilian biomes. Mycotaxon 133(1): 209–209 https://doi.org/10.5248/133.209

Leal PL, Siqueira O, Stürmer SL (2013) Switch of tropical Amazon forest to pasture affects taxonomic composition but not species abundance and diversity of arbuscular mycorrhizal fungal community. Applied Soil Ecology 71:72–80

Maia LC, de Carvalho Júnior AA, Cavalcanti LH et al. (2015a) Diversity of Brazilian Fungi. Rodriguésia 66(4)

Maia RS, Vasconcelos SS, Carvalho CJR (2015b) Soil phosphorus fractions and mycorrhizal symbiosis in response to the availability of moisture and nutrients at a secondary forest in eastern Amazonia. Acta Amaz. 45:3

Marinho F, Silva GA, Ferreira ACA, Veras JSN, Souza NMF, Goto BT, Maia LC, Oehl F (2014) Bulbospora minima, a new genus and a new species in the Glomeromycetes from semi-arid Northeast Brazil. Sydowia 66: 313–323

Marinho F, da Silva IR, Oehl F, Maia LC (2018) Checklist of arbuscular mycorrhizal fungi in tropical forests. https://doi.org/10.12905/0380.sydowia70-2018–0107

Matias SR, Pagano MC, Muzzi FC, Oliveira CA, Carneiro AA, Horta SN, SCOTTI MR (2009) Effect of rhizobia, mycorrhizal fungi and phosphate-solubilizing microorganisms in the rhizosphere of native plants used to recover an iron ore area in Brazil?. European Journal of Soil Biology, v. 46, p. 10.1016

Mello CMA, da Silva IR, Pontes JS, Goto BT, da Silva GA, Maia LC (2012) Diversity of arbuscular mycorrizal fungi in an area of Caatinga, PE, Brazil. Acta Botanica Brasilica 26(4): 938–943

Mello CMA, Silva GA, Assis DMA, Pontes JS, Ferreira ACA, Leao MPC, Vieira HEE, Maia LC, Oehl F (2013) *Paraglomus pernambucanum* sp. nov. and Paraglomus bolivianum comb. nov., and biogeographic distribution of Paraglomus and Pacispora. Journal of Applied Botany and Food Quality 86: 113–125

Moreira M, Baretta D, Tsai SM, Cardoso EJBN (2006) Spore density and root colonization by arbuscular mycorrhizal fungi in preserved or disturbed Araucaria angustifolia (Bert.) O. Ktze. ecosystems. Sci Agric. 63: 380–385

Moreira M, Baretta D, Tsai SM, Cardoso EJBN (2009) Arbuscular mycorrhizal fungal communities in native and in replanted Araucaria forest. Sci Agr. 66: 677–684

Moreira M, Baretta D, Tsai SM, Gomes-da-Costa SM, Cardoso EJBN. (2007a) Biodiversity and distribution of arbuscular mycorrhizal fungi in Araucaria angustifolia forest. Sci Agr 64: 393–399

Moreira M, Nogueira MA, Tsai S M, Gomes-da-Costa S M, Cardoso EJBN (2007b) Sporulation and diversity of arbuscular mycorrhizal fungi in Brazil Pine in the field and in the greenhouse. Mycorrhiza 17: 519–526

Moreira M, Zucchi MI, Gomes JE, Tsai SM, Alves-Pereira A, Cardoso EJBN (2016) *Araucaria angustifolia* aboveground roots presented high arbuscular mycorrhizal fungal colonization and diversity in the Brazilian Atlantic Forest. Pedosphere 26(4): 561–566

Moreira-Souza M, Trufem SFB, Gomes-da-Costa SM, Cardoso EJBN (2003) Arbuscular mycorrhizal fungi associated with *Araucaria angustifolia* (Bert.) O. Ktze. Mycorrhiza 13: 211–215

Myers N, Mittermeier RA, Mittermeier CG, Fonseca GA, Kent J (2000) Biodiversity hotspots for conservation priorities. Nature 403: 853–858

Oehl F, Sieverding E, Ineichen K, Ris EA, Boller T, Wiemken A (2005) Community structure of arbuscular mycorrhizal fungi at different soil depths in extensively and intensively managed agroecosystems. New Phytol. 165: 273–283

Oki Y, Goto BT, Jobim K, Rosa LH, Ferreira MC, Coutinho ES et al. (2016) Arbuscular Mycorrhiza and Endophytic Fungi in Ruspestrian Grasslands. pp 157–179. In: Ecology and Conservation of Mountaintop grasslands in Brazil. Fernandes GW(Ed.) Springer, Switzerland.

Öpik M, Moora M, Liira J, Zobel M (2006) Composition of root-colonizing arbuscular mycorrhizal fungal communities in different ecosystems around the globe. Journal of Ecology 94: 778–790

Pagano MC, Scotti MR (2009) A survey of the arbuscular mycorrhiza occurrence in *Paepalanthus bromelioides* and *Bulbostylis* sp. in rupestrian fields, Brazil. Micologia Aplicada International 21: 1–10

Pagano MC, Scotti MR (2010) Effect of Phosphorus Fertilization on Arbuscular Mycorrhizal Colonization of Zeyheria tuberculosa a Native Species in Brazil?s Forest. Middle-East Journal of Scientific Research, v. 6, p. 604–611

Pagano MC, Cabello MN, Scotti MR (2010) Arbuscular mycorrhizal colonization and growth of *Eremanthus incanus* Less. in a highland field. Plant Soil Environ., 56, (9): 412–418
Pagano MC, Cabello MN (2012) Mycorrhizas in natural and restored riparian zones. In: Pagano MC, editor Mycorrhiza: Occurrence and Role in Natural and Restored Environments. Nova Science Publishers, Hauppauge, NY. ISBN: 978-1-61209-226-3.
Pagano MC, Zandavalli RB, Araújo FS (2013) Biodiversity of arbuscular mycorrhizas in three vegetational types from the semiarid of Ceará State, Brazil. Applied Soil Ecology 67: 37–46
Pagano MC (2016) Recent Advances on Mycorrhizal Fungi. 1. ed. London: Springer, 2016. v. 1. 150
Pereira CMR, da Silva DKA, Ferreira ACA, Goto BT, Maia LC (2014) Diversity of arbuscular mycorrhizal fungi in Atlantic forest areas under different land uses. Agriculture, Ecosystems and Environment 185: 245–252
Pontes JS, Oehl F, Pereira CD, Machado CTT, Coyne D, da Silva DKA, Maia LC (2017) Diversity of arbuscular mycorrhizal fungi in the Brazilian's Cerrado and in soybean under conservation and conventional tillage Applied Soil Ecology 117–118: 178–189
Ratter, J. A.; Ribeiro, J. F., Bridgewater, S. 1997. The Brazilian cerrado vegetation and threats to its biodiversity. Annals of Botany 80(3): 223–230.
Rizzini CT (1997) Tratado de fitogeografia do Brasil. Âmbito Cultural, Rio de Janeiro.
Santos OM, Oliveira NC, Novais RF (1995) Observações preliminares sobre fungos micorrízicos vesículo-arbusculares em plantas crescendo em dunas na Bahia. Revista Ceres 42: 191–202
Santos RS, Barreto-Garcia PAB, Scoriza RN (2018) Mycorrhizal fungi and litter as indicators of the edge effect in a fragment seasonal forest. Ciência Florestal 28: 324–335
Silva CF, Pereira MG, Silva EMR, Correia MEF, Saggin-Júnior OJ (2006) Fungos micorrízicos arbusculares em áreas no entorno do Parque Estadual da Serra do Mar em Ubatuba (SP). Revista Caatinga 19: 1–10
Silva DKA, Souza RG, Velez BAA, da Silva GA, Oehl F, Maia LC (2015a) Communities of arbuscular mycorrhizal fungi on a vegetation gradient in tropical coastal dunes. Appl. Soil Ecol. 96: 7–17. https://doi.org/10.1016/j.apsoil.2015.06.009.
Silva DKA, Coutinho FP, Escobar IEC, Souza RG, Oehl F, da Silva GA, Cavalcante UMT, Maia LC (2015b) The community of arbuscular mycorrhizal fungi in natural and revegetated coastal areas (Atlantic Forest) in northeastern Brazil. Biodivers. Conserv. 24(9): 2213–2226 https://doi.org/10.1007/s10531-015-0968-7.
Silva IR, da Silva, DKA, de Souza FA, Oehl F, Maia LC (2017) Changes in arbuscular mycorrhizal fungal communities along a river delta island in northeastern Brazil. Acta Oecologica 79: 8–17
Souza RG, Silva DK, Mello CMA et al. (2013) Arbuscular mycorrhizal fungi in revegetated mined dunes. Land Degradation and Development 24(2)
Stürmer SL, Bellei MM (1994) Composition and seasonal variation of spore populations of arbuscular mycorrhizal fungi in dune soils on the island of Santa Catarina, Brazil. Canadian Journal of Botany 72: 359–363
Stürmer SL, Klauberg Filho O, de Queiroz MH, Mendonça MM (2006) Occurrence of arbuscular mycorrhizal fungi in soils of early stages of a secondary succession of Atlantic Forest in South Brazil. Acta Bot Brasil 20 (3):513–521
Stürmer SL, Bellei MM (2011) Composition and seasonal variation of spore populations of arbuscular mycorrhizal fungi in dune soils on the island of Santa Catarina, Brazil. Canadian Journal of Botany 72(3): 359–363
Stürmer SL, Siqueira JO (2006) Diversity of Arbuscular Mycorrhizal Fungi in Brazilian Ecosystems. In: Moreira FMS, Siqueira JO, Brussaard L (eds) Soil Biodiversity in Amazonian and Other Brazilian Ecosystems. CABI Publishing, London, 206–236
Stürmer SL, Stürmer R, Pasqualini D (2013) Taxonomic Diversity and community structure of arbuscular mycorrhizal fungi (Phylum *Glomeromycota*) in three maritime sand dunes in Santa Catarina State, South Brazil. Fungal Ecology 6: 27–36
Teixeira AFS, Kemmelmeier K, Marascalchi MN, Stürmer SL, Carneiro MAC, Moreira FMS (2017) Arbuscular mycorrhizal fungal communities in an iron mining area and its surround-

ings: Inoculum potential, density, and diversity of spores related to soil properties. Ciência e Agrotecnologia, 41(5): 511–525
Trufem SFB, Malatinszky SMM, Otomo HS (1994) Fungos micorrízicos arbusculares em rizosferas de plantas do litoral arenoso do Parque Estadual da Ilha do Cardoso, SP, Brasil. Acta bot. bras. 8: 219–229
Trufem SB (1995) Aspectos ecológicos de fungos micorrízicos arbusculares na rizosfera de plantas de restinga da Ilha do Cardoso, SP, Brasil. Revta. Brasil. Bot. 18: 51–60
Valyi K, Mardhiah U, Rillig MC, Hempel S (2016) Community assembly and coexistence in communities of arbuscular mycorrhizal fungi. The ISME Journal 10: 2341–2351
Vieira LC, da Silva DKA, de Melo MAC, Escobar IEC, Oehl F, da Silva GA (2019) Edaphic factors influence the distribution of arbuscular mycorrhizal fungi along an altitudinal gradient of a tropical mountain, in press
Whitcomb S, Stutz JC (2007) Assessing diversity of arbuscular mycorrhizal fungi in a local community: role of sampling effort and spatial heterogeneity. Mycorrhiza 17: 429–437
Zangaro W, Rostirola LV, de Souza PB, Alves RA, Lescano LEAM, Rondina ABL, Nogueira MA, Carrenho R (2013) Root colonization and spore abundance of arbuscular mycorrhizal fungi in distinct successional stages from an Atlantic rainforest biome in southern Brazil. Mycorrhiza 23: 221–233
Zangaro W, Assis RL, Rostirola LV, Souza PB, Gonçalves MC, Andrade G, Nogueira MA (2008) Changes in arbuscular mycorrhizal associations and fine root traits in sites under different plant successional phases in southern Brazil, Mycorrhiza, 19:37–45.

Chapter 10
Mycorrhizas in Central Savannahs: *Cerrado* and *Caatinga*

Jadson B. Moura and Juliana S. R. Cabral

10.1 Introduction

10.1.1 Central Savannahs: Cerrado and Caatinga

Savannah is the name given to biomes that have a phytophysiognomy composed predominantly of low vegetation and small tree species (Veloso et al. 1991). The Cerrado and Caatinga, which together constitute 35% of the Brazilian territory (Vieira 2001; Beuchle et al. 2015) (Fig. 10.1), are considered savannahs since their phytophysiognomies are composed of herbaceous plants, small shrubs and grasses, and spaced tree species. These phytophysiognomies have instances of xeromorphism or adaptations to arid environments, such as trees and shrubs with tortuous branches, thick barks, hardened leaves with shiny surfaces and covered also by trichomes, and flower blooms occurs in the dry season.

The Cerrado (Fig. 10.2) is the second largest Brazilian biome after the Amazon, extending over an area of 2,045,064 km^2 that covers eight states of Central Brazil: Minas Gerais, Goiás, Tocantins, Bahia, Maranhão, Mato Grosso, Mato Grosso do Sul, Piauí and Distrito Federal (Hunke et al. 2015). It is crossed by three of the largest hydrographic basins in South America (SA) and has regular rainfall indexes that provide it with a great biodiversity. The Cerrado is nowadays considered the last agricultural frontier of the Americas (Braz et al. 2004; Klink and Machado 2005).

The predominant soil class in this biome is the oxisol type, which is a deep soil with low natural fertility and phosphorus levels, high iron contents, aluminium oxides and acids with intense weathering (Santos et al. 2013).

J. B. Moura (✉)
Evangelical Faculty of Goianésia – FACEG, Goiás, Brazil

J. S. R. Cabral
Institute of Higher Education of Rio Verde – IESRIVER, Goiás, Brazil

M. C. Pagano, M. A. Lugo (eds.), *Mycorrhizal Fungi in South America*, Fungal Biology, https://doi.org/10.1007/978-3-030-15228-4_10

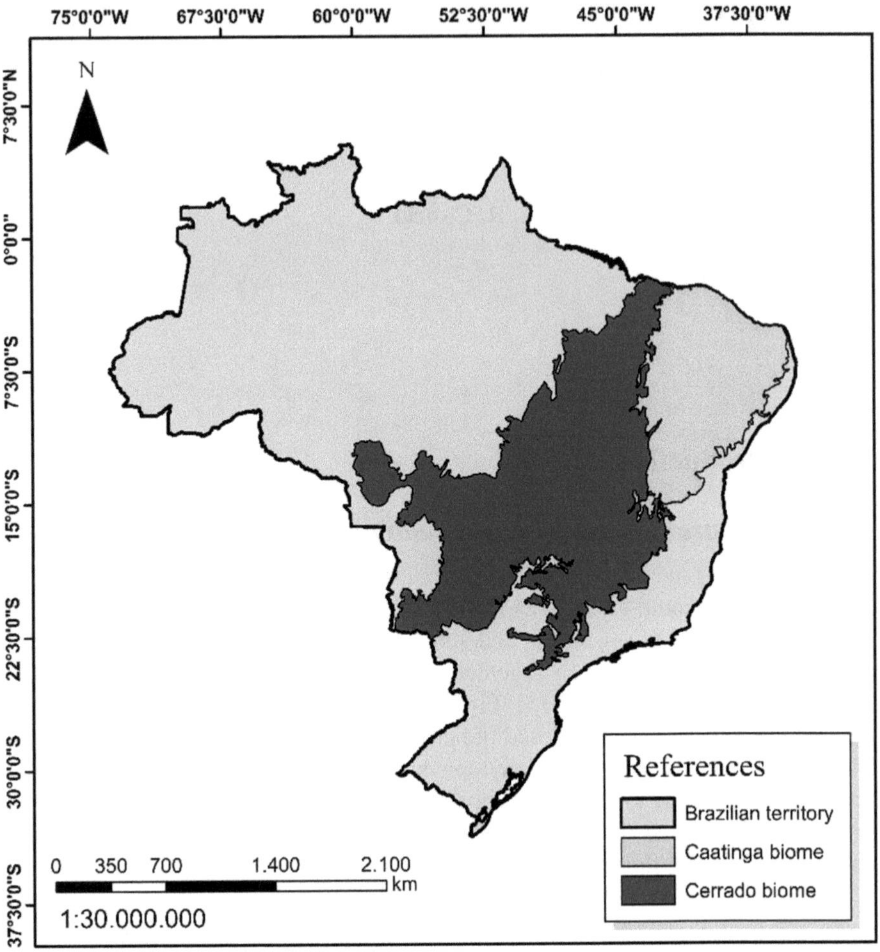

Fig. 10.1 Distribution of the Cerrado and Caatinga Biomes in the Brazilian territory

The Cerrado has one of the largest biodiversity on the planet because it is a transitional biome which is in direct geographical contact with other important South American biomes such as; Amazonia, Caatinga, Mata Atlântica, Pantanal and Bolivian Chaco (Taber et al. 1997; Klink and Machado 2005).

The Caatinga (Fig. 10.3) is a semi-arid biome and the main one of the Brazilian Northeast, covering nine states: Ceará, Bahia, Rio Grande do Norte, Paraíba, Pernambuco, Alagoas, Sergipe, southwest of Piauí and Maranhão, and northern Minas Gerais (Vasconcelos et al. 2017) and an area of approximately 900,000 km^2 that represent almost the 54% of the northeastern region as well as the 11% of the Brazilian territory.

Fig. 10.2 *Cerrado* area of the Chapada dos Veadeiros National Park, Goiás, Brazil

Fig. 10.3 *Caatinga* area in Nova Fazenda property, Desterro, Paraíba

The Caatinga biodiversity is fundamental to the sustainability of this biome. In addition to the herbaceous and shrub species, as well as the Cerrado, the Caatinga is composed of many thorny and aculeate species (such as Cactaceae and Bromeliaceae) and deciduous species that lose their leaves at the beginning of the dry season (Santos et al. 2017; Lima and Rodal 2010).

Due to the dry climate with scarce and irregular precipitations concentrated in the months of January to May, the little weathered soils are predominantly litholic, alic and dystrophic, of medium texture and poorly developed, varying from shallow to very shallow (Mello et al. 2012; Vasconcelos et al. 2017).

10.2 Mycorrhizae in the Cerrado and Caatinga

Central American savannahs are environments that naturally offer adverse abiotic conditions for plant growth and development, with low phosphorus level and limited water regime. In these environments, the plants depend directly on the performance of mycorrhizal fungi to survive such conditions, turning the association between fungi and plants as an important resilience factor to stressful situations (Thomazini 1974; Porcel and Ruiz- Lozano 2004; Hunke et al. 2015; Moura et al. 2017).

The different surveys carried out on different types of Cerrado soils show that arbuscular mycorrhizal fungi (AMF) are associated to a large number of native plants of the region including grasses, legumes and tree species such as "pequi" and "buriti" (Miranda 2008).

10.2.1 Arbuscular Mycorrhizal Fungi in the Central American Savannahs

Arbuscular mycorrhizal fungi (AMF) have recently received a new systematic status that ranges from phylum (Glomeromycota) to sub-phylum (Glomeromycotina) (Spatafora et al. 2016). Together with this advance, new families and genera have been described (Błaszkowski et al. 2018; Symanczik et al. 2018). Moreover, 3 classes, 16 families, 41 genera and 300 species have recently been officially reporting (Błaszkowski et al. 2018).

The average density of AMF species in the Cerrado varies from 25 to 50 spores per 50 cm^3 of soil. In the Caatinga area, there is a strong variation in the number of these fungi propagules due to plant community and soil chemical composition variations, which could contain high phosphorus levels (Mello et al. 2012).

Mimosa tenuiflora, a native plant of the Caatinga flora, is an important mycorrhizal fungi community host in the Brazilian semi-arid regions. In soils were *M. tenuiflora* was developed, 18 species of AMF were identified. Spores of the genera

Acaulospora, *Claroideoglomus*, *Dentiscutata*, *Entrophospora*, *Funneliformis*, *Gigaspora*, *Glomus*, *Racocetra*, *Rhizoglomus,* and *Scutellospora* were highly produced in the dry season (Mello et al. 2012; Souza et al. 2016). In the Caatinga area, in the state of Pernambuco, Mello et al. (2012) identified 16 taxa of AMF with a predominance of the genus *Glomus* represented by 7 species. These authors also reported for the first time in Brazil the presence of the species *Pacispora boliviana*, which was believed to be only found in Bolivia (Oehl and Sieverding 2004).

A new AMF species, named *Ambispora brasiliensis*, was identified in the Cerrado region by Goto et al. (2008) in the state of Minas Gerais. Further, Silva et al. (2008) found a new species in the northeast of Brazil, described under the name *Scutellospora pernambucana.* Moreover, de Pontes et al. (2017) identified *Acaulospora spinulifera* as a new species in Cerrado areas, the Atlantic forest, and in soybean crops in addition to an isolated new species, *Scutellospora alterata,* found in the northeast region of Caatinga (de Pontes et al. 2013). Thus, 67% of the 79 species of AMF found in Brazilian biomes were identified in the Cerrado, this could be mainly due to the regional soil conditions (Moreira et al. 2006; Moura 2015).

10.2.2 Ectomycorrhizal Fungi in the Central American Savannahs

There are few recorded instances of ectomycorrizal fungi associations in tropical savannahs, except for i) those found in some southeastern Asian tropical forests where plants of the Dipterocarpaceae family predominate in association with a large number of fungal species (Oliveira and Giachini 1999), and ii) for some leguminous species found in African savannahs (Bâ et al. 2012). In Brazil, there are reports of this association in the Amazon forest (Singer and Araujo 1979), in the Atlantic forest (Ferreira et al. 2013) and in *Eucalyptus* forests in the southern region of the country (Bertolazi et al. 2010).

Thomazini (1974) was a pioneer in the surveying of ectomycorrhizal fungi species in the Cerrado region. In addition, Casagrande (1985, 1986, 1987) found these fungal associations with "ata brava" *(Duquetia furfuraceae*), "carobinha" *(Jacaranda decurrens*), "guabiroba-do-Cerrado" (*Campomanesia coerulea*), "sene-do-campo" *(Cassia cathartica)* and "pata-de-vaca-do-Cerrado" *(Bauhinia holophylla).*

The hypogeal fungal species *Scleroderma polyrhizum* was found associated with "pequi" rhizosphere (*Caryocar brasiliense*) (Baseia and Milanez 2000). Underground fungi are more difficult to locate and hence to identify and classify. Consequently, the reported existence of this fungal growth habit associated with the Cerrado plants opens a new horizon for further research studies.

In the Caatinga region, due to the low density of arboreal species, there are no confirmed reports of fungi that form ectomycorrhizal associations with plants.

Baseia and Galvão (2002) looked for ectomycorrhizal associations among fungi of the phylum Basidiomycota in the rhizosphere of plants of the Caatinga but they did not reported the presence of symbiotic associations.

10.3 Agroecosystems in American Savannahs and Mycorrhizae

The different soil systems directly influence the rhizosphere mycorrhizal community. In general, this community is more abundant in agroecosystems cultivated under conservation systems than in fallow soils, intensive cultivations with phytosanitary products and systemic fungicides, or in crops with low or no dependence on symbiotic associations with mycorrhizal fungi (Miranda 2008). The highest abundance of AMF occurs in tropical systems, and the highest rates of colonization were observed in tropical savanna plants (Treseder and Cross 2006).

10.3.1 Arbuscular Mycorrhizal Fungi in Agroecosystems of American Savannahs

The AMF community is usually high in cultivated systems, especially in those that adopt conservation systems such as no-tillage or with reduced use of agrochemicals (Miranda and Miranda 2011). Ferreira et al. (2013) observed a lower spore density and AMF colonization in no-tillage systems and deforested areas than in riparian forests, forest edges, and pastures. Moreover, the authors also found a predominance of the families Acaulosporaceae, Glomeraceae, and Gigasporaceae. The maintenance of high amounts and diversity of AMF species was observed (de Pontes et al. 2013) in a cowpea (*Vigna unguiculata*) bean planting system, with organic fertilizer usage, emphasizing the importance of the myco-trophic crop species in agricultural production systems.

Proper soil and plant management are fundamental to benefit the mycorrhizal community and the symbiotic associations, especially in areas where the community is quantitatively and qualitatively low. The cultivation system used, although there is no great specificity between host and fungus, is the major influential factor in the occurrence and abundance of mycorrhizal fungi in the soil (Bever et al. 1996).

In the Cerrado region, where the climate is well defined with a dry and a rainy season, it can be observed that the spores abundance in the native soil is low and inferior to that in the cultivated soil where there is an increased in the AMF abundance. As regards to the spore number, it increases along with soil moisture at the beginning of the rainy season whereas it decreases in the upper soil layers in the dry period (Howeler et al. 1987).

Besides no-tillage and reduced use of agrochemicals, another practice of common management in the cultivated areas of the Cerrado is the correction of soil pH with the application of limestone, which positively influences the soil mycorrhizal community by gradually increasing the spore number. Even though Miranda and Miranda (2011) verified the gradual increase of spore number in an area that suffered pH correction; limestone overdoses have also been proved to cause a decrease in spore density. In addition to pH correction, the application of phosphorus, a scarce nutrient in most Cerrado soils, could be also beneficial to the mycorrhizal fungi community (Miranda 2008).

The mostly adopted planting methods in the region are the conventional system, which makes use of soil tillage, and the no-tillage system which does not have tillage and aims at constant vegetation cover. Moura (2015) obtained similar values of spores density and mycorrhizal community composition when comparing conventional and conservationist systems applied to wheat, sugarcane, brave beans, maize, corn, crotalaria, and brachiaria. Oehl et al. (2004) verified two different farming systems, conventional and organic farming, they report that AMF spore abundance and species diversity was significantly higher in the organic than in the conventional systems.

10.3.2 The Use of Ectomycorrhizae in Agroecosystems under the Cerrado Soil

Ectomycorrhizal fungi, unlike AMF, are characterized by their high fungus-host specificity, having a predilection for arboreal species. Genera such as *Eucalyptus*, widely cultivated in the Cerrado region, are among the species capable of forming symbiotic associations with ectomycorrhizal fungi (Miranda 1986).

Giachini et al. (2000) found in cultivated soils the genera *Chondrogaster, Descomyces, Hysterangium, Pisolithus* and *Setchelliogaste*r associated with *Eucalyptus* spp., and the genera *Amanita, Rhizopogon, Lactarius,* and *Suillus* associated with cultivated *Pinus*. In the Amazon forest region, Singer and Araujo (1979) verified the presence of symbiotic associations in the dense litter layer. Further, Pagano and Scotti (2008) demonstrated that the cultivation of *Eucalyptus camaldulensis* has a variation in its mycorrhizal colonization with a predominance of AMF associations (with *Glomus* sp.) while *Eucalyptus grandis* has a predominance of ectomycorrhizal colonization with low spore density, showing the different mycorrhizal dependence among these species.

In the southern region of Bahia in the Cerrado soil, positive results were observed when adopting the inoculation to establish *Pinus* seedlings with *Thelephora terrestris* and *Pisolithus tinctorius* (Filho and Krugner 1980).

10.4 Final Considerations

All things considered, among all the planting practices adopted in the tropical savannahs, the choice of crop species, irrigation system and agrochemicals used have a much stronger impact on the mycorrhizal community than the soil types or planting system adopted. Water availability is also a limiting factor in these regions, the water regime is divided into dry and rainy annual seasons, that also influence the mycorrhizal communities.

References

Bâ, A. M., Duponnois, R., Moyersoen, B., & Diédhiou, A. G. (2012). Ectomycorrhizal symbiosis of tropical African trees. Mycorrhiza, 22(1), 1–29.

Bertolazi AA, Canton GC, Azevedo IG, Cruz ZMA, Soares DNES, Conceição JM, Santos WO, Ramos AC (2010) O papel das ectomicorrizas na biorremediação dos metais pesados no solo. Nat on-line 8:24–31

Baseia IG, Galvão TC de O (2002) Some interesting Gasteromycetes (Basidiomycota) in dry areas from northeastern Brazil. Acta Bot Brasil 16:1–8

Baseia IG, Milanez AI (2000) First record of *Scleroderma polyrhizum* Pers. (Gasteromycetes) from Brazil. Acta Bot Brasil 14(2):181–184

Beuchle R, Grecchi RC, Shimabukuro YE, Seliger R, Eva HD, Sano E, Achard F (2015) Land cover changes in the Brazilian Cerrado and Caatinga biomes from 1990 to 2010 based on a systematic remote sensing sampling approach. Appl Geogr 58:116–127

Bever JD, Morton JB, Antonovics J, Schultz PA (1996) Host-Dependent Sporulation and Species Diversity of Arbuscular Mycorrhizal Fungi in a Mown Grassland. Br Ecol Soc 84:71–82

Błaszkowski J, Kozłowska A, Niezgoda P, Goto BT, Dalpé Y (2018) A new genus, *Oehlia* with *Oehlia diaphana* comb. nov. and an emended description of *Rhizoglomus vesiculiferum* comb. nov. in the Glomeromycotina. Nova Hedwigia 107:501–518

Braz SP, Urquiaga S, Alves B, Boddey RM (2004) Degradação de pastagens, matéria orgânica do solo e a recuperação do potencial produtivo em sistemas de baixo input tecnológico na região dos cerrados. Seropédica, Embrapa

Casagrande LIT (1985) Micorriza em plantas de Cerrado: *Duquetia furfuraceae* (St. Hil.) & Hook. Rev Agric (Brazil) 60(2):107–110

Casagrande LIT (1986) Um caso de micorriza ectendotrófica em planta de Cerrado. Rev Agric (Brazil) 61:27–33

Casagrande LIT (1987) Ectomicorrizas e endomicorrizas em *Jacaranda decurrens* Cham. Rev Agric (Brazil) 62:229–237

de Pontes JS de, Sánchez-Castro I, Palenzuela J, Maia LC, Silva GA da, Oehl F (2013) *Scutellospora alterata*, a new gigasporalean species from the semi-arid Caatinga biome in Northeastern Brazil. Mycotaxon 125:169–181

de Pontes JS de, Santos VM, Pereira CD, Alves da Silva G, Maia LC, Oehl F (2017) *Acaulospora spinulifera*, a new arbuscular mycorrhizal fungal species from the Brazilian Cerrado and Atlantic Rain forest. Nova Hedwigia 105:219–229

Ferreira AJ, Wartchow F, Cortez VG (2013) *Limacella ochraceolutea* (Agaricomycetes) in the Atlantic forest of southern Brazil. Field Mycol 14:64–67

Filho MT, Krugner TL (1980) Formação de ectomicorrizas e crescimento de mudas de *Pinus caribaea* var. *bahamensis* em solo de viveiro infestado artificialmente com *Thelephora terrestris* e *Pisolithus tinctorius* no Litoral Sul da Bahia. IEPF 21:21–37

Giachini AJ, Oliveira VL, Castellano MA, Trappe JM (2000) Ectomycorrhizal fungi in *Eucalyptus* and *Pinus* plantations in Southern Brazil. Mycologia 92:1166–1177
Goto BT, Moreira M, Tsai SM, Cardoso EJBN, Maia LC (2008) Notas sobre *Acaulospora bireticulata* Rothwell &Trappe e primeiro registro de *Acaulospora koskei* Blask. para o Brasil. Acta Bot Brasilica 22:583–587
Howeler RH, Sieverding E, Saif S (1987) Practical aspects of mycorrhizal technology in some tropical crops and pastures. Plant Soil Interfaces Interact. Dordrecht, Springer Netherlands, p 249–283
Hunke P, Mueller EN, Schröder B, Zeilhofer P (2015) The Brazilian Cerrado: assessment of water and soil degradation in catchments under intensive agricultural use. Ecohydrology 8:1154–1180
Klink CA, Machado RB (2005) A conservação do Cerrado brasileiro. Megadiversidade 1(1):147–155
Lima ALA, Rodal MJN (2010) Phenology and wood density of plants growing in the semi-arid region of Northeastern Brazil. J Arid Environ 74:1363–1373
Mello CMA de, Silva IR da, de Pontes JS de, Goto BT, Silva GA da, Maia LC (2012) Diversidade de fungos micorrízicos arbusculares em área de Caatinga, PE, Brasil. Acta Bot Brasilica 26:938–943
Miranda JCC de (1986) Utilização das Micorrizas na Agricultura. Brasilia, Brazil
Miranda JCC de (2008) Cerrado: Micorriza arbuscular: ocorrência e manejo. Planaltina, Embrapa Cerrados, p 169
Miranda JCC de, Miranda LN de (2011) Contribuição da micorriza arbuscular para a produtividade e sustentabilidade nos sistemas de produção com plantio direto e convencional no Cerrado. Planaltina, DF: Embrapa Cerrados
Moreira FMS, Siqueira JO, Brussaard L (2006) Soil biodiversity in Amazonian and other Brazilian ecosystems. CABI 2006. Micorrizas. In: Moreira FMS, Siqueira JO (org) Microbiol e Bioquim do Solo. 2nd edition, p 729
Moura JB de (2015) Diversidade e colonização micorrízica em diferentes usos do solo no Cerrado. UNB
Moura JB de, Valentim NM, Ventura MVA, Junior WGV (2017) Taxa de colonização micorrízica sob diferentes sistemas de cultivo no cerrado em cana-de-açúcar. Faculdade de Tecnologia e Ciencias 2:60–66
Oehl F, Sieverding E (2004) *Pacispora*, a new vesicular arbuscular mycorrhizal fungal genus in the Glomeromycetes. J Appl Bot 78:72–82
Oehl F, Sieverding E, Mäder P, Dubois D, Ineichen K, Boller T, Wiemken A (2004) Impact of long-term conventional and organic farming on the diversity of arbuscular mycorrhizal fungi. Oecologia 138:574–583
Oliveira VL, Giachini AJ (1999) Ecologia e Aplicação de Ectomicorrizas. In: Siqueira OJ, Moreira FMS, Lopes AS, Guilherme LRG, Faquim V, Furtini Neto AE, Carvalho JG (eds) Inter-relação Fertil Biol do Solo e Nutr Plantas. Lavras, UFLA, p 818
Pagano MC, Scotti MR (2008) Arbuscular and ectomycorrhizal colonization of two *Eucalyptus* species in semiarid Brazil. Mycoscience 49(6):379–384
Porcel R, Ruiz-Lozano JM (2004) Arbuscular mycorrhizal influence on leaf water potential, solute accumulation, and oxidative stress in soybean plants subjected to drought stress. J Exp Bot 55:1743–1750
Santos HG dos, Jacomine PKT, Anjos LHC dos, Oliveira VÁ de, Lumbreras JF, Coelho MR, Almeida JA de, Cunha TJ (2013) Sistema brasileiro de classificação de solos. Embrapa, Brasilia, Brazil
Santos W de S, Henriques IGN, Santos W de S, Ramos GG, Vasconcelos G dos S, Vasconcelo ADM (2017) Análise florística-fitossociológica e potencial madeireiro em área de Caatinga submetida a manejo florestal. Agropecuária Científica no Semiárido 13:203–211
Silva DKA da, Freitas N de O, Cuenca G, Maia LC, Oehl F (2008) *Scutellospora pernambucana*, a new fungal species in the Glomeromycetes with a diagnostic germination orb. Mycotaxon 106:361–370

Singer R, Araujo I de J da S (1979) Litter decomposition and ectomycorrhiza in Amazonian forests. Acta Amaz 9:25–42

Souza BR, Moura JB, Oliveira TC, Ramos MLG, Lopes Filho LC (2016) Arbuscular Mycorrhizal fungi as indicative of soil quality in conservation systems in the region of vale do São Patrício, Goiás. Int J Curr Res IN 8:43307–43311

Spatafora JW, Chang Y, Benny GL, Lazarus K, Smith ME, Berbee ML, Bonito G, Corradi N, Grigoriev I, Gryganskyi A, James TY, O'Donnell K, Roberson RW, Taylor TN, Uehling J, Vilgalys R, White MM, Stajich JE (2016) A phylum-level phylogenetic classification of zygomycete fungi based on genome-scale data. Mycologia 108:1028–1046

Symanczik S, Al-Yahya'ei MN, Kozłowska A, Ryszka P, Błaszkowski J (2018) A new genus, *Desertispora*, and a new species, *Diversispora sabulosa*, in the family Diversisporaceae (order Diversisporales, subphylum Glomeromycotina). Mycol Prog 17:437–449

Taber A, Navarro G, Arribas MA (1997) A new park in the Bolivian Gran Chaco – an advance in tropical dry forest conservation and community-based management. Oryx 31:189–198

Thomazini LI (1974) Mycorrhiza in plants of the 'Cerrado'. Plant Soil 41:707–711

Treseder KK, Cross A (2006) Global Distributions of Arbuscular Mycorrhizal Fungi. Ecosystems 9:305–316

Vasconcelos ADM, Henriques IGN, Souza MP de, Santos W de S, Santos W de S, Ramos GG (2017) Caracterização florística e fitossociológica em área de Caatinga para fins de manejo florestal no município de São Francisco-PI. Agropecuária Científica no Semiárido 13:329–337

Veloso HP, Rangel Filho ALR, Lima JCA (1991) Classificação da Vegetação Brasileira Adaptada a um Sistema Universal. CDDI, organizador. Rio Janeiro, IBGE, Dep. Recur. Nat. e Estud. Ambient. Rio de Janeiro: fundação instituto brasileiro de geografia e estatistica

Vieira FA (2001) Micorrizas na Região do Cerrado. Universidade de Brasilia, Brazil, p 28

Chapter 11
Structure and Diversity of Arbuscular Mycorrhizal Fungal Communities Across Spatial and Environmental Gradients in the Chaco Forest of South America

Gabriel Grilli, Nicolás Marro, and Lucía Risio Allione

11.1 Introduction

The Gran Chaco Americano represents the largest seasonally dry forest ecosystem in South America (SA). This biogeographical region encompasses part of Argentina, East Bolivia, a little of southeastern Brazil, and western Paraguay, occupying more than 1,200,000 km^2 (Zak et al. 2008; Caldas et al. 2015). In central Argentina, the Chaco forest may be classified, according to precipitation patterns, into eastern humid to sub-humid Chaco and into western dry semi-arid to arid Chaco (Hoyos et al. 2013). Recently, Oyarzabal et al. (2018) distinguished 7 units of vegetation in the Chaco region of central Argentina according to physiognomic and floristic features: the Wetlands, Sub-humid Chaco, Arid Chaco, Semiarid Chaco, Chaco Serrano, Mountain grassland and "Salinas" (Salt flats) (Fig. 11.1). Therefore, the Chaco forest has a wide variety of environments, each with high levels of biodiversity (Bucher and Huszar 1999; Nori et al. 2016) and providing important ecosystemic services such as biodiversity maintenance, clean water supply, carbon sequestration and soil erosion protection (Kauffman et al. 2003; Caldas et al. 2015). Regrettably, as in many ecosystems around the globe, the land use changes are profoundly affecting the biodiversity present in this region (e.g. Cagnolo et al. 2006; Longo et al. 2014; Grilli et al. 2012, 2017; Verga et al. 2017). Biodiversity studies

G. Grilli (✉) · N. Marro
Laboratorio de Micología. Instituto Multidisciplinario de Biología Vegetal, FCEFyN (CONICET-Universidad Nacional de Córdoba), Córdoba, Argentina
e-mail: ggrilli@imbiv.unc.edu.ar

L. Risio Allione
Laboratorio de Micología, Diversidad e Interacciones Fúngicas (MICODIF), Área de Ecología, Facultad de Química, Bioquímica y Farmacia, Universidad Nacional de San Luis, San Luis, Argentina

IMIBIO-CONICET, Universidad Nacional de San Luis, San Luis, Argentina

M. C. Pagano, M. A. Lugo (eds.), *Mycorrhizal Fungi in South America*, Fungal Biology, https://doi.org/10.1007/978-3-030-15228-4_11

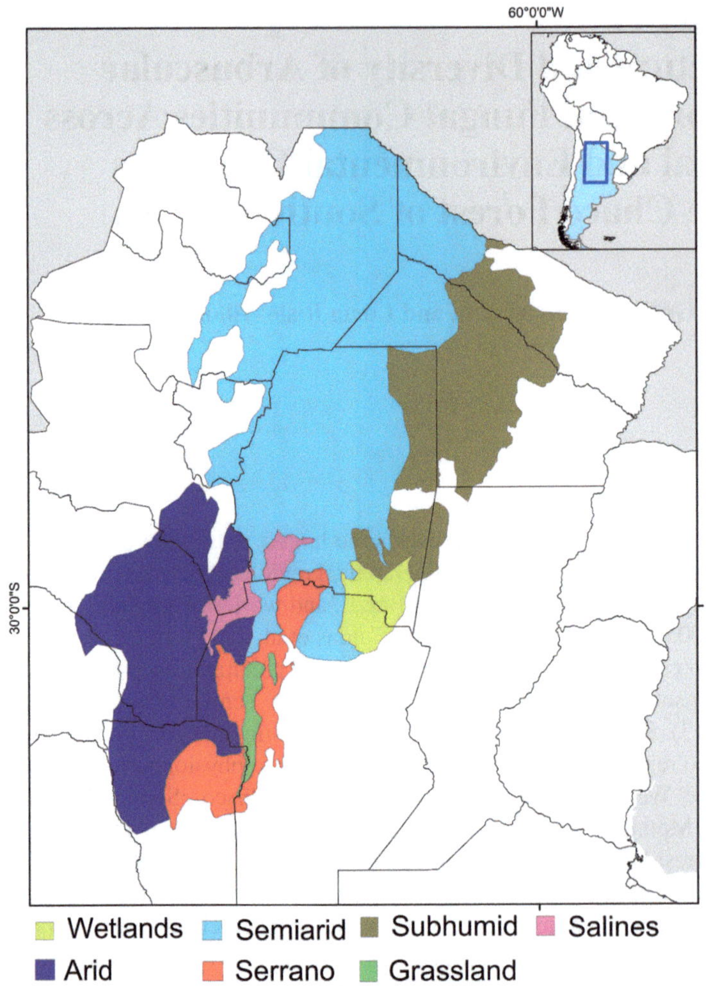

Fig. 11.1 Vegetation units at Chaco forest in Argentina (Oyarzabal et al. 2018)

conducted in this region have been mainly focused on aboveground organisms (i.e. mainly plants; e.g. Zak et al. 2004; Cagnolo et al. 2006; Cabido et al. 2018) while a lower number of studies considered belowground organisms (e.g. Lugo et al. 2005; Urcelay et al. 2009; Becerra et al. 2014; Grilli et al. 2013; Soteras et al. 2016). Therefore, information about how belowground communities are spatially structured across vegetation units in the Chaco forest is still scarce.

Regarding belowground organisms, the arbuscular mycorrhizal fungi (AMF) are one of the main elements and most widespread fungal type in soil microbiota. The AMF establish a symbiotic association with the majority of land plants, providing them with major access to soil nutrients in return for carbon compounds (Smith and Read 2008). The AMF were traditionally identified on the basis of resistance structures such as asexual spores (i.e. morphospecies). Lately, improvement in DNA sequencing techniques (i.e. Next Generation Sequencing) allowed the identification of uncultured environmental communities (Öpik et al. 2009), enabling the study of variation in the composition of AM fungal communities across spatial, environmental and biotic gradients.

The study of how communities assemble in nature has been one of the major paradigms in ecology since the beginning of the discipline (Preston 1948; Pielou 1975; Tokeshi 1990; McGill et al. 2006; Götzenberger et al. 2012). Accordingly, several theories have arisen in order to explain the community structure of living organisms. Among them, the deterministic niche-based processes predict niche differentiation by inter-specific competition as the main force structuring the community (Gause 1934; Silvertown 2004), while stochastic neutral theories affirm that demographic stochasticity and dispersal limitation structure the communities (Hubbell 2001). In general, these theories have been tested on macro-organisms (reviewed in Götzenberger et al. 2012), and to a minor extent on micro-organisms (Dumbrell et al. 2010; Kivlin et al. 2011; Verbruggen et al. 2012). Phylogenetic analysis might be helpful to prove deterministic and stochastic processes in AMF communities (Egan et al. 2017). If we consider that functional traits in AMF might be phylogenetically constrained (Powell et al. 2009), we could expect that niche-based mechanisms exclude species with similar traits due to inter-specific competition and therefore, enhance the phylogenetic dispersion of AM fungal communities (i.e. phylogenetic overdispersion). On the contrary, neutral mechanisms such as dispersal limitation or deterministic such as habitat filtering reduce the phylogenetic distance between the AMF communities (i.e. phylogenetically clustered). Regarding the Chaco forest, the forces structuring AM fungal communities in the vegetation units remain poorly understood.

Therefore, the aims of this chapter include reviewing the data available in the literature concerning AMF communities in order to (i) analyze the relationship of AMF taxonomic diversity with spatial and environmental variables across ecosystems in the Chaco forest, and to (ii) evaluate the phylogenetic diversity and the structure (i.e. clustered or overdispersed) of AMF communities at the Chaco forest ecoregions compared to local, regional and global species pools.

11.2 Taxonomic Diversity of AMF Communities: Relationship with Spatial and Environmental Variables of Vegetation Units at the Chaco Forest

We have analyzed the morphospecies diversity occurrence along spatial and environmental variables in the vegetation units (Oyarzabal et al. 2018) of the Chaco phytogeographic province. We searched in Google Scholar for studies performed in the specified ecosystem using the terms: "arbuscular mycorrhiza", "chaco/chaquean forest" and "soil nutrients". After filtering the search, we found nine studies performed in four of the seven vegetation units of the Chaco forest that matched our terms. The studies found were assigned to vegetation units as follows: two to Salinas, one to Arid Chaco, two to Chaco Serrano and four to Mountain grassland. No study was found for the Wetlands, Sub-humid Chaco and Semi-arid Chaco (Table 11.1).

We considered the datasets in the studies as independent for the analysis of AMF diversity between vegetation units. To test the differences in AMF richness, we fitted generalized linear models using "vegetation unit" as the fixed factor and random intercept "plant host/source" with *glmer.nb()* function in package lme4 in R (Bates et al. 2015). The relationship between similarity of AMF communities and the euclidean distance between study sites was assessed with a distance decay of similarity analysis (Mantel test).

Arbuscular mycorrhizal fungi richness between vegetation units showed significant differences (Fig. 11.2a). The fixed factor "vegetation unit" and the random term "host/source" were significant for the model fit (df = 1, Chisq = 37.72, $P < 0.0001$). The morphospecies richness decrease in the direction Chaco Serrano > Mountain grassland = Salinas > Arid Chaco. The Chaco Serrano showed the highest richness (mean ± SD =16.37 ± 3.07) which was four times higher than the one in the Arid Chaco (mean ± SD = 4.49 ± 3.26; z = −3.54, $P < 0.0001$), while the Mountain grassland and Salinas' richness showed no significant differences between them (z = −0.199, $P = 0.84$).

The results showed a significant distance-decay relationship of the AMF community similarity with increased geographic distance. The Mantel test (r = 0.33; P = 0.001) showed a significant spatial correlation between the geographic and AMF community matrices. Similar results in distance decay of similarity were observed in AMF communities at local (Dumbrell et al. 2010; Davison et al. 2012) and regional scale (Lekberg et al. 2007; van der Gast et al. 2011) and with empirical evidence on a global scale (Davison et al. 2015).

To analyze the diversity of AMF communities at different geographical scales we performed Principal Components of Neighbour Matrices (PCNM). Therefore, spatial relationships of the biological response were evaluated at multiple scales (Borcard and Legendre 2002). We used the function *pcnm()* from package vegan (Oksanen et al. 2018) to calculate the PCNMs variables. The wider spatial scale is represented by the first PCNM eigenvector (PCNM1) and decrease while the PCNMs eigenvalue increases. Only positive and significant (alpha = 0.05) PCNM variables were selected. To evaluate the relationship of AMF communities with the edaphic data, we selected soil variables that were present in all the studies (Nitrogen, Phosphorus and pH). The altitude (m) of sampling points in the dataset was obtained with the *getData()* function from package raster (Hijmans et al. 2017). In addition, we downloaded climatic data from Worldclim database (Fick and Hijmans 2017) for all the study sites. In particular, we gathered information about mean annual temperature and precipitation ("Temp", "Prec"). Arbuscular mycorrhizal fungi community dissimilarity distance matrix was constructed with the *vegdist()* function based on Sorensen index. The response and explanatory variables (PCNMs,

Table 11.1 Studies carried out in the Chaco forest at Central Argentina

Study	latitude	longitude	Site	Host/Source	Plant type/ source	Vegetation unit
Becerra et al. (2014)	−29.45	−64.30	Salinas Ambargasta	*Allenrolfea patagonica*	Shrub	Salinas
	−29.45	−64.30	Salinas Ambargasta	*Atriplex argentina*	Shrub	Salinas
	−29.45	−64.30	Salinas Ambargasta	*Hetterostachys ritteriana*	Shrub	Salinas
	−29.45	−64.30	Salinas Ambargasta	*Suaeda divaricata*	Shrub	Salinas
	−29.73	−64.52	Salinas Grandes	*Allenrolfea patagonica*	Shrub	Salinas
	−29.73	−64.52	Salinas Grandes	*Atriplex argentina*	Shrub	Salinas
	−29.73	−64.52	Salinas Grandes	*Hetterostachys ritteriana*	Shrub	Salinas
	−29.73	−64.52	Salinas Grandes	*Suaeda divaricata*	Shrub	Salinas
Lugo et al. (2005)	−30.53	−64.71	Cruz del Eje	*Larrea divaricata*	Shrub	Arid
	−30.53	−64.71	Cruz del Eje	*Neobouteloua lophostachya*	Herb	Arid
	−30.53	−64.71	Cruz del Eje	*Sporobolus pyramidatus*	Herb	Arid
	−30.53	−64.71	Cruz del Eje	*Trichloris crinita*	Herb	Arid
Longo et al. (2014)	−31.04	−64.32	Agua de Oro	Soil	Soil	Serrano
	−31.12	−64.28	Salsipuedes	Soil	Soil	Serrano
	−31.77	−64.47	La Serranita	Soil	Soil	Serrano
	−31.50	−64.59	Cuesta Blanca	Soil	Soil	Serrano
	−31.30	−64.50	Bialet Massé	Soil	Soil	Serrano
	−31.04	−64.32	Agua de Oro	Soil	Soil	Serrano
	−31.12	−64.28	Salsipuedes	Soil	Soil	Serrano
	−31.77	−64.47	La Serranita	Soil	Soil	Serrano
	−31.50	−64.59	Cuesta Blanca	Soil	Soil	Serrano
	−31.30	−64.50	Bialet Massé	Soil	Soil	Serrano
Cofré et al. (2012)	−29.74	−64.53	Salinas Grandes	*Atriplex cordobensis*	Shrub	Salinas
Lugo et al. (2003)	−31.33	−64.75	Pampa de Achala	*Alchemilla pinnata*	Herb	Mountain grassland
	−31.33	−64.75	Pampa de Achala	*Briza subaristata*	Herb	Mountain grassland
	−31.33	−64.75	Pampa de Achala	*Deyeuxia hieronymi*	Herb	Mountain grassland
	−31.33	−64.75	Pampa de Achala	*Poa stuckertii*	Herb	Mountain grassland
	−31.33	−64.75	Pampa de Achala	*Eragrostis lugens*	Herb	Mountain grassland
	−31.33	−64.75	Pampa de Achala	*Sorghastrum pellitum*	Herb	Mountain grassland

(continued)

Table 11.1 (continued)

Study	latitude	longitude	Site	Host/Source	Plant type/ source	Vegetation unit
Menoyo et al. (2009)	−31.42	−64.78	Pampa de Achala	*Polylepis australis*	Tree	Mountain grassland
	−31.62	−64.82	Pampa de Achala	*Polylepis australis*	Tree	Mountain grassland
Soteras et al. (2014)	−31.97	−64.93	Los Molles	*Polylepis australis*	Tree	Mountain grassland
	−31.97	−64.93	Los Molles	*Polylepis australis*	Tree	Mountain grassland
	−31.97	−64.93	Los Molles	*Deyeuxia hieronymi*	Tree	Mountain grassland
Urcelay et al. (2009)	−31.50	−64.58	Cuesta Blanca	Soil	Soil	Serrano
Soteras et al. (2015)	−31.38	−64.80	Los Gigantes	*Polylepis australis*	Tree	Mountain grassland
	−31.97	−64.93	Los Molles	*Polylepis australis*	Tree	Mountain grassland
	−31.73	−64.78	Pampa de Achala	*Polylepis australis*	Tree	Mountain grassland

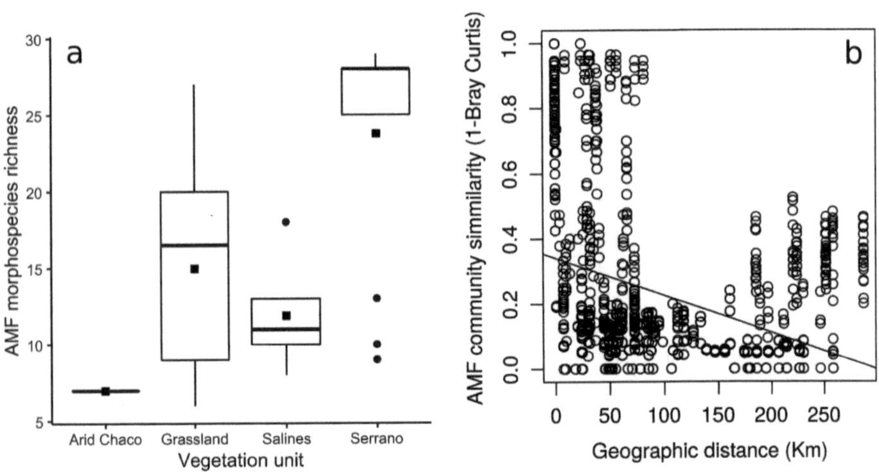

Fig. 11.2 AMF morphospecies (**a**) mean richness in vegetation units of the Chaco forest and (**b**) community distance-decay analisys (Mantel test) of similarity with geographic distance (Km)

edaphic and environmental) were standardized with the *decostand()* function in order to center (means ~ 0) and scale (SD = 1) the data.

Variation in AMF communities composition (presence-absence) was partitioned among explanatory variables (PCNMs, environmental and edaphic) with a distance based redundancy analysis (db-RDA, Legendre and Anderson 1999) using the *capscale()* function from package vegan that allows the analysis of another distance than the euclidean, such as dissimilarity distance. The most significant variables (i.e. geographic, environmental and edaphic) were selected with a permutational (999 permutations) forward selection model procedure performed with the *ordiR2step()* function based on the adjusted R^2 and p-value (Blanchet et al. 2008).

The results of the forward selection model showed that AMF community composition was being significantly structured by four geographic variables (in order of importance, PCNM1: F = 20.68, $P = 0.001$, PCNM2: F = 34.4, $P = 0.001$, PCNM6: F = 3.33, $P = 0.015$, PCNM3: F = 4.93, $P = 0.002$), environmental variables (altitude: F = 37.2, $P = 0.001$; precipitation: 32.49, $P = 0.001$; temperature: F = 18.25, $P = 0.001$) and edaphic variables (pH: F = 2.44, $P = 0.048$; Nitrogen: F = 4.78, $P = 0.001$, Phosphorus: ns).

The distance based redundancy analysis showed that the complete model with the selected environmental variables (PCNMs geographic, climatic and edaphic) explained the 41% of the AMF community variation ($R_a^2 = 0.79$, df = 9, F = 17.61, $P = 0.001$). In particular, the geographic variables explained the 29% ($R_a^2 = 0.52$, df = 4, F = 14.8, $P = 0.001$, Fig. 11.3a), the climatic variables explained the 22% ($R_a^2 = 0.4$, df = 3, F = 12.5, $P = 0.001$; Fig. 11.3b) and the edaphic explained the 13% ($R_a^2 = 0.22$, df = 2, F = 8.3, $P = 0.001$; Fig. 11.3c).

Our results suggest that variation in AMF communities respond to spatial structure at a wider scale according to PCNMs variables, and to environmental and edaphic variables, consistent with previous evidence (Dumbrell et al. 2010; Soteras et al. this book; Davison et al. 2015; Powell and Rillig 2018; Kotilínek et al. 2017).

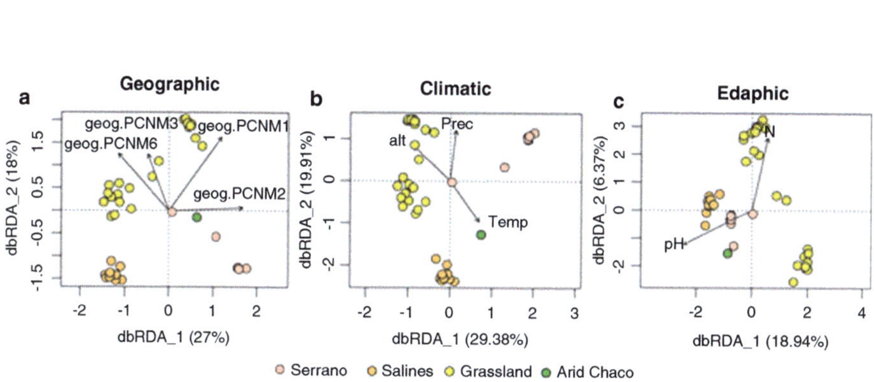

Fig. 11.3 Arbuscular mycorrhizal fungi communities ordination in vegetation units of the Chaco forest considering mean morphospecies richness and community distance-decay analysis (Mantel test) of similarity with geographic distance (Km). Distance based redundancy analysis (dbRDA) of AMF communities considering (**a**) Geographic, (**b**) Climatic and (**c**) Edaphic variables. References: the length and direction of the vectors represent the strength and direction of the relationship between the variables and community composition

11.3 Phylogenetic Diversity and Structure of AMF Communities in the Chaco Forest

In order to analyze the phylogenetic diversity and structure of AMF communities, we gathered studies that evaluate AMF molecular diversity in any of the 7 units of vegetation in the Chaco forest of central Argentina. We circumscribed our study by adhering to the Virtual Taxa (VT) proposed by Öpik et al. (2010) since it is the most extended and quality-controlled database of AMF molecular diversity. To identify those studies that match our search we used Google Scholar with the following terms: "arbuscular mycorrhiza", "ribosomal small-subunit (SSU)" and "Chaco/chaquean forest". The search yielded a total of two studies performed in only two of the Chaco regions: "Mountain grasslands" (Davison et al. 2015) and "Chaco Serrano" (Grilli et al. 2015). Despite the low number of studies, both datasets yielded a total of 137 VT of AMF in 75 root samples of five plant species. To assess the sampling efficacy in both datasets we constructed species accumulation curves (Fig. 11.4) with the function *speccacum()* in the R package vegan (Oksanen et al. 2018). Virtual taxa richness between regions was compared with the *glmer.nb()* function from package lme4 to control overdispersion. The model was specified by using "region" as a fixed factor and "plant host species" as a random factor. To visualize the variation in AMF community composition (i.e. presence-absence) between regions we used non-metric multidimensional scaling (NMDS, package vegan) using Bray-Curtis dissimilarity distance. Further, multivariate ANOVA with permutations (PERMANOVA) was used to assess variation in AMF community composition between regions. Since PERMANOVA is also sensitive to multivariate dispersion (i.e. analogous to the homogeneity of variance) we used the *betadisper()* function from package vegan (Anderson 2001) to test for the possibility of differences arising from within-group dispersion rather than from a compositional change in the community.

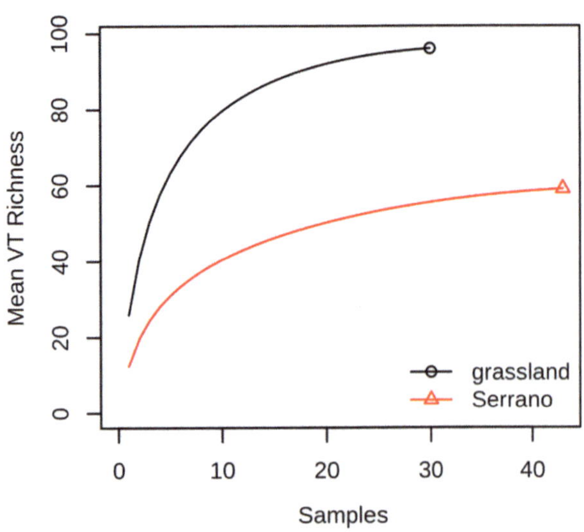

Fig. 11.4 Species accumulation curves of VT richness in Chaco Serrano and Mountain grassland in the Chaco forest

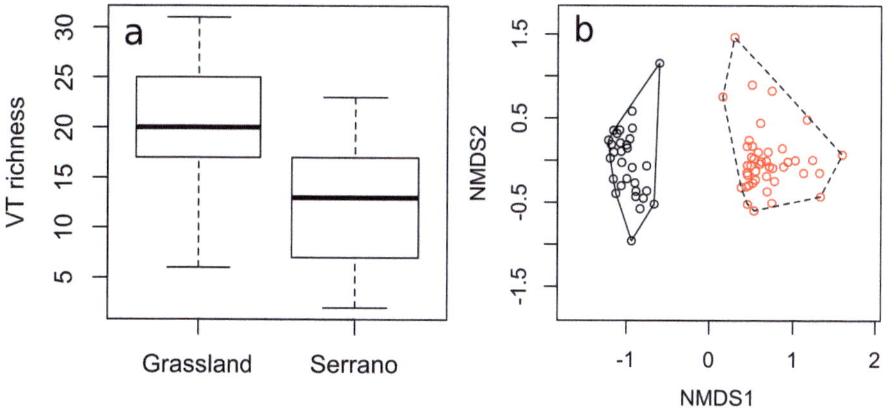

Fig. 11.5 The (**a**) mean AMF VT richness (Mean ± SD) and (**b**) the variation of AMF community composition (NMDS, stress =) in the Mountain grassland and Chaco Serrano. References: black circle: Mountain grassland, red circle: Chaco Serrano

The VT richness was higher (df = 1, $\chi^2 = 26.9$, $P \leq 0.0001$) in the Mountain grassland (mean ± SD = 20.36 ± 6.2) than in the Chaco Serrano (mean ± SD = 12.18 ± 5.9; Fig. 11.5a). In addition, the AMF community composition (presence-absence) differed between the Mountain grassland and the Chaco Serrano (PERMANOVA pseudo-F = 44.37, $P = 0.001$; Fig. 11.5b). We found no differences between Chaco regions when considering multivariate dispersion (i.e. analogous to the homogeneity of variance) (F = 0.13, $P = 0.72$), suggesting that variation in AMF communities composition arises from between-group variability rather than from within-group variability. These results are consistent with global evidence of AMF community richness being higher in grasslands than in forest ecosystems (Davison et al. 2015).

Regarding the phylogenetic structure of AMF communities, we found that AMF communities in the Mountain grasslands and Chaco Serrano showed a lower mean pairwise phylogenetic distance than expected by chance compared to local (Córdoba Province), regional (South America) and global species pool (Table 11.2). Therefore, these AM fungal communities were phylogenetically clustered compared to local, regional and global phylogenies suggesting a dispersal limitation or habitat filtering structuring the community (Dumbrell et al. 2010). These findings are in contradiction with Grilli et al. (2015) probably because more data sets of AMF species pool of these regions (i.e. local and regional) are available (Öpik et al. 2013; Davison et al. 2015). As regards the Mountain grasslands, there is evidence of clustered phylogenetic diversity in mountain elevated AMF communities (Egan et al. 2017; Kotilínek et al. 2017).

Table 11.2 Mean pairwise phylogenetic distance (MPD) of AMF communities in the Mountain grassland and Chaco Serrano compared with random MPD of (**a**): global, (**b**): regional and (**c**): local species pool

	[a]Ntaxa	MPD. obs	MPD. rand	MPD. sd	MPD.obs. rank	MPD. obs.z	MPD. obs.p	runs
a- global VT pool								
Grassland	93	0.16	0.28	0.03	1	−3.69	**0.001**	999
Serrano	58	0.14	0.28	0.04	1	−3.22	**0.001**	999
b- South America VT pool								
Grassland	93	0.49	0.61	0.04	5	−2.88	**0.005**	999
Serrano	58	0.40	0.61	0.06	1	−3.58	**0.001**	999
c- Córdoba VT pool								
Grassland	93	0.90	1.016	0.03	2	−3.65	**0.002**	999
Serrano	58	0.80	1.014	0.05	1	−4.26	**0.001**	999

aReferences: Ntaxa: taxa number, MPD.obs: observed MPD, MPD.rand: randomized MPD, MPD.sd: standard deviation of randomized MPD, MPD.obs.rank: rank of MPD, MPD.obs.z: z value, MPD.obs.p: p values, runs: number of randomizations.

11.4 Final Considerations

All things considered, general data about AM fungal communities remain scarce at the Chaco forest. Nevertheless, we could glimpse some general patterns of diversity and structure of AM fungal communities in the vegetation units of this ecosystem. In this chapter, the complementation of phylogenetic metrics and traditional taxon-based approaches to analyze diversity allowed a more powerful insight to disentangle the AM fungal community structure in ecosystems (Egan et al. 2017). In general, a variation of AM fungal communities might be determined by the spatial configuration and environmental conditions. In particular, the main neutral force that structures the AM fungal community composition might be the dispersal limitation (Kivlin et al. 2011) together with underlying deterministic niche based processes such as habitat filtering due to edaphic properties (i.e. pH and N) and environmental variables (i.e. Alt, Prec, Temp). Evidence of stochastic underlying deterministic processes structuring AM fungal communities has been studied in other ecosystems (Chave 2004). Conversely, it appears that in the Chaco forest niche-based models (edaphic and environmental variables) affecting the AM fungal communities are secondary to dispersal limitation (neutral process). However, whether future studies of AMF communities in this region might change this pattern is unknown.

Acknowledgments This work was supported by the Consejo Nacional de Investigaciones Científicas y Técnicas (CONICET), FONCYT, Universidad Nacional de Córdoba and Universidad Nacional de San Luis. We are grateful to María del Rosario Iglesias for helping with the map edition.

References

Anderson MJ (2001) A new method for non-parametric multivariate analysis of variance. Austral Ecol 26(1):32–46

Bates D, Mächler M, Bolker B, Walker S (2015) Fitting linear mixed-effects models using lme4. J Stat Softw 67(1):1–48

Becerra A, Bartoloni N, Cofré N, Soteras F, Cabello M (2014) Arbuscular mycorrhizal fungi in saline soils: Vertical distribution at different soil depth. Braz J Microbiol 45(2):585–594

Blanchet FG, Legendre P, Borcard D (2008) Forward selection of explanatory variables. Ecology 89:2623–2632

Borcard D, Legendre P (2002) All-scale spatial analysis of ecological data by means of principal coordinates of neighbour matrices. Ecol Model 153(1–2):51–68

Bucher E, Huszar P (1999) Sustainable management of the Gran Chaco of South America: ecological promise and economic constraints. J Environ Manage 57:99–108

Cabido M, Zeballos SR, Zak M, Carranza ML, Giorgis M A, Cantero JJ, Acosta AT (2018) Native woody vegetation in central Argentina: Classification of Chaco and Espinal forests. Appl Veg Sci 21(2):298–311

Caldas M, Goodin D, Sherwood S, Campos Krauer J, Wisely S (2015) Landcover change in the paraguayan chaco: 2000–2011. J Land Use Sci 10:1–18

Cagnolo L, Cabido M, Valladares G (2006) Plant species richness in the Chaco Serrano Woodland from central Argentina: ecological traits and habitat fragmentation effects. Biol Conserv 132(4):510–519

Chave J (2004) Neutral theory and community ecology. Ecol Lett 7(3):241–253

Cofré N, Becerra A, Nohura E, Soteras F (2012). Arbuscular mycorrhizae and dark-septate endophytes on *Atriplex cordobensis* in saline sites from Argentina. J Agric Technol 8:2201–2214

Davison J, Öpik M, Zobel M, Vasar M, Metsis M, Moora M (2012) Communities of arbuscular mycorrhizal fungi detected in forest soil are spatially heterogeneous but do not vary throughout the growing season. PLoS One 7(8): e41938

Davison J, Moora M, Öpik M, Adholeya A, Ainsaar L, Bâ A, Burla S, Diedhiou AG, Hiiesalu I, Jairus T, Johnson NC, Kane A, Koorem K, Kochar M, Ndiaye C, Pärtel M, Reier Ü, Saks Ü, Singh R, Vasar M, Zobel M (2015) Global assessment of arbuscular mycorrhizal fungus diversity reveals very low endemism. Science 127(6251):970–973

Dumbrell AJ, Nelson M, Helgason T, Dytham C, Fitter AH (2010) Relative roles of niche and neutral processes in structuring a soil microbial community. ISME J 4(3):337–345

Egan CP, Callaway RM, Hart MM, Pither J, Klironomos J (2017) Phylogenetic structure of arbuscular mycorrhizal fungal communities along an elevation gradient. Mycorrhiza 27(3):273–282

Fick SE, Hijmans RJ (2017) WorldClim 2: new 1 km spatial resolution climate surfaces for global land areas. Int J Climatol 37(12):4302–4315

Gause GF (1934) Experimental analysis of Vito Volterra's mathematical theory of the struggle for existence. Science 79(2036):16–17

Götzenberger L, de Bello F, Bråthen KA, Davison J, Dubuis A, Guisan A, Lepš J, Lindborg R, Moora M, Pärtel M, Pellissier L, Pottier J, Vittoz P, Zobel K, Zobel M (2012) Ecological assembly rules in plant communities approaches, patterns and prospects. Biol Rev 87(1):111–127

Grilli G, Urcelay C, Galetto L (2012) Forest fragment size and nutrient availability: complex responses of mycorrhizal fungi in native–exotic hosts. Plant Ecol 213:155–165

Grilli G, Urcelay C, Galetto L (2013) Linking mycorrhizal fungi and soil nutrients to vegetative and reproductive ruderal plant development in a fragmented forest at central Argentina. Forest Ecol Manag 310:442–449

Grilli G, Urcelay C, Galetto L, Davison J, Vasar M, Saks Ü, Jairus T, Öpik M (2015) The composition of arbuscular mycorrhizal fungal communities in the roots of a ruderal forb is not related to the forest fragmentation process. Environ Microbiol 17:2709–2720

Grilli G, Longo S, Huais PY, Pereyra M, Verga EG, Urcelay C, Galetto L (2017) Fungal diversity at fragmented landscapes: synthesis and future perspectives. Curr Opin Microbiol 37:161–165

Hijmans RJ, van Etten J, Cheng J, Mattiuzzi M, Sumner M, Greenberg JA (2017) Raster: Geographic Data Analysis and Modeling. R package version 2.3–33; 2016

Hoyos L, Cingolani A, Zak M, Vaieretti M, Gorla D, Cabido M (2013) Deforestation and precipitation patterns in the arid Chaco forests of central Argentina. Appl Veg Sci 16:260–271

Hubbell SP (2001) The unified neutral theory of species abundance and diversity. Princeton University Press, Princeton

Kauffman J, Steele M, Cummings D, Jaramillo V (2003) Biomass dynamics associated with deforestation, fire, and conversion to cattle pasture in a Mexican tropical dry forest. Forest Ecol Manag 176:1–12

Kivlin SN, Hawkes CV, Treseder KK (2011) Global diversity and distribution of arbuscular mycorrhizal fungi. Soil Biol Biochem 43(11):2294–2303

Kotilínek M, Hiiesalu I, Košnar J, Šmilauerová M, Šmilauer P, Altman J, Dvorský M, Kopecký M, Doleža J (2017) Fungal root symbionts of high altitude vascular plants in the Himalayas. Scientific REPortS 7:6562. DOI:https://doi.org/10.1038/s41598-017-06938-x

Legendre P, Anderson MJ (1999) Distance based redundancy analysis: testing multispecies responses in multifactorial ecological experiments. Ecol Monograph 69 (1):1–24

Lekberg YLVA, Koide RT, Rohr JR, Aldrich-Wolfe L, Morton JB (2007) Role of niche restrictions and dispersal in the composition of arbuscular mycorrhizal fungal communities. J Ecol 95(1):95–105

Longo S, Nouhra E, Goto B, Berbara R, Urcelay C (2014) Effects of fire on arbuscular mycorrhizal fungi in the Mountain Chaco Forest. Forest Ecol Manag 315:86–94

Lugo MA, Gonzalez Maza E, Cabello MN (2003) Arbuscular mycorrhizal fungi in a mountain grassland II: Seasonal variation of colonization studied, along with its relation to grazing and metabolic host type. Mycologia 95 (3):407–415

Lugo MA, Anton AM, Cabello MN (2005) Arbuscular mycorrhizas in the *Larrea divaricata* scrubland of the arid 'Chaco', Central Argentina. J Agric Technol 1(1):163–178

McGill BJ, Maurer BA, Weiser, MD (2006) Empirical evaluation of neutral theory. Ecology 87(6):1411–1423

Menoyo E, Renison D, Becerra A (2009) Arbuscular mycorrhizas and performance of *Polylepis australis* trees in relation to livestock density. Forest Ecol Manag 258:2676–2682

Nori J, Torres R, Lescano J, Cordier J, Periago M, Baldo D (2016) Protected areas and spatial conservationpriorities for endemic vertebrates of the Gran Chaco, one of the most threatened ecoregions of the world. Divers Distrib 22:1212–1219

Oksanen J, Blanchet F, Kindt R, Legendre P, Minchin P, O'Hara R (2018) Vegan: Community Ecology Package. R package vegan, vers. 2.2–1; 2015

Öpik M, Metsis M, Daniell TJ, Zobel M, Moora M (2009) Large-scale parallel 454 sequencing reveals host ecological group specificity of arbuscular mycorrhizal fungi in a boreonemoral forest. New Phytol 184(2):424–437

Öpik M, Vanatoa A, Vanatoa E, Moora M, Davison J, Kalwij JM, Reier U, Zobel M (2010) The online database MaarjAM reveals global and ecosystemic distribution patterns in arbuscular mycorrhizal fungi (Glomeromycota). New Phytol 188(1) 223–241

Öpik M, Zobel M, Cantero JJ, Davison J, Facelli JM, Hiiesalu I, Jairus T, Kalwij JM, Koorem K, Leal ME, Liira J, Metsis M, Neshataeva V, Paal J, Phosri C, Põlme S, Reier Ü, Saks Ü, Schimann H, Thiéry O, Vasar M, Moora M (2013) Global sampling of plant roots expands the described molecular diversity of arbuscular mycorrhizal fungi. Mycorrhiza 23:411–430

Oyarzabal M, Clavijo J, Oakley L, Biganzoli F, Tognetti P, Barberis I, Maturo H, Aragón R, Campanello P, Prado D, Oesterheld M, León R (2018) Unidades de vegetación de la Argentina. Ecol Austral 28:040–063

Pielou EC (1975) Ecology diversity. John Wiley and Sons.New York, p 165

Powell JR, Parrent JL, Hart MM, Klironomos JN, Rillig MC, Maherali H (2009) Phylogenetic trait conservatism and the evolution of functional trade-offs in arbuscular mycorrhizal fungi. Proc R Soc Lond B Biol Sci 276(1676):4237–4245

Powell JR, Rillig MC (2018) Biodiversity of arbuscular mycorrhizal fungi and ecosystem function. New Phytol 220 (4):1059–1075

Preston FW (1948) The commonness, and rarity, of species. Ecology 29(3):254–283
Silvertown J (2004) Plant coexistence and the niche. Trends Ecol Evol 19(11):605–611
Smith S, Read D (2008) Mycorrhizal Symbiosis, third ed. Academic, London
Soteras F, Renison D, Becerra A (2014) Restoration of high altitude forests in an area affected by a wildfire: *Polylepis australis* Bitt. seedlings performance after soil inoculation. Trees-Struct Funct 28:173–182
Soteras F, Grilli G, Cofré MN, Marro N, Becerra A (2015) Arbuscular mycorrhizal fungal composition in high montane forests with different disturbance histories in central Argentina. Appl Soil Ecol 85:30–37
Soteras F, Coutinho Moreira B, Grilli G, Pastor N, Carneiro Mendes F, Ruela Mendes D, Renison D, Megumi Kasuya MC, de Souza FA, Becerra A (2016) Arbuscular mycorrhizal fungal diversity in rhizosphere spores versus roots of an endangered endemic tree from Argentina: is fungal diversity similar among forest disturbance types?. Appl Soil Ecol 98:272–277
Tokeshi M (1990). Niche apportionment or random assortment: species abundance patterns revisited. J Anim Ecol 59(3):1129–1146
Urcelay C, Díaz S, Gurvich D, Chapin F, Cuevas E, Domínguez LS (2009) Mycorrhizal community resilience in response to experimental plant functional type removals in a woody ecosystem. J Ecol 97:1291–1301
van der Gast CJ, Gosling P, Tiwari B, Bending GD (2011) Spatial scaling of arbuscular mycorrhizal fungal diversity is affected by farming practice. Environ Microbiol 3(1):241–249
Verbruggen E, van Der Heijden MG, Weedon JT, Kowalchuk GA, Röling WF (2012) Community assembly, species richness and nestedness of arbuscular mycorrhizal fungi in agricultural soils. Mol Ecol 21(10):2341–2353
Verga EG, Sánchez Hümöller HL, Peluc SI, Galetto L (2017) Forest fragmentation negatively affects common bird species in subtropical fragmented forests. Emu-Austral Ornithol 117(4):359–369
Zak MR, Cabido M, Hodgson JG (2004) Do subtropical seasonal forests in the Gran Chaco, Argentina, have a future?. Biol Cons 120(4):589–598
Zak MR, Cabido M, Cáceres D, Díaz S (2008) What drives accelerated land cover change in central Argentina? Synergistic consequences of climatic, socioeconomic, and technological factors. J Environ Manage 42(2):181–189

Chapter 12
Southern Highlands: Fungal Endosymbiotic Associations

Mónica A. Lugo and Eugenia Menoyo

12.1 Introduction

South American Highlands ecosystems comprise different ecoregions and phytogeographic areas including in the Andean and Chaco regions. Along the Andes are stretched the High Andean region, Costal Peruvian Andean Desert, North Andean Páramo, Atacama Desert, Subandean Patagonia, Magellanic Andes, and Puna. The Highlands of the Chaco region are represented by transitional biomes between the Cerrado, Caatinga and Mata Atlantica, the Brazilian Altiplano with rupestrian grasslands or fields, the Prepuna in Bolivia and Northwest of Argentina, and the High grasslands or Mountain grasslands in Central Argentina (Morrone 2001a; Cabido et al. 2010; Oyarzabal et al. 2018).

These Highland ecosystems present unique biodiversity, climate, geographic position, geologic origin, and biogeography. Accordingly, most of them are considered the main biodiversity hotspots (Myers et al. 2000; Cardoso da Silva and Bates 2002; Madriñán et al. 2013; Young et al. 2015). Further, some of these Highlands constitute true "sky islands" due to the physical separation between the mountains to the surrounded ecosystems by a desertic area. Consequently, this separation fostering endemism, vertical migration of species and relict populations. The complex dynamics of the species richness that can be found in the sky islands draw attention to biogeography as well as biodiversity studies (Hughes and Atchison 2015).

M. A. Lugo (✉)
Biological Sciences, National University of San Luis, Grupo MICODIF (Micología, Diversidad e Interacciones Fúngicas)/IMIBIO (Instituto Multidisciplinario de Investigaciones Biológicas)-CONICET-CCT SL, San Luis, San Luis, Argentina

E. Menoyo
National University of San Luis, Grupo MICODIF (Micología, Diversidad e Interacciones Fúngicas)-UNSL/IMASL (Instituto de Matemática Aplicada San Luis)-CONICET, San Luis, Argentina

M. C. Pagano, M. A. Lugo (eds.), *Mycorrhizal Fungi in South America*, Fungal Biology, https://doi.org/10.1007/978-3-030-15228-4_12

The advance of the agronomic frontier over natural environments and its usual consequent deforestation has increased together with the cropland expansion, fertilizers use, high soil salinization, water table rise, water soil degradation, energy use and common pollution (Foley et al. 2011), all of these having strongest negative effects in arid and semiarid zones (Manuel-Navarrete et al. 2007; Viglizzo and Jobbágy 2010). Therefore, it becomes necessary to contemplate the conservation of fragile, unique and diverse systems such as the South American Highlands ecosystems.

The mycorrhizal fungi are symbiotic plant root colonizers abreast of other fungal endophytes and are important components of soil communities. They link the biotic and abiotic soil ecosystem components and contribute to geochemical cycles, plant nutrition and growth, providing essential ecosystem services (Gianinazzi et al. 2010) such as soil health (Bradford 2014) and global carbon dynamics intervention (Averill et al. 2014). Different interactions occur between plants and symbiotic fungi as ectomycorrhizas (ECM), ericoid mycorrhizas (ER), orchid mycorrhizas (OM), arbuscular mycorrhizas (AM) and fungal root endophytes as fine root endophytes (FRE), coarse root endophytes (CRE), and dark septate endophytes (DSE). These interactions are essential to the host plant survival, mainly due to the acquired different resource acquisition strategies that may differ among plant species and ecosystems (Smith and Read 2008, Walker et al. 2018b). In the world, Highlands floras are diverse and consequently, these ecosystems also include different mycorrhizal types and fungal root endophytes associated (Smith and Read 2008). Thus, in alpine ecosystems ECM, AM are abundant above treeline, as also DSE (Oehl and Körner 2014) whereas ER are common in heathlands and AM are predominant in grasslands (Smith and Read 2008). In general, AM fungi (AMF) occur in soils that have low organic material content, high nitrogen content and low phosphorus availability (Allen 1991; Read 1991). Dark septate endophytes are found in a wide range of habitats since they are more tolerant to harsh conditions than AMF due to their melanized hyphae, which are highly resistant to drought and heat stress (Redman et al. 2002; Knapp et al. 2012), two prevailing conditions in the South American Highlands.

This Chapter reviews and discusses mycorrhizal and root endophytic fungi research focused on root colonization in arid and semiarid Highlands of South America (SA). The areas analyzed comprise Andean and Chaco mountain ecosystems placed above the treeline in at least 700 m of elevation, and the low elevation area for the southern Andes, with minor annual mean rain than 1000 mm. Further, the Highlands of SA have been contrasted to worldwide knowledge to provide new perspectives for future research considering global change.

12.2 Highlands Ecosystems in South America

Different conceptual frameworks were used to define biological areas with ecological and sociological values such as biomes, ecoregions, hotspots, phytogeographical and biogeographical domains, kingdoms, districts, and provinces among others, and each conceptual frameworks with their own methods, scales, properties, utilities and applications (Matteucci et al. 2016). As regards defining South American Highland ecosystems, the classification with a biogeographic approach proposed by Morrone (2001a, 2006) has been taken into account. This author characterizes and classifies the biogeographic regions by abiotic features of these areas based on their geological and climatic history while including in his classification of biotic characters not only animal diversity (birds, insects, mammals), which is usually included in most of the biogeographic known systems and applied internationally with a conservationist view, but also vegetation and fungi diversity. Thus, his classification system is the most complete framework for the mycorrhizal association analysis in the Highlands of SA. Further, this approach is complemented by the ecoregion definitions of Morello et al. (2018) that include a soil characterization of these areas which is crucial information to mycorrhizal studies together with vegetation data, social utilities, productivity, and protected areas. Moreover, the vegetation communities characterization together with phytogeographic and floristic information (Cabrera and Willink 1980; Oyarzabal et al. 2018; Martínez Carretero 1995; Martínez Carretero et al. 2016) were also considered.

The hypothesis of the dual origin of the SA continent has been supported by cladistic, biogeographic and panbiographics studies. This hypothesis relates southern SA to the southern temperate areas of the Austral kingdom while tropical SA is related to African and North American regions because of their similar animals, plants, and fungi. Further, this compound origin was also addressed to the Andean region which has been intimately related to the southern temperate areas and also to the Neotropical region that has been closely related to the Old World tropics (Morrone 2004, 2006 and references therein). Therefore, the North Andean Páramo has been originated from another pangeographic area different from the one that originated the rest of the southern Andean region, creating greatly different fauna, flora, soils, and fungal associations and diversity. Moreover, the Prepuna and Andean regions are considered biogeographically as the South American transition zones (Morrone 2006) which are extended along the Highlands of the Andes, western Venezuela, northern Chile, and west Central Argentina including all the arid Highlands of SA such as the North Andean Páramo, Coastal Peruvian Desert, Puna, Atacama, and Monte regions (Morrone 2006, 2004; Posadas et al. 1997). Transition zones are areas with an overlap of organisms from different biotic elements with historically and ecologically different origins, these zones are useful to infer their original biota and sister areas with diverse biogeographies. Moreover, transition zones may present a low or high biodiversity with evolutionary importance as representative areas of biotic interactions (Morrone 2006; Ruggiero and Ezcurra 2003), and therefore they are crucial to the study of mycorrhizal and fungal root endophytic associations.

The Chacoan subregion includes the biogeographic provinces of Caatinga, Cerrado, Chaco and Pampa (Morrone 2001a). The rupestrian grasslands are located along a transition zone between Caatinga, Cerrado and Mata Atlántica and it includes mountains above 700 m from the Brazilian Espinhaço Range. As regards the Prepuna region, authors such as Oyarzabal et al. (2018) consider it a part of the

Chacoan domain, whereas others such as Morrone (2006, 2004) and Posadas et al. (1997) include it in the Andean grouping. "Pampa de Achala" is a highland plateau with grasslands and “tabaquillo” (*Polylepis australis*) forests located in the wet ravines of the Argentinian Sierras Grandes above 2000 m in Central Argentina in the Arid Chaco (Oyarzabal et al. 2018) with vegetal communities that have patchy structures of dry-grasslands, tall-grasslands and turfs in relation with the type of rock substrate of the landscape (Cabido 1985; Cabido et al. 1987, 1991).

Meanwhile, the Andean Highlands of SA are from the northern part of the continent since the North Andean Páramo, which is a biogeographical province located in the high mountains of Venezuela, Colombia, Ecuador and Perú, above the montane treeline of 3000 m (Morrone 2001a, 2006, 2014) and reaching up to the snowline at 4500–4800 m. The soils are young with very low phosphorus content, acidic availability and organic matter accumulation. There are extreme daily temperature fluctuations in the air and soil. Thus, the Páramo vegetation (Monasterio 1980) lives under highly stressful conditions including grazing pressure (Montilla et al. 1992; Barnola and Montilla 1997), agriculture, farming and mining (Vásquez et al. 2015). The Páramo is regarded as a sky island or continental biogeographic island on mountain tops, because of the recent orogeny of the northern Andes during the successive folding and glaciation effects during the Miocene and Quaternary. These geomorphological events created an “archipelago of islands” on mountain tops with biogeographical barriers that generated a rich diversity of niches that have promoted the diversification of the organisms inhabiting them (Peyre et al. 2018 and reference therein). Thus, the Páramo is considered a hotspot with the world’s fastest evolving and important biodiversity (Myers et al. 2000; Madriñán et al. 2013; Young et al. 2015) and the richest tropical highland flora, which includes ca. 5000 plant species (Sklenář et al. 2014; Peyre et al. 2018 and reference therein) with characteristic plant taxa (Morrone 2001a) such as the ferns *Dicksonia stuebelii* (Dicksoniaceae) and *Muhlenbergia cleefi* (Poaceae), the typical “frailejones” (Asteraceae: Espeletiinae) which are more than 144 species distributed among the genera *Carramboa*, *Coespeletia*, *Espeletia*, *Espeletiopsis*, *Libanothamnus*, *Paramiflos*, *Ruilopezia* and *Tamania* (Diazgranados and Barber 2017); *Draba arauquensis* (Brassicaceae), *Gunnera antioquensis* and *G. caucana* (Gunneraceae), *Aragoa* (Scrophulariaceae) and *Passiflora truxillensis* (Passifloraceae). Aside from this huge biodiversity and high endemism that reaches to 60% of its flora (Peyre et al. 2018 and reference therein), the Páramo is further known for providing essential ecosystem services such as water provision, climate regulation and carbon stocking (Buytaert et al. 2011; Farley et al. 2013; Peyre et al. 2018 and reference therein). The Andean Páramo is considered a highly vulnerable region of high irreplaceability and has been included in the global biodiversity conservation priorities (Brooks et al. 2006).

The Peruvian Coastal Desert is a narrow ribbon along the Pacific Ocean coast from the north of Perú to the north of Chile (Cabrera and Willink 1980; Morrone 2001a). The vegetation is scant, there are only permanent communities on the riversides and near the sea; the arborescent Cactaceae are abundant between the elevations of 1500–3000 m, among which shrubs and herbs grow with the rains (Cabrera and Willink 1980). The dominant plant species are *Acacia macrantha*, *Caesalpinia tinctoria*, *Diplostephium tacorense*, *Franseria fruticosa*, *Inga feuillei*, *Kaegeneckia lanceolata*, *Lemaireocereus cartwrightianus*, *L. laetus*, *Neoraimondia macrostibas*, *Paspalum vaginatum*, *Prosopis chilensis*, *P. limensis*, *Salicornia ambigua*, *Schinus areira*, *Tillandsia latifolia*, *T. purpurea*, *T. straminea* and *Trichocereus peruvianus* (Cabrera and Willink 1980; Morrone 2001a, b). This Andean biogeographic province is threatened by overgrazing, the alteration of the river flow regimes and the firewood collection (Dinerstein et al. 1995).

The Atacama Desert is in the northern extreme of Chile (Morrone 2001a, b) and extends along 1000 km of the Pacif coast, with elevations from 900–1000 m, raising to mean values of 3000 m–5000 m. It is one of the driest and possibly oldest deserts in the world with annual precipitations of 0.1–2.3 mm, with mean annual rains of 12 mm representing an extreme habitat for life on Earth which is an analog for life in dry conditions on Mars (McKay et al. 2003; Clarke 2006). The vegetation in this area is scarce; however, there are rich communities as an oasis, such as the "lomas" formations along the hills, which are supported by the mists that are formed during winter due to the currents of the Pacific Ocean. The Atacama Desert is also the scenario of the desert flowering, an amazing and particular phenomenon which occurred when the El Niño events have promoted heavy rains of more than 15 mm (Pliscoff et al. 2017). The biogeographic province of Atacama is characterized by the angiosperms taxon *Chuquiraga ulicina* (Asteraceae) (Morrone 2001b); however, its flora is diverse (Morong 1891), with 980 Chilean native plant species which represent the 54.3% of the endemic species from Chile and 119 species naturalized (Letelier et al. 2008) with 9.6% of these endemic species in the Red List (Squeo et al. 2008). In this environment, there are different vegetation formations and a high number of endemic flora species (Pliscoff et al. 2017). The Atacama is one bioma threatened by overgrazing of domestic livestock, the alteration of rivers flow, the firewood extraction (Dinerstein et al. 1995; Morrone 2001a, b) and, mining of copper-gold ore deposits (Clarke 2006).

The Andes, High Andean region, was considered as one of the highly vulnerable regions of strongly irreplaceable which has been included in the global biodiversity conservation priorities (Brooks et al. 2006). It occupies the high peaks of the "Cordillera de los Andes", above 4400–6000 m in its northern parts to 500 m in the southernmost area in the Argentinian state of Tierra del Fuego (Cabrera 1976; Morello et al. 2018), it is along the Precordillera in this southern area, at low elevations which also represent the Andean region as the Magellanic Andean regions and the Subandean Patagonia (Morrone 2001a). The uplift of the Andean Cordillera occurred 100 million years ago (mya) in the Cretaceous; 40 mya later, the elevation of the land occurred due to the accumulation of sand and silt from the swamps formed by subduction of the Nazca plate below the continental plate of SA at 5.5 mya. Along the time this uplift was followed by successive uplifts and volcanic events, further, the pressure effects on the Nazca plate uplift by the South American continental plate molded the Andes that is still is in an active geomorphogenesis (Morello et al. 2018). The climate is cold and dry, with strong winds, precipitation in the form of snow or hail at any time of the year. In the summits, they present eternal snows. The average monthly temperatures are below zero for more than half of the year; its heliophany is high and the thermal amplitude is very large. The annual precipitations are between 100 to 200 mm. There is intense solar radiation. Glaciers are found in the region and the irrigation water comes from the thaws. The soils are very poor and little evolved or immature; predominantly rocky, stony or sandy soils, generally loose, shallow with rocky subsoil and incipient evolution with the bare rock representing up to 86% of the surface in the High Andes (Morello et al. 2018; Cabrera and Willink 1980). The plant species which habit High Andean are more than 300 (Ferreyra et al. 1998; Ferreyra and Grigera 2002; Oyarzabal et al. 2018), and among them endemic and microendemic species which have to be conserved (Ferreyra and Grigera 2002). The vegetation is a combination of species with characteristics associated with xerophitism extreme, at low temperatures, and in the wind, with graminoid form isolated, in low and compact, circular or semilunar stands. The dicotyledons are characterized by a large underground development, small leaves, presence of resins, protected stomata and trichomes. Creeping shrubs and the plants in cushion or in plates attached to soil are common (Cabrera 1976), with the vegetation unit of low steppe of *Senecio algens* and *Oxalis compacta* which form cushions, with *Jaborosa laciniata*, *Nastanthus ventosus*, *Calandrinia* spp; the open graminoid steppe, represented by an association of *Festuca orthophylla*, *Festuca chrysophylla*, *Poa gymnantha*, *Stipa speciosa* or *Pappostipa vaginata*, *Pappostipa frigida, Nassella mucronata, Deyeuxia cabrerae*, among others; the shrub steppe and the semi-desert of lichens, and the fertile plain of Poaceae, Juncaceae and Cipereceae (Cabrera 1976; Martínez Carretero et al. 2016; Morello et al. 2018; Oyarzabal et al. 2018).

The Puna is a highland plateau which extends from Southern Perú to Northwestern Argentina (Martínez Carretero 1995; Morrone 2001a; Oyarzabal et al. 2018), limited by the Cordillera Real to the east and the Andes to the west, between 2000–4400 m. Its climate is dry and cool all year round with a large thermal range, intense solar radiation, and scarce precipitations in the form of snow, hail, or summer rain with a dry season in winter and an annual precipitation of 100–400 mm in summer (Cabrera and Willink 1980) but also 32 mm (Ruthsatz and Movia 1975; Ruthsatz 1977). Solar radiation is intense and relative air humidity is low (10–15%), existing large thermal differences along the day (16–20 °C). The resulting climate is of a desertic type. The annual mean temperature oscillates between 8.5–9.5 °C (Cabrera and Willink 1980), during summer, the monthly mean temperature is near 6 °C (Ruthsatz and Movia 1975; Ruthsatz 1977). The vegetation consists in shrubby steppes and grasses with scattered trees of *Polylepis tomentella* ("Quiñuales") in the moist ravines (Cabrera and Willink 1980; Martínez Carretero et al. 2016; Renison et al. 2013). The plants present adaptations to the lack of water during large part of the year, low temperatures and grazing, such as deep roots (*Adesmia schickendanzii*), storage of water (*Tephrocactus* spp.), small leaves or photosynthetic stems (*Fabiana denudata*), spines (*Aloysia salsoloides*), or growth in cushion at ground level (*Werneria aretioides, Azorella compacta*). The shrubs steppe of *Fabiana densa* and *Baccharis boliviensis* is the community most widespread in the Puna province, with low coverage of herbs and the presence of other shrub species such as *Adesmia horrida*, *Aloysia salsoloides*, and *Tetraglochin cristatum*. In addition, there are endemic species (Martínez Carretero 1995). The graminose steppe is in the higher areas, there is also azonal communities (e.g., steppe of halophytes, steppe of psammophilous plants, vega, among others) (Oyarzabal et al. 2018). The soils are superficial and immature, very poor in organic matter, sandy and rocky (Cabrera 1976), and, Entisol and Aridisol type (Martínez Carretero 1995). The Puna is threatened by agriculture, overgrazing of domestic livestock (llamas, goats, and sheep), fires and firewood collection (Dinerstein et al. 1995; Morrone 2001a). Furthermore, Puna ecoregion was considered as vulnerable by its conservation degree since 2002 (Olson and Dinerstein 2002); moreover, in the northwest of Argentina, there are strong degraded areas in Puna considered as hotspots (Navone and Bosio 2008–2009).

On the other hand, the Chacoan Highlands of SA are included into the "Chaco" or "Chaqueña" subregion which occupies the north and center of Argentina, south of Bolivia, west and center of Paraguay, and center and northwest of Brazil (Morrone 2001a, 2004; Prado 1993ab; Prado and Gibbs 1993). These Chaco Highlands are included in the huge Chaco biogeographic province, which covers low elevation plains and mountains with a continental climate, warm, with mainly summer precipitation between 500 mm–1200 mm from the west to the east, and the temperature average is 20–23 °C. The predominant vegetation type of Chaco is the deciduous xerophytic forest but there are also palms, savannahs, halophilic steppes, among others vegetation communities. The Chaco region is very disturbed by the forestry exploitation, livestock, and agriculture. The most important tree species is the "quebracho colorado", used to extract tannin and to manufacture poles and sleepers for railways; because of its very slow growth there is hardly any reforestation and in many areas, the quebracho has been practically exterminated. The livestock has destroyed the primitive graminoid herbaceous layer, determining the invasion of thorny shrubs and bromeliads, and, these areas are used for agricultural purposes with many areas which have been dismantled to grow cotton, tobacco or sugar cane (Cabrera and Willink 1980). In this context, Highlands of Chaco are favorable areas to be included in conservation plans due to their restricted access and the characteristics of their plants communities.

In Brazil, Chacoan Highlands are located in southeastern and northeastern, named the "campos rupestres" are rupestrian grasslands or rupestrian fields, considered an important biodiversity hotspot of the Cerrado (Cardoso da Silva and Bates 2002; Overbeck et al. 2015) especially for mycorrhizal fungi and fungal endophytes (Oki et al. 2016). The climate is a typical subtropical seasonal with a dry season during winter and a rainy season during summer (Fiaschi and Piradi 2009 and references therein). Among the Espinhaço Plateau, at 700–2000 m of elevation, harbors old growth grasslands with ca. 6000 plants species with high endemisms among its diverse habitats (Veldman et al. 2015), which leads to its unique vegetation composed of hundreds of endemic plant species (Giullieti and Pirani 1988; Fiaschi and Pirani 2009; Echternacht et al. 2011). The main vegetation type at higher elevations within the Espinhaço Range is the "campos rupestres", these grasslands are growing on rocky or sandy soil mostly of quartzite origin, but occasionally of arenite origin, placed above 700 m, mostly at elevations of 1000–1400 m (Coutinho et al. 2015); with acidic and poor in nutrients quartzitic soils but in the southern portion the soils are mostly ferruginous. These rupestrian grasslands include diverse habitats with different substrates, floristic compositions, the presence of nude rocks or sandy sediments, and soils with dry or humid periods. Its peculiar vegetation type is associated with a mosaic of bare bedrocky and white sands and has been formed as a consequence of recent diversification events at 3–4.7 mya (Echternacht et al. 2011 and references therein). In these highlands, the family with the greatest number of endemic species is Eriocaulaceae (Rapini et al. 2008), with clades involving *Viguiera* (Asteraceae), *Microlicieae* (Melastomataceae) and *Minaria* (Apocynaceae). The phylogenetic studies of the endemic flora of the Cerrado have proposed some angiosperm groups as candidates to study the possible recent radiations of this system, such as *Eremanthus*, *Lychnophora*, *Richterago* (Asteraceae); *Encholirium* (Bromeliaceae); *Kielmeyera* (Clusiaceae); *Eriocaulaceae*, *Pseudotrimezia* (Iridaceae); *Eriope* (Lamiaceae); *Chamaecrista*, *Mimosa* (Fabaceae); *Diplusodon* (Lythraceae); *Byrsonima* (Malpighiaceae); *Microlicia*, *Trembleya* (Melastomataceae); *Sauvagesia* (Ochnaceae); *Declieuxia* (Rubiaceae); *Barbacenia* and *Vellozia* (Velloziaceae) (Fiaschi and Piradi 2009 and references therein; Echternacht et al. 2011 and references therein).

The Prepuna is another arid Highland of the Chaco biogeographic province, extended along the Central to Norwest of Argentina, from Jujuy to La Rioja states for some authors (Cabrera 1976; Cabrera and Willink 1980; Oyarzabal et al. 2018) or to the northern Mendoza states to others (Morrone 2001ab), and along also areas of Chuquisaca, Tarija and Potosí states of Bolivia (López 2000; Oyarzabal et al. 2018). Argentinian Prepuna occupies the dry eastern slopes and ravines of the Sub-Andean mountains and the "Sierras Pampeanas" in the northwest of Argentina, with elevations of 2000–3400 m in the northern area, and 1000–3000 m the south extreme. The precipitations are less than 200 mm in the year and occurred mainly in summer. In general, the presence of the Prepuna in Argentina is determined not only by the elevation but very particularly by the disposition and orientation of the ravines (Cabrera 1976). The vegetation presents adaptations to the lack of water as in Cactaceae, plants in the form of a cushion, presence of small leaves and aphyllous growth forms. The typical vegetation is the shrub steppe with small trees (*Prosopis ferox*, *Gochnatia glutinosa*, *Senna crassiramea*, *Aphyllocladus spartioides*, *Cercidium andicola*, *Zuccagnia punctata*, among others). There is a characteristic vegetation unit of large columnar cacti *Trichocereus atacamensis* or *T. tarijensis* above 3000 m; further, a lower layer of creeping and globular Cactaceae (*Airampoa ayrampo*, *Tunilla tilcarensis*, *Parodia maassii*, *P. stuemeri*), and grasses (*Digitaria californica*, *Munroa argentina*, *Jarava leptostachya*, *J. media*, *Eragrostis andicola*) (Aagesen et al. 2009). The Prepuna presents also a steppe with terrestrial Bromeliaceae, many of them forming cushions, which occupies the rocky and steep slopes, dominated by *Deuterocohnia brevifolia*, *D. lorentziana*, *D. digitata*, *Tillandsia virescens*, *Puya castellanosii* and *P. dyckioides* (Cabrera 1976). Although Argentinian Prepuna has a distinctive physiognomy, it lacks endemic species that define it (Aagesen et al. 2009, 2012; Oyarzabal et al. 2018), in contrast to Bolivian

Prepuna which has its own endemic species (López 2000; López and Beck 2002). The Bolivian Prepuna has similar characteristic than Argentinian ones but located in the Andes of southern Bolivia between 2300–3300 m. Soils in the Prepuna region are mountain soils, immature, stony, sandy and very permeable (Cabrera 1976; López 2000). The floristic composition of Bolivian Prepuna includes three hundred and twenty-four species, with two hundred and ninety-nine native and numeorus endemic species. Further, considering its species composition the Prepuna mainly consists of a southern South American floristic composition originated in the subtropical dry Chaco forests and is also influenced by the Andean Puna region, in which predominate species exclusive to the whole Prepuna region (Argentinian and Bolivian) as well as Bolivian endemic (most of them exclusive to the Bolivian Prepuna). These two Prepuna regions share many genera and even some species (López 2000; López and Beck 2002; Oyarzabal et al. 2018). The Prepuna as a whole is a biogeographical transition zone, with important value for biogeographic inferences and for its biodiversity (Morrone 2006). There is no detailed information about the threatened plant species to Argentinian Prepuna (Dinerstein et al. 1995); however, Bolivian Prepuna has almost 50% of native and endemic species, this feature highlights the importance for its conservation (López and Beck 2002).

Among Chaco biogeographic region, the "Pastizales de Altura" or tall grasslands are above 1500 m or Grasslands of "Stipeas and Festuceas", they are graminoid steppes placed higher than the Chaco forests, and are characterized by a zonal community and another azonal community composed by hygrometic cespitose grasses and forbs predominantly of the tribes Stipae and Festucea (Cabrera 1976; Oyarzabal et al. 2018). Among these Highlands, the flat areas filled by modern sedimentent are named "Pampas" (Cabido et al. 2010). These grasslands are along the edges of hills and mountains and constitute the last vegetation floor of the "Sierras Pampeanas" and Sub-Andean Mountains, between 1700–1900 m above sea level and arise over 2000 m in Pampa de Achala, Sierras Grandes in Córdoba state (Cabido et al. 2010; Martínez Carretero et al. 2016) but also are found in Sierras de San Luis and the east of Catamarca states (Cabrera 1976). The Pastizales de Altura main floristic composition of the zonal community is shaped by Poaceae as *Nassella filiculmis*, *N. niduloides*, *Nassella tenuissima* among others (Stipea) or *Festuca hieronymi*, *F. lilloi*, *Poa stuckertii*, *Deyeuxia hieronymi*, and others of the Festucea. As an azonal community, there is a high prairie of hydrophytic areas, in places where moisture is accumulated and herbs and forbs appear such as *Lachemilla pinnata*, *Eleocharis pseudoalbibracteata*, *Carex gayana*. Further, the populations of the huge grass *Cortaderia selloana* are common where the water flows. The ravines present little and small forests of *Polylepis australis* ("tabaquillo" or "queñoa") and *Maytenus boaria* ("maitén" or "orco molle"), which are distributed along areas protected of grazing and fire by the rock outcrops (Cabido et al. 2010). These tabaquillo forests also present a smaller stratum of shrubs of *Berberis hieronymi*, *Clinopodium gilliesii*, *Clinopodium odorum*, *Gaultheria poeppigii*, *Baccharis tucumanensis*, *Baccharis flabellata*, *Escallonia cordobensis*, *Heterothalamus alienus* and *Colletia spinosissima*, and ferns such as *Blechnum pennamarina*, *Cystopteris fragilis*, *Woodsia montevidensis*, *Pleopeltis pinnatifida*, *Polystichum montevidense*, *Pellaea ternifolia*, *Elaphoglossum gayanum* and *Elaphoglossum lorentzii* (Cabido et al. 2010). In particular, the Pampa de Achala, is a granitic plateau at 2250 m above sea level, in "Sierras de Córdoba", the climate is temperate with cold dry winters (average temperature 5 °C) and short cool summers (average temperature 11.4 °C), the average rainfall is 850 mm during spring-summer, frost may occur at almost any time, occasionally snow may fall in winter and spring (Díaz et al. 1994). The vegetation is a climatically determined grassland traditionally subjected to pastoral use (cattle, sheep, and horse). The plant communities and soil characteristics were described by Cabido et al. (1987, 2010). The floristic features are detailed in Cabido (1985). The soils have light acid pH and high degradation velocity, the texture varies from loam to clay-loam; the soils related to the "*Deyeuxia* grassland" is classified as Humic Cambisol/Cumulic Haplumbrept, Haplic Phaeozem/Entic Hapludoll, and Haplic Phaeozem/Fluventic Hapludoll (Cabido et al. 1987). Furthermore, in the highland grasslands of the Sierras de Córdoba, logging, fire and browse of tabaquillo sprouts have confined these forests to ravines and protected areas as the "Quebrada del Condorito" National Park (Cabido et al. 2010).

12.3 Mycorrhizal associations in South American Highlands

In South American Highlands, the fungal endosymbiotic associations above the treeline were recorded in Andean and Chaco biogeographic regions. In Andean Highlands, AMF diversity in Puna is low in richness and spores abundance, and decrease with the increase of the elevation between 3320–3850 m; with predominance of the glomoid and its sporocarpic species on the acaulosporoid and scutellosporoid species (Lugo et al. 2008) without effect of the host photosynthetic pathways (C_3 or C_4) on AMF diversity. Further, in Magellanic/Patagonian Andean region, sites from 800 to 2005 m of elevation from shrublands to forest with different habitat features as precipitations, soils, and plants communities have shown also low AMF species richness and abundance, with *Acaulospora laevis* and its family (Acaulosporaceae) as dominant taxa at all sites in these AMF communities; which were grouped according to habitat with certain structural patterns of the AMF community, that in some cases were common with other highlands environments instead others were exclusive to this Andean region (Velázquez et al. 2016). In contrast, in the Chaco Highlands, AMF diversity was higher (Lugo and Cabello 2002) than in the Andean region. Thus, in Pampa de Achala at lower elevation of 2250 m, AMF communities composition and richness showed a relationship with a low host preference, with different AMF taxa associated to C_3 or C_4 grasses (Lugo and Cabello 2002). In the rupestrian grasslands of the Espinhaço Range, AMF diversity also have shown variation with soil propierties along an altitudinal gradient from 800 to 1400 m above sea level, where the pattern of AMF diversity with the elevation increase (Coutinho et al. 2015) was similar to Andean Regions (Lugo et al. 2008); further, in the rupestrian grasslands at ca. 1200 m the diversity of AMF and its communities structure were found related more to the heterogeneity of habitats and their soil physicochemical features than to the soil chemical characteristics and the plant species richness (Carvalho et al. 2012). Moreover, for a more detailed and deeper review of the AMF diversity among the different ecosystems of SA see also Chapters of Becerra et al., Cofré et al., Grilli et al., and Soteras et al. in this Book.

In this Chapter, we have focused on plants root colonization by fungal endophytic associations. Thus, it was revised our unpublished data and published data (Table 12.1) available online of root colonization in arid and semiarid Highlands above treeline. The articles were searched online for Google Scholar and Scopus, in total only 25 publications were found in these topics. Thus, the root colonization by fungal endosymbiotic associations was reviewed Highlands above treeline along 12 Ecoregions from northern to southern SA (Table 12.1) representing to the Andean region by the Páramo, High Andean, Puna, Central Andean, Magellanic Andean region, and the Cerrado and Chaco regions for the Chaco Highlands. There are no reports or publications either for root colonization by fungal endosymbionts, and mycorrhizal colonization or for mycorrhizal fungi and spores for the Atacama and Costal Peruvian Andean Deserts.

The main proportion of colonization data from plants of the South American Highlands are from the Andean region (74%) and less information is recorded from the Chaco (26%) (Figs. 12.1, 12.2). The regions with the main proportion of plant species studied were Venezuelan Páramo (23%), the Argentinian Puna (20%), the High grasslands of the Argentinian Chaco (16%) and the Andes of Central Chile (11%) (Figs. 12.1, 12.2).

The root colonization of 205 plant taxa was considered from publications and unpublished results of the Chapter´s authors. These plant taxa are included in a total of 42 plant families, mostly Poaceae (35%) and Asteraceae (20%), and the families scarcely represented in the studied areas were Apiaceae, Aspleniaceae, Berberidaceae, Blechnaceae, Calyceraceae, Campanulaceae, Caryophyllaceae, Chenopodiaceae, Crassulaceae, Cunoniaceae, Cyperaceae, Dryopteridaceae, Ephedraceae, Ericaceae, Eriocaulaceae, Fabaceae, Gentianaceae, Geraniaceae, Hydrophyllaceae, Hypericaceae, Iridaceae, Juncaceae, Lamiaceae, Lycopodiaceae, Malvaceae, Melastomataceae, Orchidaceae, Oxalidaceae, Polygonaceae, Polypodiaceae, Portulacaceae, Ranunculaceae, Rosaceae, Scrophulariaceae, Solanaceae, Valerianaceae, Verbenaceae, Violaceae, Xyridaceae and Winteraceae. Along Andean and Chaco Highlands, the families studied in common were Apiaceae, Asteraceae, Cyperaceae, Gentianaceae, Lycopodiaceae, Orchidaceae, Poaceae, Polygonaceae, Rosaceae, and Scrophulariaceae, with a predominance of Poaceae, 39% and 24% respectively (Table 12.1, Fig. 12.3a). Among South American Highlands, the habits of the majority of plants were herbs (42%), graminoid (36%) and shrub (12%) (Fig. 12.3b) with the vast majority (73%) of native or endemic plants (13%), only a low number species were exotic (7%) (Fig. 12.3c).

The distribution of mycorrhizal associations and fungal endophytic colonization in plants roots of Highlands studied are detailed in Fig. 12.4a. Different categories were observed including AM, DSE, ER and their dual (AM-DSE, DSE-OM) and triple associations (AM-DSE-ER). Nevertheless, the most frequent association was AM (43%) and dual association AMF-DSE (40%). Then, the prevalent fungal endo-

Table 12.1 List of plant species studied in Andean and Chaco Highlands. The data were obtained from publications and unpublished data of Chapter´s authors. The species name, its families, habits (fern, herb, graminoid, shrub, succulent, tree), distribution status (native, exotic, endemic), AMF root colonization (AMF), DSE root colonization (DSE), orchid mycorrhiza, ericoid mycorrhiza and location (Andean or Chaco regions) were included

Plant species	Family	Habit	Status	AMF	DSE	[c]Location/Source
Muehlenbergia ligularis	Poaceae	g	Native	x	nd	Andean: Venezuelan Páramo[2]
Agrostis trichodes	Poaceae	g	Native	x	nd	Andean: Venezuelan Páramo[2]
Festuca australis	Poaceae	g	Endemic	x	nd	Andean: Venezuelan Páramo[2]
Aciachne pulvinata	Poaceae	g	Native	x	nd	Andean: Venezuelan Páramo[2]
Trisetum irazuense	Poaceae	g	Native	x	nd	Andean: Venezuelan Páramo[2,13]
Eleocharis acicularis	Cyperaceae	g	Native	x	nd	Andean: Venezuelan Páramo[2]
Carex albolutescens	Cyperaceae	g	Native	-	nd	Andean: Venezuelan Páramo[2]
Lachemillia sp.	Rosaceae	-	-	x	nd	Andean: Venezuelan Páramo[2]
Lucilia venezuelensis	Asteraceae	h	Native	x	nd	Andean: Venezuelan Páramo[2]
Oritrophium paramense	Asteraceae	h	Native	x	nd	Andean: Venezuelan Páramo[2]
Hypericum brathys	Asteraceae	h	Native	x	nd	Andean: Venezuelan Páramo[2]
Taraxacum oficinale	Asteraceae	h	Exotic	x	nd	Andean: Venezuelan Páramo[2]
Geranium sp.	Geraniaceae	-	-	x	nd	Andean: Venezuelan Páramo[2]
Rumex acetosella L.	Polygonaceae	h	Exotic	x	nd	Andean: Venezuelan Páramo[2,13]
Sysyrinchium sp.	Iridaceae	-	-	x	nd	Andean: Venezuelan Páramo[2]
Hypericum laricifolium	Hypericaceae	sh	Native	x	nd	Andean: Venezuelan Páramo[2]
Espeletia shultzii	Asteraceae	h	Endemic	x	nd	Andean: Venezuelan Páramo[2,13]
Espeletia floccosa	Asteraceae	h	Endemic	x	nd	Andean: Venezuelan Páramo[2]
Pseudognaphalium moritzianum	Asteraceae	h	Native	x	nd	Andean: Venezuelan Páramo[2]
Bidens andicola	Asteraceae	h	Native	x	nd	Andean: Venezuelan Páramo[2]
Hinterhubera ericoides	Asteraceae	sh	Endemic	x	nd	Andean: Venezuelan Páramo[2]

(continued)

Table 12.1 (continued)

Plant species	Family	Habit	Status	AMF	DSE	cLocation/Source
Blakiella bartsiaefolia	Asteraceae	h	Native	x	nd	Andean: Venezuelan Páramo[2]
Conyza lasseriana	Asteraceae	h	Endemic	-	nd	Andean: Venezuelan Páramo[2]
Lucilia radians	Asteraceae	h	Native	x	nd	Andean: Venezuelan Páramo[2]
Hypochoeris setosus	Asteraceae	h	Native	x	nd	Andean: Venezuelan Páramo[2]
Stipa philipii	Poaceae	g	Native	x	nd	Andean: Venezuelan Páramo[2]
Poa petrosa	Poaceae	g	Native	x	nd	Andean: Venezuelan Páramo[2]
Luzula racemosa	Juncaceae	h	Native	x	nd	Andean: Venezuelan Páramo[2]
Arenaria sp.	Caryophyllaceae	-	-	x	nd	Andean: Venezuelan Páramo[2]
Echeverria venezuelensis	Crassulaceae	s	Native	x	nd	Andean: Venezuelan Páramo[2,13]
Lobelia ternera	Campanulaceae	h	Native	x	nd	Andean: Venezuelan Páramo[2]
Bacharis prunifolia	Asteraceae	sh	Native	x	nd	Andean: Venezuelan Páramo[13]
Gnaphalium paramorum	Asteraceae	h	Native	x	nd	Andean: Venezuelan Páramo[13]
Gnaphalium purpureum	Asteraceae	h	Native	x	nd	Andean: Venezuelan Páramo[13]
Stevia elatior	Asteraceae	h	Native	x	nd	Andean: Venezuelan Páramo[13]
Senecio formosus	Asteraceae	h	Native	x	nd	Andean: Venezuelan Páramo[13]
Geranium sp.	Geraniaceae	-	-	x	nd	Andean: Venezuelan Páramo[13]
Poa annua	Poaceae	g	Exotic	x	nd	Andean: Venezuelan Páramo[13]
Agrostis jahnii	Poaceae	g	Native	x	nd	Andean: Venezuelan Páramo[13]
Hypericum laricoides	Hypericaceae	sh	Native	x	nd	Andean: Venezuelan Páramo[13]
Orthosanthus chimborasencis	Iridaceae	h	Native	x	nd	Andean: Venezuelan Páramo[13]
Lupinus meridanus	Fabaceae	h	Native	x	nd	Andean: Venezuelan Páramo[13]
Acaena cylindrostachya	Rosaceae	h	Native	x	nd	Andean: Venezuelan Páramo[13]

(continued)

Table 12.1 (continued)

Plant species	Family	Habit	Status	AMF	DSE	[c]Location/Source
Acaena elongata	Rosaceae	sh	Native	x	nd	Andean: Venezuelan Páramo[13]
Lachemilla fulvescens	Rosaceae	h	Native	x	nd	Andean: Venezuelan Páramo[13]
Lachemilla hirta	Rosaceae	h	Native	x	nd	Andean: Venezuelan Páramo[13]
Lachemilla verticilata	Rosaceae	h	Native	x	nd	Andean: Venezuelan Páramo[13]
Calamagrostis effusa	Poaceae	g	Native	x	nd	Andean: Colombian Páramo[14]
Espeletia grandiflora	Asteraceae	h	Native	x	nd	Andean: Colombian Páramo[14]
Espeletia corymbosa	Asteraceae	h	Native	x	nd	Andean: Colombian Páramo[14]
Weinmannia tomentosa	Cunoniaceae	t	Native	x	nd	Andean: Colombian Páramo[14]
Drimys granadensis	Winteraceae	t	Native	x	nd	Andean: Colombian Páramo[14]
Chusquea scandens	Poaceae	g	Native	x	nd	Andean: Colombian Páramo[14]
Asplenium castaneum	Aspleniaceae	h	Native	-	-	Andean: Peruvian High Andean[16]
Perezia coerulescens	Asteraceae	h	Native	-	x	Andean: Peruvian High Andean[16]
Senecio sp.	Asteraceae	-	-	x	x	Andean: Peruvian High Andean[16]
Werneria sp.	Asteraceae	-	-	x	x	Andean: Peruvian High Andean[16]
Mnioides sp.	Asteraceae	-	-	x	x	Andean: Peruvian High Andean[16]
Werneria orbignyana	Asteraceae	h	Native	-	x	Andean: Peruvian High Andean[16]
Xenophyllum rosenii	Asteraceae	sh	Native	-	x	Andean: Peruvian High Andean[16]
Ephedra rupestris	Ephedraceae	sh	Native	-	-	Andean: Peruvian High Andean[16]
Erodium cicutarium	Geraniaceae	h	exotic	-	-	Andean: Peruvian High Andean[16]
Calamagrostis antoniana	Poaceae	g	Native	x	-	Andean: Peruvian High Andean[16]
Calamagrostis ovata	Poaceae	g	Native	-	-	Andean: Peruvian High Andean[16]
Astragalus arequipensis	Fabaceae	h	Native	x	x	Andean: Peruvian High Andean[16]

(continued)

Table 12.1 (continued)

Plant species	Family	Habit	Status	AMF	DSE	[c]Location/Source
Lupinus aridulus	Fabaceae	h	Native	x	x	Andean: Peruvian High Andean[16]
Lycopodium (Huperzia) sp.	Lycopodiaceae	f	-	x	x	Andean: Peruvian High Andean[16]
Nototriche sulphurea	Malvaceae	h	Native	x	-	Andean: Peruvian High Andean[16]
Bartsia pumila	Scrophulariaceae	h	Endemic	x	x	Andean: Peruvian High Andean[16]
Valeriana pycnantha	Valerianaceae	h	Native	x	x	Andean: Peruvian High Andean[16]
Adesmia spinosissima	Fabaceae	sh	Native	x	x	Andean: Bolivian High Andean[19]
Bacharis incarum	Asteraceae	sh	Native	x	x	Andean: Bolivian High Andean[1,19]
Chersodoma jodopappa	Asteraceae	sh	Native	x	x	Andean: Bolivian High Andean[19]
Chuquiraga atacamensis	Asteraceae	sh	Native	x	x	Andean: Bolivian High Andean[19]
Parastrephia lepidophylla	Asteraceae	sh	Native	x	x	Andean: Bolivian High Andean[19]
Parastrephia quadrangularis	Asteraceae	sh	Native	x	x	Andean: Bolivian High Andean[19]
Fabiana densa	Solanaceae	sh	Native	x	x	Andean: Bolivian High Andean[19]
Junelia seriphioides	Verbenaceae	sh	Native	x	x	Andean: Bolivian High Andean[19]
Lampayo castellani	Verbenaceae	sh	Native	x	x	Andean: Bolivian High Andean[19]
Nassella publiflora	Poaceae	g	Native	x	x	Andean: Bolivian High Andean[19]
Jarava leptostachya	Poaceae	g	Native	x	x	Andean: Bolivian High Andean[19]
Stipa plumosa	Poaceae	g	Native	x	x	Andean: Bolivian High Andean[19]
Chenopodium quinoa	Chenopodiaceae	h	Native	x	x	Andean: Bolivian High Andean[19]
Bromus catharticus	Poaceae	g	Native	x	x	Andean: Argentinian Puna[9]
Polypogon interruptus	Poaceae	g	Native	x	x	Andean: Argentinian Puna[9]
Vulpia myuros f. megalura	Poaceae	g	Exotic	x	x	Andean: Argentinian Puna[9]
Calamagrostis sp.	Poaceae	g	-	-	nd	Andean: Argentinian Puna[8]

(continued)

Table 12.1 (continued)

Plant species	Family	Habit	Status	AMF	DSE	[c]Location/Source
Calamagrostis breviaristata	Poaceae	g	Native	x	nd	Andean: Argentinian Puna[8]
Calamagrostis trichodonta	Poaceae	g	Native	x	nd	Andean: Argentinian Puna[8]
Danthonia annableae	Poaceae	g	Native	x	x	Andean: Argentinian Puna[*9]
Danthonia boliviensis	Poaceae	g	Native	x	x	Andean: Argentinian Puna[9]
Chascolytrum subaristatum	Poaceae	g	Native	-	nd	Andean: Argentinian Puna[8]
Festuca humilior	Poaceae	g	Native	x	x	Andean: Argentinian Puna[*9]
Festuca rigescens	Poaceae	g	Native	x	nd	Andean: Argentinian Puna[8]
Hordeum muticum	Poaceae	g	Native	x	x	Andean: Argentinian Puna[9]
Jarava plumosula	Poaceae	g	Endemic	x	x	Andean: Argentinian Puna[9]
Koeleria praeandina	Poaceae	g	Endemic	x	x	Andean: Argentinian Puna[9]
Nassella meyeniana	Poaceae	g	Native	x	nd	Andean: Argentinian Puna[8]
Piptochaetium indutum	Poaceae	g	Native	x	nd	Andean: Argentinian Puna[8]
Poa calchaquiensis	Poaceae	g	Native	x	nd	Andean: Argentinian Puna[8]
Poa laetevirens	Poaceae	g	Native	x	x	Andean: Argentinian Puna[9]
Poa lilloi	Poaceae	g	Native	x	nd	Andean: Argentinian Puna[8]
Poa pratensis	Poaceae	g	Exotic	-	nd	Andean: Argentinian Puna[8]
Poa superata	Poaceae	g	Endemic	x	nd	Andean: Argentinian Puna[8]
Trisetum spicatum	Poaceae	g	Native	x	nd	Andean: Argentinian Puna[8]
Aristida adscensionis	Poaceae	g	Native	x	x	Andean: Argentinian Puna[9]
Bouteloua barbata	Poaceae	g	Native	x	x	Andean: Argentinian Puna[9]
Bouteloua simplex	Poaceae	g	Native	x	x	Andean: Argentinian Puna[9]
Eragrostis nigrican	Poaceae	g	Native	x	x	Andean: Argentinian Puna[9]

(continued)

Table 12.1 (continued)

Plant species	Family	Habit	Status	AMF	DSE	[c]Location/Source
E. nigricans var. punensis	Poaceae	g	Native	x	x	Andean: Argentinian Puna[*9]
E. mexicana subsp. virescens	Poaceae	g	Native	x	x	Andean: Argentinian Puna[9]
Microchloa indica	Poaceae	g	Native	x	x	Andean: Argentinian Puna[9]
Aristida asplundii	Poaceae	g	Native	x	x	Andean: Argentinian Puna[9]
Cynodon dactylon	Poaceae	g	Exotic	x	x	Andean: Argentinian Puna[9]
Cynodon dactylon var. biflorus	Poaceae	g	Exotic	x	nd	Andean: Argentinian Puna[8]
Eragrostis sp	Poaceae	g	-	x	nd	Andean: Argentinian Puna[8]
Eragrostis lugens	Poaceae	g	Native	-	nd	Andean: Argentinian Puna[8]
Muhlenbergia rigida	Poaceae	g	Native	x	-	Andean: Argentinian Puna[9]
Gentiana prostrata	Gentianaceae	h	Native	x	x	Andean: Argentinian Puna[15]
Gentianella helianthemoides	Gentianaceae	h	Native	x	-	Andean: Argentinian Puna[15]
Digitaria californica	Poaceae	g	Native	x	nd	Andean: Argentinian Puna[5]
Polypogon monspeliensis	Poaceae	g	Exotic	x	nd	Andean: Argentinian Puna[5,6]
Stipa speciosa	Poaceae	g	Endemic	x	nd	Andean: Argentinian Puna[5,6]
Trichloris crinita	Poaceae	g	Native	x	nd	Andean: Argentinian Puna[5,6]
Polypogon interrupus	Poaceae	g	Native	x	nd	Andean: Argentinian Puna[6]
Puccinellia frigida	Poaceae	g	Native	x	x	Andean: High Andean Hypersaline[17]
Azorella madreporica	Apiaceae	sh	Native	x	x	Andean: Chilean High Andean[3]
Laretia acaulis	Apiaceae	h	Native	x	x	Andean: Chilean High Andean[3]
Pozoa coriacea	Apiaceae	h	Native	x	x	Andean: Chilean High Andean[3]
Chaetanthera lycopodioides	Asteraceae	h	Native	-	-	Andean: Chilean High Andean[*3]
Erigeron andicola	Asteraceae	h	Native	x	x	Andean: Chilean High Andean[3]

(continued)

Table 12.1 (continued)

Plant species	Family	Habit	Status	AMF	DSE	[c]Location/Source
Nassauvia lagascae	Asteraceae	h	Native	-	x	Andean: Chilean High Andean[3]
Perezia carthamoides	Asteraceae	h	Native	-	-	Andean: Chilean High Andean[3]
Senecio bustillosianus	Asteraceae	sh	Endemic	x	x	Andean: Chilean High Andean[3]
Senecio francisci	Asteraceae	sh	Native	x	x	Andean: Chilean High Andean[3]
Taraxacum officinale	Asteraceae	h	Exotic	x	x	Andean: Chilean High Andean[3]
Nastanthus agglomeratus	Calyceraceae	h	Native	x	x	Andean: Chilean High Andean[3]
Cerastium arvense	Caryophyllaceae	h	Exotic	x	x	Andean: Chilean High Andean[3]
Adesmia sp.	Fabaceae	h	-	-	-	Andean: Chilean High Andean[3]
Phacelia secunda	Hydrophyllaceae	h	Native	x	x	Andean: Chilean High Andean[3]
Oxalis compacta	Oxalidaceae	h	Native	x	x	Andean: Chilean High Andean[3]
Hordeum comosum	Poaceae	g	Native	x	x	Andean: Chilean High Andean[3]
Calandrinia caespitosa	Portulacaceae	h	Native	-	-	Andean: Chilean High Andean[3]
Montiopsis sericea	Portulacaceae	h	Endemic	-	x	Andean: Chilean High Andean[3]
Barneoudia major	Ranunculaceae	h	Native	x	x	Andean: Chilean High Andean[3]
Acaena pinnatífida	Rosaceae	h	Native	x	x	Andean: Chilean High Andean[3]
Melosperma andicola	Scrophulariaceae	sh	Native	-	x	Andean: Chilean High Andean[3]
Viola atropurpurea	Violaceae	h	Endemic	x	x	Andean: Chilean High Andean[3]
Viola philippii	Violaceae	sh	Endemic	-	-	Andean: Chilean High Andean[3]
Deschampsia flexuosa	Poaceae	g	Exotic	x	x	Andean: Magellanic steppe[4]
Poa rigidifolia	Poaceae	g	Endemic	x	x	Andean: Magellanic steppe[4]
Gavilea australis	Orchidaceae	h	Endemic	-	-[a]	Andean: Magellanic steppe[22]
Gavilea lutea	Orchidaceae	h	Endemic	-	-[a]	Andean: Magellanic steppe[22]

(continued)

Table 12.1 (continued)

Plant species	Family	Habit	Status	AMF	DSE	[c]Location/Source
Codonorchis lessonii	Orchidaceae	h	Endemic	-	-[a]	Andean: Magellanic steppe[22]
Gentianella magellanica	Gentianaceae	h	Endemic	x	x	Andean: Subandean Patagonia[15]
Paepalanthus bromelioides	Eriocaulaceae	h	Native	x	nd	Chaco: Rupestrian fields[20,23]
Bulbostylis sp.	Cyperaceae	g	-	x	nd	Chaco: Rupestrian fields[20,23]
Eremanthus incanus	Asteraceae	t	Native	x	nd	Chaco: Rupestrian fields[23]
Centrosema coriaceum	Fabaceae	h	Native	x	nd	Chaco: Rupestrian fields[23]
Pavonia viscosa	Malvaceae	h	Native	x	nd	Chaco: Rupestrian fields[23]
Tibouchina multiflora	Melastomataceae	t	Native	x	nd	Chaco: Rupestrian fields[23]
Syngonanthus elegans	Eriocaulaceae	h	Endemic	x	nd	Chaco: Rupestrian fields[24]
Loudetiopsis chrysothrix	Poaceae	g	Native	x	nd	Chaco: Rupestrian fields[24]
Xyris sp.	Xyridaceae	h	Native	x	nd	Chaco: Rupestrian fields[24]
Bulbophyllum weddelii	Orchidaceae	h	Native	-	x[a]	Chaco: Rupestrian fields[25]
Epidendrum dendrobioides	Orchidaceae	h	Native	-	x[a]	Chaco: Rupestrian fields[25]
Maxillaria acicularis	Orchidaceae	h	Native	-	x[a]	Chaco: Rupestrian fields[25]
Oncidium gracile	Orchidaceae	h	Native	-	x[a]	Chaco: Rupestrian fields[25]
Pleurothallis teres	Orchidaceae	h	Endemic	-	x[a]	Chaco: Rupestrian fields[25]
Prosthechea vespa	Orchidaceae	h	Native	-	x[a]	Chaco: Rupestrian fields[25]
Sophronitis milleri	Orchidaceae	h	Native	-	x[a]	Chaco: Rupestrian fields[25]
Sarcoglottis sp.	Orchidaceae	h	Native	-	x[a]	Chaco: Rupestrian fields[25]
Digitaria swalleniana	Poaceae	g	Native	x	-	Chaco: Argentinian Prepuna[11]
Lachemilla pinnata	Rosaceae	h	Native	x	x	Chaco: High grasslands[7,12]
Briza subaristata	Poaceae	g	Native	x	nd	Chaco: High grasslands[7]

(continued)

Table 12.1 (continued)

Plant species	Family	Habit	Status	AMF	DSE	[c]Location/Source
Deyeuxia hieronymi	Poaceae	g	Native	x	x	Chaco: High grasslands[7,12]
Poa stuckertii	Poaceae	g	Endemic	x	x	Chaco: High grasslands[7,12]
Eragrostis lugens	Poaceae	g	Native	x	nd	Chaco: High grasslands[c6,7]
Sorghastrum pellitum	Poaceae	g	Native	x	nd	Chaco: High grasslands[c7]
Agrostis glabra	Poaceae	g	Endemic	x	nd	Chaco: High grasslands[c10]
Muhlenbergia peruviana	Poaceae	g	Native	x	nd	Chaco: High grasslands[c10]
Nassella nidulans	Poaceae	g	Endemic	x	nd	Chaco: High grasslands[c10]
Vulpia myurus	Poaceae	g	Exotic	x	nd	Chaco: High grasslands[10]
Polystichum montevidense	Dryopteridaceae	f	Native	x	-	Chaco: High grasslands[12]
Eryngium agavifolium	Apiaceae	h	Native	x	x	Chaco: High grasslands[12]
Oreomyrrhis andicola	Apiaceae	h	Native	x	x	Chaco: High grasslands[12]
Achyrocline satureioides	Asteraceae	sh	Native	x	x	Chaco: High grasslands[12]
Gamochaeta americana	Asteraceae	h	Native	x	x	Chaco: High grasslands[12]
Hypochaeris radicata	Asteraceae	h	Exotic	x	x	Chaco: High grasslands[12]
Berberis hieronymi	Berberidaceae	sh	Native	x	x	Chaco: High grasslands[12]
Blechnum penna-marina	Blechnaceae	f	Native	x	x	Chaco: High grasslands[12]
Carex fuscula	Cyperaceae	g	Native	x	x	Chaco: High grasslands[12]
Gaultheria poeppigii	Ericaceae	sh	Native	x	x[b]	Chaco: High grasslands[12,18]
Gentianella achalensis	Gentianaceae	h	Native	x	x	Chaco: High grasslands[12]
Lepechinia meyenii	Lamiaceae	h	Native	x	x	Chaco: High grasslands[12]
Satureja odora	Lamiaceae	sh	Native	x	x	Chaco: High grasslands[12]
Huperzia saururus	Lycopodiaceae	f	Native	x	x	Chaco: High grasslands[12]

(continued)

Table 12.1 (continued)

Plant species	Family	Habit	Status	AMF	DSE	[c]Location/Source
Festuca tucumanica	Poaceae	g	Native	x	x	Chaco: High grasslands[12]
Polypodium bryopodum	Polypodiaceae	f	Native	x	x	Chaco: High grasslands[12]
Duchesnea indica	Rosaceae	h	Exotic	x	x	Chaco: High grasslands[12]
Polylepis australis	Rosaceae	sh/t	Native	x	x	Chaco: High grasslands[12]
Bartsia crenoloba	Scrophulariaceae	h	Native	x	x	Chaco: High grasslands[12]
Aa achalensis	Orchidaceae	h	Endemic	-	x[a]	Chaco: High grasslands[21]
Gentianella multicaulis	Gentianaceae	h	Endemic	x	x	Chaco: High grasslands[15]
Gentianella parviflora	Gentianaceae	h	Endemic	-	x	Chaco: High grasslands[15]

Habits: (f) fern, (h) herb, (g) graminoid, (sh) shrub, (s) succulent, (t) tree; AMF root colonization: (x) presence, (-) absence; DSE root colonization: (x) presence, (-) absence, (nd): non determined; [a]orchid mycorrhiza; [b]ericoid mycorrhiza. [c]Location/Source: (1) Angulo-Veizaga and Garcia-Apaza (2014), (2) Barnola and Montilla (1997), (3) Casanova-Katny et al. (2011), (4) García et al. (2012), (5) Lugo et al. (1995), (6) Lugo et al. (1997), (7) Lugo et al. (2003), (8) Lugo et al. (2012), (9) Lugo et al. (2018), (10) Lugo unpublished, (11) Lugo unpublished, (12) Menoyo et al. (2007), (13) Montilla et al. (1992), (14) García Romero et al. (2004), (15) Salvarredi et al. (2010), (16) Schmidt et al. (2008), (17) Silvani et al. (2013), (18) Urcelay (2002), (19) Urcelay et al. (2011), (20) Pagano and Scotti (2009), (21) Fracchia et al. (2014a), (22) Fracchia et al. (2014b), (23) Pagano and Cabello (2012), (24) Costa et al. (2016), (25) Nogueira et al. (2005). The Location/Source numbers are also placed at its respective geolocations in the Fig. 12.2

symbiotic associations was AM, which AMF colonization represented in the 83% of roots of the species while only 17% was not colonized for these fungi (Fig. 12.4b). Plants that were not colonized by AMF were found within several families as Aspleniaceae, Asteraceae, Cyperaceae, Ephedraceae, Fabaceaea, Gentianaceae, Geraniaceae, Orchidaceae, Poaceae, Portulacaceae, Scrophulariaceae, and Violaceae; however, those families also showed presence of AMF colonization depending of the species studied, only Aspleniaceae, Ephedraceae, Orchidaceae and Portulacaceae never was colonized by AMF (Table 12.1). The DSE colonization was recorded in 48% of the total plant taxa studied in Highlands, while only 9% did not present this type of colonization (Fig. 12. 4c). Thus, there is a remaining 43% of plants without information of the presence of colonization by DSE. In addition, as also occurred for AMF colonization, the plants without DSE colonization were representatives of several families as Aspleniaceae, Asteraceae, Dryopteridaceae, Ephedraceae, Fabaceaea, Gentianaceae, Geraniaceae, Malvaceae, Orchidaceae, Poaceae, Portulacaceae, and Violaceae; however, those families also showed presence of DSE colonization depending of the species studied, only Aspleniaceae,

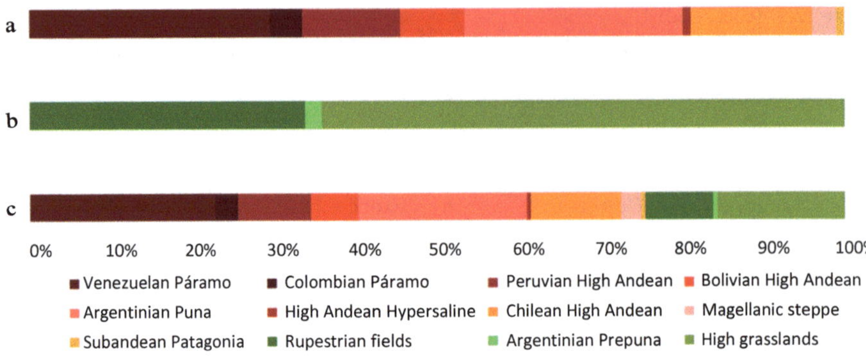

Fig. 12.1 Percentage of plant species studied along the different Highlands of South America (**a**) Andean Highlands, (**b**) Chaco Highlands, (**c**) Total data

Dryopteridaceae, Ephedraceae, Geraniaceae, and Malvaceae never were colonized by DSE (Table 12.1).

Furthermore, a triple association of ER, AM, and DSE was also recorded in *Gaultheria poeppiggi* DC (Ericaceae) in Chaco Highlands of Pampa de Achala (Table 12.1).

In regard to mycorrhizal associations and fungal endophytic colonization in plants roots, in Andean environments 9 Ecoregions (Table 12.1) were reported and revised considering the plant species analyzed (Figs. 12.1, 12.2) such as the Venezuelan Páramo (30%), the Argentinian Puna (27%), the Andes of Central Chile (15%) and Peruvian High Andean (12%), these plants species are widespread in 33 families, with the main proportion of species studied included in Poaceae (39%) and Asteraceae (25%) (Fig. 12.3a). The majority of the plants observed were herbaceous (39%) and graminoid (40%), also including few ferns, shrubs, succulents and trees (Fig. 12.3b); which were majority native (72%) or endemic (12%) plants, with few exotic species (8%) (Fig. 12.3c). Arbuscular mycorrhizal colonization was recorded in the 85% of the plant species and the 43% presented DSE colonization; however, the 47% of the total plant species were not analyzed for the presence of colonization by DSE (Figs. 12.4b, c).

Along Highlands of the Chaco region, 3 Ecoregions (Table 12.1) were considered to be tall grasslands such as Rupestrian fields of Cerrado region, Prepuna and Pampa de Achala grasslands (Figs. 12.1, 12.2) that were represented by the main species numbers (64%) among 20 families of plants, mostly Poaceae (24%) (Fig. 12.3a). As in the Andean highlands, the habits of the majority of the plants studied were herbaceous (52%) and graminoid (26%) including the lesser percentage of ferns, shrubs, and tree (Fig. 12.3b). The great number of species were native (76%) or endemic plants (16%), with a scarce number of exotic species (6%) (Fig. 12.3c). The 80% of the species presented AMF colonization and the 64% of plants were colonized by DSE while the remaining 32% were not analyzed for the presence of DSE colonization (Figs. 12.4b, c).

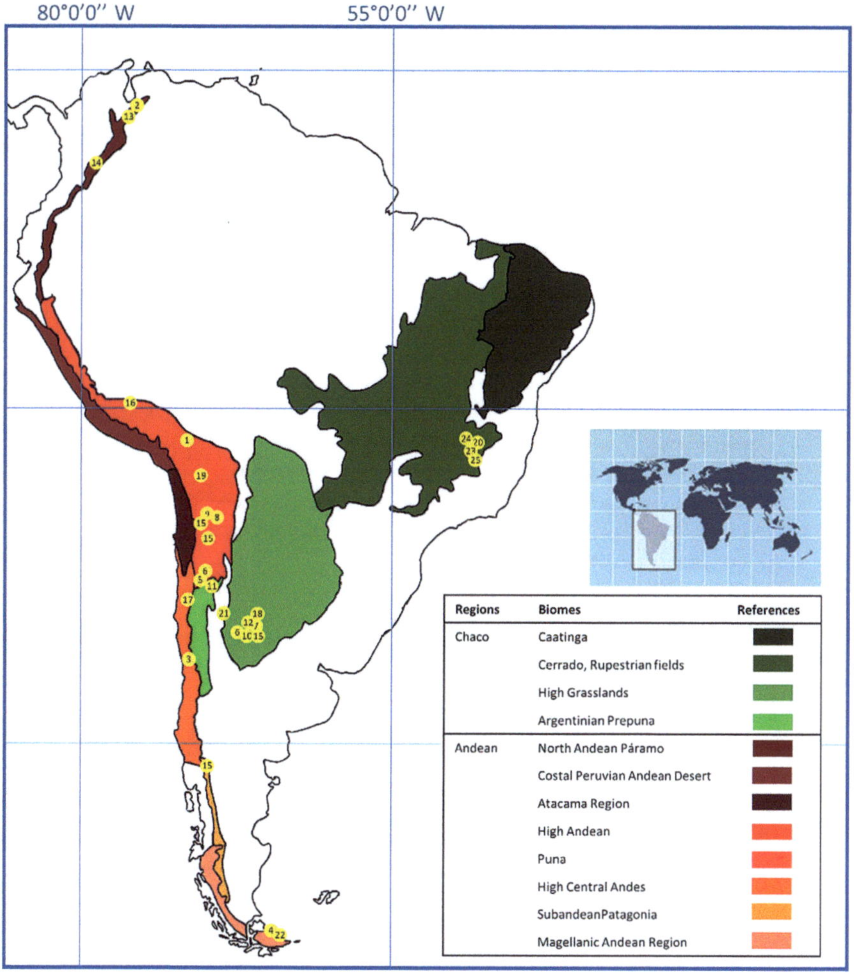

Fig. 12.2 Map of the root symbiosis distribution in Highlands of South America. The Location/Source numbers are detailed in the foot note of the Table 12.1 (Map author: Hebe J. Iriarte – IMIBIO Institute)

The glomalean fungal endophytes colonization in roots of Highlands of Andean (Fig. 12.5) and Chaco regions (Fig. 12.6) were fine root endophytes (FRE) (Figs. 12.5a, 12.6a) and coarse root endophytes (CRE); the coarse and medium hyphae were found in both Andean (Fig. 12.5b) and Chaco regions (Fig. 12.6b) forming colonization from *Arum* (Fig. 12.5c, Fig. 12.6c) to *Paris* (Fig. 12.5c) types, and a CRE with a particular *Paris* type (Fig. 12.6b) which was presented in roots of Gentianaceae also in both ecorigions. Furthermore, DSE colonization was also observed in the roots of Andean (Fig. 12.5d) and Chaco Gentianaceae (Fig. 12.6d) and Poaceae (Fig. 12.6e).

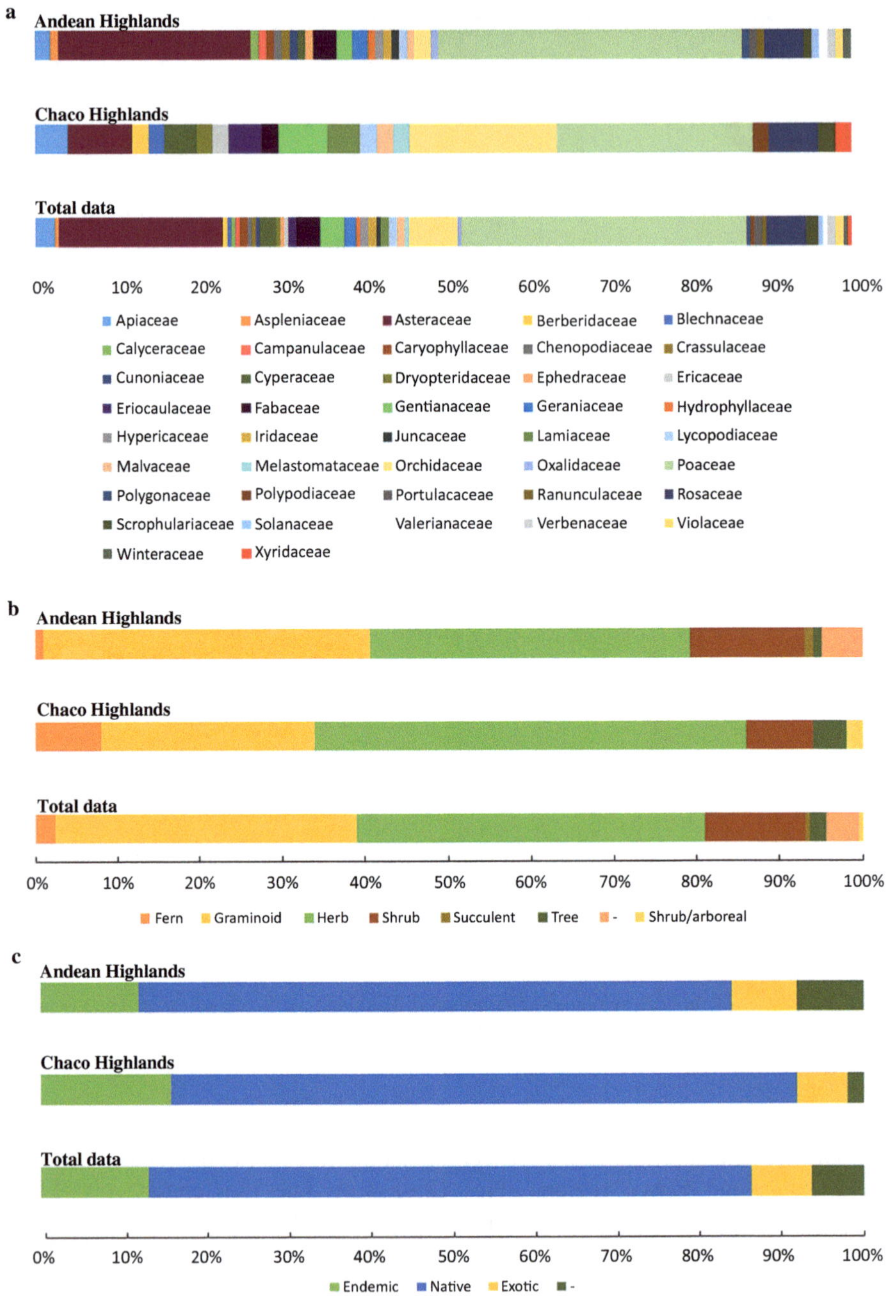

Fig. 12.3 Percentage of plant species studied along the different Highlands of South America considering (**a**) Family, (**b**) Habits, (**c**) Status

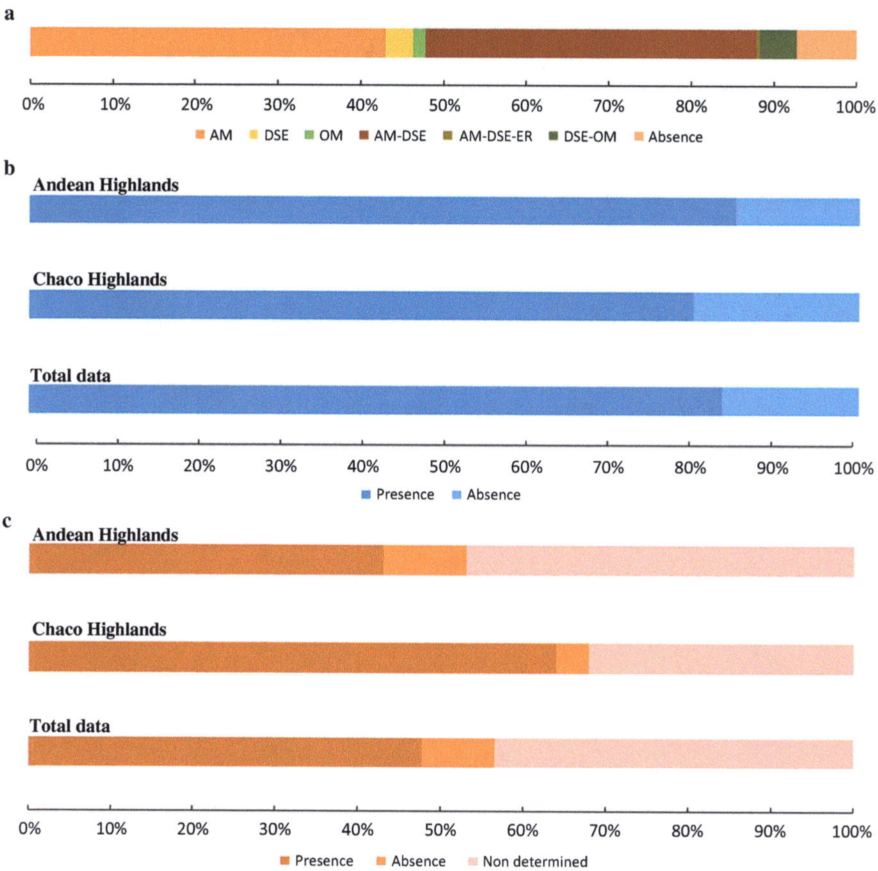

Fig. 12.4 Percentage of plant species associated to different fungal endophytes and mycorrhizal fungi along the Highlands of South America (**a**) Percentage of total hosts species by different root fungal associations, (**b**) Percentage of hosts species colonized by AMF, (**c**) Percentage of hosts species colonized by DSE. References: AM: Arbuscular mycorrhizas, DSE: Dark septate endophytes, OM: Orchid mycorrhizas, ER: Ericoid mycorrhizas

The frequency of plant species colonized by each family was analyzed at each environment studied, showing differences between the Ecoregions. Thus, some combinations of well represented families were found in Andean and Chaco Highlands (Table 12.1, Fig. 12.3a). It is important to take into account in Highlands, the effects of global warming on plants community, with nutrient induced loss of plant diversity such as the graminoid or sedge promotion that could have an impact on the fungal communities of soils, especially in the AMF (Wahl and Spiegelberg 2016).

Ectomycorrhizal associations were absent in Highlands of SA neither Andean nor Chaco regions. The ECM has been known in alpine environments in roots of trees, few shrubs and herbaceous plants such as *Betula* spp. (Betulaceae), *Salix* spp. (Salicaceae), *Dryas* spp. (Rosaceae) and *Kobresia* spp. (Cyperaceae) (Gardes and

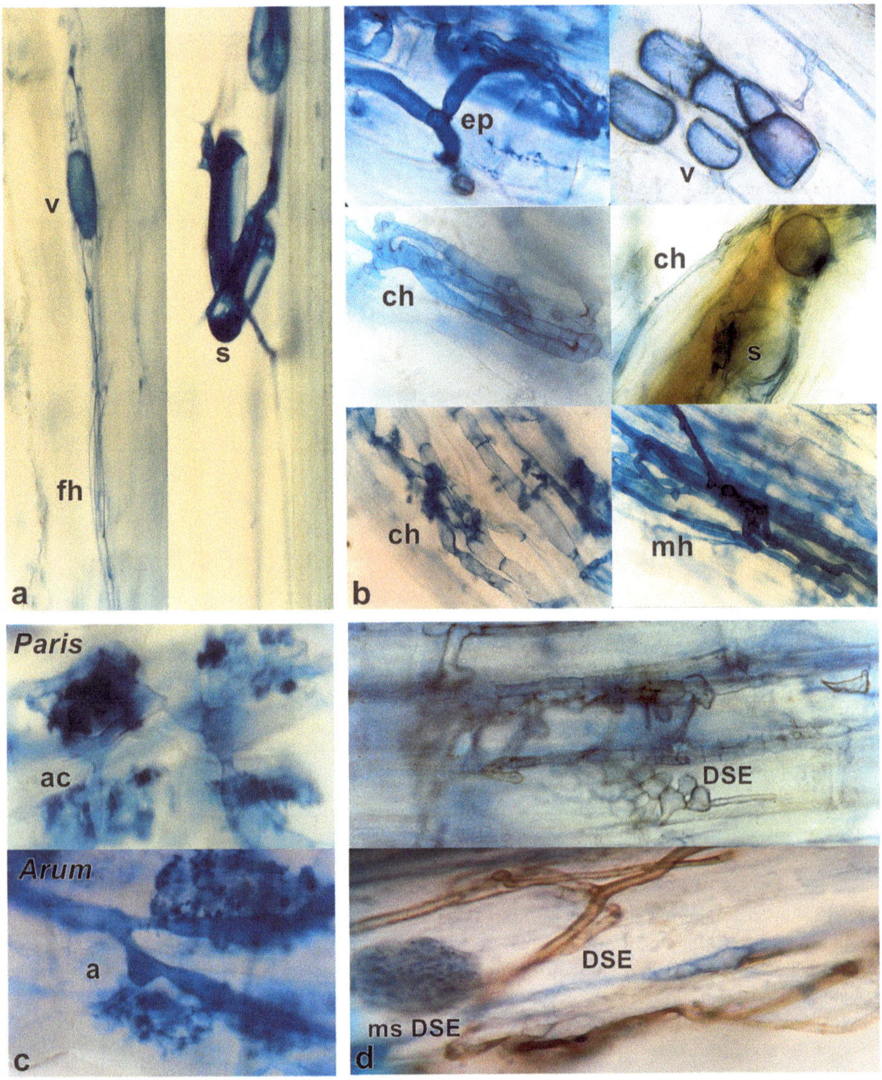

Fig. 12.5 Root colonization types in Poaceae of the South American Highlands of Puna, Andean region (**a**) fine root endophytes (FRE), (**b**) coarse (CRE) and medium glomalean root endophytes, (**c**) CRE forming *Arum* and *Paris* type colonization, (**d**) DSE in the roots. *References: meaning of letters on the images: a, arbuscule; ac, arbusculated coil; ch, coarse hypha; ep, entry point; fh, fine hypha; mh, medium hypha; ms, microsclerotia; v, vesicle; s, spore (Photo-credit: M. A. Lugo)

Dahlberg 1996; Smith and Read 2008). Although these families and potential ectomycorrhizal hosts plants were revised in this Chapter, there are no ECM records yet among South American Highlands.

It is important to note that *Gaultheria poeppigii* is the unique citation of ER along the South American Highlands, also for triple associations in native plants

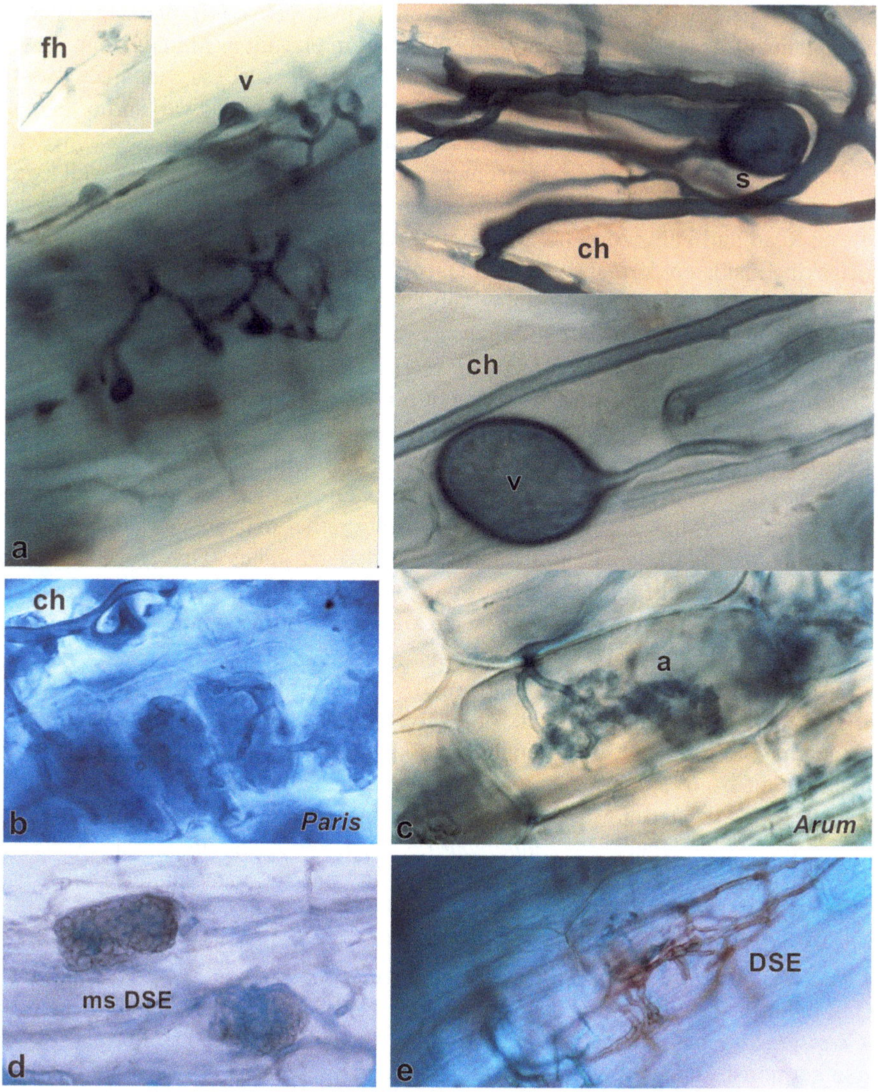

Fig. 12.6 Root colonization types of the South American Pampa de Achala Highlands of Chaco regions (**a**) fine root endophytes (FRE) in roots of Poaceae, (**b**) coarse root glomalean endophyte (CRE) with *Paris* type colonization in roots of Gentianaceae, (**c**) CRE in roots of Poaceae with *Arum* type colonization, (**d**): DSE in roots of Gentianaceae, (**e**) DSE in roots of Poaceae *References: a, arbuscule; ch, coarse hypha; fh, fine hypha; ms, microsclerotia; v, vesicle; s, spore (Photo-credit: M. A. Lugo)

roots. This Ericaceae growth in Pampa de Achala, in the Chaco Highland in Córdoba mountains of Argentina, between 1700 and 2884 m elevation (Menoyo et al. 2007; Urcelay 2002). Plants belonging to Ericales (e.g. Ericaceae) are taxa commonly distributed in cold-dominated environments, co-occurring with dwarf shrubs in high areas named heathlands (Kohn and Stasovski 1990; Read 1991; Smith and Read 2008). The heathlands are also high environments of the world which are well kwon in regard to its ericoid mycorrhizas diversity and functionality (Gardes and Dahlberg 1996; Read and Pérez Moreno 2003; Smith and Read 2008). Ericoid colonization occurring inside epidermal cells of typical ericaceous hair-like roots, where hyphae of the fungal symbionts are forming "coils" in most ericaceous arctic and alpine plants (Peterson et al. 2004), alpine dwarf plant productivity also have been positively correlated with ericoid infection level. Further, fungal symbionts forming ericoid mycorrhizas are anamorphic forms of Ascomycota and Basidiomycota, and/or DSE and/or AMF (Jumpponen and Trappe 1998; Smith and Read 2008). Moreover, the triple colonization of *G. poeppiggi* by *Rhizoctonia*-like, *Phialocephala*-like and AMF has been considered as an evolutionary feature of this species, with the capability to form AM as a new evolutionary novelty rather than an ancestral character regarding to *Gaultheria* is an evolutionarily derived genus (Urcelay 2002). The study of this type of fungal associations and the native Ericaceae and other families among Ericales is an important gap in the knowledge of the symbiotic associations for Highlands of SA.

Another exceptional record for South American Highlands was the orchid mycorrhizal associations in roots of twelve terrestrial native species. In Central Argentina, *Aa achalensis* Schltr. habits in high grasslands above treeline at Sierra de Velasco in Chaco region. This orchid is an endemic and endangered species, which has been associated to five fungal strains of two basidiomycetous fungi of *Rhizoctonia*-like related to *Thanatephorus cucumeris*, and three ascomycetous fungi belonging to *Phialophora graminicola* and one to an uncultured Pezizaceae (Fracchia et al. 2014a). In ruprestian grassland of the Cerrado, eight Brazilian native species (*Bulbophyllum weddelii*, *Epidendrum dendrobioides*, *Maxillaria acicularis*, *Oncidium gracile*, *Pleurothallis teres*, *Prosthechea vespa*, *Sophronitis milleri*, *Sarcoglottis* sp.) were assocciated with DSE and *Rhizoctonia*-like fungi such as *Ceratorhiza*, *Epulorhiza*, and *Rhizoctonia* (Nogueira et al. 2005). Further, in the southernmost Andean region in Tierra del Fuego, three native austral orchid species *Gavilea australis* (Skottsb.) MN Correa, *G. lutea* (Pers.) M.N. Correa and *Codonorchis lessonii* (Brongn.) Lindl. were associated to four fungal basidiomycetous endophytes belonging to two strains of Ceratobasidiaceae, *Tulasnella calospora,* and *Ceratobasidium albasitensis*. Strikingly, *G. australis* is an endangered and endemic austral orchid species which has been shown a low mycorrhizal specificity with its endophytic fungi (Fracchia et al. 2014b). Thus, this new knowledge of fungal symbionts of *G. australis* could be applied in conservation efforts of this endangered orchid; moreover, these researching approaches might be positive applications also to more threatened host species.

The AMF and DSE were the predominant symbionts associated in the roots of Highland plants. The AMF were the fungal endosymbionts that colonized the largest number of hosts, followed by the DSE. Although in the world mountain environments occupy ca. a 25% of the Earth surface (Körner 1999, 2007) and the presence of AMF colonization in the highlands has been reported at elevations as high as 4545 m in extreme conditions of the Swiss Alps (Oehl and Körner 2014), the research studies in these high ecosystems are still scarce (Wahl and Spiegelberg 2016), especially in SA as it has been shown in this Chapter. Strikingly, the AMF colonization had already been previously reported at 5250 m elevation for the High Andes of Perú (Schmidt et al. 2008). Thus, the presence of AMF in the South American host plants surpassing in height to those in the Northern Hemisphere alpine environments except for the reports from Himalaya Mountains in India, with AMF found in elevations up to 5800 m, although the host plants grow up to 6150 m elevations, showing that AMF presence and colonization is more compelled by extreme conditions than by host presence; instead, DSE were present throughout from 3400 to 6150 m (Kotilínek et al. 2017). Further among Andean region, AMF colonization was recorded at 4314 m at Puna Highlands (Lugo et al. 2012), where has been shown an inverse relationship between altitude and root colonization as it was also found in the Perú Andean Highlands (Schmidt et al. 2008) and at 4123-4260 m in the hypersaline sites in High Andean wetlands (Silvani et al. 2013). However, in Puna, the effect of altitude on AMF colonization seems to be more related to the photosynthetic pathway (C_3, C_4) of grasses hosts than to life cycles (annual, perennial). These Andean sites are the most extreme places showing evidence of well-established AMF communities in SA (Schmidt et al. 2008; Lugo et al. 2008, 2012; Silvani et al. 2013), comparable also with AMF communities in alpine and mountain grasslands around the world. In Andean region, AMF colonization has been found also along an extent altitudinal range from the Northern Andean Páramo, High Bolivian Andes, High Central Andes, mean elevation sites of Puna to 400–700 m at the lowest elevation in Andean Southermost extreme, in Tierra del Fuego (Barnola and Montilla 1997; Montilla et al. 1992; García Romero et al. 2004; Urcelay et al. 2011; Angulo-Veizaga and García-Apaza 2014; Salvarredi et al. 2010; Lugo et al. 1995, 1997; García et al. 2012).

The roots colonization by the AMF and the DSE was well represented for the Andean and Chaco Highlands. The AMF and DSE had a slightly higher percentage of hosts colonized in Andes or Chaco, repectively. This differential behavior of AMF vs. DSE between Andean and Chaco could be due to diverse factors such as environmental (soil nutrients, climate, elevation, among others) and biological (host species, phylogenetic relationships, biotic stressors, species competition, among others) conditions. The first distinctly differential factor between Andes and Chaco are the soils types and its nutrients availability. The mycorrhizal fungi and fungal root endophytes associations can function as resource acquisition strategies in host plants, and may differ among species and ecosystems as a function of their fungal symbionts (ECM, ER, OR, AMF, DSE or its combinations), as each fungal group may access different sources of growth-limiting nutrients (Smith and Read 2008).

It is well known, AMF associations occur often in environments with poor soils with low organic material content, high nitrogen content and low phosphorous availability (Allen 1991; Read 1991). Ericoids and ECM are indispensable for their host plants which are obligate mycotrophs; in contrast, AMF can form associations with plant species that can live with or without mycorrhizal fungi as facultative mycotrophs, colonizing plant roots without offering any apparent benefit (Trappe 1987). Moreover, AMF colonization can function as mycorrhizal (AM) or not when the mycorrhizal structures formed into the roots are only hyphae and vesicles but the arbuscules (the host-fungus interchanging nutrients structures) are absent, this non mycorrhizal colonization was defined as "glomalean fungus colonization" (GFC) by Brundrett and Tedersoo (2018). Further, AMF can present two other root colonizing types commonly in graminoid and herb hosts, they are "fine" (FRE) and "coarse" (CRE) root endophytes. The FRE type of colonization has been attributed to *Glomus tenuis* which also was considered as an increased colonizer in alpine highlands (Read and Haselwandter 1981; Walker et al. 2018a and reference therein). Recently, FRE have been included in the sub-phylum Mucoromycotina (Orchard et al. 2017), rather than Glomeromycota where were placed before together with CRE; actually, the FRE named in the past as *Glomus tenue* (Greenall) I. R. Hall was renamed as *Planticonsortium tenue* (Greenall) C. Walker et D. Redecker and included in Mucoromycotina, instead the CRE fungi remain in Glomeromycota (Walker et al. 2018a and reference therein). The CRE and FRE colonization types were present even in Andean and in Chaco Highlands (Schmidt et al. 2008 and this Chapter). In general, it has been accepted that FRE colonization is increased with the elevation (Wahl and Spiegelberg 2016 and references therein). The coarse hyphae (CRE) and fine endophyte (FRE) colonization in roots of Poaceae in the Andean Puna and Chaco Highlands were reported in this Chapter (Figs. 12.5, 12.6). Furthermore, CRE and FRE colonization have been detected in diverse hosts belonging to the families Asteraceae, Fabaceae, Poaceae, Scrophulariaceae and Valeraniaceae in Andean Highlands of Perú (Schmidt et al. 2008). Although CRE and FRE colonization types were found along Andean and Chaco Highlands (Schmidt et al. 2008 and this Chapter), is lacking yet the information of this type of colonization patterns, because there are only presence/absence records of them in SA.

It has been showed that DSE may replace the function of AMF at high elevations or latitudes, such as in the high Arctic tundra and in European and North American alpine communities (Read and Haselwandter 1981; Trappe 1987; Kohn and Stasovski 1990; Gardes and Dahlberg 1996). In high latitude and elevation environments of the Northern Hemisphere (Read and Haselwandter 1981; Kohn and Stasovski 1990), the AMF colonization seems to be scarcity versus DSE colonization although it causes remain poorly understood (Gardes and Dahlberg 1996). However, these studies have been reported some explanations for the reduction in AMF colonization such as high fertility in high cold areas instead of poor soils conditions that benefit AMF proliferation; presence of nival zones where vegetation is scant and the host roots availability is low hinder AMF spread by root-to-root contact given by their biotrophic nutrition, and its soils with few AMF spores, resulting in a scant presence of fungal propagules in highlands. In turn, DSE colonization typically is improved in fertile areas with low availability of roots, probably due to DSE have the capacity to decompose organic matter in the absence of host plant roots and have a relatively high tolerance to extreme conditions. Moreover, plants such as the Cyperaceae, which host DSE but are not associated with AMF, often are abundant in highlands.

Although DSE and AMF can co-occur in roots, there is no consensus on whether these interactions are competitive, facilitative or amensalistic (Ruotsalainen and Eskelinen 2011). However, in Bolivian Andean Highlands, Urcelay et al. (2011) suggested that in an environment characterized by aridity, cold temperatures, and nutrient-poor soils, the relative colonization by AMF vs. DSE, rather than the total colonization by AMF or DSE per se, better predict the functional implications of the fungal-root symbiosis. Further, in Puna AMF and DSE patterns of root colonization shifted as a function of elevation in most grass species, but in general, these trends differ from previous studies in the Northern Hemisphere. However, the variation among sites in AMF and DSE colonization could not be explained by different elevations of sites, instead, other environmental factors as microenvironments features might exert a strong influence on AMF and DSE colonization. Moreover, both AMF and DSE may have established synergistic and beneficial associations with hosts in Puna harsh ecosystems (Lugo et al. 2018).

In the Northern Hemisphere, particularly in Arctic regions, organic soils tussock in tundra and in alpine environments, Cyperaceae and Juncaceae are abundant, usually as dominant species in the plant communities, and they are frequently non-mycorrhizal. The prevalence of cyperaceous graminoid plants in these ecosystems may entail the scarcity of colonization by AMF. However, in the same environment types Cyperaceae and Juncaceae usually for association in its roots with DSE and FRE, and they are colonizing roots of these graminoid more often than AMF in high elevation when harsh climatic conditions are the prevailing (Walker et al. 2018b and references therein). However, one more time in Highlands of SA, Cyperaceae and Juncaceae showed a different pattern of colonization compared to the Northern Hemisphere; thus, in Andean Venezuelan Páramo the Cyperaceae and Juncaceae at 380 m were colonized by AMF, and DSE colonization was still not studied (Barnola and Montilla 1997). Instead, in Chaco Highlands any species of Juncaceae were not studied in the highlands above treeline; meanwhile, the Cyperaceae species were associated to both AMF and DSE in Chaco Highlands at 2190 m (Menoyo et al. 2007) and only AMF colonization was studied and recorded at the rupestrian fields of the Serra do Cipó at 600-900 m (Pagano and Scotti 2009; Pagano and Cabello 2012).

Several plants families were inconsistently colonized by AMF as Asteraceae, Cyperaceae, Fabaceaea, Gentianaceae, Geraniaceae, Poaceae, Scrophulariaceae, and Violaceae; that is, these families showed AMF colonization depending on the species studied (Table 12.1). The same lack of a trend of colonization at host family level occurred for DSE along the families Asteraceae, Fabaceaea, Gentianaceae, Orchidaceae, Poaceae, Portulacaceae, and Violaceae. Further, Aspleniaceae, Dryopteridaceae, Ephedraceae, Geraniaceae, Malvaceae families were not colonized by DSE (Table 12.1). Then, a general pattern associated with phylogenetic relationships could not be found for AMF and DSE colonization since the same family presented colonized and non-colonized species. However, many plants families recorded associated to AMF and DSE in South American Highlands had already been reported mostly as only AM, AM and ECM or non mycorrhizal (NM) in different mountains environments around the world (e.g. Apiaceae, Berberidaceae, Blechnaceae, Caryophyllaceae, Lamiaceae, Lycopodiaceae, Oxalidaceae, Polypodiaceae, Portulaceaea, Scrophulareaceae, Solanaceae, Valerianaceae among others) (Wang and Qiu 2006; Brundrett 2009; Brundrett and Tedersoo 2018). Behind, also the South American family Calyceraceae is reported in this Chapter associated to both AMF and DSE. Moreover, several families reported associated dually to fungal symbionts in Andean were different than in Chaco Highlands.

In other hand, in several families such as Chenopodiaceae and Cyperaceae the mycorrhizal status the type of associations established with fungal symbionts on their roots have been a controversial issue due to involved many NM plants but also have been reported associated to AMF (Wang and Qiu 2006 and references therein; Brundrett 2009 among others). Moreover, the majority of angiosperms are associated with symbiotic fungi forming AM (Brundrett 2009). In the Poaceae, 99.6% of the species studied are AM symbionts (Wang and Qiu 2006), constituting an AM group in Poales (Brundrett 2009). However, it is important to note that most of the studied plant species are mainly from the Northern Hemisphere and Eurasia, and only a few of them are from SA. In South American Highlands, these families were NM, AM, DSE or AM-DSE in Cyperaceae and AM-DSE in Chenopodiaceae. It has been proposed (Brundrett and Tedersoo 2018 and references therein) that hosts with multifunctional roots (e.g. ERN and OM, ECM and AM, etc.) are able to suffer morphological and taxonomical diversification in short period of evolutionary time, particularly in poor soils of arid environments. Thus, the plants of South American Highlands which have mostly multifunctional roots forming dual (e.g. AM and DSE) and triple (e.g. OM, AM and DSE) associations could be subject of quick evolution.

Although, more studies could be done in South American Highlands to arrive at conclusive results in this issue. Futhermore, we did not find any report of mycorhizal or fungal endophyte associations with bryophytes in South American Highlands.

In South American Highlands, the plant roots mainly were associated forming AM (43%) and dual associations with AMF-DSE (40%) followed by dual DSE-OM associations (4.4%), DSE colonization (3.3%), OM (1.5%), triple colonization by AM-DSE-ER (0.5%), and 7.3% of non mycorrhizal (NM) plants. Worldwide, it has been proposed that the majority of vascular plants are mycorrhizal, forming 72% AM, 2.0% ECM, 1.5% ECM and 10% OM; in contrast, 8% were reported as NM, whereas 7% have inconsistent NM-AM associations, so-called GFC (Brundrett and Tedersoo 2018). Further, the NM plants and the plants associated with GFC are mainly nutritional or habitat specialists such as carnivores, parasites, hydrophytes, and epiphytes (Brundrett 2009; Brundrett and Tedersoo 2018). Moreover, NM plants are most inhabiting arid, disturbed and harsh habitats, also arid and alpine environments (Brundrett and Tedersoo 2018 and references therein). Further, erroneous determinations of the mycorrhizal status may be due to a misinterpretation of the GFC colonization as a really functional mycorrhizic association (Brundrett and Tedersoo 2018 and references therein). Therefore, the patterns of root colonization by symbiotic fungi in South American Highlands seems to be similar of worldwide.

12.4 Conclusion

In this Chapter, the needs for more information to known, understand, conserve and management of South American Highlands have been highlighted. Several thematic vacant areas and gaps of mycorrhizal information detected are ECM in shrubs and herbs, ericoid mycorrhizas, orchidoid mycorhizal, mycorhizal and endophitic fungal associations with bryophytes, molecular approach to study mycorrhizal association and fungal symbionts are urgently required for AMF and DSE especially inside roots, and also in soils. Further, the examination and use of arbuscular mycorrhizas morphological classification of the colonization types such as coarse (CRE) and fine root endophytes (FRE), glomalean but not mycorrhizal colonizazion (GFC) and *Arum-Paris* type continuum in different ecosystems have been scantly mentioned in the bibliography and are required as necessary for future studies to functional approaches of this root associations.

Throughout this Chapter, we have been able to appreciate the diversity of symbiotic associations with fungal endophytes and mycorrhizal fungi in the roots of predominantly native plants, and endemic of South American Highlands. Thus, it has been shown that the colonization patterns in these Highlands are different of those of the Northern Hemisphere, and these differences involved either host families, type of fungal associations established in host roots, and colonization types. Furthermore, the number of endemic and native plants studied is still scarce; data included in this Chapter comprise around to 2% of vegetation in Highlands of the Andean region and 0.8% in Chaco region. It is also important to note that the symbiotic association studies of Páramo and Rupestrian field vegetation are particularly scarce, principally considering the great diversity of plant species and the high number of endemic species that they present, with publications of these South American regions covering only 1% and 0.3% of them, respectively. In future research, it could be necessary to include and emphasize the study of the endemic species, which are of fundamental importance for the management and conservation of these high ecosystems, to evaluate also the presence of multifunctional associations in the roots that could be useful to measure the functional aspects of root symbiosis. The common presence of multifunctional root associations in Highlands that are under the effect of rapid evolutionary processes, behind to the marked disturbances and perturbations caused by the advance of the agricultural frontiers across the continent, and global warming effects on fungal symbiosis in roots claim for rapid results in these issues to maintain these biodiversity hotspots Highlands.

Acknowledgements The authors are especially grateful for his collaboration to the Biol. Esteban M. Crespo in the field works in the Highlands grasslands, Puna and Prepuna, and to the Microb. Hebe J. Iriarte (Research Technical Assistant of CONICET) for her assistance with the elaboration and improvement of the Figures. Likewise, we would like to express our gratitude to the researchers Drs. Marta Cabello and Laura Domínguez, who were pioneers and propellers of mycorrhizal fungi research of the native ecosystems in Argentina. In addition, ML wants to thank especially to her worthy director, Dr. Ana Anton for having trained her not only in the knowledge of the plants but also by the integral academic training received from her. This work was financially supported by PROICO 02-2718 (FQByF-UNSL), and both authors are staff researchers from CONICET.

References

Aagesen L, Bena MJ, Nomdedeu S, Panizza A, López RP, Zuloaga FO (2012) Areas of endemism in the southern central Andes. Darwiniana 50:218–251

Aagesen L, Szumik CA, Zuloaga FO, Morrone O (2009) Quantitative biogeography in the South America highlands – recognizing the Altoandina, Puna and Prepuna through the study of Poaceae. Cladistics 25:295–310

Allen MF (1991) The ecology of mycorrhizae. In: De Barnes RSK, Birks HJB, Connor EF, Harper JL, Paine R (eds). Cambridge University Press, Cambridge, New York, Port Chester, Melbourne, Sydney

Angulo-Veizaga WV, García-Apaza E (2014) *Baccharis incarum* and fungus Arbuscular Mycorrhizal symbiotic relationship for land fallow in the Bolivian highland. CienciaAgro 3:51–58

Averill C, Turner BL, Finzi AC (2014) Mycorrhiza-mediated competition between plants and decomposers drives soil carbon storage. Nature 505:543–545

Barnola LG, Montilla MG (1997) Vertical distribution of mycorrhizal colonization, root hairs, and belowground biomass in three constrasting sites from the tropical high mountains, Mérida, Venezuela. Arct Antarct Alp Res 29:206–212

Bradford MA (2014) Good dirt with good friends. Nature 505:486–487

Brooks TM, Mittermeier RA, da Fonseca GAB, Gerlach J, Moffmann M, Lamoreux JF, Mittermeier CG, Pilgrim JD, Rodrigues RR (2006) Global biodiversity conservation priorities. Science 313:58–61

Brundrett MC (2009) Mycorrhizal associations and other means of nutrition of vascular plants: understanding the global diversity of host plants by resolving conflicting information and developing reliable means of diagnosis. Plant Soil 320:37–77

Brundrett MC, Tedersoo L (2018) Evolutionary history of mycorrhizal symbioses and global host plant diversity. New Phytologist 220:1108–1115

Buytaert W, Cuesta-Camacho F, Tobón C (2011) Potential impacts of climate change on the environmental services of humid tropical alpine regions. Global Ecology and Biogeography 20(1):19–33

Cabido MR, Carranza ML, Acosta ATR, Paez S (1991) Contribución al conocimiento fitosociológico del Bosque Chaqueño Serrano en la provincia de Córdoba, Argentina. Phytocoenologia 19:547–566

Cabido MR (1985) Las comunidades vegetales de la Pampa de Achala, Sierras de Córdoba, Argentina. Documents phytosociologiques 9:431–443

Cabido MR, Breimer R, Vega G (1987) Plant communities and associated soil types in a high plateau of the Córdoba mountains, Central Argentina. Mountain Res Develop 7:25–42

Cabido M, Giorgis M, Tourn. 2010. Guía para una excursión botánica en las Sierras de Córdoba. Boletín de la Sociedad Argentina de Botánica 45(1–2):209–219

Cabrera AL (1976) Territorios fitogeográficos de la República Argentina. Enciclopedia Argentina de Agricultura y Jardinería, 2da. Edición, II (1). ACME, Argentina

Cabrera AL, Willink A (1980) Biogeografía de América Latina. OEA, Washington, DC

Cardoso da Silva JM and Bates JM (2002) Biogeographic Patterns and Conservation in the South American Cerrado: A tropical savanna hotspot. BioScience 52:225–233
Carvalho F, Souza FA, Carrenho R, Moreira FMS, Jesus EC, Fernandes GW (2012) The mosaic of habitats in the high-altitude Brazilian rupestrian fields is a hotspot for arbuscular mycorrhizal fungi. Appl Soil Ecol 52:9–19
Casanova-Katny MA, Torres-Mellado GA, Palfner G, Cavieres LA (2011) The best for the guest: high Andean nurse cushions of *Azorella madreporica* enhance arbuscular mycorrhizal status in associated plant species. Mycorrhiza 21:613–622
Clarke JDA (2006) Antiquity of aridity in the Chilean Atacama Desert. Geomorphology 73:101–114
Costa HAO, Stürmer SL, Ragonezi C, Grazziotti PH, Fonseca Santos Grazziotti DC, de Barros Silva E (2016) Species richness and root colonization of arbuscular mycorrhizal fungi in *Syngonanthus elegans*, an endemic and threatened species from the Cerrado domain in Brazil. Ciência e Agrotecnologia 40:326–336
Coutinho ES, Fernandes GW, Berbara RLL, Valério HM, Goto BT (2015) Variation of arbuscular mycorrhizal fungal communities along an altitudinal gradient in rupestrian grasslands in Brazil. Mycorrhiza 25:627–638
Díaz S, Acosta A, Cabido M (1994) Grazing and the phenology of flowering and fruiting in a montane grassland in Argentina: a niche approach. Oikos 70: 287–295
Diazgranados M, Barber JC (2017) Geography shapes the phylogeny of frailejones (Espeletiinae Cuatrec., Asteraceae): a remarkable example of recent rapid radiation in sky islands. PeerJ, https://doi.org/10.7717/peerj.2968
Dinerstein ED, Olson M, Graham DJ, Webster AL, Primm SA, Bookbinder MP, Ledec YG (1995) Una evaluación del estado de conservación de las ecorregiones terrestres de América Latina y el Caribe. World Bank, Washington, DC
Echternacht L, Trovóa M, Oliveira CT, Pirania JR (2011) Areas of endemism in the Espinhaço Range in Minas Gerais, Brazil. Flora 206:782–791
Farley KA, Bremer LL, Harden CP, Hartsig J (2013) Changes in carbon storage under alternative land uses in biodiverse Andean grasslands: implications for payment for ecosystem services. Conservation Letters 6(1):21–27
Ferreyra M, Cingolani A, Ezcurra C, Bran D (1998) High-Andean vegetation and environmental gradients in northwestern Patagonia. Journal Vegetation Science 9:307–316
Ferreyra M, Grigera D (2002) Conservación de la vegetación altoandina en el Parque Nacional Nahuel Huapi. Jornadas Regionales Desarrollo y Ciencia en Áreas de Montaña, Universidad Nacional del Comahue
Fiaschi P, Pirani J (2009) Review of plant biogeographic studies in Brazil. Journal of Systematics and Evolution 47(5):477–496
Foley JA, Ramankutty N, Brauman KA, Cassidy ES, Gerber JS, Johnston M, Mueller ND, O'Connell C, Ray DK, West PC, Balzer C, Bennett EM, Carpenter SR, Hill J, Monfreda C, Polasky S, Rockström J, Sheehan J, Siebert S, Tilman D, Zaks DPM (2011) Solutions for a cultivated planet. Nature 478:337–342
Fracchia S, Silvani V, Flachsland E, Terada G, Sede S (2014a) Symbiotic seed germination and protocorm development of *Aa achalensis* Schltr., a terrestrial orchid endemic from Argentina. Mycorrhiza 24:35–43
Fracchia S, Aranda-Rickert A, Flachsland E, Terada G, Sede S (2014b) Mycorrhizal compatibility and symbiotic reproduction of *Gavilea australis*, an endangered terrestrial orchid from south Patagonia. Mycorrhiza 24:627–634
García I, Mendoza R, Pomar MC (2012) Arbuscular mycorrhizal symbiosis and dark septate endophytes under contrasting grazing modes in the Magellanic steppe of Tierra del Fuego. Agriculture, Ecosystems and Environment 155:194–201
García Romero JF, García Fernandéz DC, Correa de Restrepo M (2004) Incidencia de las Micorrizas Arbusculares y Vesículos Arbusculares como estrategia adaptativa de especies de páramo y selva Altoandina, Cordillera Oriental de Colombia. Colombia forestal 8:43–59

Gardes M, Dahlberg A (1996) Mycorrhizal diversity in arctic and alpine tundra: an open question. New Phytologist 133:147–157

Gianinazzi S, Gollotte A, Binet MN, van Tuinen D, Redecker D, Wipf D (2010) Agroecology: the key role of arbuscular mycorrhizas in ecosystem services. Mycorrhiza 20(8):519–530

Giulietti AM, Pirani JR (1988) Patterns of geographic distribution of some plant species from the Espinhaço Range, Minas Gerais and Bahia, Brazil. In: Vanzolini PE, Heyer WR (eds) Proceedings of a Workshop on Neotropical Distribution Patterns. Academia Brasileira de Ciências, Rio de Janeiro, Brazil, 1987

Hughes CE, Atchison GW (2015) The ubiquity of alpine plant radiations: from the Andes to the Hengduan Mountains. New Phytologist 207:275–282

Jumpponen A, Trappe JM (1998) Dark septate endophytes: a review of facultative biotrophic root-colonizing fungi. New Phytologist 140:295–310

Knapp DG, Pintye A, Kovács GM (2012) The dark side is not fastidious-Dark septate endophytic fungi of native and invasive plants of semiarid sandy areas. PloS ONE 7:e32570

Kohn LM, Stasovski E (1990) The mycorrhizal status of plants at Alexandra Fiord, Ellesmere Island, Canada, a High Arctic site. Mycologia 82:23–35

Körner C (1999) Alpine plant life. Springer-Verlag Heidelberg, New York

Körner C (2007) The use of "altitude" in ecological research.Trends Eco. Evol 22:569–574

Kotilínek M, Hiiesalu I, Košnar J, Šmilauerová M, ŠmilauerP, Altman J, Dvorský M, Kopecký M, Doleža J (2017) Fungal root symbionts of high altitude vascular plants in the Himalayas. Scientific REPortS 7:6562. https://doi.org/10.1038/s41598-017-06938-x

Letelier L, Squeo FA, Arancio G, Marticorena A, Muñoz-Schick M, Arroyo MTK, León-Lobos P, Montecinos S, Gutiérrez JR (2008) Diversidad vegetal de la Región de Atacama, Chile. In: Squeo FA, Arancio G, Arancio G (eds) Libro Rojo de la Flora nativa y de los Sitios Prioritarios para su Conservación. Ediciones Universidad de La Serena, Chile, p 123–135

López RP (2000) La prepuna boliviana. Ecología en Bolivia 34:45–70

López RP, Beck S (2002) Phytogeographical affinities and life form composition of the Bolivian Prepuna. Candollea 57:77–96

Lugo MA, Domínguez de Toledo L, Anton AM (1995) *Sclerocystis sinuosa* (Glomales, Zygomycetes) en cuatro *Poaceae* Argentinas. Kurtziana 24:145–152

Lugo MA, Domínguez de Toledo L, Anton AM (1997) Seis especies de Glomales (Zygomycetes) en *Poaceae* Argentinas. I. Kurtziana 25:187–204

Lugo MA, Cabello MN (2002) Native arbuscular mycorrhizal fungi (AMF) from mountain grassland (Córdoba, Argentina) I. Seasonal variation of fungal spore diversity. Mycologia 94 (4):579–586

Lugo MA, González Maza ME, Cabello MN (2003) Arbuscular mycorrhizal fungi (AMF) from native Argentinian -South American- mountain grassland II. Seasonal variation of colonization and its relation with grazing and metabolic host type. Mycologia 95:407–415

Lugo MA, Ferrero M, Menoyo E, Estévez MC, Siñeriz F, Anton AM (2008) Arbuscular mycorrhizal fungi and rhizospheric bacteria diversity along an altitudinal gradient in South American Puna grassland. Microb Ecol 55:705–713

Lugo MA, Menoyo E, Risio Allione L, Negritto MA, Henning JA, Anton AM (2018) Arbuscular mycorrhizas and dark septate endophytes associated with grasses from the Argentine Puna. Mycologia 110:654–665

Lugo MA, Negritto MA, Jofré M, Anton A, Galetto L (2012) Colonization of native Andean grasses by arbuscular mycorrhizal fungi in Puna: a matter of altitude, host photosynthetic pathway and host life cycles. FEMS Microbiol Ecol 81:455–466

Madriñán S, Cortés AJ, Richardson JE (2013) Páramo is the world's fastest evolving and coolest biodiversity hotspot. Frontiers in Genetics 4:1–6

Martínez Carretero E (1995) La Puna Argentina: delimitación general y división en distritos florísticos. Bol Soc Argent Bot 31:27–40

Martínez Carretero E, Faggi AM, Fontana JL, Aceñolaza P, Gandullo R, Cabido M, Iriart D, Prado D, Roig FA, Eskuche U (2016) Prodromus sinsistemático de la República Argentina y una breve introducción a los estudios fitosociológicos. Bol Soc Argent Bot 51:469–549

Matteucci SD, Silva ME, Rodríguez AF (2016) Clasificaciones de la tierra: ¿provincias fitogeográficas, ecorregiones o paisajes?. Fronteras 14:1–16

McKay CP, Friedmann EI, Gómez-Silva B, Cáceres-Villanueva L, Andersen DT, Landheim R (2003) Temperature and moisture conditions for life in the extreme arid region of the Atacama Desert: Four years of observations including the El Niño of 1997–1998. Astrobiology 3 (2):393–406

Menoyo E, Becerra AG, Renison D (2007) Mycorrhizal associations in *Polylepis* woodlands of Central Argentina. Can J Bot 85:526–531

Monasterio M (1980) Las formaciones vegetales de los páramos de Venezuela. In: Monasterio M (ed) Estudios Ecológicos en los Páramos Andinos. Editorial de la Universidad de Los Andes, Mérida, p 93–158

Montilla M, Herrera RA, Monasterio M (1992) Micorrizas vesiculo-arbusculares en parcelas que se encuentran en sucesión-regeneración en los Andes tropicales. Suelo y Planta 2:59–70

Morello J, Matteucci SD, Rodríguez AF, Silva ME (2018) Ecorregiones y complejos Ecosistémicos argentines, 2da ed, FADU, GEPAMA, Orientación Gráfica Editora, Argentina

Morong T (1891) The Flora of the Desert of Atacama. Bulletin of the Torrey Botanical Club 18 (2):39–48

Morrone JJ (2001a) Biogeografía de América Latina y el Caribe. M&T–Manuales & Tesis SEA, vol. 3. Zaragoza

Morrone JJ (2001b) Toward a formal definition of the Paramo-Punan subregion and its provinces. Rev Mus Argent Cienc Nat 3:1–12

Morrone JJ (2004) Panbiogeografía, componentes bióticos y zonas de transición. Revista Brasileira de Entomologia 48(2):149–162

Morrone JJ (2006) Biogeographic areas and transition zones of Latin America and the Caribbean Islands based on panbiogeographic and cladistics analyses of the entomofauna. Annu Rev Entomol 51:467–494

Morrone JJ (2014) Biogeographical regionalisation of the Neotropical region. Zootaxa 3782:1–110

Myers N, Mittermeier RA, Mittermeier CG, da Fonseca GAB, Kent J (2000) Biodiversity hotspots for conservation priorities. Nature 403:853–858

Manuel-Navarrete D, Gallopín GC, Blanco M, Díaz-Zorita M, Ferraro DO, Herzer H, Laterra P, Murmis MR, Podestá GP, Rabinovich J, Satorre EH, Torres F, Viglizzo EF (2007) Multi-causal and integrated assessment of sustainability: the case of agriculturization in the Argentine Pampas. Environ Dev Sustain 11:621–638

Nogueira RE, Liparini Pereira O, Megumi Kasuya MC, da Silva Lanna MC, Pimentel Mendonça M (2005) Fungos micorrízicos associados a orquídeas em campos rupestres na região do Quadrilátero Ferrífero, MG, Brasil. Acta Bot Bras 19:417–424

Navone S, Bosio MJ (2008–2009) Región PUNA. Centro de Investigación y Aplicación de la Teledetección CIATE, Facultad de Agronomía – Universidad de Buenos Aires, Reports 2008–2009, http://www.desertificacion.gob.ar/wp-content/uploads/2013/06/Informe-SP-Puna-2008-2009

Oehl F, Körner C (2014) Multiple mycorrhization at the coldest place known for Angiosperm plant life. Alpine Botany 124:193–198

Oki Y, Goto BT, Jobim K, Rosa LH, Costa Ferreira M, Silva Coutinho E, de Azevedo Xavier JH, Fernanda Carvalho, de Souza Moreira FM, de Souza FA, Louro Berbara RL, Fernandes GW (2016) Arbuscular Mycorrhiza and Endophytic Fungi in Ruspestrian Grasslands. In: GW Fernandes (ed) Ecology and Conservation of Mountain top Grasslands in Brazil, Springer, Switzerland. https://doi.org/10.1007/978-3-319-29808-5-8, p 157–179

Olson DM, Dinerstein E (2002) The global 200: priority ecoregions for global conservation. Ann Missuri Bot Garden 89:199–224

Orchard S, Hilton S, Bending GD, Dickie IA, Standish RJ, Gleeson DB, Jeffery RP, Powell JR, Walker C, Bass D, Monk J, Simonin A, Ryan MH (2017) Fine endophytes (*Glomus tenue*) are related to Mucoromycotina, not Glomeromycota. New Phytologist 213:481–485

Overbeck GE, Vélez-Martin E, Scarano FR, Lewinsohn TM, Fonseca CR, Meyer ST, Muller SC, Ceotto P, Dadalt L, Durigan G, Gislene G, Gossner MM, Guadagnin DL, Lorenzen K, Jacobi CM, Weisser WW, Pillar VD (2015) Conservation in Brazil needs to include non-forest ecosystems. Diversity Distrib 21:1455–1460

Oyarzabal M, Clavijo J, Oakley L, Biganzoli F, Tognetti P, Barberis I, Maturo HM, Aragón R, Campanello PI, Prado D, Oesterheld M, León RJC (2018) Unidades de vegetación de la Argentina. Ecología Austral 28:40–63

Pagano MC, Scotti MR (2009) A survey of the arbuscular mycorrhiza occurrence in *Paepalanthus bromelioides* and *Bulbostylis* sp. in rupestrian fields, Brazil. Micologia Aplicada Internacional 21:1–10

Pagano MC, Cabello MN (2012) Occurrence of mycorrhizas in Highland fields. In: Pagano MC (ed) Mycorrhiza: Occurrence in Natural and Restored Environments. NOVA Science Publishers Inc., New York, p 87–98

Peterson RL, Massicotte HB, Melville L (2004) Mycorrhizas: anatomy and cell biology. NRC Research Press and CABI Publishing, Ottawa and Wallingford

Peyre G, Balslev H, Font X (2018) Phytoregionalisation of the Andean páramo. PeerJ PeerJ 6:e4786. https://doi.org/10.7717/peerj.4786

Pliscoff P, Zanetta N, Hepp J, Machuca J (2017) Efectos sobre la flora y vegetación (del evento de precipitación extremo de agosto 2015 en Alto Patache, Desierto de Atacama, Chile. Revista de Geografía Norte Grande 68:91–103

Posadas PE, Estévez JM, Morrone JJ (1997) Distributional patterns and endemism areas of vascular plants in the Andean subregion. Fontqueria 48:1–10

Prado DE (1993a) What is the Gran Chaco vegetation in South America? I. A review. Contribution to the study of flora and vegetation of the Chaco. V. Candollea 48:145–172

Prado DE (1993b) What is the Gran Chaco vegetation in South America? II. A redefinition. Contribution to the study of flora and vegetation of the Chaco. VII. Candollea 48:615–629

Prado DE, Gibbs PE (1993) Patterns in species distribution in the dry seasonal forest of South America. Annals Missouri Botanical Garden 80:902–927

Rapini A, Ribeiro PL, Lambert S, Pirani JR (2008) A flora dos campos rupestres da Cadeia do Espinhaço. Megadiversidade 4:15–23

Read DJ (1991) Mycorrhizas in ecosystems. Experientia 47:376–391

Read DJ, Haselwandter K (1981) Observations on the mycorrhizal status of some alpine plant communities. New Phytologist 88:341–352

Read DJ, Pérez Moreno J (2003) Mycorrhizas and nutrient cycling in ecosystems – a journey towards relevance? New Phytologist 157:475–492

Redman RS, Sheehan KB, Stout RG, Rodriguez RJ, Henson JM (2002) Thermotolerance generated by plant/fungal symbiosis. Science 298:1581

Renison D, Cuyckens GAE, Pacheco S, Guzmán GF, Grau HR, Marcora P, Robledo G, Cingolani AM, Dominguez J, Landi M, Bellis L, Hensen I (2013) Distribución y estado de conservación de las poblaciones de árboles y arbustos del género *Polylepis* (Rosaceae) en las montañas de Argentina. Ecología Austral 23:27–36

Ruggiero RM, Ezcurra C (2003) Regiones y transiciones biogeográficas: complementaridad de los análisis en biogeografía histórica y ecológica. In: Morrone JJ, Llorente Bousquets J (eds) Una Perspectiva Latinoamericana de la Biogeografía de México. Prensas Científicas, UNAM, México DF, p 141–154

Ruotsalainen AL, Eskelinen A (2011) Root fungal symbionts interact with mammalian herbivory, soil nutrient availability and specific habitat conditions. Oecologia 166:807–817

Ruthsatz B (1977). Pflanzengesellschaften und ihre Lebensbedingungen in den Andinen Halbwüsten Nordwest-Argentiniens. J. Cramer, Vaduz

Ruthsatz B, Movia CP (1975) Relevamiento de las estepas andinas del noreste de la Provincia de Jujuy, Argentina. Fundación para la Educación, la Ciencia y la Cultura, Buenos Aires

Salvarredi LA, Crespo EM, Menoyo E, Filippa EM, Barboza GE, Lugo MA (2010) Micorrizas arbusculares y endófitos septados oscuros en gentianaceae nativas de la Argentina. Bol Soc Argent Bot 45:223–229

Schmidt SK, Sobieniak-Wiseman LC, Kageyama SA, Halloy SRP, Schadt CW (2008) Mycorrhizal and dark-septate fungi in plant roots above 4270 meters elevation in the Andes and Rocky Mountains. Arct Antarct Alp Res 40:576–583

Silvani V, Rothen C, Rodríguez MA, Cisneros G, Godeas A, Aranda-Rickert A, Fracchia S (2013) Fungal root colonization of *Puccinellia frigida* (Phil.) Johnston, a dominant grass species inhabiting the margins of high-altitude hypersaline Andean wetlands. Aquatic Botany 108:26–32

Sklenář P, Hedberg I, Cleef AM (2014) Island biogeography of tropical alpine floras. Journal of Biogeography 41(2):287–297. https://doi.org/10.1111/jbi.12212

Smith SE, Read DJ (2008) Mycorrhizal symbiosis. Academic press, Cambridge, UK

Squeo FA, Arroyo MTK, Marticorena A, Arancio G, Muñoz-Schick M, Negritto M, Rojas G, Rosas M, Rodríguez R, Humaña AM, Barrera E, Marticorena C (2008) Catálogo de la Flora Vascular de la Región de Atacama. In: Squeo FA, Arancio G, Arancio G (eds) Libro Rojo de la Flora nativa y de los Sitios Prioritarios para su Conservación. Ediciones Universidad de La Serena, Chile, p 67–120

Trappe JM (1987) Phylogenetic and ecologic aspects of mycotrophy in the Angiosperms from and Evolutionary standpoint. In: Safir GR (ed) Ecophysiology of VA Mycorrhizal Plants. CRC Press, Boca Raton, FL, p 5–25

Urcelay C, Acho J, Joffre R (2011) Fungal root symbionts and their relationship with fine root proportion in native plants from the Bolivian Andean highlands above 3700 m elevation. Mycorrhiza 21:323–330

Urcelay C (2002) Co-ocurrence of three fungal root symbionts in *Gaultheria poeppiggi* DC in Central Argentina. Mycorrhiza 12:89–92

Vásquez DL, Balslev H, Sklenář P (2015) Human impact on tropical-alpine plant diversity in the northern Andes. Biodiversity and Conservation 24(11):2673–2683

Velázquez MS, Stürmer SL, Bruzone C, Fontenla S, Barrera M, Cabello M (2016) Occurrence of arbuscular mycorrhizal fungi in high altitude sites of the Patagonian Altoandina region in Nahuel Huapi National Park (Argentina). Acta Botanica Brasilica 30 (4):521–531

Veldman JW, Overbeck GE, Negreiros D, Mahy G, Le Stradic S, Fernandes GW, Durigan G, Buisson E, Putz FE, Bond WJ (2015) Where tree planting and forest expansion are bad for biodiversity and ecosystem services. BioScience 347:484–485

Viglizzo EF, Jobbágy E (eds) (2010) Expansión de la Frontera Agropecuaria en Argentina y su Impacto Ecológico-Ambiental. Ministerio de Agricultura, Ganadería y Pesca de la Nación. Ediciones INTA

Wahl A-L, Spiegelberg T (2016) Arbuscular mycorrhizal fungi in changing mountain grassland ecosystems- a challenge for research. Botany 94:435–458

Walker C, Gollotte A, Redecker D (2018a) A new genus, *Planticonsortium* (Mucoromycotina), and new combination (*P. tenue*), for the fine root endophyte, *Glomus tenue* (basionym *Rhizophagus tenuis*). Mycorrhiza 28:213–219

Walker C, Harper CJ, Brundrett MC, Krings M (2018b) Looking for Arbuscular Mycorrhizal Fungi in the Fossil Record: An Illustrated Guide. In: Krings M, Cúneo NR, Harper CJ, Rothwell G (eds) Transformative Paleobotany. AP, Elsevier, p 481–517

Wang B, Qiu Y-L (2006) Phylogenetic distribution and evolution of mycorrhizas in land plants. Mycorrhiza 16:299–363

Young BE, Josse C, Stern M, Vasconez S, Olander J, Smyth R, Zador M, Sánchez de Lozada A, Comer PJ, Moull K, Echavarría M, Hak J (2015) Hotspot de biodiversidad de los Andes tropicales. In: Resumen técnico del perfil del ecosistema. Washington, D.C.: NatureServer & EcoDecisión

Chapter 13
Arbuscular Mycorrhizal Fungal Communities of High Mountain Ecosystems of South America: Relationship with Microscale and Macroscale Factors

Florencia Soteras, Eugenia Menoyo, Gabriel Grilli, and Alejandra G. Becerra

13.1 Introduction

South America harbors one of the main hotspots of diversity, the high mountain ecosystems, despite only accounting for a quarter of the Earth's land surface (Myers et al. 2000; Barry 2008; La Sorte and Jetz 2010; Hoorn et al. 2013). Several plants, birds, and macrofungal species show endemism in the high mountain of many regions of South America (Fjeldså and Kessler 1996; Myers et al. 2000; Robledo et al. 2006). These ecosystems comprise natural watersheds, providing several ecosystem services such as hydrological regime regulation, soil protection, and conservation of biodiversity (Grêt-Regamey et al. 2012). Mountain habitats show distinctive abiotic conditions that differentiate them from lowlands (Barry 2008). For instance, temperature decrease in average 6 °C per each km in elevation also influenced by latitude (Barry 2008). Generally, the studies in mountain ecosystems have been focused on aboveground diversity (plants, animals and macrofungi) (Robledo and Renison 2010; Castillo et al. 2017; Nouhra et al. 2018; Quintero and Jetz 2018), but little is known about soil communities (Lugo and Cabello 2002;

F. Soteras (✉)
Laboratorio de Ecología Evolutiva y Biología Floral, IMBIV, CONICET, Universidad Nacional de Córdoba, Córdoba, Argentina
e-mail: fsoteras@conicet.gov.ar

E. Menoyo
Grupo de Estudios Ambientales, IMASL, CONICET, Universidad Nacional de San Luis, San Luis, Argentina

G. Grilli · A. G. Becerra
Laboratorio de Micología, IMBIV, CONICET, and Cátedra de Diversidad Biológica I, Facultad de Ciencias Exactas, Físicas y Naturales, Universidad Nacional de Córdoba, Córdoba, Argentina

M. C. Pagano, M. A. Lugo (eds.), *Mycorrhizal Fungi in South America*, Fungal Biology, https://doi.org/10.1007/978-3-030-15228-4_13

Becerra et al. 2009; Menoyo et al. 2009; Geml et al. 2014; Soteras et al. 2016). Among them, arbuscular mycorrhizal fungi (AMF) are ubiquitous root symbionts in the Glomeromycota that form an obligate root symbiosis with great part of land plants (Schüßler et al. 2001; Spatafora et al. 2006). Despite the large diversity of host plants (ca. 200,000 species), there has just been identified in average 250 AMF morphological taxa (hereafter "morphospecies"), and no correlation between plant species and AMF richness has been globally found (Bever et al. 2001; Tedersoo et al. 2014).

The vast majority of the AMF taxa occur in nearly every climatic zones and continents (Davison et al. 2015). Last studies have postulated that a recent dispersion is the main factor shaping the cosmopolitan distribution of the most of the AMF taxa (Davison et al. 2015). However, these fungi are differentially affected by soil characteristics (Smith and Read 2008). In addition, different host species are colonized by particular AMF present in their rhizosphere (Senés-Guerrero and Schüßler 2016; Soteras et al. 2016), although there is a lack of a global positive correlation with plant richness. As plants and terrestrial animals, AMF taxa richness has been evidenced to correlate negatively with latitude (Hillebrand 2004; Davison et al. 2015), but different from ectomycorrhizal fungi (Tedersoo et al. 2014) and other soil microorganisms (Bardgett and Van Der Putten 2014). In addition, variables such as precipitation and temperature through the alteration of soil moisture, locally affect AMF richness (Davison et al. 2015). South America comprises diverse high mountain ecosystems, from low latitude tropical to high latitude temperate, where different local conditions also influence AMF communities (Matus et al. 2014).

Taxa of AMF could be grouped by their functional characteristics that are phylogenetically constrained (Hart and Reader 2002; Maherali and Klironomos 2007). Thereby, members of Gigasporaceae produce extensive extra-radical mycelia, sporulate lately in the growing season, and provide high nutritional benefits to hosts. On the other hand, Glomeraceae mainly colonize intraradically, produce spores early, and provide less nutritional benefits to hosts. Finally, Acaulosporaceae represents an intermediate colonization strategy, producing low biomass inside and outside the roots, and being highly resistant to soil acidity and low temperatures (Hart et al. 2001; Hart and Reader 2002). Accordingly, and based on the competitor-stress tolerant-ruderal framework of Grime (1979), Gigasporaceae are considered as "competitor", Glomeraceae as "ruderal" and Acaulosporaceae as "stress tolerant" (Chagnon et al. 2013).

In this chapter we reviewed and re-analyzed the data of the studies performed at high mountain ecosystems of South America to evaluate the variation of AMF morphospecies richness and composition of AMF communities in relation to micro- and macro-scale factors. Particularly, we hypothesized that high mountain forests harbor different richness and composition of AMF communities due to changes in microscale (host species, pH, N, P) and macroscale factors (latitude, temperature, and precipitation) rather than similar AMF communities as expected from its cosmopolitan distribution.

13.2 Arbuscular Mycorrhizal Fungi in the High Mountain Ecosystems of South America

Traditionally, the studies of AMF diversity were based on the morphological characteristics and ontogeny of the asexual spores (Smith and Read 2008). The advance of DNA-based methods improved the taxonomic identification of non-sporulating and AMF species. This kind of studies are very scarce in South America even more in mountain ecosystems (Soteras et al. 2016; Senés-Guerrero and Schüßler 2016). Therefore, we only considered the morphological diversity of AMF in high mountain ecosystems of South America. We compiled published studies searching in Google Scholar articles containing the following combination of terms: "arbuscular mycorrhizal" AND "high mountain" OR "Andean". We reviewed all the studies performed at mountain sites at around 1200 meters above sea level focusing on "highlands" *sensu* Barry (2008) that identified AMF spores morphologically. Following this procedure, we obtained in total 12 studies: 6 from Brazil, 5 from Argentina and 1 from Chile (Fig. 13.1, Table 13.1). Considering all of them, 168 AMF morphospecies were identified.

13.3 Arbuscular Mycorrhizal Fungi Richness Versus Macroscale and Microscale Factors

To disentangle the relationship of AMF richness with microscale and macroscale factors we fitted generalized linear models (GLM) with the *glm()* function as implemented in the R environment with Poisson error distribution and identity logarithmic link function (R Core Team 2018). When overdispersion was detected the standard errors were corrected using a quasi-GLM model (Zuur et al. 2009). Microscale factors included: host species or vegetation type and soil characteristics as pH, N and P content, obtained from the studies when available. Macroscale factors included: latitude, mean annual temperature (in degree Celsius multiplied by 10) and mean annual precipitation from MERRAclim (Vega et al. 2017a), available in the DRYAD database (Vega et al. 2017b).

Vegetation type or host species showed significant differences in AMF rhizospheric richness (Fig. 13.2). Mountain ecosystems in Brazil (savanna forest, quartz gravel field dominated by *Vellozia* sp., and rocky outcrops of Cerrado and Atlantic Forest) showed the highest AMF richness. This result is probably due to the dominance of AMF in hot and seasonal environments (van der Heijden et al. 2008). For the contrary, the lowest AMF richness was observed in successional temperate forests of *N. pumilio* (Fig. 13.2). Generally, in temperate forests, where nutrient availability is low and the organic form is present in litter and humus, predominate the colonization by ectomycorrhizal decomposer fungi (Matus et al. 2014). In consequence, ectomycorrhizal fungi are responsible for almost the 80% of the N acquired by plants of temperate and boreal ecosystems (van der Heijden et al. 2008). As in *N.*

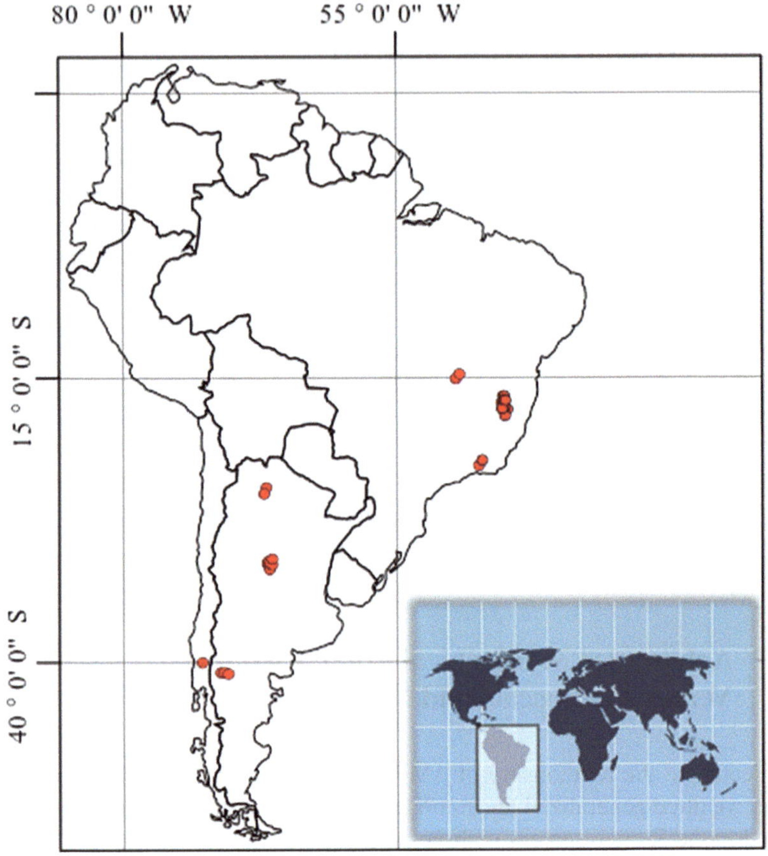

Fig. 13.1 Map showing the location of the high mountain ecosystems included in this study

pumilio forests, we found that reforested *Araucaria* forests of Brazil also showed a very low AMF diversity. In this study, rhizosphere soil samples were taken from reforested areas with *A. angustifolia* (8–12 years old) and *Pinus elliotti* plants (Moreira-Souza et al. 2003). Several studies have described changes in AMF communities associated with exotic plant invasion (Mummey and Rillig 2006). The very low AMF richness in this ecosystem compared with 19 other mountain hosts and ecosystems support the evidence that exotic plant species might negatively influence on soil AMF communities.

Arbuscular mycorrhizal fungi richness related to microscale (pH, N and P content) and macroscale (latitude, mean annual temperature and mean annual precipitation) factors are shown in Fig. 13.3. AMF richness was negatively related to pH ($t = 2.049$, $P = 0.046$, Fig. 13.3a), positively to N ($t = 3.003$, $P = 0.006$, Fig. 13.3b), but not significant relationship was observed with P ($t = 0.236$, $P = 0.81$, Fig. 13.3c). In addition, a negative relationship was observed of AMF richness with latitude in absolute numbers ($t = -4.015$, $P < 0.001$, Fig. 13.3d), and a positive relationship

Table 13.1 Summary of the studies performed in high mountain ecosystems of different sites of South America. Sites are ordered by increasing latitude, and information about treatment of the study, vegetation type or rhizosphere host or dominant plant, altitude, soil texture, P and N is provided

Site	Latitude	Longitude	Treatment	Vegetation type/Rhizosphere host / dominant plant species	Altitude (m)	Soil texture	P (ppm)	N (%)
Brazil[a1]	15°36′6.36″S	47°43′3.92″W	Cerrado CS-I	Savanna forest	1100	Sandy clay loam	2.4	–
	15°35′34.04″S	47°44′12.09″W	Cerrado CS-II	Savanna forest	1100	Sandy clay loam	1.7	–
	15°35′53.74″S	47°42′24.93″W	Cerrado CS-III	Savanna forest	1100	Sandy clay loam	1.4	–
Brazil[a2]	18°12′21.1″S	43°33′47.6″W	Soberbo stream	Rupestrian grassland: *Syngonanthus elegans*, *Loudetiopsis chryssothrix*, *Xyris* sp.	1368	–	5.8	–
	17°55′02.9″S	43°35′53.74″S	National Park “sempre-vivas”		1310	–	2.4	–
Brazil[a3]	19°16′ 50.2″S	43°35′ 27.7″W	Sandy bogs	*Lagenocarpus rigidus*	1158	Sandy loam	2	0.7
	19°16′ 54.4″S	43° 35′ 29″W	Peat bogs	*Axonopus siccus*	1146	Sandy loam	3	1.5
	19°17′ 15.2″S	43°35′ 39.2″W	Rocky outcrops	*Trachupogon spicatus*	1163	Sandy loam	2	0.9
	19°17′ 04.1″S	43°35′ 37.7″W	Quartz gravel field	*Vellozia* sp.	1192	Sandy loam	3	0.8
	19°16′ 57.7″S	43°35′ 40.0″W	Cerrado	*Schizachyrium tenerum* Nees.	1173	Clay loam	2	1.7

(continued)

Table 13.1 (continued)

Site	Latitude	Longitude	Treatment	Vegetation type/Rhizosphere host / dominant plant species	Altitude (m)	Soil texture	P (ppm)	N (%)
Brazil[a4]	19°15′50.6″S	43°35′10.3″W	Cerrado	Rocky outcrop and Cerrado *sensu stricto*	1000	Sandy	1.16	–
	19°13′56.5″S	43°34′34.8″W		Rocky outcrop and Cerrado *sensu stricto*	1100	Sandy	1.15	–
	19°17′43.0″S	43°33′17.4″W		Rocky outcrop	1200	Sandy	2.71	–
	19°17′49.6″S	43°35′28.2″W		Rupestrian grassland	1300	Sandy	1.08	–
	19°16′59.3″S	43°32′08.9″W		Rupestrian grassland	1400	Sandy	2.38	–
Brazil[a5]	22°44′ S	45°30′W	Native *Araucaria* forests	*Podocarpus lambertii*, *Ilex paraguariensis*, *Clethra scabra*, *Weinmannia piannata*, *Cryptocarya aschersoniana*, *Prunus myrtifolia*, *Symplocus aegrota*, *Drymys winterii*	1674	–	10	–
			Reforested *Araucaria* forests	*Araucaria agustifolia* and *Pinus eliotii*	1674	–	4.5	–
Brazil[a6]	23°19′31″S	45°05′02″W	Atlantic forest	*Euterpe edulis* Mart., *Cecropia glaziovii* Snethl., *Guapira opposita* (Vell.) Reitz, *Bathysa australis* (A.St.-Hil.) Benth. & Hook., *Mollinedia schottiana* (Spreng.) Perkins, *Coussarea* sp., *Myrcia spectabilis* DC.	1000	Sandy clay loam	4.8	–
Yungas of Argentina[a7]	26°58′S	65°45′W	Quebrada del Portugués	*Alnus acuminata* Kunth.	2187	Sandy loam	13.75	2.22
	27°43′S	65°54′W	Narváez range	*Alnus acuminata* Kunth.	1820	Loam	9.73	3.65

Site	Latitude	Longitude	Treatment	Vegetation type/Rhizosphere host / dominant plant species	Altitude (m)	Soil texture	P (ppm)	N (%)
Central Argentina[a8,9,10]	31°58′S	64°56′W	Los Molles	*Polylepis australis* Bitt.	1800–2000	Sandy loam	56.86	0.94
	31°23′S	64°48′W	Los Gigantes	*Polylepis australis* Bitt.	1800–1900	Sandy loam	33.23	0.59
	31°44′S	64°47′W	Santa Clara	*Polylepis australis* Bitt.	2000–2200	Sandy loam	34.73	0.72
	31° 25′S	64° 47′W	Los Gigantes	*Polylepis australis* Bitt.	2140	Sandy loam	10.17	0.12
	31° 37′S	64° 49′W	Quebrada del Condorito national park	*Polylepis australis* Bitt.	2190	Sandy loam	16.3	0.11
	31° 20′S	64° 45′W	Mountain grassland	*Briza subaristata* Lam., *Deyeuxia hieronymi* (Hack.) Türpe, *Poa stuckertii* (Hack.) Parodi, *Eragrostis lugens* Nees., *Sorghastrum pellitum* (Hack.) Parodi, *Alchemilla pinnata*	2250	Loam to clay loam	–	–

(continued)

Table 13.1 (continued)

Site	Latitude	Longitude	Treatment	Vegetation type/Rhizosphere host / dominant plant species	Altitude (m)	Soil texture	P (ppm)	N (%)
Chile[a11]	40°47′S	72°12′W	Forest (site 1)	*Nothofagus pumilio*	1150	Silt loam	18.28	24
			Forest (site 2)				17.49	22
			Forest (site 3)				16.70	21
			Crater (site 4)	Successional forest: *Bacharis nivalis* Schultz Bip., *Senecio bipontinii* Wedd., *Pernettya pumila* (L.F.) Hook., *Quianchamalium chilense* Lam.	1273	Silt loam	16.16	47
			Crater (site 5)				14.06	40
			Crater (site 6)				11.97	34
			Disturbed (site 7)	*N. Pumilio*	1050	Silt loam	8.32	13
			Disturbed (site 8)				8.12	12
			Disturbed (site 9)				7.92	12
Patagonia Argentina [a12]	41°16′12″S	71°18′16″W	Chalhuaco Hill	*Chiliotrichum rosmarinifolium* Less.	1629	Sandy loam	26.40	0.33
	41°10′20″S	71°18′56″W	Catedral Hill	*Armenia maritima* (Mill.) Wild.	1886	Sandy loam	1.1	0.07
	41°10′33″S	71°49′04″W	Tronador Hill	*A. maritima, Baccharis magellanica* (Lam.) Pers., *B. empetrifolia* Lam., *C. rosmarinifolium, Nassauvia revoluta* Don., *Quinchamalium chilense, Senecio bipontinii* Wedd.	1904	Sandy loam	1.9	0.08

[a]References: [1]Souza de Pontes et al. (2017), [2]Orlandi Costa et al. (2016), [3]de Carvalho et al. (2012), [4]Coutinho et al. (2015), [5]Moreira-Souza et al. (2003), [6]Bonfim et al. (2016), [7]Becerra et al. (2011), [8]Soteras et al. (2015), [9]Menoyo et al. (2009), [10]Lugo and Cabello (2002), [11]Marín et al. (2016), [12] Velázquez et al. (2016)

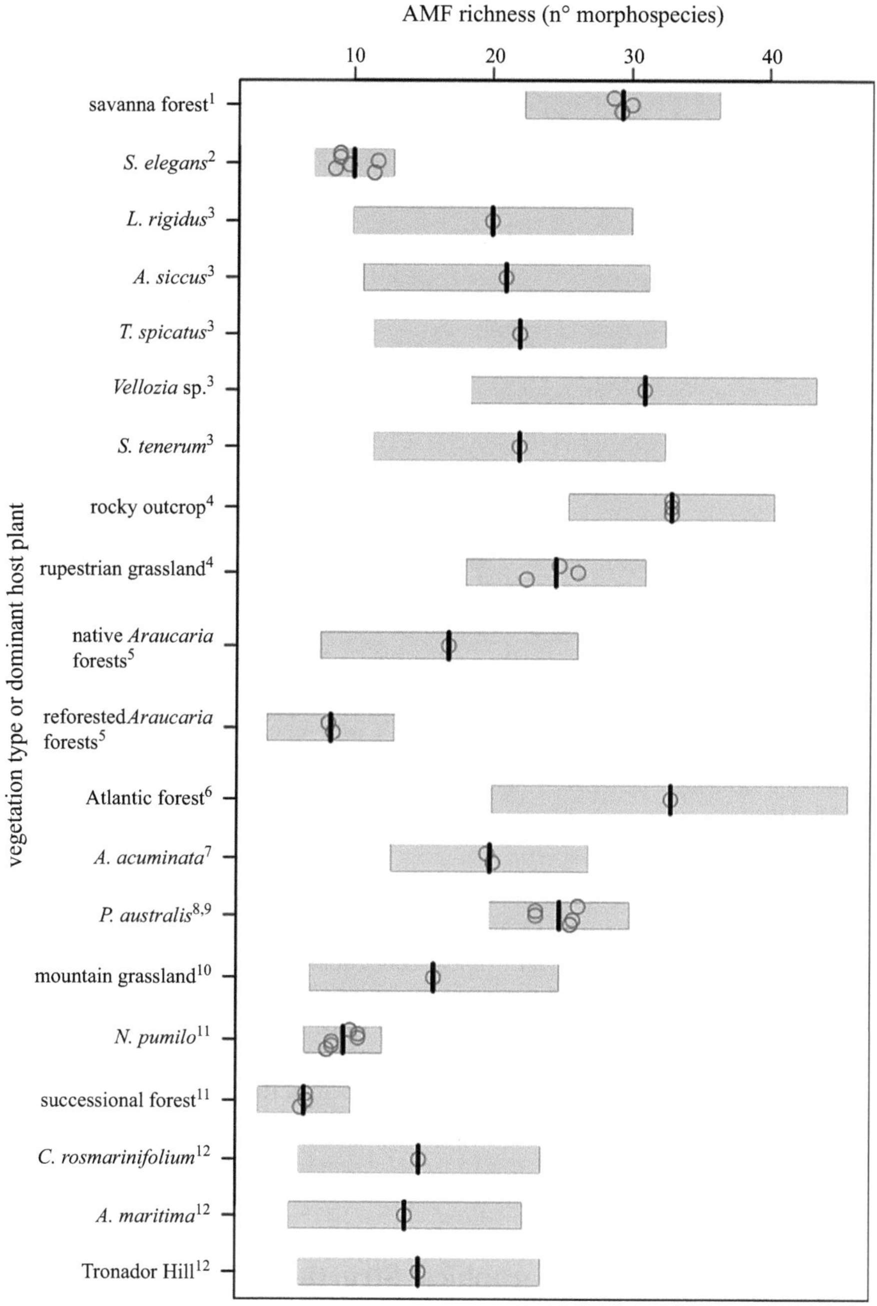

Fig. 13.2 AMF richness related to vegetation type or dominant host plant (ordered by increasing latitude)

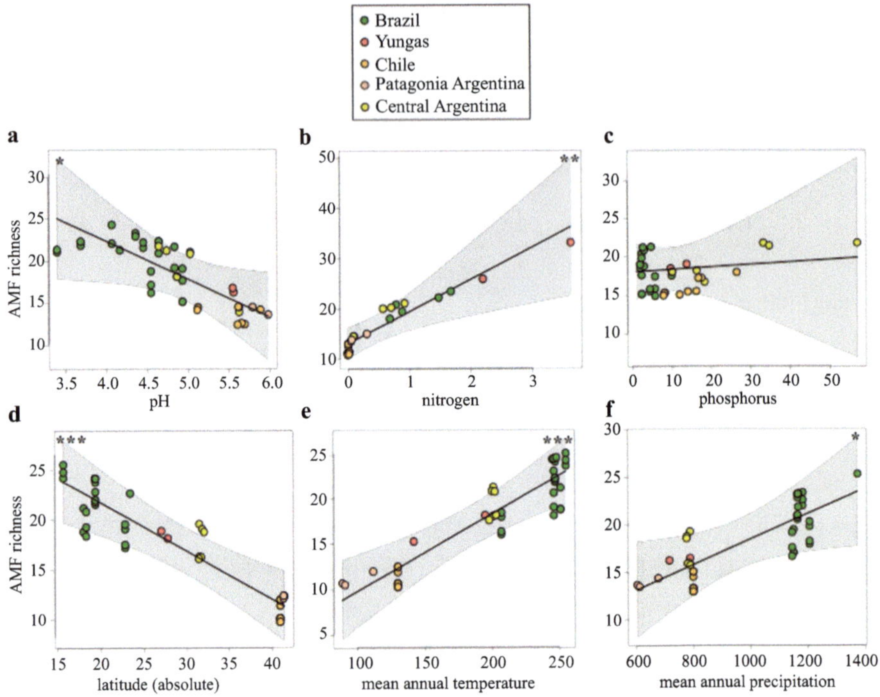

Fig. 13.3 AMF richness related to microscale (pH, N and P content) and macroscale (latitude, mean annual temperature and mean annual precipitation) factors. Asterisks indicate significant relationship according to the GLM (*** P < 0.001, ** P < 0.01, * P < 0.05). Points color represents sampling sites

with both mean annual temperature (t = 4.191, P < 0.001, Fig. 13.3e) and precipitation (t = 2.137, P = 0.039, Fig. 13.3f). AMF communities of high mountain showed high richness at lower latitudinal tropical ecosystems, where seasonal changes of solar radiation, day length and temperature are small (Barry 2008). These ecosystems showed the lowest pH and intermediate N values. The same latitudinal pattern was observed for global AMF richness studies (Davison et al. 2015), plants and animals (Hillebrand 2004), but not for ectomycorrhizal fungi which are associated with specific forest types (Tedersoo et al. 2014).

13.4 Arbuscular Mycorrhizal Fungi Communities' Composition: Geographical Structure and Relationship with Macroscale Factors

In order to evaluate the variation on AMF community composition in relation to different geographical scales and macroscale factors, we first constructed principal coordinates of neighbor matrices (PCNM). The PCNM variables allow to detect if the biological response (i.e. AMF community composition) is associated with different spatial structures along the study area. We obtained six geographical variables able to detect the spatial structure of the data at all scales encompassed by the sampling design (Borcard and Legendre 2002; Borcard et al. 2004). The order of the PCNM variables follows a progression from larger to smaller spatial scales (Borcard et al. 2004). For each response data model, the most significant PCNM variables were chosen by permutational forward model selection and ensuring that the adjusted R^2 of the reduced models did not exceeded the adjusted R^2 of the global models. The AMF community composition (presence-absence) was partitioned among the selected geographical variables and macroscale factors (latitude, mean annual temperature and mean annual precipitation) using distance-based redundancy analysis (db-RDA), with *capscale()* function from R package vegan (Legendre and Andersson 1999; Oksanen et al. 2018). The dissimilarity distance between pairs of AMF morphospecies was estimated using the Sorensen index. The variation explained by geographical variables and macroscale factors was determined by the automatic selection of variables using forward model choice on adjusted R^2 with 999 permutations using the *ordiR2step()* function. In this procedure, the variables that best fit the data are sequentially selected and added to the final model. The analyses were performed using the vegan package in R. The significance among centroids of sites was assessed with the *envfit()* function of the vegan package after 999 permutations. To determine whether the significant effects were attributed to either differences of multivariate site (between group variability) or to dispersion (within group variability) we used the *betadisper()* function of vegan. Microscale factors were not included in this analysis due to missing data in some sites. The Yungas, Cerrado and Soberbo stream from Brazil (Orlandi Costa et al. 2016; Souza de Pontes et al. 2017) were discarded from the db-RDA analysis due to significant effect of within heterogeneity, which avoids the possibility to differentiate the effects of multivariate dispersion from the compositional change among sites.

Four geographical variables were significantly structuring AMF communities (ordered in increasing importance for final model fit: PCNM1: $F = 17.737$, $P = 0.002$; PCNM3: $F = 11.047$, $P = 0.002$; PCNM4: $F = 4.779$, $P = 0.002$; and PCNM2: $F = 2.697$, $P = 0.018$). The three macroscale variables significantly structured AMF community composition of each site (latitude: $F = 19.899$, $P = 0.002$; mean annual precipitation: $F = 92.853$, $P = 0.002$; and mean annual temperature: $F = 5.532$, $P = 0.002$) being kept in the final model. Site differences in relation to their AMF community was associated 25% with both geographical and macroscale factors ($R^2 = 0.72$, pseudo-$F = 12.32$, $P = 0.001$, Fig. 13.4a), 21% with only geographical factors ($R^2 = 0.59$, pseudo-$F = 12.35$, $P = 0.001$, Fig. 13.4b), and 19% with only macroscale factors ($R^2 = 0.55$, pseudo-$F = 13.63$, $P = 0.001$, Fig. 13.4c).

According to the analysis derived from the db-RDA, the AMF community differed significantly among sites ($r^2 = 0.94$, $P < 0.001$, Fig. 13.4a). At a wider scale (represented by PCNM1; associated with db-RDA1: $r^2 = 0.95$, $P = 0.001$), latitude, precipitation and temperature were highly related to differences between Brazil and Chile in their AMF community composition. This is in concordance with global studies of AMF biogeography that showed influences of temperature and precipitation on AMF root colonizing composition (Öpik et al. 2013; Davison et al. 2015).

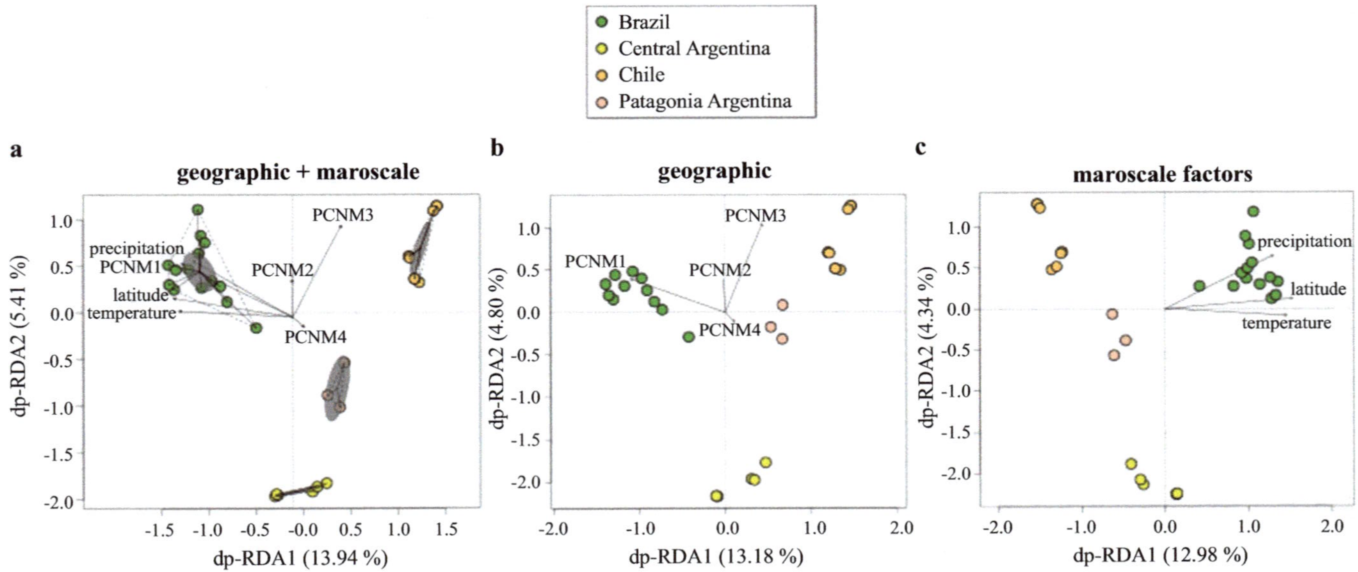

Fig. 13.4 Distance-based redundancy analysis (db-RDA) of localities-based of AMF presence-absence community composition. Arrows indicate the direction of the maximum change in geographical (PCNMs) and macroscale factors (latitude, mean annual temperature and precipitation); **a** complete model: ellipses represent the 95% confidence dispersion around localities centroids, lines connect replicates within localities to their centroids; **b** geographical model, and **c** climatic model

At coarse scales (mainly represented by PCNM3; associated with db-RDA2: $r^2 = 0.66$, P = 0.001) Central Argentina and Patagonia Argentina differentiated in their AMF community composition mainly due to the differences in host species (PERMANOVA: F = 12.54, $r^2 = 0.77$, P = 0.001), soil pH (PERMANOVA: F = 10.53, $r^2 = 0.47$, P = 0.001) and N content (PERMANOVA: F = 7.678, $r^2 = 0.39$, P = 0.001). Several studies provide evidence that the distribution of AMF can be affected by host species, pH and total N (Koske 1987; Johnson et al. 1992; Egerton-Warburton et al. 2004).

To evaluate the strength of association of sampling sites, and vegetation type or dominant host with AMF morphospecies, an indicator species analysis was applied using the *indval()* function of the R package labdsv (Dufrene and Legendre 1997; Roberts 2013). Two species were significantly associated with Brazil, nine with Yungas, eleven with Central Argentina, three with Chile, and six with Patagonia. Of the 20 vegetation types and dominant hosts, six AMF morphospecies were significantly associated with savanna forest, one with *A. siccus*, one with *T. spicatus*, four with rocky outcrop, one with rupestrian grassland, one with *A. acuminata*, four with *P. australis*, three with *N. pumilio*, and one with successional forest. A meta-analysis of global distribution patterns of root-colonizing AMF also demonstrated different type of ecosystems hosting different assemblages of AMF morphospecies (Öpik et al. 2006).

13.5 Relationship Between AMF Functional Richness and Abiotic Characteristics

Arbuscular mycorrhizal fungi were grouped into three functional groups according to their traits (*sensu* Chagnon et al. 2013): "ruderal-Glomeraceae" (Claroideoglomeraceae + Glomeraceae + Pacisporaceae + Diversisporaceae), "stress-tolerant-Acaulosporaceae" (Acaulosporaceae + Ambisporaceae + Entrophosporaceae + Archaeosporaceae), and "competitor-Gigasporaceae". To determine the relationship among AMF functional groups with microscale and macroscale factors we fitted generalized linear models (GLM) with the *glm()* function as implemented in the R environment with Poisson error distribution and identity or logarithmic, in the case of Gigasporaceae, link function. When overdispersion was detected, the standard errors were corrected using a quasi-GLM model.

Glomeraceae and Gigasporaceae families were negatively associated with pH (t = 3.685, P < 0.001; t = 2.785, P = 0.009; respectively). Meanwhile, and contrary to previous evidence (Veresoglou et al. 2012), Acaulosporaceae did not show a significant relationship with pH (t = 0.747, P = 0.460). Glomeraceae and Gigasporaceae showed higher morphospecies richness in soils with pH between 3.5 and 5.0, and Acaulosporaceae from 5.0 to 6.0. Contrary to Glomeraceae, sporulation of Acaulosporaceae is promoted in acidic soils, but its members also occur on higher pH soils (Clark 1997). Only Gigasporaceae showed a significant and positive association with N ($t = 5.106$, $P < 0.001$), and a negative association with P ($t = 2.038$, $P = 0.048$). Meanwhile, Glomeraceae and Acaulosporaceae did not show a significant relationship with any of these variables (Fig. 13.5). In P- limited ecosystems with high N availability, host plants may select AMF taxa with extensive hyphal networks that forage P effectively, such as Gigasporaceae (Egerton-Warburton et al. 2007). This is because excess in N availability is expected to improve plant photosynthesis thus making the availability of C for transfer to AMF symbionts less costly for the plant (Johnson 2010). Nonetheless, evidence that increase N availability reduce the occurrence of AMF taxa with greater P benefit (i.e. Gigasporaceae) has been also documented (Treseder et al. 2018).

Among macroscale factors, Glomeraceae and Gigasporaceae showed a negative significant relationship with latitude ($t = 4.450$, $P < 0.001$; $t = 5.180$, $P < 0.001$; respectively), and a positive association with mean annual temperature ($t = 5.302$, $P < 0.001$; $t = 3.902$, $P < 0.001$; respectively) and precipitation ($t = 2.779$, $P = 0.008$; $t = 3.815$, $P < 0.001$; respectively). However, Acaulosporaceae did not show significant association with any of these variables (Fig. 13.5). Gigasporaceae members are adapted to live in stable ecosystems (de Souza et al. 2005), and highly dependent on precipitation (Veresoglou et al. 2012) as observed here.

13.6 Conclusions

High mountain ecosystems of South America differed in their AMF communities due to macroscale and microscale factors, revealing indicator AMF morphospecies associated with either sampling site or vegetation type or host identity. This is in line with global molecular studies of AMF, which evidenced patchily distributed AMF communities (Öpik et al. 2010, 2013), although contrary to an AMF taxa cosmopolitan distribution (Davison et al. 2015). As stated by Davison et al. (2015), several high mountain ecosystems of South America remain unexplored thus making our results probably related to low sampling effort. However, it is important to take into account that these authors presented global patterns of molecularly identified AMF species considering only four records among grassland and successional forests at South America thus probably losing the patchily structure of AMF communities of high mountain ecosystems. The AMF richness relationships with micro and macroscale factors were mainly due to Glomeraceae and Gigasporaceae responses to these variables. At higher scales, tropical and temperate ecosystems differentiated in their AMF community composition due to macroscale factors as latitude, precipitation and temperature. At lower scales, soil characteristics and host species became the most relevant factors in differentiating AMF community composition of sites. High mountain ecosystems of South America comprise a particular environment in which AMF communities could not be framed in a cosmopolitan pattern but rather they adjust to their own pattern associated with specific conditions of the highlands.

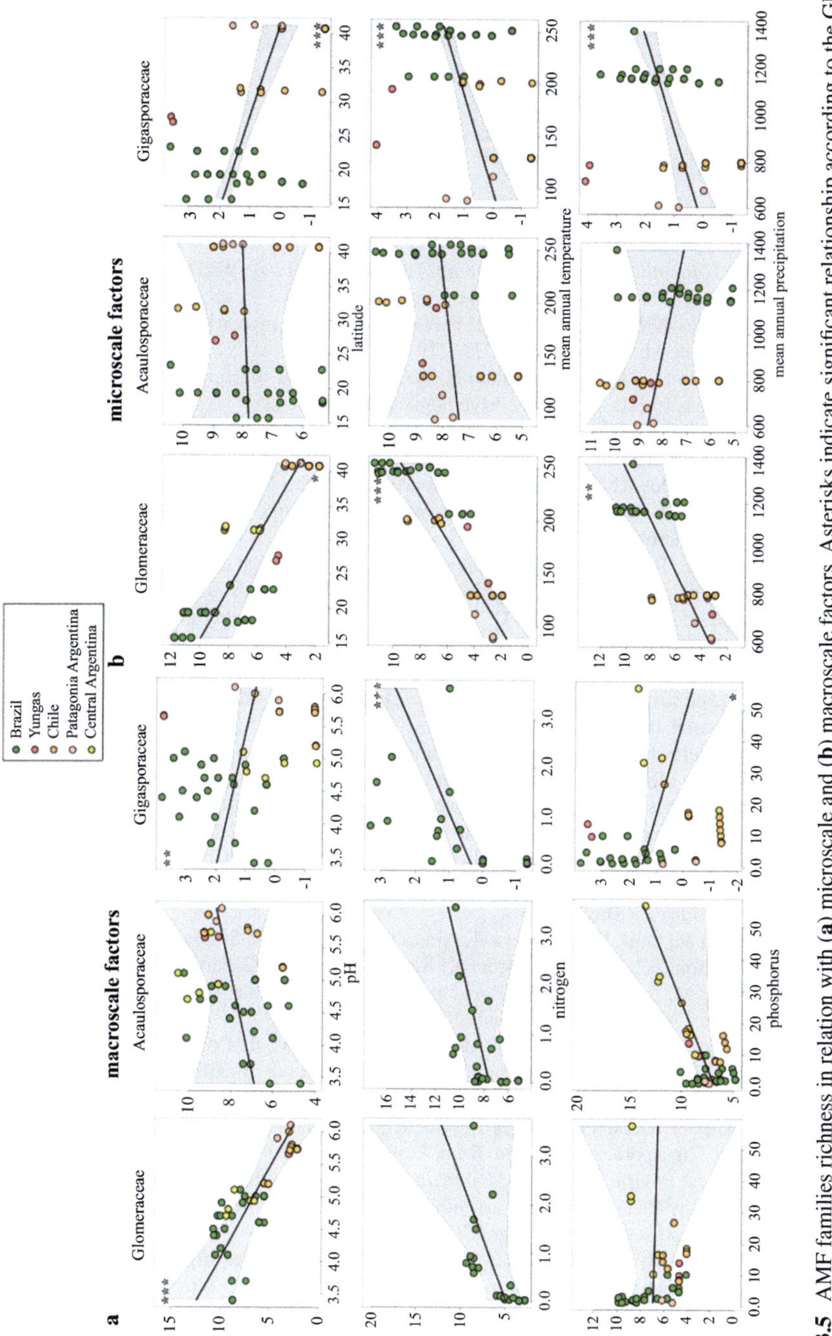

Fig. 13.5 AMF families richness in relation with (**a**) microscale and (**b**) macroscale factors. Asterisks indicate significant relationship according to the GLM (*** $P < 0.001$, ** $P < 0.01$, * $P < 0.05$). Points color represents sampling sites

Acknowledgements This work was financially supported by FONCyT (BID PICT 438–2006 granted to A.B., BID 2015 PICT 338 granted to F.S), and Idea Wild Foundation. All authors are staff researchers from CONICET.

References

Bardgett RD, Van Der Putten WH (2014) Belowground biodiversity and ecosystem functioning. Nature 515(7528):505–511

Barry RG (2008) *Mountain weather and climate*. Third edit, *Mountain Weather and Climate, Third Edition*. Third edit. New York: Cambridge University Press

Becerra AG, Cabello MN, Bartoloni NJ (2011) Native arbuscular mycorrhizal fungi in the Yungas forests, Argentina. Mycologia 103(2):273–279

Becerra AG, Cabello M, Zak MR, Bartoloni N (2009) Arbuscular mycorrhizae of dominant plant species in Yungas forests, Argentina. Mycologia 101(5):612–621

Bever JD, Schultz PA, Pringle A, Morton JB (2001) Arbuscular mycorrhizal fungi: More diverse than meets the eye, and the ecological tale of why. BioScience 51(11):923–931

Bonfim JA, Vasconcellos RLF, Gumiere T, de Lourdes Colombo Mescolotti D, Oehl F, Nogueira Cardoso EJB (2016) Diversity of arbuscular mycorrhizal fungi in a Brazilian Atlantic Forest toposequence. Microb Ecol 71(1):164–177

Borcard D, Legendre P (2002) All scale spatial analysis of ecological data by means of principal coordinates of neighbor matrices. Ecol Model 153:51–68

Borcard D, Legendre P, Avois-Jacquet C, Tuomisto H (2004) Dissecting the spatial structure of ecological data at multiple scales. Ecology 85(7):1826–1832

Castillo DD, Carrasco JC, Quevedo LA, Ricaurte CB, Gavilanes AV, Borz SA (2017) Diversity, composition and structure of Andean High Forest in Ecuador, South America. Bull Transilvania University Braşov 10(59):1–16

Chagnon P-L, Bradley RL, Maherali H, Klironomos JN (2013) A trait-based framework to understand life history of mycorrhizal fungi. Trends Plant Sci 18(9):484–491

Clark RB (1997) Arbuscular mycorrhizal adaptation, spore germination, root colonization, and host plant growth and mineral acquisition at low pH. Plant Soil 192:15–22

Coutinho ES, Fernandes GW, Berbara RLL, Valério HM, Goto BT (2015) Variation of arbuscular mycorrhizal fungal communities along an altitudinal gradient in rupestrian grasslands in Brazil. Mycorrhiza 25(8):627–638

Davison J, Moora M, Öpik M, Adholeya A, Ainsaar L, Bâ A, Burla S, Diedhiou AG, Hiiesalu I, Jairus T, Johnson NC, Kane A, Koorem K, Kochar M, Ndiaye C, Pärtel M, Reier Ü, Saks Ü, Singh R, Vasar M, Zobel M (2015) Global assessment of arbuscular mycorrhizal fungus diversity reveals very low endemism. Science 127(6251):970–973

de Carvalho F, de Souza FA, Carrenho R, de Souza Moreira FM, da Conçeição Jesus E, Wilson Fernandes G (2012) The mosaic of habitats in the high-altitude Brazilian rupestrian fields is a hotspot for arbuscular mycorrhizal fungi. Appl Soil Ecol 52:9–19

de Souza F, Dalpé Y, Declerck S, de la Providencia I, Séjalon-Delmas N (2005) Life History Strategies in Gigasporaceae: Insight from Monoxenic Culture. In: Declerck S, Fortin JA, Strullu DG (eds) *In Vitro Culture of Mycorrhizas*. Springer, Berlin, Heidelberg, p 73–91

Dufrene M, Legendre P (1997) Species assemblages and indicator species: the need for a flexible asymmetrical approach. Ecol Monogr 67(3):345–366

Egerton-Warburton LM, Allen EB, Allen MF (2004) Conservation of mycorrhizal fungal communities under elevated atmospheric CO2 and anthropogenic nitrogen deposition. In: Sivasithamparama K, Dixon KW, Barret RL (eds) *Microorganisms in plant conservation and biodiversity*. Springer, Netherlands, p 19–43

Egerton-Warburton LM, Johnson NC, Allen EB (2007) Mycorrhizal community dynamics following nitrogen fertilization: a cross-site test in five grasslands. Ecol Monogr 77(4):527–544

Fjeldså J, Kessler M (1996) *Conserving the biological diversity of Polylepis woodlands of the highland of Peru and Bolivia. A contribution to sustainable natural resource management in the Andes*. Copenhagen, Denmark: Nordeco

Geml J, Pastor N, Fernandez L, Pacheco S, Semenova T, Becerra AG, Wicaksono CY, Nouhra ER (2014) Large-scale fungal diversity assessment in the Andean Yungas forests reveals strong community turnover among forest types along an altitudinal gradient. Mol Ecol 23(10):2452–2472

Grêt-Regamey A, Brunner SH, Kienast F (2012) Mountain ecosystem services: Who cares?. Mount Res Dev 32(S1):S23–S34

Grime JP (1979) *Plant Strategies and Vegetation Processes*. John Wiley & Sons, Chichester, New York, Brisbane,Toronto

Hart MM, Reader RJ (2002) Taxonomic basis for variation in the colonization strategy of arbuscular mycorrhizal fungi. New Phytol 153:335–344

Hart MM, Reader RJ, Klironomos JN (2001) Life-history strategies of arbuscular mycorrhizal fungi in relation to their successional dynamics. Mycologia 93(6):1186–1194

van der Heijden MGA, Bardgett RD, van Straalen NM (2008) The unseen majority: soil microbes as drivers of plant diversity and productivity in terrestrial ecosystems. Ecol Lett 11(3):296–310

Hillebrand H (2004) On the Generality of the Latitudinal Diversity Gradient. Am Nat 163(2):192–211

Hoorn C, Mosbrugger V, Mulch A, Antonelli A (2013) Biodiversity from mountain building. *Nature Geoscience*. Nature Publishing Group, a division of Macmillan Publishers Limited. All Rights Reserved, 6, p 154

Johnson NC (2010) Resource stoichiometry elucidates the structure and function of arbuscular mycorrhizas across scales. New Phytol 185(3):631–647

Johnson NC, Tilman D, Wedin D (1992) Plant and soil controls on mycorrhizal fungal communities. Ecology 73(6):2034–2042

Koske RE (1987) Distribution of VA mycorrhizal fungi along a latitudinal temperature gradient. Mycologia 79(1):55–68

La Sorte FA, Jetz W (2010) Projected range contractions of montane biodiversity under global warming. Proc R Soc Biol Sci Ser B 277(1699):3401–3410

Legendre P, Andersson MJ (1999) Distance-based redundancy analysis: Testing multispecies responses in multifactorial ecological experiments. Ecol Mon 69(1):1–24

Lugo MA, Cabello MN (2002) Native arbuscular mycorrhizal fungi (AMF) from mountain grassland (Cordoba, Argentina) I. Seasonal variation of fungal spore diversity. Mycologia 94(4):579–586

Maherali H, Klironomos JN (2007) Influence of phylogeny on fungal community assembly and ecosystem functioning. Science 316(5832):1746–1748

Marín C, Aguilera P, Cornejo P, Godoy R, Oehl F, Palfner G, Boy J (2016) Arbuscular mycorrhizal assemblages along contrasting Andean forests of Southern Chile. J soil Sci plant Nutr 16(4):916–929

Matus F, Rumpel C, Neculman R, Panichini M, Mora ML (2014) Soil carbon storage and stabilisation in andic soils: A review. Catena 120:102–110

Menoyo E, Renison D, Becerra AG (2009) Arbuscular mycorrhizas and performance of *Polylepis australis* trees in relation to livestock density. For Ecol Manag 258(12):2676–2682

Moreira-Souza M, Trufem SFB, Gomes-Da-Costa SM, Cardoso EJ (2003) Arbuscular mycorrhizal fungi associated with *Araucaria angustifolia* (Bert.) O. Ktze. Mycorrhiza 13(4):211–215

Mummey DL, Rillig MC (2006) The invasive plant species *Centaurea maculosa* alters arbuscular mycorrhizal fungal communities in the field. Plant Soil 288(1–2):81–90

Myers N, Mittermeier RA, Mittermeier CG, da Fonseca GAB, Kent J (2000) Biodiversity hotspots for conservation priorities. Nature 403(6772):853–858

Nouhra E, Soteras F, Pastor N, Geml J (2018) Richness, species composition and functional groups in Agaricomycetes communities along a vegetation and elevational gradient in the Andean Yungas of Argentina. Biodivers Conserv 27(8):1849–1871

Oksanen J, Blanchet FG, Friendly M, Kindt R, Legendre P, McGlinn D, Minchin PR, O'Hara RB, Simpson GL, Solymos P, Stevens MHH, Szoecs E, Wagner H (2018) vegan: Community Ecology Package. R package version 2.5–2. https://CRAN.R-project.org/package=vegan

Öpik M, Moora M, Liira J, Zobel M (2006) Composition of root-colonizing arbuscular mycorrhizal fungal communities in different ecosystems around the globe. Journal of Ecology 94(4):778–790

Öpik M, Vanatoa A, Vanatoa E, Moora M, Davison J, Kalwij JM, Reier U, Zobel M (2010) The online database MaarjAM reveals global and ecosystemic distribution patterns in arbuscular mycorrhizal fungi (Glomeromycota). New Phytol 188(1):223–241

Öpik M, Zobel M, Cantero JJ, Davison J, Facelli JM, Hiiesalu I, Jairus T, Kalwij JM, Koorem K, Leal ME, Liira J, Metsis M, Neshataeva V, Paal J, Phosri C, Põlme S, Reier Ü, Saks Ü, Schimann H, Thiéry O, Vasar M, Moora M (2013) Global sampling of plant roots expands the described molecular diversity of arbuscular mycorrhizal fungi. Mycorrhiza 23(5):411–430

Orlandi Costa HA, Stürmer SL, Ragonezi C, Grazziotti PH, Fonseca Santos Grazziotti DC, de Barros Silva E (2016) Species richness and root colonization of arbuscular mycorrhizal fungi in Syngonanthus elegans, an endemic and threatened species from the Cerrado domain in Brazil. Ciência Agrotecnologia 40(3):326–336

Quintero I, Jetz W (2018) Global elevational diversity and diversification of birds. Nature 555:246–250

R Core Team (2018) R: A language and environment for statistical computing. R Foundation for Statistical Computing, Vienna, Austria. http://www.R-project.org.

Roberts DW (2013) labdsv: ordination and multivariate analysis for ecology package. Version 1.6–1. http://cran.r-project.org/web/packages/labdsv

Robledo GL, Renison D (2010) Wood-decaying polypores in the mountains of central Argentina in relation to *Polylepis* forest structure and altitude. Fungal Ecology 3(3):178–184

Robledo G, Urcelay C, Domínguez L, Rajchenberg M (2006) Taxonomy, ecology, and biogeography of polypores (Basidiomycetes) from Argentinian *Polylepis* woodlands. Can J Bot 84(10):1561–1572

Schüßler A, Schwarzott D, Walker C (2001) A new fungal phylum, the Glomeromycota: phylogeny and evolution. Mycol Res 105(12):1413–1421

Senés-Guerrero C, Schüßler A (2016) A conserved arbuscular mycorrhizal fungal core-species community colonizes potato roots in the Andes. Fungal Divers 77(1):317–333

Smith SE, Read D (2008) Mycorrhizal symbiosis. Academic Press, Great Britain

Soteras F, Grilli G, Cofré MN, Marro N, Becerra A (2015) Arbuscular mycorrhizal fungal composition in high montane forests with different disturbance histories in central Argentina. Appl Soil Ecol 85:30–37

Soteras F, Coutinho Moreira B, Grilli G, Pastor N, Carneiro Mendes F, Ruela Mendes D, Renison D, Megumi Kasuya MC, de Souza FA, Becerra A (2016) Arbuscular mycorrhizal fungal diversity in rhizosphere spores versus roots of an endangered endemic tree from Argentina: Is fungal diversity similar among forest disturbance types?. Appl Soil Ecol 98:272–277

Souza de Pontes JS, Oehl F, Donizete Pereira C, Torres de Toledo Machado C, Coyne D, Alves da Silva DK, Maia LC (2017) Diversity of arbuscular mycorrhizal fungi in the Brazilian's Cerrado and in soybean under conservation and conventional tillage. Appl Soil Ecol 117–118:178–189

Spatafora JW, Sung G-H, Johnson D, Hesse C, O'Rourke B, Serdani M, Spotts R, Lutzoni F, Hofstetter V, Miadlikowska J, Reeb V, Gueidan C, Fraker E, Lumbsch T, Lücking R, Schmitt I, Hosaka K, Aptroot A, Roux C, Miller AN, Geiser DM, Hafellner J, Hestmark G, Arnold AE, Büdel B, Rauhut A, Hewitt D, Untereiner WA, Cole MS, Scheidegger C, Schult M, Sipman H, Schoch CL (2006) A five-gene phylogeny of Pezizomycotina. Mycologia 98(6):1018–1028

Tedersoo L, Bahram M, Põlme S, Kõljalg U, Yorou NS, Wijesundera R, Ruiz L. V, Vasco-Palacios AM, Thu PQ, Suija A, Smith ME, Sharp C, Saluveer E, Saitta A, Rosas M, Riit T, Ratkowsky D, Pritsch K, Põldmaa K, Piepenbring M, Phosri C, Peterson M, Parts K, Pärtel K, Otsing E, Nouhra E, Njouonkou AL, Nilsson RH, Morgado LN, Mayor J, May TW, Majuakim L, Lodge DJ, Lee SS, Larsson K-H, Kohout P, Hosaka K, Hiiesalu I, Henkel TW, Harend H, Guo L,

Greslebin A, Grelet G, Geml J, Gates G, Dunstan W, Dunk, Drenkhan R, Dearnaley J, De Kesel A, Dang T, Chen X, Buegger F, Brearley FQ, Bonito G, Anslan S, Abell S, Abarenkov K (2014) Global diversity and geography of soil fungi. Science 346(6213): 1256688–1–10

Treseder KK, Allen EB, Egerton-Warburton LM, Hart MM, Klironomos JN, Maherali H, Tedersoo, L (2018) Arbuscular mycorrhizal fungi as mediators of ecosystem responses to nitrogen deposition: A trait-based predictive framework. J Ecol 106(2):480–489

Vega GC, Pertierra LR, Olalla-Tárraga MÁ (2017a) MERRAclim, a high-resolution global dataset of remotely sensed bioclimatic variables for ecological modelling. Scientific Data 4:1–12

Vega GC, Pertierra L, Olalla-Tárraga M (2017b) *Data from: MERRAclim, a high-resolution global dataset of remotely sensed bioclimatic variables for ecological modelling. Dryad Digital Repository*

Velázquez MS, Stürmer SL, Bruzone C, Fontenla S, Barrera M, Cabello M (2016) Occurrence of arbuscular mycorrhizal fungi in high altitude sites of the Patagonian Altoandina region in Nahuel Huapi National Park (Argentina). Acta Bot Brasilica 30(4):521–531

Veresoglou SD, Chen B, Rillig MC (2012) Arbuscular mycorrhiza and soil nitrogen cycling. Soil Biol Biochem 46:53–62

Zuur AF, Ieno NI, Walker NJ, Saveliev AA, Smith MG (2009) *Mixxed effects models and extensions in ecology with R*, Gail M, Krickeberg K, Samet JM, Tsiatis A, Wong W (eds). Springer, Verlag, New York

Chapter 14
Mycorrhizas in the South American Mediterranean-Type Ecosystem: Chilean Matorral

Patricia Silva-Flores, Ana Aguilar, María José Dibán, and María Isabel Mujica

14.1 Introduction

The five Mediterranean-type ecosystems (MTEs, singular: MTE) in the world are climatically characterized with warm-dry summers and cool-wet winters (Rundel and Cowling 2013). These ecosystems are located in California, central Chile, the Mediterranean Basin, the Cape Region of South Africa, and southwestern and south Australia (Dallman 1998; Rundel and Cowling 2013). A remarkable feature of the MTEs is the fact that they occupy, in total, less than 3% of the Earth's surface and contain almost 50,000 species of vascular plants, which correspond to 20% of the world's known species (Cowling et al. 1996; Rundel and Cowling 2013). Also, many of the plant species are endemic (Cowling et al. 1996) and, at the same time,

P. Silva-Flores (✉)
Centro de Estudios Avanzados en Fruticultura (CEAF), Santiago, Chile

Departamento de Botánica, Universidad de Concepción, Concepción, Chile

Micófilos ONG, Concepción, Chile
e-mail: psilvaf@ceaf.cl

A. Aguilar
Centro Regional de Innovación Hortofrutícola de Valparaíso (CERES), Quillota, Chile

Pontificia Universidad Católica de Valparaíso, Valparaíso, Chile

M. J. Dibán
Micófilos ONG, Concepción, Chile

Departamento de Ciencias Ecológicas, Universidad de Chile, Santiago, Chile

Instituto de Ecología y Biodiversidad (IEB), Ñuñoa, Chile

M. I. Mujica
Instituto de Ecología y Biodiversidad (IEB), Ñuñoa, Chile

Departamento de Ecología, Pontificia Universidad Católica de Chile, Santiago, Chile

M. C. Pagano, M. A. Lugo (eds.), *Mycorrhizal Fungi in South America*, Fungal Biology, https://doi.org/10.1007/978-3-030-15228-4_14

they are threatened by several human-related factors (Underwood et al. 2009). All these characteristics have placed the MTEs as biodiversity hotspots (Myers et al. 2000), which means that they are priorities for conservation. Due to this situation, much research has been done in MTEs, mostly in plants (Dallman 1998), but also some attention has been paid to animals (Rundel and Cowling 2013). However, the soil microbiological biodiversity has been overlooked, despite the fact that they are increasingly being recognized as key players in the restoration of degraded ecosystems (Harris 2009).

One of the most important microorganisms in the soil are the mycorrhizal fungi (MF) which form symbiotic relations with the roots of approximately 90% of known plant species (Brundrett and Tedersoo 2018). These mycorrhizal symbiotic relations occurs in almost all ecosystems (Read 1991; Read and Perez-Moreno 2003; Read et al. 2004) and as four main types: arbuscular mycorrhiza (AM), ectomycorrhiza (EcM), orchid mycorrhiza (OrM) and ericoid mycorrhiza (ErM) (Brundrett and Tedersoo 2018). The MF uptake nutrients from soil and supplies to the plant partner in return for carbohydrates (Smith and Read 2008) and lipids (Jiang et al. 2017; Keymer et al. 2017; Luginbuehl et al. 2017). Consequently, the mycorrhizal symbiosis, significantly influences plant fitness, as well as several ecosystem processes such as carbon, nitrogen and phosphorous cycles, regulation of plant diversity, soil aggregation and seedling survival (van der Heijden et al. 2015). Thus, because of the importance of MF to plants, they cannot be ignored in the efforts to preserve ecosystems as MTEs. However, regarding MF in the MTEs, scarce research has been performed. In fact, in an ISI Web of Knowledge search (status May 2018), it is possible to find only 512 publications in a 10 year span regarding this topic. Moreover, in the last XIV MEDECOS and XIII AEET Consortium Meeting held in Spain in February 2017 that gathered 538 participants (Arista et al. 2017), there were only fifteen investigations dealing with mycorrhiza in MTEs – most of them from the Mediterranean Basin (Álvarez-Garrido et al. 2017; Benito Matías et al. 2017; Parker et al. 2017; Pérez-Izquierdo et al. 2017; Rincón Herranz et al. 2017; Romero Munar et al. 2017; Verdú 2017; Calviño-Cancela et al. 2017; Dias et al. 2017; Gil-Martínez et al. 2017; Hernández-Rodríguez et al. 2017; López García et al. 2017; Marañón et al. 2017; Mediavilla et al. 2017; Navarro-Fernández et al. 2017). Recently, researchers have started to focus on MF in the central Chile MTE. Due to this, in this chapter we aim to highlight and compile the arising and existing knowledge on mycorrhizas of central Chile MTE. This chapter will focus on basic and applied research on different mycorrhizal types, as well as on the detection of knowledge gaps and proposals for future research directions.

14.2 Mycorrhizas in Central Chile Mediterranean-Type Ecosystem

The unique South American Mediterranean-type ecosystem is located from 30° to 36° South Latitude in the western portion of the continent (Armesto et al. 2007). It is surrounded by the Atacama Desert in the north, by the Pacific Ocean in the west, by the Andes in the east and by the evergreen-deciduous temperate forests in the south. This topography, which in turn produces antagonistic radiation/moistness systems and a highly variable soil system, generates environmental gradients that have produced highly rich plant communities (Armesto et al. 2007). Also, several types of plant communities occur here, which altogether are known as **Chilean matorral** (Armesto et al. 2007).

From a recent study it was possible to extract the proportions of the mycorrhizal types from up to 1576 native plant species (from a total of 1591) of the Chilean matorral (Silva-Flores et al. unpublished work). On that study, the proportions of mycorrhizal types were calculated at three resolution levels (Fig. 14.1). The first one calculated the proportions using an empirical approach (Bueno et al. 2018), i.e. using published in peer-reviewed literature in where the mycorrhizal type of a plant species was assessed through empirical methods (Fig. 14.1a). With this approach it was possible to assess the mycorrhizal type of only 17.3% of the plant species in the Chilean matorral (Fig. 14.1a). The other two levels of resolution used to assess the mycorrhizal type of the plant species, calculated the proportions using a taxonomic approach (Bueno et al. 2018), i.e. extrapolating a certain plant species mycorrhizal type to a taxonomically and phylogenetically related plant species. Extrapolating information from plant species to plant genus (Fig. 14.1b) and subsequently to plant family (Fig. 14.1c), allowed to assess the mycorrhizal type of 78% and 99% (at genus and family level of resolution respectively) of the plant species of the Chilean matorral. From the results of that study was possible to extract that the AM type is in higher proportion relatively to the other mycorrhizal types, independent of the level of resolution (Fig. 14.1). The other mycorrhizal types proportions vary depending on the level of resolution. In order to learn the exact proportions of all mycorrhizal types in the Chilean matorral more empirical data are urgently needed, since

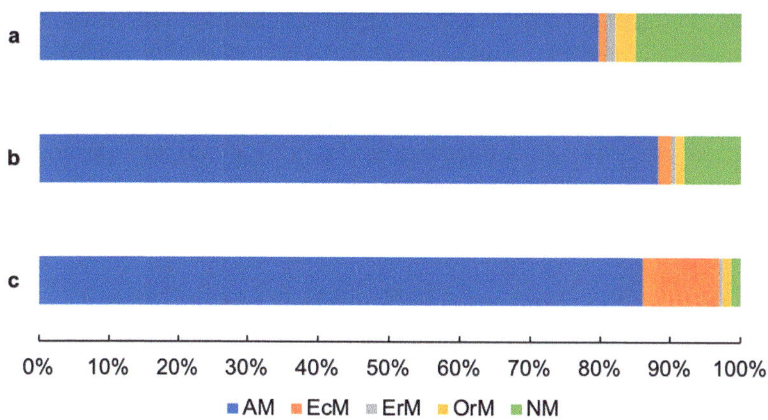

Fig. 14.1 Proportions of mycorrhizal types in the Chilean matorral. Proportions were calculated from: (**a**) empirical data of plant species, (**b**) extrapolation from plant species to genus, and (**c**) extrapolation from plant genus to family. AM: arbuscular mycorrhiza, EcM: ectomycorrhiza, ErM: ericoid mycorrhiza, OrM: orchid mycorrhiza and NM: non-mycorrhiza

it has been recently showed the inaccuracy of the taxonomic approach to describe plant mycorrhizal types (Bueno et al. 2018). To learn the mycorrhizal type distribution patterns in plant species can indicate the relationship strength between plant communities and mycorrhizas (Moora 2014), potentially regulating both the mycorrhizal and plant community (Neuenkamp et al. 2018). Consequently, this information might be useful in developing tools for restoration of degraded ecosystems as the Chilean MTE.

It is important to highlight from the above-mentioned research (Silva-Flores et al. unpublished work) that even with the extrapolation of mycorrhizal type data from plant species to plant families, there are still 17 plant species in the Chilean matorral where the mycorrhizal type is absolutely unknown. Two of these species are in the Chilean national regulation of classification of species according to their conservation status. The species *Berberidopsis corallina* Hook. f. (Berberidopsidaceae) and *Gomortega keule* (Molina) Baill. (Gomortegaceae) are both classified as endangered; thus, it would be essential to clarify their mycorrhizal type, mycorrhizal abundance and other aspects on mycorrhizal biology in order to assess the relative importance of the MF in the recovery of these plant species.

14.3 Arbuscular Mycorrhiza in the Chilean Matorral

Arbuscular mycorrhizal fungi (AMF) are found as root symbionts in 72% of land plant species (Brundrett and Tedersoo 2018). This symbiosis, known as AM, occurs between the roots of certain plants and the hyphae of fungi from the Phylum Glomeromycota (Tedersoo et al. 2008). With the currently available research, it is known that the AM symbiosis is the more frequent across the plant species of the Chilean matorral (Fig. 14.1) and consequently probably a key component for this ecosystem. Despite this, there is scarce information on the topic, probably due the lack of awareness. In fact, to our knowledge, there are only five published studies regarding basic research on AM symbiosis related to the Chilean matorral (Casanova-Katny et al. 2011; Torres-Mellado et al. 2012; Marín et al. 2017; Benedetti et al. 2018; Silva-Flores et al. 2019) and two on applied research (Curaqueo et al. 2010, 2011).

14.3.1 Basic Research on AM Symbiosis in the Chilean Matorral

One study explored the unknown mycorrhizal type of 10 plant species of the Amaryllidaceae family: *Gilliesia curicana*, *G. graminea*, *G. montana*, *Miersia chilensis*, *M. leporina*, *M. myodes*, *M. tenuiseta*, *Solaria atropurpurea*, *S. miersioides* and *Speea humilis* (Torres-Mellado et al. 2012). The study found that all the plant species had an association with the AM type with a mean colonization percentage of 45%. The study has an impact for conservation strategies of those plant species since all of them are either vulnerable or endangered. The authors concluded that because the plants have a high mean mycorrhization level, they should be highly AM dependent and thus the associated AMF should be considered in conservation strategies as plant propagation. The study also suggests that the next step should be the identification of AMF species associated to the plant species in order to successfully use it in conservation programs.

In the same line, another investigation also explored the mycorrhizal type of 23 Andean plant species, 18 of which belonged to the AM type (Casanova-Katny et al. 2011). This was done in order to show that cushion-associated plants had a higher AM colonization in comparison with the same plant species growing in bare soil.

One study explored the factors affecting AMF communities in ten Chilean *Nothofagus* forests (Marín et al. 2017), included three sampling plots that according to Armesto et al. (2007) can be considered as part of the Chilean matorral. One plot was from *Reserva Natural Los Ruiles* and was dominated by *N. alesandrii* (P1). The other two plots were from *Parque Nacional La Campana*, one plot dominated by *N. macrocarpa* (P2) and the second by *Luma apiculata* and *Peumus boldus* (P3). In that study, AMF communities were studied through the analysis of spores extracted from soil samples. Only *Glomus* sp. was found in P1 and P2, whilst in P3 was also present together with *G. diaphanum.* Consequently, P3 resulted to have a higher diversity than P1 and P2, whilst the community composition of P1 and P2 were similar, but also different from P3. The content in the soil of plant available phosphorus and magnesium were the main edaphic variables affecting the AM fungal community composition in those three plots. The higher spore abundance was in P2, followed by P3 and P1.

A recent published study, quantified and morphologically identified the AMF spores associated to nine sites of *P. boldus* forests (Benedetti et al. 2018). They reported a total of 23 AM fungal species considering all sites. *Funneliformis badium* was present in all sites and with a high abundance relative to the other AM fungal species. In contrast, *Septoglomus constrictum* was also present in all sites but with a low abundance. *F. mosseae, Acaulospora spinosa* and *Rhizophagus irregularis* were also frequent species considering all sites.Finally, in the VI Region of Libertador Bernardo O'Higgins, the role of biotic and abiotic factors in regulating soil AMF spore density in two sclerophyllous shrublands were explored. The results showed a strong regulation of climatic seasons on spore density in both shrublands, in contrast to plant host species that did not have an effect on soil spore density in any of the shrublands. Soil factors as clay content, electrical conductivity, soil organic matter and available phosphorus and nitrogen also affected AMF spore density (Silva-Flores et al. 2019).

There are also several ongoing studies with unpublished data or submitted results regarding AM symbiosis on the Chilean matorral. In this respect, a study performed in the *Reserva Río Los Cipreses* ecosystem (34°27′54″S 70°27′18″W) allowed scientists to have a first screening of the diversity of AMF in the upper part of the *Cachapoal* river (Aguilar et al. unpublished data). Three plant formations were studied in this ecosystem: the *Austrocedrus chilensis,* the Matorral and the Espinal.

The main AMF genera found in all three plant formations were *Glomus, Acaulospora* and *Archeospora* (Fig. 14.2). Moreover, the Matorral plant formation showed a higher species richness relative to the Espinal and *Austrocedrus chilensis* (Fig. 14.2). The tree plant formations had different species composition (Fig. 14.2). Also, a high level of AMF colonization was observed in the roots of all three plant formations; *Austrocedrus chilensis* (90%), the Matorral (75%) and the Espinal (65%). Finally, a positive correlation was observed between the diversity of AMF and edaphic factors, such as nitrogen and phosphorus concentration, available and exchangeable potassium and soil organic matter percentage.

Another study performed in the VI Region of *Libertador Bernardo O'Higgins* in central Chile, aimed to assess the mycorrhizal type of the dominant plant species of the sclerophyllous shrubland plant formation (Silva-Flores et al. submitted). It was possible to stablish that *P. boldus*, *Kageneckia oblonga*, *Escallonia pulverulenta*, *Quillaja saponaria* and *Cryptocarya alba* were all AM plant species. In *Lithrea caustica* it was possible to observe AMF hyphae only in the surface of roots thus further analyses are recommended to accurately assess whether is an AM plant.

Finally, another study explored the variation of the molecular AMF community of the sclerophyllous shrubland in relation with host plant species, compartment (root or soil), physico-chemical soil factors and seasons (Silva-Flores et al. unpublished data). So far, a richness of 153 virtual taxa (VT) has been found. Also, the main AMF genera found were *Glomus, Claroideoglomus* and *Paraglomus*. AMF richness was regulated by host plant species, while AMF community composition was regulated by seasons, host plant species, soil compartment (root or soil) and some physico-chemical soil factors.

All the studies above described indicate that AMF are highly present in the Chilean matorral and consequently playing an important role on this MTE. However, all this research is in a descriptive stage and further studies quantifying the AMF

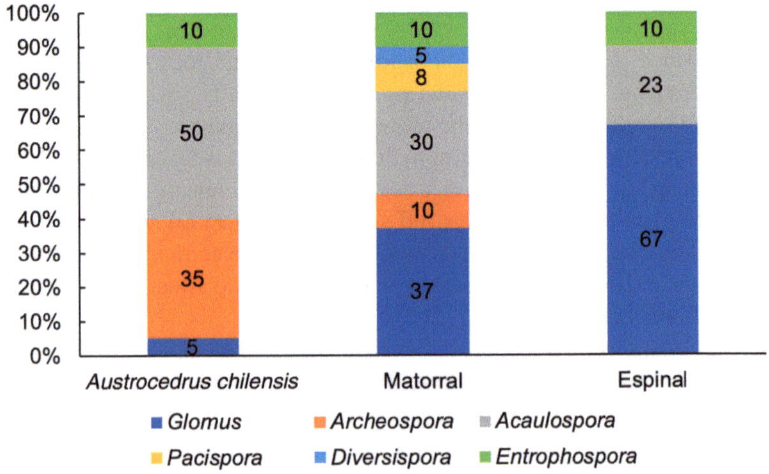

Fig. 14.2 Percentage of AMF spore genera in 100 g of soil in each plant formation studied

contribution to ecosystem processes should be performed, in order to develop efficient conservation strategies for this MTE.

14.3.2 Applied Research on AMF in the Chilean Matorral

Two studies related to AMF and applications of it have been performed in the Chilean Mediterranean agroecosystems. The first one evaluated the effect of no tillage and conventional tillage on soil organic matter, arbuscular mycorrhizal hyphae and soil aggregates (Curaqueo et al. 2010). Tillage affected the quantity and quality of soil organic matter, as well as AMF activity, glomalin content, and soil aggregation. No tillage produced higher values of hyphal length and glomalin production in contrast to conventional tillage. Thus, no tillage favors soil aggregation and consequently contribute to the stability of organic matter of the Mediterranean agroecosystems. The second study explored the effect of conventional tillage and no tillage for 6 and 10 years on AMF propagules (spore density and total and active fungal hyphae) and glomalin content (Curaqueo et al. 2011). AMF propagules and glomalin content resulted to be higher in a 6 year no tillage system compared with a conventional tillage system and 10 years no tillage system, suggesting that the positive effects of no tillage system for 6 year vanished after certain time.

Finally, Aguilar et al. (unpublished data) studied the effect of two different agricultural managements (organic vs conventional) on the diversity of AMF present in the soil of Mediterranean Chilean vineyards. The morphological analysis from spores of grapevine rhizospheric soil resulted in a total of twelve morphospecies of AMF (Fig. 14.3). Organic management had a higher species richness (11) compared to conventional management (10). Also, the species composition was different between managements. The organic management was composed by 2 exclusive species (*Acaulospora* sp. and *Pacispora scintillans*) and 9 shared with conventional management, while conventional had 1 exclusive (*Claroideoglomus etunicatum*) (Fig. 14.3). Finally, a molecular analysis showed that the three most common colonizers of grapevine roots, independent of the management, were *Funneliformis verruculosum*, *Septoglomus constrictum* and an unknown *Septoglomus* sp. This study provides valuable information since identification of AMF species have the potential for being used in sustainable management practices to improve grapevine production in the Mediterranean region.

The investigation on AMF in Mediterranean agroecosystems shows that AM symbiosis is also important and contributes to a better performance of the productive systems (Curaqueo et al. 2010, 2011). However, more detailed studies are needed in order to assess their role as a provider of ecological services in, for instance, sustainable agriculture (Johansson et al. 2004).

Finally, the recognition of the crucial role of AMF in the central Chilean MTE is needed in order to protect the diversity of AMF populations as well as the vegetation diversity. Additionally, it is also important to considerer the relationships between AMF and other microorganisms (e.g. PGPR and rhizobia).

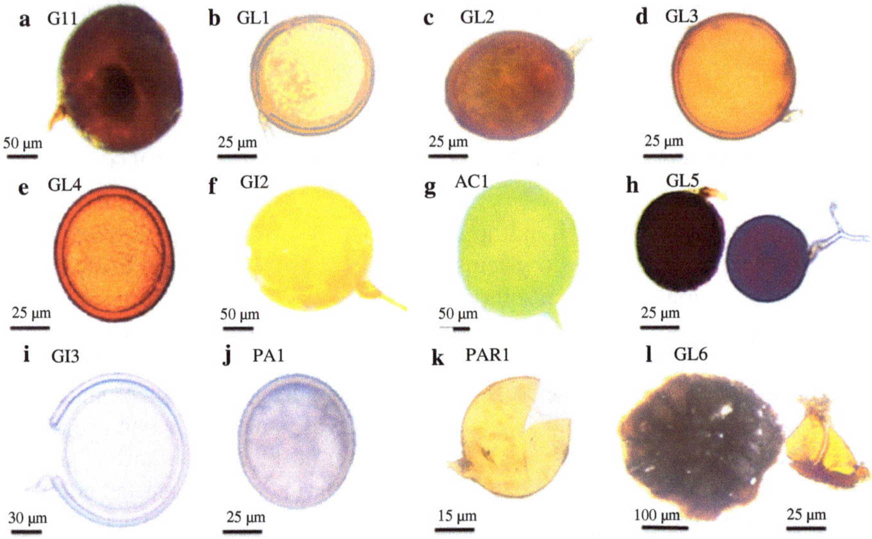

Fig. 14.3 AMF spore morphotypes from soil of ten Chilean grapevine valleys. (**a**) GI1 *Scutellospora* sp., (**b**) GL1 *Funeliformis verruculosum*, (**c**) GL2 Uncultured *Septoglomus,* (**d**) GL3, *Claroideoglomus etunicatum***, (**e**) GL4 Uncultured *Septoglomus,* (**f**) GI2 *Gigaspora* sp., (**g**) AC1 *Acaulospora* sp.*, (**h**) GL5 *Septoglomus constrictum,* (**i**) GI3 *Cetrospora gilmorei,* (**j**) PA1 *Pacispora scintillans**, (**k**) PAR1 *Paraglomus* sp. and (**l**) GL6 *Sclerocystis* sp. *species exclusively from soils with organic management, ** species exclusively from soils with conventional management

14.4 Ectomycorrhiza in the Chilean Matorral

In Chile, *Nothofagus* is the only native plant genus documented as EcM (Garrido 1988). It has 10 species, where six of them can be found in the Chilean matorral through altitudinal gradient replacement. *N. macrocarpa* inhabits in small, relictual and disjunct populations in the top of Coastal Mountain range, forming the northern limit of genus distribution (Alcaras 2010). In contrast, *N. obliqua*, *N. glauca*, *N. alessandri, N. alpina*, and *N. dombeyi* inhabit the altitudinal intermediate zones.

Several studies on *Nothofagus* forests in central Chile revealed the presence of 43 species of ectomycorrhizal fungi (Fig. 14.4), divided in 3 Orders, 9 Families, and 13 Genera (Singer 1969; Moser and Horak 1975; Horak 1980; Garrido 1985, 1988). *Cortinarius* is the dominant fungal genus, comprising 56% of the total species richness (Table 14.1). In addition, these forests are characterized by a high degree of fungal endemism, being 44% of the fungal species endemic to Chile, and 42% are endemic of the South American *Nothofagus* forest, comprising central-south of Chile and southwest of Argentina (Niveiro and Albertó 2012, 2013, 2014; Romano and Lechner 2013; Romano et al. 2017). Thus, a total of 86% of EcM fungal species found in the Chilean matorral are endemic of Nothofagus forests in southern South America (Table 14.1).

Fig. 14.4 Some native EcM species: (**a**) *Austropaxillus statuum*, (**b**) *Cortinarius austroturmalis*, (**c**) *C. magellanicus*, and (**d**) *Descolea antarctica*

Currently, macromycetes are being studied in forests dominated by *N. macrocarpa* in two locations: *Cerro El Roble* (33°00′S, 71°00′W) and *Reserva Natural Altos de Cantillana* (33°52′S, 71°00′W). We have found 17 ectomycorrhizal fungi corresponding to 10 species of the genus *Cortinarius*, two species of *Inocybe*, two species of *Laccaria*, one *Hebeloma* species, one *Amanita* species and one *Paxillus* species (Dibán et al. unpublished work). Although they are in the process of taxonomic determination, none of them coincide with those described in the literature for *N. macrocarpa* (Singer 1969; Moser and Horak 1975; Garrido 1985). Thereby, this study increases documented EcM fungal richness for *N. macrocarpa* to 28 species in total.

Most of the species records previously mentioned are based on the presence of ectomycorrhizal species through fruiting bodies (Singer 1969; Moser and Horak 1975; Garrido 1985), with few studies confirming the presence of the species in the roots (Garrido 1988). Thus, in mixed forests with the presence of two or more *Nothofagus* species, it makes it difficult to interpret which ectomycorrhizal species is associated with which host species. In addition, some fungal genera are both, ectomycorrhizal and saprotrophs (e.g. *Ramaria* spp.) (Tedersoo et al. 2008). Thus, in studies based only on fruiting bodies, there is no certainty whether *Ramaria* spp. are forming EcM associations or not. Consequently, one of the challenges in the

Table 14.1 List of ectomycorrhizal fungal species in four localities of the Chilean matorral, and its distribution. En = endemic, Ch = Chile, Ar = Argentina. S1: *Cerro El Roble* (33°00′S, 71°00′W), S2: *Altos de Vilches* (35°36′S, 71°12′W), S3: *Reserva Forestal El Maule* (35°50′S, 72°31′W) and S4: *Pilén* (35°57′S, 72°25′W)

Species	Distribution	S1	S2	S3	S4
Amanita diemii Singer	Ch, Ar	0	1	1	0
Amanita merxmuelleri Bresinsky & Garrido	En Ch	0	0	1	0
Amanita umbrinella Gilb. Et Clel.	Gondwanic	0	1	0	0
Austropaxillus boletinoides (sing.) Bresinsky & Jarosch	En Ch, Ar	0	0	1	0
Austropaxillus statuum (Speg.) Bresinsky & Jarosch	En Ch, Ar	0	1	1	0
Boletus araucarianus Garrido	En Ch	0	0	1	0
Boletus putidus Horak	En Ch	0	1	0	0
Butyriboletus loyo Philippi	En Ch	0	1	0	0
Cortinarius albocanus (Horak & Moser) Peintner & Moser	Gondwanic	0	1	1	0
Cortinarius albocinctus Moser	En Ch, Ar	1	0	0	0
Cortinarius amoenus (Moser & Horak) Garnier	En Ch, Ar	0	0	1	0
Cortinarius argillohygrophanicus Moser & Horak	En Ch, Ar	1	0	0	0
Cortinarius aridus Moser	En Ch	1	0	0	0
Cortinarius austroturmalis Moser & Horak	En Ch, Ar	0	1	1	1
Cortinarius austroturmalis var. austroturmalis	En Ch	1	0	0	0
Cortinarius brevisporus Moser	En Ch	1	0	0	0
Cortinarius cauquenensis Garrido	En Ch	0	0	1	0
Cortinarius coigue Garrido	En Ch	0	0	1	0
Cortinarius columbinus Moser & Horak	En Ch, Ar	1	0	0	1
Cortinarius darwinii Spegazzini	En Ch, Ar	0	1	0	0
Cortinarius elaiotus Moser & Horak	En Ch	0	0	0	1
Cortinarius gracilipes Moser	En Ch	0	0	0	1
Cortinarius hualo Garrido	En Ch	0	0	1	0
Cortinarius magellanicus Spegazzini	Native	0	0	1	0
Cortinarius maulensis Moser	En Ch, Ar	0	0	0	1
Cortinarius pachynemeus Moser	En Ch	1	0	0	0
Cortinarius paguentus Garrido & Horak	En Ch	0	0	1	0
Cortinarius roblemaulicola Garrido & Horak	En Ch, Ar	0	0	1	0
Cortinarius teraturgus Moser	Gondwanic	1	0	0	0
Cortinarius teresae (Garrido) Garnier	En Ch	0	0	1	0
Cortinarius tumidipes Moser	En Ch, Ar	0	1	0	0
Cortinarius viridurifolius Moser	En Ch	0	0	0	1
Descolea antarctica Singer	En Ch, Ar	1	1	1	0
Inocybe neuquenensis Singer	En Ch, Ar	0	0	1	0
Laccaria ohiensis (Mont.) Singer	Broad	0	0	1	0
Paxillus aff involutus (Batsch ex Fr.) Fr.	Broad	1	0	0	0

(continued)

Table 14.1 (continued)

Species	Distribution	S1	S2	S3	S4
Russula austrodelica Singer	En Ch	0	1	1	0
Russula nothofaginea Singer	En Ch, Ar	0	0	1	0
Stephanopus vilchensis Garrido & Horak	En Ch	0	1	0	0
Tricholoma cortinatellum Singer	En Ch, Ar	0	1	1	0
Tricholoma fagnani Singer	En Ch, Ar	0	0	1	0
Tricholoma fusipes Singer	En Ch, Ar	0	0	1	0
Zelleromyces alveolatus (sing. & Sm.) Trappe, Lebel & Castellano	En Ch	1	0	0	0
	TOTAL	**11**	**12**	**20**	**6**

study of EcM in the Chilean matorral is to combine the taxonomy of fruiting bodies together with direct observation of the roots, and to sequence the described species to extend the genetic database. Another challenge is to increase sampling locations, especially in forests of *N. alessandri* and *N. macrocarpa*, that are scarcely sampled and they are both classified as endangered species (Benoit 1989).

14.5 Orchid Mycorrhiza in the Chilean Matorral

The Orchidaceae family forms an exclusive type of mycorrhiza, called the orchid mycorrhiza (OrM). In this association, orchids interact with a polyphyletic group of life-free saprophytic fungi called *Rhizoctonia* that includes fungi from three basidiomycetes families: Tulasnellaceae, Ceratobasidiaceae and Sebacinaceae (Dearnaley et al. 2012). In addition to the exchange of nutrients and carbon between fungi and adult plants (Cameron et al. 2006), OrM are crucial for orchid germination and seedling development. Orchid seeds are extremely small and lack of energy reserves (Arditti and Ghani 2000), so they require associating with MF that provide the nutrients and carbon needed to germinate (Rasmussen 2002). This process, known as symbiotic germination, is one of the defining characteristics of Orchidaceae (Givnish et al. 2016) and it means that all orchids are mycoheterotrophic (MHT) at least in one stage of their life. Most orchids are autotrophic at adulthood, but there are some species that remain fully MHT throughout life (Leake 1994). Furthermore, some green orchids species present a third nutrition mode called partial MHT or mixotrophy, in which they obtain carbon from MF and from photosynthesis (Selosse and Roy 2009; Hynson et al. 2013).

Although Orchidaceae has a tropical center of diversity, it shows a considerable secondary diversity outside tropical regions (Dressler 1981). This is the case of Mediterranean ecosystems, where the scarce orchid flora of Southern California is an exception compared to Mediterranean Australia, Chile, south Africa and southern Europe (Bernhardt 1995).

There are 42 orchid species that show continuous or partial distribution through the Chilean matorral (Novoa et al. 2015), all of them terrestrial and photosynthetic. Little is known about OrM associations in Chile (Herrera et al. 2019), and even less in the Chilean matorral, however the evidence available shows these orchids form associations mainly with fungi form the families Tulasnellaceae and Ceratobasidiaceae (Herrera et al. 2017). These findings support observations from other Mediterranean zones that show that Tulasnellaceae and Ceratobasidiaceae fungi are the main associates in a high number of orchid species (Girlanda et al. 2011; Jacquemyn et al. 2015). Interestingly, there is no record of Sebacinaceae forming OrM in Chilean matorral. This result agrees with other studies in Southern South America (Fracchia et al. 2014a, b) but differs from observations in Mediterranean Basin (Girlanda et al. 2011). Further studies assessing the distribution of OrM fungi in soils would help to elucidate the causes of the lack of Sebacinaceae. Nevertheless, more research on other orchid species including more populations is needed to confirm this pattern.

Another exciting observation is the variation in the degree of specificity among orchid species of Chilean matorral. Specificity ranges from generalist associations like in *Chloraea longipetala* (Herrera et al. 2017) and *Bipinnula fimbriata* (Steinfort et al. 2010) to more specialists, as observed in *Chloraea gavilu* (Herrera et al. 2017). This agrees with variation on mycorrhizal specificity observed among orchids species from Mediterranean Australia (Bonnardeaux et al. 2007; Swarts and Dixon 2009). Additionally, variation in specificity among populations of the same species was observed in *Bipinnula fimbriata* and *B. plumosa*, which was related to changes in soil nutrient availability (Mujica et al. 2016). In the last decade, it has been an increasing effort to identify OrM in Chilean matorral (Herrera et al. 2019). However, further studies are required to expand this knowledge and to allow comparisons between Mediterranean climates. For example, to our knowledge, there is no evaluation of nutritional modes of Chilean matorral orchids, while mixotrophy has been detected in Mediterranean Basin orchids (Liebel et al. 2010; Girlanda et al. 2011). This is particularly interesting considering that this nutritional mode might be more frequent in green orchids than previously thought (Gebauer et al. 2016). There is a lot to be done in the study of Chilean matorral OrM, especially bearing in mind that most of Chilean orchids are endemic and insufficiently known or in some degree of threat (Novoa et al. 2015; Herrera et al. 2019); and mainly considering that knowledge on OrM is crucial for successful strategies in orchid conservation (Batty et al. 2002; Swarts and Dixon 2009).

14.6 Final Considerations

Mycorrhizal research in the Chilean matorral is evidently scarce. However, an emerging interest is arising from several researchers – mainly in AM, EcM and OrM. South America, in general, with their contrasting mycorrhizal patterns in comparison with the northern hemisphere, climatic conditions and other features have the potential for new, interesting discoveries (Bueno et al. 2017), and, of course, the South American MTE is not distant to this option.

AMF research in the Chilean MTE requires the increase of sampling efforts in all plant communities that constitute the Chilean matorral – with both complementary morphological and molecular approaches – not only from the soil compartment, but also from the direct observations of roots. Studies should be done in order to promote the conservation of AMF with their respective plant hosts. Also, more research is needed to encourage sustainable agriculture since most of the plants of productive interest have AM. The Chilean MTE is under high agriculture pressure thus, conservation and production should find an equilibrium; and, through AM symbiosis research, this aim could be reached.

Ectomycorrhizal research, is based mainly on fruiting bodies. Thus, here also direct morphological observation of roots is needed, as well as the use of molecular approaches in order to increase the knowledge – at least in terms of diversity.

Orchid mycorrhizal research has been focused on the fungal diversity associated with this symbiosis. However, more studies are needed in this respect since the orchids of the Chilean matorral are endemic and many are threatened; thus, OrM research will aid orchid conservation.

It is important to highlight the lack of studies in ErM not only for the Chilean Matorral, but also at a national level. Thus, an urgent call is made regarding this mycorrhizal type.

Mycorrhizal research in the Chilean MTE and, in general, in South America is emerging and filling basic knowledge gaps through ecological diversity studies. However, in the future, the integration of physiological studies in order to quantify the contribution of mycorrhizas to ecosystem processes will be necessary as well as the use of molecular approaches to understand the mechanism of the ecological patterns that we are finding. Lastly, it is worth mentioning that the Chilean matorral is a biodiversity hotspot; thus, all the mycorrhizal knowledge will be useful for conservation purposes, as well as the restoration of already degraded plant communities of this ecosystem that are constantly submitted to anthropic negative pressure.

Acknowledgements Patricia Silva-Flores was funded by the National Doctorate Grant N° 21140639 of CONICYT and CONICYT Regional/CEAF/R08I1001. P.S.F. also thanks the support of the Roberto Godoy regular FONDECYT 1190642. Ana Aguilar was funded by the National Doctorate Grant N° 21120047 and N° 81150505 of CONICYT and VI Scientific Research Fund of Pacific Hydro SA. A.A. also thanks the support of the regular postdoctoral 2018 grant of the Pontificia Universidad Católica de Valparaíso. María José Dibán was funded by Luis Felipe Hinojosa FONDECYT 1150690 and AFB170008. M.J.D also thanks to Dr. Götz Palfner, co-supervisor of Master Thesis, specifically in guiding taxonomic identification of some species. María Isabel Mujica thanks to CONICYT for the National Doctorate Grant N° 21151009.

References

Alcaras C (2010) Caracterización de una población de *Nothofagus macrocarpa* (A.DC.) Vasq. & Rodr., en sector Granizo del Parque Nacional La Campana. Trabajo de Titulación. Universidad Austral de Chile

Álvarez-Garrido L, Hortal S, Viñegla B, Carreira de la Fuente JA (2017) Plant-soil interactions as modulatory mechanism of adaptive capacity to global change in relict conifer forests. In: Abstract book of the XIV MEDECOS & XIII AEET meeting. Human driven scenarios for evolutionary and ecological changes. Asociación Española de Ecología Terrestre, Madrid, p 266

Arditti J, Ghani AKA (2000) Numerical and physical properties of orchid seeds and their biological implications. New Phytol 145:367–421. https://doi.org/10.1046/j.1469-8137.2000.00587.x

Arista A, Arroyo J, Berjano R, Jiménez-Lobato V, Jiménez-López J, Lloret F, López-Jurado J, Márquez-Corro JI, Olmedo-Vicente E, Rodríguez-Castaneda NL, Sánchez M, Simón-Porcar VI, Vilà M, Picó FX (2017) Present and future of ecological and evolutionary research in Mediterranean-type ecosystems: Conclusions from the last International Mediterranean Ecosystems Conference. Am J Bot 104:1777–1782. https://doi.org/10.3732/ajb.1700367

Armesto J, Arroyo M, Hinojosa L (2007) The Mediterranean environment of central Chile. In: Veblen TT, Young KR, Orme AR (eds) The physical geography of South America. Oxford University Press, Inc., New York, p 184–199

Batty AL, Dixon KW, Brundrett MC, Sivasithamparam K (2002) Orchid conservation and mycorrhizal associations. In: Sivasithamparam K, Dixon KW, Barrett RL (eds) Microorganisms in Plant Conservation and Biodiversity. Kluwer Academic Publishers, Dordrecht, p 195–226

Benedetti S, Balocchi F, Hormazábal M (2018) Arbuscular mycorrhizal fungi (AMF) linked to *Peumus boldus* natural formation in Central Chile. Gayana Bot 75:431–437

Benito Matías LF, Álvarez Lafuente A, Peñuelas JL (2017) *Tuber melanosporum* improves freezing tolerance in *Quercus faginea* and *Q. ilex* seedlings. In: Abstract book of the XIV MEDECOS & XIII AEET meeting. Human driven scenarios for evolutionary and ecological changes. Asociación Española de Ecología Terrestre, Madrid, p 266

Benoit IL (1989) Libro Rojo de la Flora Terrestre de Chile (Primera Parte). CONAF, Santiago de Chile

Bernhardt P (1995) Biogeography and Floral Evolution in the Geoblasteae (Orchidaceae). In: Arroyo MTK, Zedler PH, Fox MD (eds) Ecology and Biogeography of Mediterranean Ecosystems in Chile, California, and Australia. Springer, New York, p 116–134

Bonnardeaux Y, Brundrett M, Batty A, Dixon K, Koch J, Sivasithamparam K (2007) Diversity of mycorrhizal fungi of terrestrial orchids: compatibility webs, brief encounters, lasting relationships and alien invasions. Mycol Res 111:51–61. https://doi.org/10.1016/j.mycres.2006.11.006

Brundrett MC, Tedersoo L (2018) Evolutionary history of mycorrhizal symbioses and global host plant diversity. New Phytol. https://doi.org/10.1111/nph.14976

Bueno CG, Marín C, Silva-Flores P, Aguilera P, Godoy R (2017) Think globally, research locally: emerging opportunities for mycorrhizal research in South America. New Phytol 215:1306–1309. https://doi.org/10.1111/nph.14709

Bueno CG, Gerz M, Zobel M, Moora M (2018) Conceptual differences lead to divergent trait estimates in empirical and taxonomic approaches to plant mycorrhizal trait assignment. Mycorrhiza. https://doi.org/10.1007/s00572-018-0869-1

Calviño-Cancela M, Santolamazza S, Durán M, Neumann M (2017) On the ecological integration of *Eucalyptus globulus* in NW Spain: new pollination and mycorrhizal interactions with local birds and fungi. In: Abstract book of the XIV MEDECOS & XIII AEET meeting. Human driven scenarios for evolutionary and ecological changes. Asociación Española de Ecología Terrestre, Madrid, p 266

Cameron DD, Leake JR, Read DJ (2006) Mutualistic mycorrhiza in orchids: Evidence from plant-fungus carbon and nitrogen transfers in the green-leaved terrestrial orchid *Goodyera repens*. New Phytol 171:405–416. doi: https://doi.org/10.1111/j.1469-8137.2006.01767.x

Casanova-Katny MA, Torres-Mellado GA, Palfner G, Cavieres LA (2011) The best for the guest: high Andean nurse cushions of *Azorella madreporica* enhance arbuscular mycorrhizal status in associated plant species. Mycorrhiza 21:613–622. https://doi.org/10.1007/s00572-011-0367-1

Cowling RM, Rundel PW, Lamont BB, Arroyo MK, Arianoutsou M (1996) Plant diversity in mediterranean-climate regions. Trends Ecol Evol 11:362–366. https://doi.org/10.1016/0169-5347(96)10044-6

Curaqueo G, Acevedo E, Cornejo P, Seguel A, Rubio R, Borie F (2010) Tillage effect on soil organic matter, mycorrhizal hyphae and aggregates in a Mediterranean agroecosystem. Rev la Cienc del suelo y Nutr Veg 10:12–21. https://doi.org/10.4067/S0718-27912010000100002

Curaqueo G, Barea JM, Acevedo E, Rubio R, Cornejo P, Borie F (2011) Effects of different tillage system on arbuscular mycorrhizal fungal propagules and physical properties in a Mediterranean agroecosystem in central Chile. Soil Tillage Res 113:11–18. https://doi.org/10.1016/j.still.2011.02.004

Dallman PR (1998) Plant Life in the World's Mediterranean Climates: California, Chile, South Africa, Australia, and the Mediterranean Basin. University of California Press Books, Berkeley

Dearnaley JDW, Martos F, Selosse M (2012) Orchid Mycorrhizas: Molecular Ecology, Physiology, Evolution and Conservation Aspects. In: Fungal Associations. Springer Berlin Heidelberg, Berlin, Heidelberg, p 207–230

Dias T, Liberati D, Munzi S, Gouveia C, Ulm F, Afonso AC, Ochoa-Hueso R, Manrique E, Sheppard L, Martins-Loução MA, Bernardes da Silva A, Cruz C (2017) Ecological independence and self-reliance of the dominant plant species loosen ecosystem integration: evidence from 7 years' of changing patterns of nitrogen pulses in a Mediterranean Basin shrubland. In: Abstract book of the XIV MEDECOS & XIII AEET meeting. Human driven scenarios for evolutionary and ecological changes. Asociación Española de Ecología Terrestre, Madrid, p 266

Dressler R (1981) The orchids: natural history and classification. Harvard University Press, Cambridge

Fracchia S, Aranda-Rickert A, Flachsland E, Terada G, Sede S (2014a) Mycorrhizal compatibility and symbiotic reproduction of *Gavilea australis*, an endangered terrestrial orchid from south Patagonia. Mycorrhiza 24:627–634. https://doi.org/10.1007/s00572-014-0579-2

Fracchia S, Silvani V, Flachsland E, Terada G, Sede S (2014b) Symbiotic seed germination and protocorm development of *Aa achalensis* Schltr., a terrestrial orchid endemic from Argentina. Mycorrhiza 24:35–43. https://doi.org/10.1007/s00572-013-0510-2

Garrido N (1985) Index Agaricalium Chilensium. J. Cramer, Vaduz

Garrido N (1988) Agaricales s.l. und ihre Mykorrhizen in den Nothofagus-Wäldern Mittelchiles. Schweizerbart Science Publishers, Stuttgart, Germany

Gebauer G, Preiss K, Gebauer AC (2016) Partial mycoheterotrophy is more widespread among orchids than previously assumed. New Phytol 211:11–15. https://doi.org/10.1111/nph.13865

Gil-Martínez M, López-García Á, Navarro-Fernández CM, et al. (2017) Understanding feedback processes between holm oak (*Quercus ilex*) and their ectomycorrhizal fungal symbionts in trace-element polluted soils in Mediterranean ecosystems. In: Abstract book of the XIV MEDECOS & XIII AEET meeting. Human driven scenarios for evolutionary and ecological changes. Asociación Española de Ecología Terrestre, Madrid, p 266

Girlanda M, Segreto R, Cafasso D, Liebel HT, Rodda M, Ercole E, Cozzolino S, Gebauer G, Perotto S (2011) Photosynthetic Mediterranean meadow orchids feature partial mycoheterotrophy and specific mycorrhizal associations. Am J Bot 98:1148–1163. https://doi.org/10.3732/ajb.1000486

Givnish TJ, Spalink D, Ames M, Lyon SP, Hunter SJ, Zuluaga A, Doucette A, Caro GG, McDaniel J, Clements MA, Arroyo MK, Endara L, Kriebel R, Williams NH, Cameron KM (2016) Orchid historical biogeography, diversification, Antarctica and the paradox of orchid dispersal. J Biogeogr 43:1905–1916. https://doi.org/10.1111/jbi.12854

Harris J (2009) Soil Microbial Communities and Restoration Ecology: Facilitators or Followers? Science 325:573–574. https://doi.org/10.1126/science.1172975

Hernández-Rodríguez M, Mediavilla O, Oria-de-Rueda J, Martín-Pinto P (2017) Ecology of fungal communities after fire in Mediterranean systems dominated by *Cistus ladanifer* L. In: Abstract book of the XIV MEDECOS & XIII AEET meeting. Human driven scenarios for evolutionary and ecological changes. Asociación Española de Ecología Terrestre, Madrid, p 266

Herrera H, Valadares R, Contreras D, Bashan Y, Arriagada C (2017) Mycorrhizal compatibility and symbiotic seed germination of orchids from the Coastal Range and Andes in south central Chile. Mycorrhiza 27:175–188. https://doi.org/10.1007/s00572-016-0733-0

Herrera H, García-Romera I, Meneses C, Pereira G, Arriagada C (2019) Orchid Mycorrhizal Interactions on the Pacific Side of the Andes from Chile. A Review. J Soil Sci Plant Nutr. https://doi.org/10.1007/s42729-019-00026-x

Horak E (1980) Agaricales y gasteromicetes secotioides. In: Flora Criptogámica de Tierra del Fuego, tomo XI, fascículo 6: Fungi, Basidiomycetes. FECYC, Buenos Aires

Hynson N, Madsen T, Selosse M, Adam IKU, Ogura-Tsujita Y, Roy M, Gebauer G (2013) The physiological ecology of mycoheterotrophic plants. In: Merckx V (ed) Mycoheterotrophy: the biology of plants living on fungi. Springer, New York, p 297–342

Jacquemyn H, Brys R, Waud M, Busschaert P, Lievens B (2015) Mycorrhizal networks and coexistence in species-rich orchid communities. New Phytol 206:1127–1134. https://doi.org/10.1111/nph.13281

Jiang Y, Wang W, Xie Q, Liu N, Liu L, Wang D, Zhang X, Yang C, Chen X, Tang D, Wang E (2017) Plants transfer lipids to sustain colonization by mutualistic mycorrhizal and parasitic fungi. Science 356:1172–1173. https://doi.org/10.1126/science.aam9970

Johansson JF, Paul LR, Finlay RD (2004) Microbial interactions in the mycorrhizosphere and their significance for sustainable agriculture. FEMS Microbiol Ecol 48:1–13. https://doi.org/10.1016/j.femsec.2003.11.012

Keymer A, Pimprikar P, Wewer V, Huber C, Brands M, Bucerius SL, Delaux PM, Klingl V, von Röpenack-Lahaye E, Wang TL, Eisenreich W, Dörmann P, Parniske M, Gutjahr C (2017) Lipid transfer from plants to arbuscular mycorrhiza fungi. Elife 6:1–33. https://doi.org/10.7554/eLife.29107

Leake J (1994) The biology of myco-heterotrophic ('saprophytic') plants. New Phytol 127:171–216. https://doi.org/10.1111/j.1469-8137.1994.tb04272.x

Liebel HT, Bidartondo MI, Preiss K, Segreto R, Stöckel M, Rodda M, Gebauer G (2010) C and N stable isotope signatures reveal constraints to nutritional modes in orchids from the Mediterranean and Macaronesia. Am J Bot 97:903–12. https://doi.org/10.3732/ajb.0900354

López García A, Gil-Martínez M, Navarro-Fernández CM, Azcón-Aguilar C, Domínguez MT, Marañón T (2017) Assessment of the recovery of functional diversity of ectomycorrhizal fungal communities in metal polluted soils. In: Abstract book of the XIV MEDECOS & XIII AEET meeting. Human driven scenarios for evolutionary and ecological changes. Asociación Española de Ecología Terrestre, Madrid, p 266

Luginbuehl LH, Menard GN, Kurup S, Van Erp H, Radhakrishnan GV, Breakspear A, Oldroyd GED, Eastmond PJ (2017) Fatty acids in arbuscular mycorrhizal fungi are synthesized by the host plant. Science 356:1175–1178. https://doi.org/10.1126/science.aan0081

Marañón T, Domínguez M, Madejón P, Navarro-Fernández CM, Gil-Martínez M, López-García Á, Murillo JM (2017) Soil functioning and ecosystem services: using trees to remediate contaminated soils. In: Abstract book of the XIV MEDECOS & XIII AEET meeting. Human driven scenarios for evolutionary and ecological changes. Asociación Española de Ecología Terrestre, Madrid, p 266

Marín C, Aguilera P, Oehl F, Godoy R (2017) Factors affecting arbuscular mycorrhizal fungi of Chilean temperate rainforests. J soil Sci plant Nutr 17, 966–984. https://doi.org/10.4067/S0718-95162017000400010

Mediavilla O, Oria de Rueda JA, Martín-Pinto P (2017) Diversity of fungal communities after a wildfire in Mediterranean pine forest is linked to vegetation replacement. In: Abstract book of the XIV MEDECOS & XIII AEET meeting. Human driven scenarios for evolutionary and ecological changes. Asociación Española de Ecología Terrestre, Madrid, p 266

Moora M (2014) Mycorrhizal traits and plant communities: Perspectives for integration. J Veg Sci 25:1126–1132. https://doi.org/10.1111/jvs.12177
Moser M, Horak E (1975) Cortinarius und nahe verwandte Gattungen in Südamerika. Beihefte zur Nov Hedwigia 52:1–628
Mujica MI, Saez N, Cisternas M, Manzano M, Armesto JJ, Pérez F (2016) Relationship between soil nutrients and mycorrhizal associations of two *Bipinnula* species (Orchidaceae) from central Chile. Ann Bot 118:149–158. https://doi.org/10.1093/aob/mcw082
Myers N, Mittermeier RA, Mittermeier CG, da Fonseca GAB, Kent J (2000) Biodiversity hotspots for conservation priorities. Nature 403:853–858. https://doi.org/10.1038/35002501
Navarro-Fernández CM, Pérez-Ramos IM, G. de la Riva E, Vera J, Roumet C, Villar R, Marañón T (2017) A functional approach to explore the drivers of mycorrhizal trait variability in Mediterranean plant communities. In: Abstract book of the XIV MEDECOS & XIII AEET meeting. Human driven scenarios for evolutionary and ecological changes. Asociación Española de Ecología Terrestre, Madrid, p 266
Neuenkamp L, Moora M, Öpik M, Davison J, Gerz M, Männistö M, Jairus T, Vasar M, Zobel M (2018) The role of plant mycorrhizal type and status in modulating the relationship between plant and arbuscular mycorrhizal fungal communities. New Phytol https://doi.org/10.1111/nph.14995
Niveiro N, Albertó E (2012) Checklist of the Argentine Agaricales I. Amanitaceae, Pluteaceae and Hygrophoraceae. Mycotaxon 119:493–494. https://doi.org/10.5248/119.493
Niveiro N, Albertó E (2013) Checklist of the Argentinean Agaricales 6. Paxillaceae, Gomphidiaceae, Boletaceae and Russulaceae. Mycotaxon 123:1–8
Niveiro N, Albertó E (2014) Checklist of the Argentine Agaricales 7. Cortinariaceae and Entolomataceae. Check List 10:72–96
Novoa P, Espejo J, Alarcon D, Cisternas M, Domínguez E (2015) Guía de campo de las orquídeas chilenas. Corporación Chilena de la Madera, Concepción
Parker I, Grove S, Haubensak KA (2017) Biotic and abiotic impacts of invasive nitrogen-fixing shrubs in the Pacific Northwest of the United States. In: Abstract book of the XIV MEDECOS & XIII AEET meeting. Human driven scenarios for evolutionary and ecological changes. Asociación Española de Ecología Terrestre, Madrid, p 266
Pérez-Izquierdo L, Zabal-Aguirre M, González-Martínez S, Verdú M, Buée M, Rincón A (2017) Fire recurrence impacts the functioning and phylogenetic structure of fungal communities in Mediterranean pine forests. In: Abstract book of the XIV MEDECOS & XIII AEET meeting. Human driven scenarios for evolutionary and ecological changes. Asociación Española de Ecología Terrestre, Madrid, p 266
Rasmussen HN (2002) Recent developments in the study of orchid mycorrhiza. Plant Soil 244:149–163. https://doi.org/10.1023/A:1020246715436
Read DJ (1991) Mycorrhizas in ecosystems. Experientia 47: 376–391
Read DJ, Leake JR, Perez-Moreno J (2004): Mycorrhizal fungi as drivers of ecosystem processes in heathland and boreal forest biomes. Can J Bot 82: 1243–1263 https://doi.org/10.1139/B04-123
Read DJ, Perez-Moreno J (2003): Mycorrhizas and nutrient cycling in ecosystems - A journey towards relevance? New Phytol 157: 475–492. https://doi.org/10.1046/j.1469-8137.2003.00704.x
Rincón Herranz A, Zabal-Aguirre M, Flores-Renteria D, González-Martínez SC, Buée M, Pérez-Izquierdo L (2017) Structural and functional responses of fungal communities to biotic and abiotic factors in Mediterranean pine forests. In: Abstract book of the XIV MEDECOS & XIII AEET meeting. Human driven scenarios for evolutionary and ecological changes. Asociación Española de Ecología Terrestre, Madrid, p 266
Romano G, Greslebin A, Lechner BE (2017) Hongos agaricoides de los bosques de Nothofagus pumilio (Chubut, Argerntina): Clave y listado de especies. Rev del Mus Argentino Ciencias Nat Nueva Ser 19:36–69. https://doi.org/10.22179/REVMACN.18.495
Romano G, Lechner BE (2013) The Cortinariaceae of Argentina' s *Nothofagus* forests. Mycotaxon 126: 1–35. https://doi.org/10.5248/126.247

Romero Munar A, Gulías J, Baraza E (2017) Ecophysiological impact of arbuscular mycorrhiza inoculation on *Arundo donax*, a promising biomass crop. In: Abstract book of the XIV MEDECOS & XIII AEET meeting. Human driven scenarios for evolutionary and ecological changes. Asociación Española de Ecología Terrestre, Madrid, p 266

Rundel PW, Cowling RM (2013) Mediterranean-Climate Ecosystems. In: Levin S (ed) Encyclopedia of Biodiversity: Second Edition, Waltham, MA: Academic, p 212–222

Selosse MA, Roy M (2009) Green plants that feed on fungi: facts and questions about mixotrophy. Trends Plant Sci 14:64–70. https://doi.org/10.1016/j.tplants.2008.11.004

Silva-Flores P, Oses I, Almada R, Molina-Montenegro MM, Palfner G (submitted) Mycorrhizal type of dominant trees in the sclerophyllous shrubland of the Mediterranean Chilean Matorral. J Soil Sci Plant Nutr

Silva-Flores P, Bueno CG, Neira J, Palfner G (2019) Physico-chemical soil factors and seasonality regulate spore bank density of Arbuscular Mycorrhizal Fungi in two sclerophyllous shrublands of the Mediterranean Chilean matorral. J Soil Sci Plant Nutr. https://doi.org/10.1007/s42729-018-0004-6

Singer R (1969) Mycoflora Australis. Nov Hedwigia Beihefte 576

Smith SE, Read DJ (2008) Mycorrhizal symbiosis. Elsevier Ltd

Steinfort U, Verdugo G, Besoain X, Cisternas MA (2010) Mycorrhizal association and symbiotic germination of the terrestrial orchid *Bipinnula fimbriata* (Poepp.) Johnst (Orchidaceae). Flora Morphol Distrib Funct Ecol Plants 205:811–817. https://doi.org/10.1016/j.flora.2010.01.005

Swarts ND, Dixon KW (2009) Terrestrial orchid conservation in the age of extinction. Ann Bot 104:543–556. https://doi.org/10.1093/aob/mcp025

Tedersoo L, Jairus T, Horton BM, Abarenkov K, Suvi T, Saar I, Koljalg U (2008) Strong host preference of ectomycorrhizal fungi in a Tasmanian wet sclerophyll forest as revealed by DNA barcoding and taxon-specific primers. New Phytol 180:479–490. https://doi.org/10.1111/j.1469-8137.2008.02561.x

Torres-Mellado G, Escobar I, Palfner G, Casanova-Katny M (2012) Mycotrophy in Gilliesieae, a threatened and poorly known tribe of Alliaceae from central Chile. Rev Chil Hist Nat 85:179–186. https://doi.org/10.4067/S0716-078X2012000200004

Underwood EC, Viers JH, Klausmeyer KR, Cox RL, Shaw MR (2009) Threats and biodiversity in the mediterranean biome. Divers Distrib 15:188–197. https://doi.org/10.1111/j.1472-4642.2008.00518.x

van der Heijden MGA, Martin FM, Selosse MA, Sanders IR (2015) Mycorrhizal ecology and evolution: the past, the present, and the future. New Phytol 205:1406–1423. https://doi.org/10.1111/nph.13288

Verdú M (2017) Plant facilitation and phylogenetics. In: Abstract book of the XIV MEDECOS & XIII AEET meeting. Human driven scenarios for evolutionary and ecological changes. Asociación Española de Ecología Terrestre, Madrid, p 266

Chapter 15
Arbuscular Mycorrhizal Symbiosis in Salt-Tolerance Species and Halophytes Growing in Salt-Affected Soils of South America

Alejandra G. Becerra, M. Noelia Cofré, and Ileana García

15.1 Introduction

Land degradation is currently recognized as one of the most important environmental problems worldwide. It has been estimated that more than 7% of the arable land is salinized and it is expected to increase up due to global change and human activities (Ruiz-Lozano et al. 2012; FAO 2015). Soil salinity, one of the major environmental factors, reduces the growth, development, and productivity of plants (Tang et al. 2015; Kalaji et al. 2016).

Arbuscular mycorrhizal fungi (AMF) are ubiquitous among a wide array of soil microorganisms inhabiting the rhizosphere and are known to exist in saline environments (Giri et al. 2003; Wang et al. 2004; García and Mendoza 2007, 2008). Arbuscular mycorrhizal fungi establish a direct physical link between plant roots and soils facilitating to acquire mineral nutrients from soils under nutrient stress conditions modifying the environment of rhizosphere, alleviating the adverse effects of salinity stress (Smith and Read 1997; Jahromi et al. 2008; Evelin et al. 2009).

Plants growing in saline soils are subjected to different physiological stresses that induce nutrient imbalance, cell organelles damage, and photosynthesis and res-

A. G. Becerra (✉)
Laboratorio de Micología, IMBIV, CONICET, and Cátedra de Diversidad Biológica I, Facultad de Ciencias Exactas, Físicas y Naturales, Universidad Nacional de Córdoba, Córdoba, Argentina
e-mail: abecerra@unc.edu.ar

M. N. Cofré
Laboratorio de Micología, IMBIV, CONICET, Universidad Nacional de Córdoba, Córdoba, Argentina

I. García
Museo Argentino de Ciencias Naturales Bernardino Rivadavia (CONICET), Buenos Aires, Argentina

M. C. Pagano, M. A. Lugo (eds.), *Mycorrhizal Fungi in South America*, Fungal Biology, https://doi.org/10.1007/978-3-030-15228-4_15

piration disruption (Juniper and Abbott 1993; Evelin et al. 2013). On AMF symbiosis the adverse effects of salinity caused inhibition of spore germination (Hirrel 1981; Juniper and Abbott 2006), hyphal growth and development (McMillen et al. 1998), as well as reduced production of arbuscules (Pfeiffer and Bloss 1988). Although increased salinity reduces AMF colonization of plant roots, the dependency of plants on arbuscular mycorrhizas (AM) is increased, indicating the significance of AMF to alleviate salinity stress on plant growth (Tian et al. 2004; Evelin et al. 2009; Miransari 2010; Porcel et al. 2012). When plant is subjected to salinity stress, absorbs less P (Munns 1993), but can alleviate this stress using different mechanisms (Al-Karaki 2000, 2006; Al-Karaki et al. 2001; Tian et al. 2004). For example, the symbiosis with "resistant AMF species" enhances leaf respiration and transpiration, increasing the exchange of carbon dioxide and water through stomatal activity in plants.

The natural vegetation growing sparsely on saline soils are called halophytes. According to Flowers and Colmer (2008), halophytes are plants that can survive and reproduce in environments where the salt concentration exceeds 200 mM of NaCl (~20 dS m−1) and constitute approximately 1% of the world's flora. Halophytes complete their life cycle under highly saline conditions (Stuart et al. 2012) and they possess special morphological and anatomical features as well as physiological processes to cope with saline environments (Hasanuzzaman et al. 2014).

Under cultivation many halophytes grow and carry high productivity without the presence of any significant salinity named "optional" or "facultative" halophytes e.g. *Atriplex* spp., *Maireana* spp., *Tamarix* spp., *Salsola* spp., *Limonium* spp., *Puccinellia* spp., etc. (Le Houérou 1993). The "obligate" or "true" halophytes need saline conditions for normal growth e.g.: *Halocnemum*, *Arthrocnemum*, *Salsola* spp., *Salicornia* spp., *Suaeda* spp., *Sarcocornia* spp. Others need no salt concentration for growing, but do better with it. They are called "preferential" halophytes, e.g. *Atriplex*, some *Maireana*, some *Tamarix*. Other non-halophytic species, i.e. which are found in nature under both saline and non-saline conditions, may be fairly tolerant to salinity. This is the case of a number of conventional cultivated species and their wild relatives which are able to grow normally, under cultivation, with an electrical conductivity (EC) soil solution of 10–15 dS m-1 or slightly above. For example: *Beta vulgaris* (beet), *Gossypium* spp. (cotton), *Cynodon* spp. (bermuda grass), *Festuca arundinacea* (tall fescue), *Agropyron elongatum* (tall wheat grass), *Medicago sativa* (alfalfa), *Medicago* spp., *Melilotus* spp. (sweet clovers), *Lotus* spp. (trefoils), *Trifolium resupinatum* (persian clover), *T. fragiferum* (strawberry clover), etc. It is known that AMF and their association with halophytes may improve plant tolerance to drought and salt (Smith and Read 2008).

Increased salinization of arable land is expected to have devastated global effects (Ruan et al. 2010; Shabala 2013). The use of halophytic plants in fodder production constitutes a useful practice in order to restore the vegetation of salt-affected areas (Yeo and Flowers 1980; O'Leary 1988). This chapter show the mycorrhizal status of halophytes species and salt-tolerance species used as a forage source for livestock in South America. Specially we made a focus in AMF associated with members of Chenopodiaceae in Salinas Grandes (Jujuy and Córdoba province) and with

Lotus tenuis (one of the most economically important naturalized legumes) in Argentinean Pampas (Buenos Aires province) of Argentina.

15.2 Soil Salinity, Halophytes, Salt Tolerance Feedstuffs and Arbuscular Mycorrhizas

Soil salinity is the salt content in the soil solution. The process of increasing the salt content is known as salinization, being one of the most important agricultural and eco-environmental problems increasing in many parts of the world (Evelin et al. 2009; Porcel et al. 2012). It can be caused due to natural salinization or topsoil salinization by human activities (land use changes and overgrazing) (Taboada et al. 2011; Bandera 2013; Di Bella et al. 2015). Examples caused by natural processes are the banks of water bodies, where fluctuating water levels and saline water evaporation over time leads to the formation of saline soil patches. By human activities, the irrigational practices for agricultural lands supplemented through saline groundwater sources.

A soil containing excess of salts impairing its productivity is called a salt-affected soil (SAS). The SAS are found mainly in the arid and semi-arid regions and can be divided into saline, saline-sodic and sodic, depending in salt amounts, type of salts, amount of sodium present and soil alkalinity. Each type of SAS will have different characteristics, which will also determine the way they can be managed. The soil map of the world (FAO-UNESCO 1974) estimated that the total area for saline soils was 397 million ha and for sodic soils was 434 million ha. The 7.6% soils in South America are salt affected (Table 15.1) (FAO-UNESCO 1971; Dudal and Purnell 1986).

In extremely saline and semi-desert environments halophytes and salt tolerant plants can growth being an essential resource in the future. Their strategies to successively grow and develop in marginal lands will be of great value for the nutrition of animals which live in those environments (El Shaer 2010).

Table 15.1 Salt-affected areas in South America (Source: Szabolcs 1979)

		Area 1000 ha		
Continent	Country	Saline/Solonchaks	Sodic/Solonetz	Total
South America	Argentina	32,473	53,139	85,612
	Bolivia	5233	716	5949
	Brazil	4141	362	4503
	Chile	5000	3642	8642
	Colombia	907	–	907
	Ecuador	387	–	387
	Paraguay	20,008	1894	21,902
	Perú	21	–	21
	Venezuela	1240	–	1240

The mycorrhizal status of many halophytes is controversial, as they belong to families like Caryophyllaceae, Chenopodiaceae and Plumbaginaceae which are frequently reported as being non-mycorrhizal (Harley and Harley 1987; Wang and Qiu 2006). Nevertheless, it has been shown that halophytes like Asteraceae and Plantaginaceae families can be intensively colonized by AMF (Harley and Harley 1987; Carvalho et al. 2001; Hildebrandt et al. 2001; Landwehr et al. 2002), protecting plants against the detrimental effects of water deficiency (Smith and Read 1997; Augé 2001; Ruiz-Lozano 2003), and alleviating salt stress symptoms (Ruiz-Lozano and Azcón 2000; Cantrell and Linderman 2001; Sharifi et al. 2007; Jahromi et al. 2008). Data on forage halophytic vegetation and its distribution in South America has been considered by Brevedan et al. (1994). In Table 15.2 we show a list of the halophytes and salt-tolerance feedstouffs mention in Brevedan et al. (2016). We included their mycorrhizal status based on local literature. Knowledge of their strategies to successively grow and develop in marginal lands will be of great value for the nutrition of animals which live in those environments (El Shaer 2010).

15.3 Mycorrhizal Symbiosis of Some Halophytes in Argentina

Salinization caused by irrigation affects 18.4 million ha in Latin America and Caribbean, particularly in Argentina, Brazil, Chile, Mexico and Perú (AQUASTAT 1997; FAO 2015). Around 85 million ha are affected by excess of salts and sodium in Argentina, and approximately 600,000 ha of irrigated soils are affected by salinity, which is the third largest area in a single country after Russia and Australia (Szabolcs 1979; Bandera 2013). In Argentina, SAS are found mainly in the arid and semi-arid regions but there are also areas naturally affected by salts in humid and sub humid climates, where salts come from groundwater. These areas are located in humid and sub-humid climates in the grasslands from the Argentinean Pampas and "Bajos Submeridionales" in Central Argentina (Lavado and Taboada 1988; Morras and Candioti 1982).

Adverse environmental conditions can negatively affect the infectivity and survival of AMF propagules (Juniper and Abbott 1993). However, AMF could survive in soil and the roots of some forage species tolerant to saline-sodic soils (Escudero and Mendoza 2005; García and Mendoza 2007, 2008). In this section of the chapter we focus in the relationship between AMF and salt tolerant feedstuffs (*Lotus* spp.) and halophytes forage shrubs of Chenopodiaceae family growing in Argentinean Pampas and "Salinas Grandes" salt flats ("Salinas" or "Salares") respectively.

Table 15.2 List of the salt tolerant and halophytes plant species cited in Brevedan et al. (2016) as valuable for grazing or browsing in South America with their geographic distribution and arbuscular mycorrhizal status (P: present, NP: not present, Nd: not determined). Ar: Argentina, Am Trop: América Tropical, Par: Paraguay

Salt tolerance/Halophytes plant species	Geographic distribution	Mycorrhizal status/ References[a]
Family Poaceae		
Agropyron scabrifolium	Ar	Nd
Agropyron elongatum	Ar	P (1)
Aristida mendocina	Ar	P (2)
Bothricloa saccharoides	Ar	Nd
Bothricloa lagonoides	Ar	Nd
Chloris gayana	Ar	P (3)
Chloris canterae	Ar	Nd
Chloris ciliata	Ar	Nd
Chloris halophila	Ar	Nd
Cenchrus ciliaris	Ar, Brazil	P (4, 5)
Diplacne uninerva	Ar	Nd
Distichlis scoparia	Ar	P (6)
Distichlis spicata	Ar, Perú, Bolivia	P (7a, 8, 9)
Distichlis australis	Ar, Perú	Nd
Distichlis humilis	Perú, Bolivia	Nd
Elymus scabrifolius	Ar	Nd
Leptochloa cloridiformis	Ar	Nd
Muhlenbergia fastigata	Perú	Nd
Panicum coloratum	Ar	P (9)
Panicum urvilleanum	Ar	Nd
Paspalum spp.	Ar	P (8, 10)
Pappophorum caespitosum	Ar	NP (2)
Pappophorum philippianum	Ar	Nd
Spartina alterniflora	Ar	NP (11)
Spartina densiflora	Ar	P (11)
Stenotaphrum secundatum	Am trop, Ar	P (8)
Family Fabaceae		
Lotus tenuis	Ar	P (8, 12)
Melilotus albus	Ar	P (13)
Melilotus officinalis	Ar	Nd
Sporobolus indicus	Ar	Nd
Sporobolus phleoides	Ar	NP (2)
Trichloris crinita	Ar	P (14)
Trichloris pluriflora	Ar	Nd
Prosopis spp.	Ar, Par, Chile, Brazil	P (3, 15)
Family Chenopodiaceae		
Atriplex spp.	Ar, Chile, Brazil	P, NP (2, 7b, 16,17,18)

(continued)

Table 15.2 (continued)

Salt tolerance/Halophytes plant species	Geographic distribution	Mycorrhizal status/ References[a]
Salicornia spp.	Ar, Perú, Bolivia	Nd
Suaeda spp.	Ar, Perú, Bolivia	Nd
Kochia spp.	Ar, Bolivia, Perú	NP (19)
Family Rosaceae		
Polylepis besseri	Chile	P (20)
Polylepis tarapacana	Chile	Nd
Family Cactaceae		
Opuntia ficus-indica	Chile, Brazil	P (5)
Opuntia cochenillifera	Brazil	Nd
Family Caryophyllales		
Sarcocornia neei	Perú	Nd
Sarcocornia perennis	Ar	Nd

[a]References: (1) Cavagnaro et al. (2014); (2) Lugo et al. (2015); (3) Mijaluk et al. (2011); (4) Di Bárbaro et al. (2018); (5) da Silva Sousa et al. (2013); (6) Pagano et al. (2011); (7) Fontenla et al. (2001a, 2001b); (8) García and Mendoza (2008); (9) Schwab et al. (2016); (10) Grigera and Oesterheld (2004); (11) Daleo et al. (2008); (12) Escudero and Mendoza (2005); (13) Hack et al. (2009); (14) Lugo et al. (2005); (15) Fracchia et al. (2009); (16) Cofré et al. (2012); (17) Soteras et al. (2012); (18) Becerra et al. (2014); (19) Schalamuk et al. (2015); (20) Hensen (1994)

15.3.1 Argentinean Pampas

In central-east of Argentina there is an extensive wetland in the sub humid portion of the Pampean plain, the Flooding Pampa (90,000 km^2) (Sala 1988), which soils affected by salts and sodium are qualified as Natraquoll and Natraqualf (US Soil Taxonomy, INTA 1977). Under such restrictive soil conditions, *Lotus tenuis* is the only legume that grows and fully colonized the flooding pampas of Argentina (Nieva et al. 2018).

Lotus tenuis is a perennial herbaceous legume appreciated by farmers due to the ability to grow in nutrient-deficient soils, and for its nutritional forage value (Mazzanti et al. 1988). Plants of *L. tenuis* seem to be tolerant of flooded conditions which commonly occur in autumn, winter and part of the spring in the Pampas (Vignolio et al. 1996, 1999). *Lotus tenuis* is highly dependent on AMF root colonization by grow at low phosphorus availability (Mendoza and Pagani 1997). In different field sites this legume presents a high colonized percentage of root length (Mendoza et al. 2000; Escudero and Mendoza 2005; García and Mendoza 2008).

Escudero and Mendoza (2005) studied over 2 years the seasonal variation in the composition of AMF communities in the rhizospheric soil from *L. tenuis* at four temperate grassland sites in the flooding Pampas. The sites represent a wide range of soil conditions (two were sodic soils, one was saline-sodic soil and the last one non-saline), hydrologic gradients, and floristic composition. At all four samples sites *Rhizophagus fasciculatum* and *R. intraradices* dominated the AMF spore communities. Spore density was highest in summer (dry season) and lowest in winter (wet season). The relative density of *R. fasciculatum* and *R. intraradices* versus two other AMF species, *Glomus* sp. and *Acaulospora* sp. had distinctive seasonal peaks. These seasonal peaks occurred at all four sites, suggesting differences among AMF species with respect to the seasonality of sporulation. This relationship occurred in spite of the broad differences among the four sites, and may suggest niche differentiation among AMF species with respect to the seasonality of sporulation. In this work, the results show the influence of many plant community and soil variables on AMF community rather than the influence of one specific dominant plant species or soil property. Consequently, there is not a clear separation between plant and soil factors on AMF sporulation in these temperate grasslands with different levels of salinity or sodicity in soil (Escudero and Mendoza 2005).

Recently, García et al. (2017) showed that AMF families and AMF colonization are a good indicator to differentiate sites by their soil characteristics (pH and Na exchangeable), and/or management (e.g. herbicide application) in the Argentinean Pampas. The AMF community was described in the rhizosphere of *L. tenuis* in halomorphic soils under different management conditions. The environments selected differ in land use: natural grasslands or *L. tenuis* promotion with glyphosate application. The soil sites present different levels of salinity and sodicity, and an increased relative frequency of *L. tenuis* in sites with glyphosate-mediated promotion. A total of twenty-two species of AMF were identified in the rhizospheric soil from *L. tenuis* at the different sites. The AMF species were grouped into seven families; four of them present the highest spore density: Glomeraceae, Claroideoglomeraceae, Acaulosporaceae and Diversisporaceae. Glomeraceae was the only family present in all sites, and particularly *Funneliformis mosseae*, was the only species described at all environments, independently of soil properties and management. Spore density data (from 2 to 26.80 spores/100 g dry soil) at the family level, soil properties and *L. tenuis* relative frequency showed that Claroideoglomeraceae spores were associated with an increase in pH and Na exchangeable, and a decrease in *L. tenuis* frequency (saline-sodic soils in natural grassland). The spores from Acaulosporaceae and Glomeraceae were associated with high *L. tenuis* frequency and a decrease in pH and Na exchangeable (non saline-non sodic soils in *L. tenuis* promotion). The Diversisporaceae family is associated with non saline-sodic soil (in *L. tenuis* promotion). Respect to AMF colonization, this fungal parameter showed a positive correlation (increased) with pH and exchangeable Na (salinity and sodicity) in grasslands, and a negative correlation (decreases) in sites with *L. tenuis* promotion. Based on these results, the AMF community in sites with *L. tenuis* promotion could be less effective in the establishment and the subsequent maintenance on root colonization. This can be explained due to the viability of the AMF spores in soil affected by the herbicide application as reported by Druille et al. (2015). The concomitant loss of plant diversity produced by the herbicide application can also affect the AMF community. Although each fungal community is adapted to the soil conditions, the different edaphic properties do not necessarily modify the proportion of root colonized by arbuscules in *L. tenuis* plants. These are agreed with previous observations on the dynamics of AMF colonization in *L. tenuis* roots in a spatial-temporal study along a hydrologic, saline and sodic gradient in grasslands of the Argentinean Pampas (García and Mendoza 2008).

Following the studies in temperate natural grasslands from the Argentinean Pampas, the seasonally dynamics of AMF root colonization, propagules and plant tissue nutrients were studied in three forage plants growing in a saline-sodic gradient (García and Mendoza 2008). The plant species were *L. tenuis*, and the grasses *Paspalum vaginatum* and *Stenotaphrum secundatum*. Soils and plant samples were collected in four sites seasonally across a topographic and saline gradient: sites 1 and 2 were on a typic Natraqualf soil (lowland sites) and sites 3 and 4 were on a typical Natraquoll soil (upland sites) (INTA-CIRN 1990) (Fig.15.1a, b). The plant communities were dominated by: *L. tenuis*, *P. vaginatum*, *Distichlis spicata*, *Eleocharis viridans* and *Cynodon dactylon* (site 1); and by *L. tenuis* (Fig.15.1c), *Bromus unioloides*, *S. secundatum* and *L. multiflorum* (site 4). The studied plant species presents a similar morphology of AMF root colonization in the adverse soil conditions. The AMF colonization was higher in *L. tenuis* than in the grass roots at all sites and seasons. The overall mean values of AMF colonization over sites and season for *L. tenuis* was 89% and for grasses was 68%. This fungal parameter was positively associated with soil water content, salinity and sodicity in *L. tenuis*, but negatively in grasses. The high levels of root colonization suggest that either the plants respond with slow root growth, the fungi colonize roots more completely or the interaction enables considerable root colonization (García and Mendoza 2008).

Fig. 15.1 The natural grassland from Buenos Aires province, (**a**) Site 1 (typic Natraqualf, the wettest site), (**b**) Site 4 (typic Natraquoll, the driest site), (**c**) *Lotus tenuis* growing in natural saline soils, (**d**) Arbuscular colonization in *L. tenuis* roots

A canonical correspondence analyses (CCA) diagram was performed by the CANOCO algorithm (Ter Braak 1987–1992) to identify the best linear combinations of soil chemical properties and concentrations of N and P in shoot and root tissues that influence AMF measurement. A positive association between EC and exchangeable Na in spring in lowland sites (sites 1 and 2) and total AM colonization and arbuscular colonization in *L. tenuis* roots (Fig. 15.1d), but a negative association between these indexes in the grass's roots (Fig. 15.2a, b). The values of total root colonization and arbuscular colonization in lowland sites (sites 1 and 2) were lower compared with the other upland sites (sites 3 and 4), suggesting that EC and exchangeable Na may affect the symbiosis more in the grasses than in the legume even when the symbiosis is functional for any combination between plant and site (Fig. 15.2a, b).

Despite the differences between the legume and the grasses, spore density in soil and AMF colonization morphology in plants vary seasonally. These seasonal effects on AM fungal variables were independent of a particular combination between plant species and soil sites, suggesting that seasonality is an important factor in regulate both spore density in soil and changes in AMF root colonization morphology in different studied plants along the saline and sodic gradient in a temperate grassland of the Argentinean Pampas (García and Mendoza 2008). AM fungi can survive and colonize plant roots, adapting to extreme saline-sodic soil conditions imposed by the environment (García and Mendoza 2007).

15.3.2 Northwestern and Central of Argentina

In northwestern of Argentina is placed the Salinas Grandes in Jujuy province at 4000 meters above sea levels with 8200 km^2 which soils are classified as typic Torripsamment (Paoli et al. 2009). The site is characterized by a medium shrub steppe dispersed and mostly growth in isolated groups, with a relatively developed herbaceous and pastures layer. Central Argentina presents some conspicuous salt flats: the Salinas Grandes and the Salinas de Ambargasta, which together occupies an area of approximately 6000 km^2. Soils of Salinas de Ambargasta and Salinas Grandes are classified as Aridisol-Orthid typic Salorthids (INTA 2003). Within these saline habitats, the distribution patterns of plant communities are defined by the salt gradient, with plant cover inversely proportional to the presence of salt. At sites where plant life is still possible, the most characteristic plant community is the halophytic shrub or "jumeal", composed of species of the Chenopodiaceae family (Cabido and Zak 1999).

The family Chenopodiaceae is represented in arid and halophytic plant communities worldwide. This family probably includes the largest number of halophytic members in comparison with other plant families. Even species of Chenopodiaceae are generally considered non-mycorrhizal (Gerdemann 1968; Hirrel et al. 1978; Mohankumar and Mahadevan 1987; Peterson et al. 1985), several species of *Atriplex* (*A. nummularia*, *A. canescens*, *A. confertifolia*, *A. gardneri*, *A. polycarpa*, *A.*

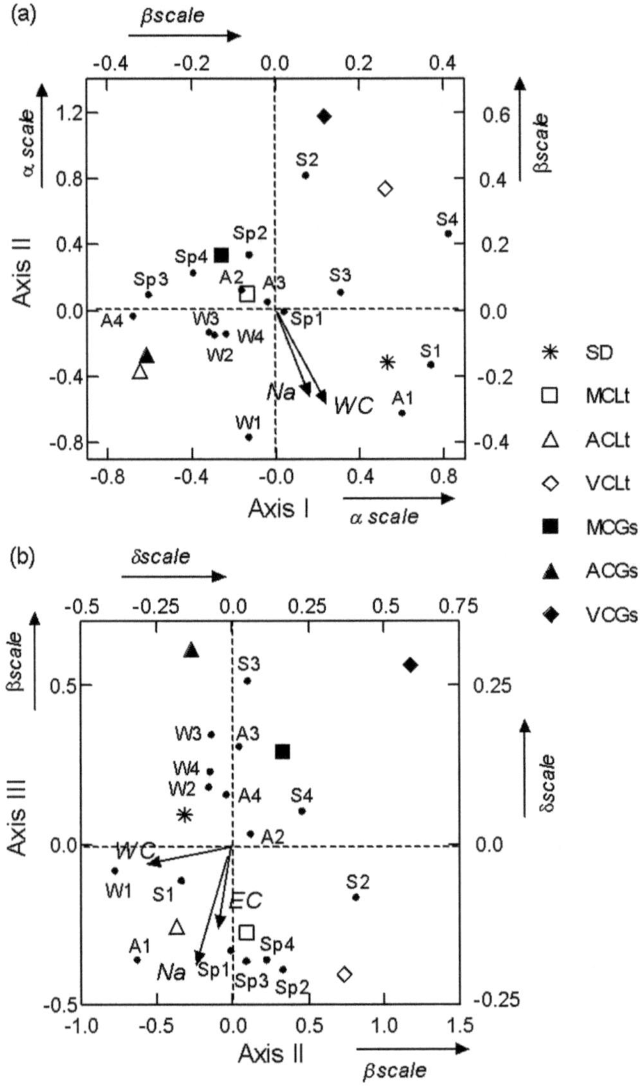

Fig. 15.2 Ordination diagram from the Canonical Correspondence Analyses (CCA) of seasonal and spatial observations based on: (**a**) AM fungal variables and (**b**) soil properties. References: *L. tenuis* (Lt), grasses (Gs), MC (total colonized root), AC (arbuscular colonization), VC (vesicle colonization), SD (spore density), Na (exchangeable sodium), WC (water content), EC (electrical conductivity), S (summer), A (autumn), W (winter) and Sp (spring). Number following each season indicates site (García and Mendoza 2008)

versicaria, *A. spinosa*, *A. lampa* and *A. argentina*), *Salicornia* sp. and *Suaeda maritima* can be colonized (Allen 1983; Kim and Weber 1985; Rozema et al. 1986; Sengupta and Chaudhuri 1990, Allen and Allen 1990; Plenchette and Duponnois 2005; Cofré et al. 2007; Soteras et al. 2009).

Fig. 15.3 Salares in the Northwestern and Central Argentina: (**a**) Salinas Grandes in Jujuy province (SGJ), (**b**) Salinas Grandes in Córdoba province (SGC), (**c**) *Atriplex cordobensis* growing in natural saline soils, (**d**) Arbuscular mycorrhizal colonization in *A. cordobensis* roots

In Argentina, *Atriplex cordobensis* (Fig.15.3c) is an endemic shrub that produce an adequate biomass for livestock all year (Aiazzi et al. 1999; Abril et al. 2000). As the mycorrhizal status of *A. cordobensis* is unknown, the presence of AMF, root colonization and AMF spore taxa were evaluated at two different salt flats ("Salinas" or "Salares"): Salinas Grandes in the Northwestern Argentina (SGJ) and Salinas Grandes in Central Argentina (SGC) (Fig. 15.3a, b) (Cofré et al. 2012). The percentage of AMF colonization ranged from 0 to 99% (Fig. 15.3d) and the AMF spore number ranged from 8.7 to 969.5 spores/100 g soil. Both fungal parameters differed between these Salares. Nine morphologically distinctive AMF species were recovered (Table 15.3, Fig. 15.4) and *Funneliformis geosporum* was the most frequent and abundant species in all sites. As stated by Bothe (2012) this AMF plays a specific role in conferring salt tolerance to plants.

Atriplex lampa ("zampa") is a shrub that constitutes a good fodder resource all year, being palatable for sheep and goats even in critical periods of drought or scarcity of food (Passera and Borsetto 1989). Their mycorrhizal status has been reported in the steppe (*sensu stricto*) and marshes, in north-west Patagonia of Argentina (Fontenla et al. 2001b). In two Salares of Central Argentina such as Salinas de Ambargasta (SA) and Salinas Grandes (SG) in Córdoba Province, Soteras et al. (2012, 2013) evaluated the vertical distribution of AMF spore and the mycorrhizal colonization in two seasons. Twenty AMF taxa were recovered in the rhizosphere, being 13 identified to species level (Table 15.3, Fig. 15.4). *Atriplex lampa* roots exhibited AMF colonization, although not arbuscules were observed. Arbuscular

Table 15.3 List of AMF morphospecies from the rhizosphere of Chenopodiaceae native tolerant plants growing in saline soils of Argentina. SGJ: Salinas Grandes de Jujuy (Northwestern Argentina), SGC: Salinas Grandes de Córdoba and SA: Salinas de Ambargasta (both Salares in Central Argentina)

Family	AMF Species	*A. cordobensis*		*A. lampa*		*A. patagonica*		*A. argentina*		*H. ritteriana*		*S. divaricata*	
		SGJ	SGC	SG	SA	SG	SA	SG	SA	SG	SA	SG	SA
Acaulosporaceae	*A. bireticulata*	*	*	*	*					*			
	A. scrobiculata	*	*	*			*	*	*	*	*	*	*
Archaeosporaceae	*Ar. undulata*			*	*		*	*		*		*	
Ambisporaceae	*Am. leptoticha*	*			*					*			
Claroideoglomeraceae	*C. claroideum*									*			
	C. etunicatum	*	*	*	*	*	*	*		*	*	*	*
	C. luteum	*	*	*	*	*	*	*	*	*	*	*	*
Diversisporaceae	*D. spurca*			*	*	*	*	*	*	*	*	*	
Glomeraceae	*F. geosporum*	*	*	*	*	*	*	*	*	*	*	*	*
	F. mosseae	*	*	*	*	*	*	*	*	*	*	*	*
	G. brohultii	*	*	*	*	*	*	*	*	*	*	*	*
	G. magnicaule			*	*	*	*			*	*	*	
	R. clarus	*	*	*	*	*	*	*	*	*	*	*	*
	R. intraradices							*		*	*	*	*
	S. aff. *constrictum*			*	*	*	*	*	*	*	*	*	*

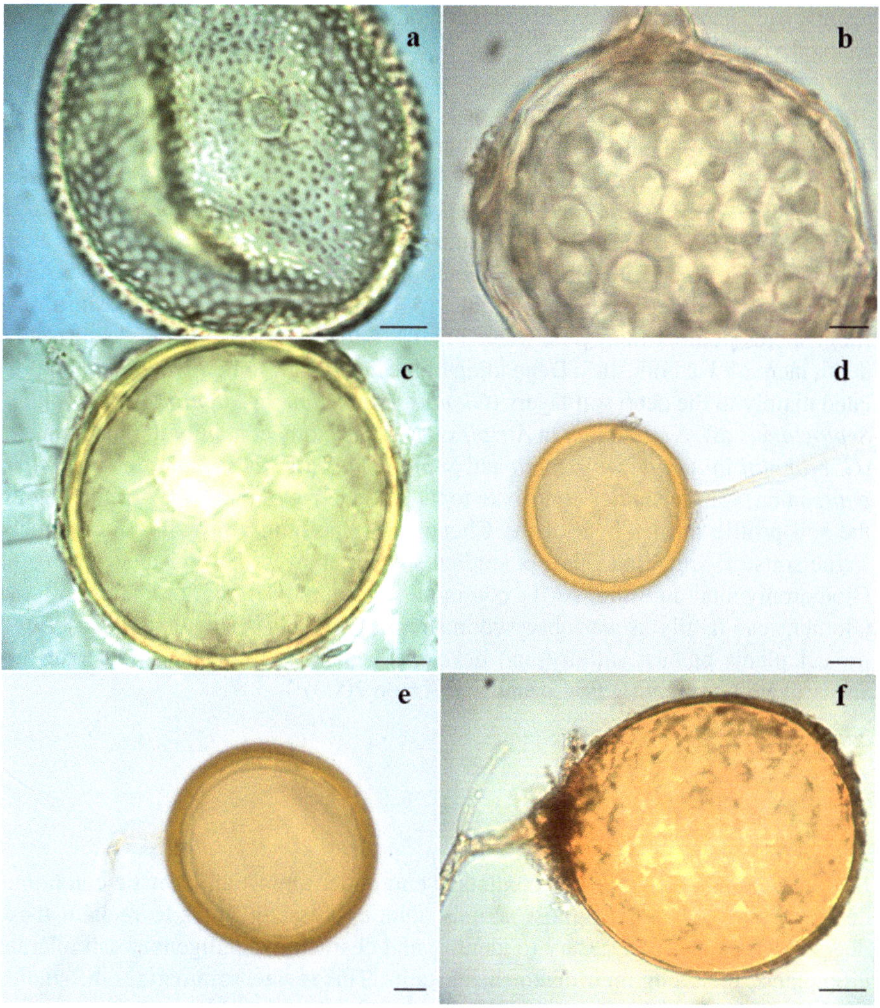

Fig. 15.4 Some of AMF morphospecies identified in saline soils of Northwestern and Central Argentina, (**a**) *Acaulospora scrobiculata*, (**b**) *Ambispora leptoticha*, (**c**) *Claroideoglomus claroideum*, (**d**) *C. etunicatum*, (**e**) *Funneliformis geosporum*, (**f**) *F. mosseae*. Scale bar a-f: 20 μm

mycorrhizal colonization varied from very low to very high (4.76–82.35%) although no significant differences were observed between seasons or soil depths. The number of AMF spores ranged between 5 and 1418/100 g dry soil with no significant differences between soil depth. The relative spore abundance in relation to the total spore number showed differences with soil depth and seasons. For *Rhizophagus clarus* during wet season in SA ranged from 0–30%, for *Septoglomus constrictum* during dry season in SG ranged from 40–60% with increasing soil depth (0–50 cm), while *Glomus brohultti* showed a decreasing relative spore abundance (20–0% for SA; 40–20% for SG) with increasing soil depth during wet season. Based on these results, and as was stated by Carvalho et al. (2001) *Glomus* spp. are the AMF more adapted to stressful conditions.

Regarding other Chenopodiaceae species (*Allenrolfea patagonica, Atriplex argentina, Heterostachys ritteriana* and *Suaeda divaricata*) Becerra et al. (2014, 2016) reported for the first time the mycorrhizal status and their vertical distribution of AMF spores in SA and SG in Córdoba Province. Low arbuscular mycorrhizal colonization was found in the studied species: 5–31% in *A. patagonica*, 2–37% in *S. divaricata*, 0–45% in *H. ritteriana* and 4–50% in *A. argentina* for both sites at all soil depths (from 0–50 cm). From a total of 19 morphologically distinctive AMF species, 15 identified to species (Table 15.3, Fig. 15.4). Arbuscular mycorrhizal fungi spores number ranged between 3 and 1162/ 100 g dry soil, and decreased as depth increased at both sites. Depending of the host plant, some AMF species sporulated mainly in the deep soil layers (*Glomus magnicaule* in *Allenrolfea patagonica, Septoglomus* aff. *constrictum* in *Atriplex argentina*), others mainly in the top layers (*G. brohultti* in *Atriplex argentina* and *Septoglomus* aff. *constrictum* in *Allenrolfea patagonica*). These studies contribute to the knowledge of the AMF diversity along the soil profile in members of the Chenopodiaceae family growing in extremely saline soils of Argentina. The six studied species showed the typical structures of Glomeromycota and the AMF community seemed to be dominated by the Glomeraceae family as was observed in stressful habitats (Wang et al. 2004). AMF protect plants against salinity and developed adaptive strategies to tolerate this stressful environment (Ruiz-Lozano and Azcón 2000).

15.4 Conclusions

Salt-affected soils occur in all continents and under almost all climatic conditions and salinity is one of the most serious abiotic stress. In order to reclaim these degraded zones, it is necessary to identify and characterize indigenous salt tolerant crop plants, including their mycorrhizal status. This review summarizes the studies carried out in South America providing information about the mycorrhizal status with salt-tolerance feedstuffs and halophytes species capable of growing in saline and/or sodic soils, used as a forage resource for livestock. Know the mycorrhizal status of these plants is a promising field and needs to be addressed in future studies.

On the other hand, the analyzed studies on *Lotus tenuis* a perennial herbaceous legume appreciated by farmers for its nutritional forage value, reveals that seasonality is an important factor in regulate both AMF spore density in soil and AMF root colonization morphology along the saline and sodic gradient. Members of Chenopodiaceae's family present AMF colonization in their roots and spores on their rhizosphere. This symbiotic relationship helps in alleviating abiotic stresses in the plant. Both in Argentinean Pampas and in Northwestern and Central of Argentina, Glomeraceae's was the dominant family observed in stressful habitats. The knowledge on halophytic and salt-tolerance plants in South America constitute a useful practice in order to restore the vegetation of salt-affected areas.

Acknowledgments This work was financially supported by Secretaría de Ciencia y Técnica - Universidad Nacional de Córdoba, Agencia de Promoción Científica y Tecnológica (PICT 438-2006) and CONICET (PIP0950). All authors are researchers from CONICET.

References

Abril A, Aiazzi M, Torres P, Argüello J (2000) Nutritional value of *Atriplex cordobensis* grown in dry Chaco of Argentina. Rev Arg Prod An 20(3–4):179–185

Aiazzi M, Abril A, Torres P, Di Rienzo J, Argüello J (1999) Seasonal variations in chemical composition of leaves and stems of *Atriplex cordobensis* (Gandoger et Stuckert), female and male plants. Phyton 65:173–178

Al-Karaki GN (2000) Growth of mycorrhizal tomato and mineral acquisition under salt stress. Mycorrhiza 10:51–54

Al-Karaki GN (2006) Nursery inoculation of tomato with arbuscular mycorrhizal fungi and subsequent performance under irrigation with saline water. Sci Hort 109:1–7

Al-Karaki GN, Hammad R, Rusan M (2001) Response of two tomato cultivars differing in salt tolerance to inoculation with mycorrhizal fungi under salt stress. Mycorrhiza 11:43–47

Allen MF (1983) Formation of vesicular-arbuscular mycorrhizae in *Atriplex gardneri* (Chenopodiaceae): seasonal response in a cold desert. Mycologia 75:773–776

Allen MF, Allen EB (1990). Carbon source of VA mycorrhizal fungi associated with Chenopodiaceae from a semiarid shrub steppe. Ecology 71:2019–2021

AQUASTAT (1997) Tablas resumen para America Latina y el Caribe. FAO. http://www.fao.org/GEO-2-199

Augé RM (2001) Water relations, drought and vesicular-arbuscular mycorrhizal symbiosis. Mycorrhiza 11:3–42

Bandera R (2013) Rehabilitación de suelos salino-sódicos: evaluación de enmiendas y de especies forrajeras. Tesis presentada para optar al título de Magíster de la Universidad de Buenos Aires, Área Recursos Naturales, p 66

Becerra A, Bartoloni J, Cofré N, Soteras F, Cabello M (2014) Arbuscular mycorrhizal fungi in saline soils: vertical distribution at different soil depths. Braz J Microbiol 45:585–594

Becerra A, Bartolini N, Cofré N, Soteras F, Cabello M (2016) Hongos micorrícico-arbusculares asociados a ambientes salinos de Córdoba. Bol Soc Arg Bot 51(1):5–13

Bothe H (2012) Arbuscular mycorrhiza and salt tolerance of plants. Symbiosis 58:1–3

Brevedan RE, Fernández OA, Villamil CB (1994) Halophytes as a resource for livestock husbandry in South America. In: Squires VR, Ayoub AT (eds) Halophytes as a Resource for Livestock and for Rehabilitation of Degraded Lands. Kluwer Acad Publ, Dordrecht, p 175–199

Brevedan RE, Fernández OA, Fioretti M, Baioni S, Busso CA, Laborde H (2016) Halophytes and Salt Tolerant Crops as a Forage Source for Livestock in South America. In: El Shaer HM, Squires VR (eds) Halophytic and Salt-Tolerant Feedstuffs Impacts on Nutrition, Physiology and Reproduction of Livestock. CRC Press. Taylor & Francis Group, Boca Ratón, p 60–78

Cabido M, Zak M (1999). Vegetación del Norte de Córdoba, Secretaría de Agricultura, Ganadería y Recursos Renovables de Córdoba, Córdoba, Argentina

Cantrell IC, Linderman RG (2001) Preinoculation of lettuce and onion with VA mycorrhizal fungi reduces deleterious effects of soil salinity. Plant Soil 233:269–281

Carvalho LM, Cacados I, Martiris-Loucao MA (2001) Temporal and spatial variation of arbuscular mycorrhizas in salt marsh plants of Tagus estuary (Portugal). Mycorrhiza 11:303–309

Cavagnaro RA, Oyarzabal M, Oesterheld M, Grimoldi AA (2014) Dinámica de crecimiento de gramíneas C_3 y C_4 en respuesta a las micorrizas y a la disponibilidad de fósforo. Lilloa 51 (Supl): 164
Cofré MN, Becerra AG, Domínguez LS (2007) Micorrizas en *Atriplex argentina*, una especie nativa forrajera de las Salinas Grandes de Córdoba. In: III Jornadas Nacionales de Flora Nativa y IV Encuentro de Cactáceas. Córdoba, Argentina. ISBN 978-987.510-079-4, p 209–217
Cofré N, Becerra A, Nouhra E, Soteras F (2012) Arbuscular mycorrhizae and dark-septate endophytes on *Atriplex cordobensis* in saline sites from Argentina. J Agric Tech 87(7):2201–2214
da Silva Sousa C, Rômulo Simões CM, Valadares de Sá Barreto Sampaio E, de Sousa Lima F, Fritz O, Costa Maia L (2013) Arbuscular mycorrhizal fungi within agroforestry and traditional land use systems in semi-arid Northeast Brazil. Acta Scientiarum 35(3):307–314
Daleo P, Alberti J, Canepuccia A, Escapa M, Fanjul E, Silliman BR, Bertness MD, Iribarne O (2008) Mycorrhizal fungi determine salt-marsh plant zonation depending on nutrient supply. J Ecol 96:431–437
Di Bárbaro G, Espeche E, Manenti L, Rizo M, Andrada H, Viale S, Batallán Morales S (2018) Asociaciones micorrícicas entre hongos nativos y plantas forrajeras cultivadas en el valle central de Catamarca. 1° Jornadas de divulgación científica y técnica Facultad de Ciencias Agrarias Universidad Nacional de Catamarca, ISBN 978-987-661-299-9, p 19–20
Di Bella CE, Rodríguez AM, Jacobo E, Golluscio RA, Taboada MA (2015) Impact of cattle grazing on temperate coastal salt marsh soils. Soil Use Manag 31:299–307
Druille M, Cabello MN, García Parisi PA, Golluscio RA, Omacini M (2015) Glyphosate vulnerability explains changes in root-symbionts propagules viability in pampean grasslands. Agric Ecos Env 202:48–55
Dudal R, Purnell MF (1986) Land Resources: salt affected soils. Reclamation and Revegetation Res 5:1–9
El Shaer H (2010) Halophytes and salt-tolerant plants as potential forage for ruminants in the Near East region. Small Rum Res 91:3–12
Escudero VG, Mendoza RE (2005) Seasonal variation of arbuscular mycorrhizal fungi in temperate grasslands along a wide hydrologic gradient. Mycorrhiza 15:291–299
Evelin H, Kapoor R, Giri B (2009) Arbuscular mycorrhizal fungi in alleviation of salt stress: a review. Ann Bot 104:1263–1280
Evelin H, Giri B, Kapoor R (2013) Ultrastructural evidence for AMF mediated salt stress mitigation in *Trigonella foenum-graecum*. Mycorrhiza 23:71–86
FAO (2015) Status of the World's report resources. Main Report, p 364–398
FAO-UNESCO (1974) Soil Map f the World. 1: 5 000 000. Volume I. Legend. United Nations Educational, Scientific and Cultural Organization, Paris, pp 59
FAO-UNESCO (1971) Soil Map f the World. 1: 5 000 000. Volume IV South America. United Nations Educational, Scientific and Cultural Organization, Paris, pp 193
Flowers TJ, Colmer TD (2008) Salinity tolerance in halophytes. New Phytol 179:945–963
Fontenla S, Puntieri J, Ocampo JA (2001a) Mycorrhizal associations in the Patagonian steppe, Argentina. Plant Soil 233:13–29
Fontenla S, Chaia E, Bustos C, Pelliza A (2001b) Microorganismos simbióticos en *Atriplex*. XXVIII Jornadas Argentinas de Botánica, La Pampa, Argentina. Bol Soc Arg Bot 36:114
Fracchia S, Aranda A, Gopar A, Silvani V, Fernandez L, Godeas A (2009) Mycorrhizal status of plant species in the Chaco Serrano Woodland from central Argentina. Mycorrhiza 19:205–214
García I, Mendoza RE (2007) Arbuscular mycorrhizal fungi and plant symbiosis in a saline-sodic soil. Mycorrhiza 17:167–174
García I, Mendoza RE (2008) Relationships among soil properties, plant nutrition and arbuscular mycorrhizal fungi-plant symbioses in a temperate grassland along hydrologic, saline and sodic gradients. FEMS Microbiol Ecol 63:359–71
García I, Cabello M, Fernández-López C, Chippano T, Mendoza R (2017) Hongos micorrícicos arbusculares en asociación con *Lotus tenuis* en ambientes halomórficos de la Cuenca del

río Salado. CONEBIOS V Ecología y Biología de Suelos. ISBN 978-987-3941-39-9. Buenos Aires. Argentina

Gerdemann JW (1968) Vesicular arbuscular mycorrhiza and plant growth. Ann Rev Phytopatol 6:397–418

Giri B, Kapoor R, Mukerji KG (2003) Influence of arbuscular mycorrhizal fungi and salinity on growth, biomass, and mineral nutrition of *Acacia auriculiformis*. Biol Fertil Soils 38:170–175

Grigera G, Oesterheld M (2004) Mycorrhizal colonization patterns under contrasting grazing and topographic conditions in flooding Pampa (Argentina). Rang Ecol Manag 57:601–605

Hack CM, Porta M, Tomei CE, Grimoldi AA (2009) Inoculación con hongos micorrícicos arbusculares y fertilización fosfatada en *Melilotus alba*. Comunicaciones Científicas y Tecnológicas. Universidad Nacional del Nordeste. CA-009

Harley JL, Harley EL (1987) A check-list of mycorrhiza in the British flora. New Phytol 105:1–102

Hasanuzzaman M, Nahar K, Alam MM, Bhowmik PC, Hossain MA, Rahman MM, Prasad MNV, Ozturk M, Fujita M (2014) Potential use of halophytes to remediate saline soils. BioMed Res Internat, Article ID 589341, 12 pages

Hensen I (1994) Estudios ecológicos y fenológicos sobre *Polylepis besseri* Hieron en la Cordillera Oriental Boliviana. Ecología Bolivia 23:21–32

Hildebrandt U, Janetta K, Fouad O, Renne B, Nawrath K, Bothe H (2001) Arbuscular mycorrhizal colonization of halophytes in Central European salt marshes. Mycorrhiza 10:175–183

Hirrel MC (1981) The effect of sodium and chloride salts on the germination of *Gigaspora margarita*. Mycology 43:610–617

Hirrel MC, Mehravaran H, Gerdemann JW (1978) Vesicular arbuscular mycorrhizae in the Chenopodiaceae and Cruciferae: do they occur? Can J Bot 56:2813–2817

INTA (1977) La Pampa Deprimida. Condiciones de drenaje de sus suelos. INTA. Departamento de Suelos. Publicación No. 154, Serie suelos, p 162, Buenos Aires

INTA-CIRN (1990) Atlas de suelos de la República Argentina. Escala 1:500.000, 1: 1.000.000. Secretaría de Agricultura, Ganadería y Pesca. Proyecto PNUD ARG 85/109, Buenos Aires

INTA (2003) Recursos Naturales de La Provincia de Córdoba: Los Suelos. Nivel de Reconocimiento 1:500.000. Secretaria de Ambiente de Córdoba, Argentina, Córdoba

Jahromi F, Aroca R, Porcel R, Ruiz-Lozano JM (2008) Influence of salinity on the in vitro development of *Glomus intraradices* and on the in vivo physiological and molecular responses of mycorrhizal lettuce plants. Microbial Ecol 55:45–53

Juniper S, Abbott L (1993) Vesicular–arbuscular mycorrhizas and soil salinity. Mycorrhiza4:45–57

Juniper S, Abbott L (2006) Soil salinity delays germination and limits growth of hyphae from propagules of arbuscular mycorrhizal fungi. Mycorrhiza16:371–379

Kalaji HM, Jajoo A, Oukarroum A, Brestic, M, Zivcak, M, Samborska A, Cetner MD, Łukasik I, Goltsev V, Ladle RJ (2016) Chlorophyll a fluorescence as a tool to monitor physiological status of plants under abiotic stress conditions. Acta Physiol Plant 38:102

Kim C-K, Weber DJ (1985) Distribution of VA mycorrhiza on halophytes on inland salt playas. Plant Soil 83:207–214

Landwehr M, Hildebrandt U, Wilde P, Nawrath K, Toth T, Biro B, Bothe H (2002) The arbuscular mycorrhizal fungus *Glomus geosporum* in European saline, sodic and gypsum soils. Mycorrhiza 12:199–211

Lavado RS, Taboada MA (1988) Water, salt and sodium dynamics in a natraquoll in Argentina. Catena 15:577–594

Le Houérou HN (1993) Salt-tolerant plants for the arid regions of the Mediterranean isoclimatic zone. In: Lieth H, Al Masoon AA (eds) Towards the rational use of high salinity tolerant plants, Vol. 1. Kluwer Academic Publisher, Dordrecht, The Netherlands, p 403–422

Lugo MA, Anton AM, Cabello MN (2005). Arbuscular mycorrhizas in the *Larrea divaricata* scrubland of the arid "Chaco", Central Argentina. Journal Agric Tech 1:163–178

Lugo MA, Reinhart KO, Menoyo E, Crespo EM, Urcelay C (2015) Plant functional traits and phylogenetic relatedness explain variation in associations with root fungal endophytes in an extreme arid environment. Mycorrhiza 25:85–95

Mazzanti A, Montes L, Minon D, Sarlangue H, Chepi C (1988) Utilización de *Lotus tenuis* en la Pampa Deprimida: resultado de una encuesta. Rev Agric Prod Anim 8:301–305

McMillen BG, Juniper S, Abbott LK (1998) Inhibition of hyphal growth of a vesicular–arbuscular mycorrhizal fungus in soil containing sodium chloride limits the spread of infection from spores. Soil Biol Biochem 30:1639–1646

Mendoza R, Pagani E (1997) Influence of phosphorus nutrition on mycorrhizal growth response and morphology of mycorrhizae in *Lotus tenuis*. J Plant Nutr 20:625–639

Mendoza R, Pagani E, Pomar MC (2000) Variabilidad poblacional de *Lotus glaber* en relación con la absorción de fósforo en suelo. Ecol Austral 10:3–14

Mijaluk A, Brandán de Weht C, García Paulucci D (2011) Micorrizas arbusculares en arbóreas nativas y en gramíneas en un sistema silvopastoril del chaco húmedo, Argentina. II Jornada sobre Ciencias del Suelo del NOA para estudiantes y jóvenes profesionales. Universidad Nacional de Tucumán, p 24

Miransari M (2010) Contribution of arbuscular mycorrhizal symbiosis to plant growth under different types of soil stress. Plant Biol 12:563–569

Mohankumar V, Mahadevan A (1987) Vesicular-arbuscular mycorrhizal association in plants of Kalakad reserve forest, India. Angew Bot 61:255–274

Morras H, Candioti L (1982) Relación entre permeabilidad, ciertos caracteres analíticos y situación topográfica de algunos suelos de los bajos submeridionales (Santa Fé). Rev Invest Agro 26:23–32

Munns R (1993) Physiological processes limiting plant growth in saline soils: some dogmas and hypotheses. Plant Cell Environ 16:15–24

Nieva AS, Bailleres MA, Llames ME, Taboada MA, Ruiz OA, Menendez A (2018) Promotion of *Lotus tenuis* in the Flooding Pampa (Argentina) increases the soil fungal diversity. Fungal Ecol 33:80–91

O'Leary JO (1988) Saline environments and halophytic crops. In: Arid land, today and tomorrow. West view Press, Boulder, Colorado, p 773–790

Pagano MC, Lugo MA, Araújo FS, Ferrero MA, Menoyo E, Steinaker D (2011) Native species for restoration and conservation of biodiversity in South America. In: Marín L, Kovač D (eds) Native Species: Identification, Conservation and Restoration. Nova Science Publishers, New York, p 1–55

Paoli HP, Volante JN, Noe YE, Vale LM, Castrillo S, Osinaga R, Chafatinos T, Nadir A (2009) Adecuación a un sistema de información geográfica del estudio "Los Suelos del NOA (Salta y Jujuy), Nadir A, Chafatinos T, 1990". Convenio INTA-UNSa. Salta: Ediciones INTA. ISBN 978-987-25050-8-0

Passera CB, Borsetto O (1989) Aspectos ecológicos de *Atriplex lampa*. Invest Agrar Prod Prot Veg 4:179–198

Peterson RL, Ashford AE, Allaway WG (1985) Vesicular-arbuscular mycorrhizal association of vascular plants on Heron Island, a great barrier reef coral ray. Aust J Bot 33:669–676

Pfeiffer CM, Bloss HE (1988) Growth and nutrition of guayule (*Parthenium argentatum*) in a saline soil as influenced by vesicular–arbuscular mycorrhiza and phosphorus fertilization. New Phytol 108:315–321

Plenchette C, Duponnois R (2005) Growth response of the saltbush *Atriplex nummularia* L. to inoculation with the arbuscular mycorrhizal fungus *Glomus intraradices*. J Arid Environ 61:535–540

Porcel P, Aroca R, Ruiz-Lozano JM (2012) Salinity stress alleviation using arbuscular mycorrhizal fungi. A review. Agron Sustain Dev 32:181–200

Rozema J, Arp W, van Diggelen J, van Esbroek M, Broekmann R, Punte H (1986) Occurrence and ecological significance of vesicular arbuscular mycorrhiza in the salt marsh environment. Acta Bot Neerlandica 35:457–467

Ruan C-J, Teixeira da Silva JA, Mopper S, Qin P, Lutts S (2010) Halophyte improvement for a salinized world. Crit Rev Plant Sci 29:329–359

Ruiz-Lozano JM (2003) Arbuscular mycorrhizal symbiosis and alleviation of osmotic stress. New perspectives for molecular studies. Mycorrhiza 13:309–317

Ruiz-Lozano JM, Azcón R (2000) Symbiotic efficiency and infectivity of an autochthonous arbuscular mycorrhizal *Glomus* sp. from saline soils and *G. deserticola* under salinity. Mycorrhiza 10:137–143

Ruiz-Lozano JM, Porcel R, Azcón R, Aroca R (2012) Regulation by arbuscular mycorrhizae of the integrated physiological response to salinity in plants: new challenges in physiological and molecular studies. J Exp Bot 63:4033–4044

Sala OE (1988) The effect of herbivory on vegetation structure. In: Werger MJA, van der Aart PJM, During HJ, Verboeven JTA (eds) Plant form and vegetation structure. SPB Academic Publishing, The Hague, The Netherlands, p 317–330

Schalamuk S, Druille M, Cabello M (2015) Arbuscular mycorrhizal fungi: Influence of agronomic practices on diversity and dynamics of colonization. In: García de Salomone I, Vázquez S, Penna C, Cassán F (eds) Rizosfera, biodiversidad y agricultura sustentable. Asoc Argent Microbiol, p 47–71

Schwab M, Di Bella CE, Casas C, Clavijo MP, Druille M, Lattanzi FA, Schaufele R, Grimoldi AA (2016) Evaluación de la incorporación de *Panicum coloratum* como fitorremediadora de suelos sódicos de la Pampa Deprimida. 39° Congreso de la Asociación Argentina de Producción Animal, PP2

Sengupta A, Chaudhuri S (1990) Vesicular arbuscular mycorrhiza (VAM) in pioneer salt marsh plants of the Ganges River delta in West Bengal (India). Plant Soil 122:111–113

Shabala S (2013) Learning from halophytes: physiological basis and strategies to improve abiotic stress tolerance in crops. Ann Bot 112:1209–1221

Sharifi M, Ghorbanli M, Ebrahimzadeh H (2007) Improved growth of salinity-stressed soybean after inoculation with pre-treated mycorrhizal fungi. J Plant Physiol 164:1144–1151

Smith SE, Read DJ (1997) Mycorrhizal symbiosis. Academic Press, San Diego

Smith SE, Read DJ (2008) Mycorrhizal symbiosis. Academic Press, London

Soteras F, Becerra A, Cofré N, Nouhra E (2009). Variación estacional de la colonización micorrícica de *Atriplex lampa* (Moq.) D. Dietr. en dos salinas de Córdoba. XXXII Jornadas Argentinas de Botánica. Bol Soc Argent Bot Córdoba, Argentina, p 130

Soteras F, Becerra A, Cofré N, Bartoloni, Cabello M (2012) Arbuscular mycorrhizal fungal species in saline environments of Central Argentina: seasonal variation and distribution of spores at different soil depth. Sydowia 64:301–311

Soteras F, Cofré N, Bartoloni J, Cabello M, Becerra A (2013) Colonización radical de *Atriplex lampa* en dos ambientes salinos de Córdoba, Argentina. Bol Soc Arg Bot 48:211–219

Stuart JR, Tester M, Gaxiola RA, Flowers TJ (2012) Plants of saline environments in AccessScience, ©McGraw-Hill Companies, Pennsylvania

Szabolcs I (1979) Review of Research on salt-affected soils. United Nations of Educational, Scientific and Cultural Organizations. Paris, ISBN 92-3-101613-X

Taboada MA, Rubio G, Chaneton EJ (2011) Grazing impacts on soil physical, chemical and ecological properties in forage production systems. In: Hatfield JL, Sauer TJ (eds) Soil management: building a stable base for agriculture. American Society of Agronomy & Soil Science Society of America, p 301–320

Tang ZS, An H, Shangguan ZP (2015) The impact of desertification on carbon and nitrogen storage in the desert steppe ecosystem. Ecol Eng 84:92–99

Ter Braak CJF (1987–1992) CANOCO—a FORTRAN program for canonical community ordination. Microcomputer Power, Ithaca, NY

Tian CY, Feng G, Li XL, Zhang FS (2004) Different effects of arbuscular mycorrhizal fungal isolates from saline or non-saline soil on salinity tolerance of plants. App Soil Ecol 26:143–148

Vignolio O, Fernández O, Maceira N (1996) Respuestas de *Lotus tenuis* y *Lotus corniculatus* (Leguminosae) al anegamiento en plantas de distintas edades. Rev Fac Agron La Plata 101:57–66

Vignolio O, Fernández O, Maceira N (1999) Flooding tolerance in five populations of *Lotus glaber* Mill. (Syn. *Lotus tenuis* Waldst. Et. Kit.). Aust J Agric Res 50:555–559

Wang B, Qiu YL (2006) Phylogenetic distribution and evolution of mycorrhizas in land plants. Mycorrhiza 16:299–363

Wang FY, Liu RJ, Lin XG, Zhou JM (2004) Arbuscular mycorrhizal status of wild plants in saline-alkaline soils of the Yellow River Delta. Mycorrhiza 14:133–137

Yeo AR, Flowers TJ (1980) Salt tolerance in the halophyte *Suaeda maritima* (L.) Dum.: evaluation of the effect of salinity upon growth. J Experiment Bot 31:1171–1183

Chapter 16
Mycorrhizal Studies in Temperate Rainforests of Southern Chile

Roberto Godoy and César Marín

16.1 Introduction

Old-growth temperate rainforests located in the mountain areas of the Southern Cone of America are often presented as global model ecosystems, as they have not been subject to chronic air pollution and have remained foristically stable throughout the Holocene (Armesto et al. 2009, 2010). In Chile, Andes and Coastal mountain ranges differ in terms of precipitation (Godoy and Oyarzún 1998; Godoy et al. 1999, 2001, 2003, 2009; Oyarzún et al. 1998, 2002, 2004, 2007, 2009, 2011; Staelens et al. 2003, 2005, 2009), and in the input of long-distance transported aerosols (Boy et al. 2014). These forests can be considered as unique, isolated biogeographic islands, as they have flora with representatives derived from Gondwanian elements, and extreme environmental, edaphic, and orographic conditions that are enhanced by seismic and volcanic activity. The Chilean Coastal mountain range served as a refugium for plants during the Last Glacial Maximum (Armesto et al. 2009), causing this area to have a high plant family endemism and a high number of isolated monotypic genera. The Coastal mountain range bedrock is highly weathered, and atmospheric nutrients coming from ocean processes have a significant influence on the biogeochemical dynamics of these forests (Boy et al. 2014). In contrast, nutrient inputs to the steep slopes of the Andes mountain range are mostly generated by young volcanic ash deposits and weathered basaltic volcanic scoria (Godoy et al. 2009).

The forests of the Chilean Coastal mountain range have developed in unique evolutionary and biogeochemical scenarios, where soil nutritional limitations and

R. Godoy (✉)
Universidad Austral de Chile, Campus Isla Teja s/n, Valdivia, Chile
e-mail: rgodoy@uach.cl

C. Marín
Universidad de O'Higgins, Av. Libertador Bernardo O Higgins 611, Rancagua, Chile

M. C. Pagano, M. A. Lugo (eds.), *Mycorrhizal Fungi in South America*,
Fungal Biology, https://doi.org/10.1007/978-3-030-15228-4_16

the dilution of marine salt aerosols have been prevalent (Armesto et al. 2009). The close connection between marine and terrestrial nutrient cycling in these forests has greatly shaped their ecosystem functioning (Hedin and Hetherington 1996; Weathers et al. 2000; Kennedy et al. 2002). Nutrient limitations make these coastal ecosystems extraordinarily sensitive to changes on biogeochemical cycles as a result of anthropogenic disturbances. In both mountain ranges there is a bedrock age gradient and therefore, a gradient of total weathering (Hedin and Hetherington 1996). The atmospheric and edaphic inputs of the Andes and Coastal mountain ranges are contrasting, thus allowing comparisons to be made at the micro-catchment scale (Oyarzún et al. 1998, 2004).

A regimen of natural disturbances maintains the population dynamics of southern Chile temperate rainforests (Godoy et al. 2009; Lara et al. 2014), which are altered by anthropogenic disturbances, resulting in genetically fragmented forest communities (Bekessy et al. 2002, 2004). Since 2005 Chile has experienced an extreme drought with a drastic drop in precipitation, causing an increased intensity and frequency of fires (Bowman et al. 2018). During the summer of 2017, fires affected 5000 km^2 of the Chilean central-southern region, affecting densely inhabited and important productive regions, as well as causing restricted access to several national parks in remote areas (Bowman et al. 2018).

16.2 Overview of Mycorrhizal Studies on Chilean Temperate Rainforests

The mycorrhizal symbiosis is one of the most common forms of mutualistic relationships, with crucial ecological and evolutionary roles on the terrestrial colonization of vascular plants (Brundrett and Tedersoo 2018). Moreover, about 92% of terrestrial plant species associate with mycorrhizal fungi (Brundrett and Tedersoo 2018). Mycorrhizal fungi improve plant survival and nutrient acquisition -mainly phosphorus and nitrogen- by creating large mycelial networks that access to both mobile and immobile forms of soil nutrients (Simard et al. 2012). Furthermore, mycorrhizal fungi influence several ecosystem processes such as plant productivity and biodiversity, soil aggregation, and carbon cycling (van der Heijden et al. 2008).

The first mycorrhizal studies on Chilean temperate rainforests classified the *Nothofagus* spp. forests as ectotrophic, and the native conifer forests as anectotrophic (Singer and Morello 1960; Singer et al. 1965; Singer 1969, 1970). Several morphoanatomical classification studies followed (Godoy and Mayr 1989; Carrillo et al. 1992; Godoy et al. 1994; Valenzuela et al. 1999, 2001), registering as many as 651 ectomycorrhizal (EM) fungi taxa exclusive to *Nothofagus* spp. (Garrido 1988), and concluding that the most abundant EM fungal orders on *Nothofagus* forests are: Boletales, Cortinariales, Gautieriales, and Russulales (Palfner and Godoy 1996a, b; Flores et al. 1997; Godoy and Palfner 1997; Palfner 2001; Nouhra et al. 2013).

Southern Chilean temperate rainforests are unique in that arbuscular mycorrhizal (AM) fungi associate with native conifers, as most of the flora, with the important exception of *Nothofagaceae* species, which are exclusively associated with EM fungi (Godoy et al. 1994; Fontenla et al. 1998; Palfner 2001, 2002; Castillo et al. 2006; Marín et al. 2016, 2017a, b, 2018a). Marín et al. (2016) registered 18 AM fungal species in three *N. pumilio* plots, which brought the number of AM fungal species described in Chile from 57 to 59 (Marín et al. 2017a). The vascular and fungal flora of southern Chile' temperate rainforests share the same climatic, geological, and evolutionary history, and the fungal flora is also characterized by a high endemism and a high number of monotypic families and genera (Palfner 2001; Marín et al. 2018b). According to descriptions and collections of EM fungi, the *Nothofagus* spp. forests of this region have a high diversity of Agaricales when compared with European *Fagus* forests (Garrido 1988; Valenzuela et al. 1999; Palfner 2001, 2002; Marín et al. 2017b). Molecular studies of soil fungi, particularly mycorrhizal fungi, are very recent in Argentinian and Chilean temperate rainforests (Nouhra et al. 2012, 2013; Tedersoo et al. 2014; Davison et al. 2015; Trierveiler-Pereira et al. 2015; Marín et al. 2017b; Truong et al. 2017, 2019).

These studies have found fungi new to science as well as pointed out vast understudied regions. Recent metagenomic studies in Chile examined soil fungi across the Andean and Coastal ranges, comparing *Nothofagus* spp. and native conifer forests, finding an inverse relationship between EM and saprotrophic fungal abundance (Marín 2018a). Another metagenomic study found more EM and saprotrophic fungi on less disturbed forests while more plant parasitic fungi were found in more disturbed forests (Marín 2017b). The survival of these forests is highly dependent on its mycorrhizal symbionts (Godoy et al. 1994; Marín et al. 2018a).

On the temperate rainforests of southern Chile, AM fungi make a significant contribution to the carbon and nitrogen cycling of soil organic matter (Etcheverría et al. 2009). This reinforces the imperative need to study the biodiversity, community composition, ecosystem roles, and eco-evolutionary parameters of the mycorrhizal symbiosis on temperate rainforests of the Southern Cone of America.

16.3 Mycorrhizal Types on Southern Chile Temperate Rainforests

We analyzed the mycorrhizal type of plant species across 17 temperate rainforest plots (30 m × 30 m) on southern Chile, conducting the plant identification at the Herbarium of Universidad de Concepción, Chile, and after the species list provided by Rodriguez et al. (2018). The mycorrhizal type was determined by analysis of the mycorrhizal colonization of roots (i.e. fixation, root staining, and microscope quantification) (Koske and Gemma 1989). Five composite soil samples from the Ah horizon were collected and thoroughly mixed (litter and organic material removed; 0–20 cm depth, aprox. 1 kg each sample). Following Sadzawka et al. (2006), analyses included soil properties known to affect plant and fungal mycorrhizal communities: pH (KCl), conductivity, total C, total N, C/N ratio, available P (Olsen P at pH 8.5), exchangeable K, Ca, and Mg (extraction with CH_3COONH_4 1 mol/L at pH 7.0), and exchangeable Al (extracted with KCl 1/mol L).

From a total of 245 vascular plant species distributed on 17 temperate rainforest plots on southern Chile, we found that 208 species (85%) have mycorrhizal associations (Table 16.1; Fig. 16.1). A total of 187 plant species associated with AM fungi, 10 plant species with ericoid (ER) mycorrhizal fungi, seven plant species with EM fungi, and four plant species associated with orchid (OR) mycorrhizal fungi. A total of 37 plant species did not form any mycorrhizal association and are considered in the literature as typical non-mycorrhizal (NM) plants, such as epiphytic ferns and broadleaf herbs, parasitic plants (Loranthaceae and Misodendraceae), and species from the families Proteaceae, Caryophyllaceae, Cyperaceae, Juncaceae, and Brassicaceae (Fig. 16.1).

From the 31 species of Pteridophytic flora found, 23 (77%) present AM symbiosis, a similar result to other latitudes (Godoy et al. 1994). The eight NM fern species were mainly epiphytic plants belonging to the family Hymenophyllaceae, showing a similar result to a study on north Patagonian forests by Fernández et al. (2005). All the native conifer species of the region associate to AM fungi, and belong to the families Araucariaceae (one species), Cupressaceae (three species), and Podocarpaceae (three species), representing an exceptional association in comparison to the Northern hemisphere conifers (i.e. Pinaceae, Taxodiaceae), that predominately associate with EM fungi (Godoy and Mayr 1989). An aspect of particular interest is the formation of root nodules in Podocarpaceae and Araucariaceae with full AM colonization (Godoy and Mayr 1989). The populations of these endemic conifers are narrow-distributed and endangered. Particularly, conifers with long life-spans as *Fitzroya cupressoides* (3600 years; Lara and Villalba 1993) and *Araucaria araucana* (1000 years; Aguilera-Betti et al. 2017), may be highly susceptible to climate change.

The autochthonous EM plants of the genus *Nothofagus* on the 17 temperate rainforest plots investigated, included seven evergreen and deciduous species (Table 16.1). Some EM fungi taxa have a wide distribution while other more specialized types occur only on isolated localities (Palfner 2001; Marín et al. 2018b). The EM fungi taxa included epigeous, secotioid, and hypogeous forms.

ER mycorrhizal associations were found on a total of 10 understory plant species of Chilean temperate rainforests: nine species on the family Ericaceae and one species (*Empetrum rubrum*) on the family Empetraceae. ER plants are common on infertile and acidic soils, characterized by a high content of recalcitrant polyphenolic compounds, leading to a very slow decomposition of soil organic matter. Instrumental to the survival of ER plants in these ecosystems, are their mycorrhizal associations, that release soil nutrients through the degradation of a wide range of complex and recalcitrant organic substrates (Smith and Read 2008). Clemmensen et al. (2015) proposed that the ER fungal biomass may contribute to the large storage of soil organic matter in older and high-altitude temperate forests, especially in the treeline under extreme environmental conditions.

Table 16.1 Mycorrhizal type for the plant species of 17 plots of temperate rainforests in southern Chile

	Forest plot																
Plant species	1	2	3	4	5	6	7	8	9	10	11	12	13	14	15	16	17
Acaena ovalifolia	–	–	–	–	–	–	AM	AM	–	–	–	–	–	–	–	–	AM
Acaena pinnatifida	AM	AM	–	–	–	–	–	–	–	–	–	–	–	–	–	–	–
Acrisione cymosa	–	–	–	–	–	–	–	AM	–	–	–	–	–	AM	–	–	–
Adenocaulon chilense	–	AM	–	–	–	–	–	–	–	–	–	–	–	–	–	AM	AM
Adesmia longipes	AM	–	–	–	–	–	–	–	–	–	–	–	–	–	–	–	–
Adesmia retusa	AM	–	–	–	–	–	–	–	–	–	–	–	–	–	–	AM	–
Adiantum chilense	–	–	–	–	–	AM	–	–	–	–	–	–	–	–	–	–	–
Adiantum sulphureum	AM	–	–	–	–	–	–	–	–	–	–	–	–	–	–	–	–
Aextoxicon punctatum	–	–	–	–	AM	–	–	–	–	–	–	–	–	AM	–	–	–
Agrostis perennans	AM	–	–	–	–	–	–	–	–	–	–	–	–	–	–	–	–
Alstroemeria aurea	–	AM	–	–	AM	AM	–	–	–	–	–	–	–	–	–	–	–
Amomyrtus luma	–	–	–	–	–	–	AM	AM	AM	–	–	–	–	AM	–	–	–
Amomyrtus meli	–	–	–	–	–	–	AM	–	AM	–	–	–	–	AM	–	–	–
Antidaphne punctulata	–	–	–	NM	–	–	–	–	NM	–	–	–	–	NM	–	–	–
Arachnitis uniflora	–	–	–	–	–	AM	–	–	–	–	–	–	–	AM	–	–	–
Araucaria araucana	–	AM	–	–	–	–	–	–	–	–	–	–	–	–	–	–	–
Aristotelia chilensis	–	–	–	–	AM	AM	AM	–	–	–	–	–	–	AM	AM	–	–
Asplenium dareoides	NM	–	–	NM	–	–	NM	NM	NM	–	–	–	–	NM	NM	–	–
Aster vahlii	–	–	–	AM	–	–	–	–	–	–	–	–	–	–	–	–	–
Asteranthera ovata	–	–	–	–	–	–	–	–	–	AM	–	–	–	–	–	AM	–
Austrocedrus chilensis	AM	–	–	–	–	–	–	–	–	–	–	–	–	–	–	–	–
Azara integrifolia	–	–	–	–	AM	–	–	–	–	–	–	–	–	–	–	–	–
Azara lanceolata	–	–	–	–	–	–	AM	AM	AM	AM	–	–	–	AM	AM	AM	–
Azara microphylla	–	AM	–	AM	–	–	–	–	–	–	–	–	–	–	–	–	–

(continued)

Table 16.1 (continued)

	Forest plot																
Plant species	1	2	3	4	5	6	7	8	9	10	11	12	13	14	15	16	17
Baccharis magellanica	–	–	–	–	–	–	AM	–	–	–	–	AM	–	–	–	–	–
Baccharis nivalis	–	AM	–	–	–	–	–	–	–	–	–	–	–	–	–	–	–
Baccharis racemosa	–	–	–	–	–	AM	AM	–	–	–	–	–	–	–	–	–	–
Baccharis sagittalis	–	–	AM	–	–	–	–	–	–	–	–	–	–	–	–	–	–
Baccharis sphaerocephala	–	–	–	–	–	–	AM	AM	AM	–	–	–	–	–	–	–	–
Berberis congestiflora	–	–	–	–	–	AM	–	–	–	–	–	–	–	–	–	–	–
Berberis darwinii	–	–	AM	–	AM	–	AM	AM	AM	–	–	–	–	–	–	–	–
Berberis microphylla	–	–	AM	–	AM	–	–	–	–	–	–	–	–	–	–	–	–
Berberis montana	AM	AM	–	–	–	–	–	–	–	–	–	–	–	–	–	–	AM
Berberis serratodentata	–	–	–	–	–	–	–	–	–	–	–	AM	–	–	–	–	AM
Berberis trigona	–	AM	–	–	–	–	–	–	–	–	–	–	AM	–	–	–	AM
Blechnum asperum	–	–	–	AM	–	AM	–	–	AM	–	–	–	–	–	AM	–	–
Blechnum blechnoides	–	–	–	–	–	–	–	–	–	–	–	–	–	AM	–	–	–
Blechnum chilense	–	–	AM	AM	–	–	AM	–	AM	AM	–	–	–	–	–	–	AM
Blechnum hastatum	–	–	–	AM	AM	AM	AM	AM	–	–	–	–	–	AM	–	–	–
Blechnum magellanicum	–	–	–	–	–	–	AM	–	AM	AM	–	AM	AM	–	–	–	–
Blechnum microphyllum	–	–	–	–	AM	–	–	–	–	–	–	–	–	–	–	–	–
Blechnum mochaenum	–	–	–	–	AM	–	–	–	–	–	–	–	–	AM	–	–	–
Blechnum penna-marina	–	AM	AM	–	–	–	–	AM	–	–	–	–	–	–	–	AM	AM
Blepharocalyx cruckshanksii	–	–	AM	AM	–	–	–	–	–	–	–	–	–	–	–	–	–
Boquila trifoliolata	–	–	–	AM	AM	AM	AM	AM	AM	AM	–	–	–	AM	–	–	–
Buddleja globosa	–	–	–	–	AM	–	–	–	–	–	–	–	–	–	–	–	–
Calandrinia ciliata	–	–	–	–	AM	–	–	–	–	–	–	–	–	–	–	–	–
Calceolaria biflora	–	AM	–	–	–	–	–	–	–	–	–	–	–	–	–	–	–
Caldcluvia paniculata	–	–	–	–	–	–	–	–	AM	–	–	–	–	AM	AM	–	–

	Forest plot																
Plant species	1	2	3	4	5	6	7	8	9	10	11	12	13	14	15	16	17
Campsidium valdivianum	–	–	–	AM	–	–	–	–	AM	AM	AM	–	AM	–	AM	–	–
Carex fuscula	–	–	NM	NM	–	–	–	–	–	–	NM	–	–	–	–	–	–
Carex sp.	–	–	NM	–	–	–	–	–	–	–	–	–	–	–	–	–	–
Centella asiatica	–	–	AM	–	–	–	–	–	–	–	–	–	–	–	–	–	–
Chascolytrum subaristatum	AM	–	–	–	–	–	–	–	–	–	–	–	–	–	–	–	–
Chloraea gaudichaudii	OR	–	–	–	–	–	–	–	–	–	–	–	–	–	–	–	–
Chloraea sp.	–	–	–	–	–	–	–	OR	–	–	OR	OR	OR	–	–	–	–
Chrysosplenium valdivicum	–	–	–	–	–	–	–	–	–	–	–	–	–	AM	–	–	–
Chusquea culeou	–	AM	AM	–	–	–	–	–	–	–	–	–	–	–	–	–	–
Chusquea montana	–	–	–	–	–	–	–	–	–	–	–	–	–	–	–	–	AM
Chusquea montana f. *nigricans*	–	–	–	–	–	–	–	–	–	–	–	–	AM	–	–	–	–
Chusquea quila	–	–	–	AM	AM	AM	AM	AM	AM	AM	AM	AM	–	AM	AM	–	–
Chusquea uliginosa	–	–	AM	–	–	–	–	–	–	–	AM	–	–	–	–	–	–
Cissus striata	–	–	–	AM	AM	AM	–	–	–	–	–	–	–	AM	–	–	–
Codonorchis lessonii	–	–	–	–	OR	–	OR	–	OR	–	–	–	–	–	–	–	OR
Coriaria ruscifolia	AM	–	–	–	–	–	–	–	–	–	–	–	–	–	–	–	–
Corynabutilon ochseni	–	–	–	–	AM	–	–	–	–	–	–	–	–	–	–	–	–
Corynabutilon vitifolium	–	–	–	–	AM	–	–	AM	–	–	–	–	–	–	–	–	–
Cynanchum pachyphyllum	–	–	–	–	AM	–	AM	AM	AM	–	–	–	–	–	–	–	–
Cyperus sp.	NM	–	–	–	–	–	–	–	–	–	–	–	–	–	–	–	–
Dasyphyllum diacanthoides	–	–	–	–	–	–	–	–	AM	AM	–	–	–	–	AM	–	–
Desfontainia fulgens	–	–	–	–	–	–	–	AM	AM	AM	–	AM	–	–	–	AM	–
Desmaria mutabilis	–	–	–	–	–	–	–	–	NM	–	–	–	–	–	–	–	–
Dioscorea brachybotrya	–	–	–	–	AM	–	–	–	–	–	–	–	–	–	–	–	–

(continued)

Table 16.1 (continued)

	Forest plot																
Plant species	1	2	3	4	5	6	7	8	9	10	11	12	13	14	15	16	17
Dioscorea bryoniifolia	–	AM	–	–	–	–	–	–	–	–	–	–	–	–	–	–	–
Discaria chacaye	–	–	AM	–	AM	–	–	–	–	–	–	–	–	–	–	–	–
Drimys andina	–	AM	–	–	–	–	–	–	–	–	–	–	–	–	–	–	AM
Drimys winteri	–	–	AM	AM	AM	–	AM	–	AM	AM	–	AM	AM	–	–	–	–
Drosera uniflora	–	–	–	–	–	–	–	–	–	–	AM	–	–	–	–	–	–
Eccremocarpus scaber	–	–	–	–	–	AM	–	–	–	–	–	–	–	–	–	–	–
Eleocharis macrostachya	–	AM	–	–	–	–	–	–	–	–	–	–	–	–	–	–	–
Elytropus chilensis	–	–	–	–	AM	–	AM	AM	AM	–	–	–	–	AM	–	–	–
Embothrium coccineum	–	NM	NM	–	–	–	NM	NM	–	–	–	NM	NM	–	–	NM	NM
Empetrum rubrum	–	–	–	–	–	–	–	–	–	–	–	–	–	–	–	–	ER
Epilobium ciliatum	–	–	–	–	–	–	–	–	–	–	–	–	–	–	–	–	AM
Ercilla volubilis	–	–	AM	–	AM	–	–	–	–	–	–	–	–	–	–	–	–
Eryngium paniculatum	AM	–	–	–	–	–	–	–	–	–	–	–	–	–	–	–	–
Escallonia alpina	–	–	–	–	–	–	–	–	–	–	–	–	AM	–	–	–	AM
Escallonia leucantha	–	–	AM	–	–	–	–	–	–	–	–	–	–	–	–	–	–
Escallonia revoluta	–	–	AM	–	–	–	–	–	–	–	–	–	–	–	–	–	–
Escallonia rosea	–	–	–	–	–	–	–	–	–	–	–	–	–	–	–	–	AM
Eucryphia cordifolia	–	–	–	–	–	–	AM	AM	–	–	–	–	–	AM	AM	–	–
Fascicularia bicolor	–	–	–	NM	NM	–	–	NM	NM	NM	–	–	–	NM	–	–	–
Festuca monticola	–	AM	–	–	–	–	–	–	–	–	–	–	–	–	–	–	–
Fitzroya cupressoides	–	–	–	–	–	–	–	–	–	–	–	AM	–	–	–	–	–
Fragaria chiloensis	–	AM	AM	–	–	–	–	AM	–	–	AM	–	–	–	–	–	–
Francoa appendiculata	–	–	–	–	–	AM	–	–	–	–	–	–	–	–	–	–	–
Fuchsia magellanica	–	–	–	–	–	–	–	–	AM	–	–	–	–	–	–	AM	–
Galium hypocarpium	–	–	–	–	AM	–	–	–	–	–	–	–	–	–	–	–	–

	Forest plot																
Plant species	1	2	3	4	5	6	7	8	9	10	11	12	13	14	15	16	17
Gamochaeta spiciformis	NM	–	–	–	–	–	–	–	–	–	–	–	–	–	–	–	–
Gaultheria caespitosa	ER	–	–	–	–	–	–	–	–	–	ER	–	–	–	–	–	–
Gaultheria insana	–	–	–	–	–	–	–	–	–	–	–	–	ER	–	–	–	–
Gaultheria mucronata	ER	–	ER	–	–	–	ER	ER	–	–	–	ER	–	–	–	–	–
Gaultheria myrtilloides	–	–	–	–	–	–	–	–	–	–	–	–	–	–	–	–	ER
Gaultheria phillyreifolia	–	ER	–	–	–	–	ER	–	–	ER	–	–	–	–	–	–	–
Gaultheria poeppigii	–	–	–	–	–	–	–	–	–	–	–	–	–	–	–	–	ER
Gaultheria poeppigii var. *linifolia*	–	–	ER	–	–	–	–	–	–	–	–	–	–	–	–	–	–
Gaultheria pumila	–	ER	ER	–	–	–	–	–	–	–	–	ER	–	–	–	–	ER
Gaultheria sp.	–	ER	–	–	–	–	–	–	–	–	–	–	–	–	–	–	–
Gavilea odoratissima	–	–	–	–	–	–	–	–	–	–	–	–	–	OR	–	–	–
Geranium robertianum	–	–	–	–	AM	–	–	–	–	–	–	–	–	–	–	–	–
Gevuina avellana	NM	–	NM	–	NM	–	NM	NM	–	–	–	–	–	NM	–	–	–
Gleichenia quadripartita	–	–	–	–	–	–	AM	–	–	–	AM	–	–	–	–	–	–
Gleichenia squamulosa	–	–	AM	–	–	–	–	–	–	–	–	AM	–	–	–	–	–
Grammitis magellanica	–	–	–	–	–	–	NM	–	–	–	–	–	–	–	–	–	NM
Greigia landbeckii	–	–	–	–	–	–	–	–	AM	–	–	–	–	–	–	AM	–
Greigia sphacelata	–	–	–	–	AM	–	AM	–	–	–	–	–	–	AM	–	–	–
Griselinia scandens	–	–	–	–	–	–	AM	–	AM	–	–	–	–	–	–	–	–
Gunnera magellanica	–	–	–	–	–	–	–	–	–	–	–	–	–	–	–	–	AM
Gunnera tinctoria	–	–	–	–	–	–	–	–	–	–	–	–	–	–	–	AM	–
Hydrangea serratifolia	–	–	–	–	–	AM	–	–	AM	–	–	–	–	AM	–	–	–
Hydrocotyle poeppigii	–	–	–	–	AM	–	–	AM	AM	–	–	–	–	AM	AM	–	–
Hymenophyllum caudiculatum	–	–	–	–	–	–	–	–	–	NM	–	NM	–	NM	–	–	–

(continued)

Table 16.1 (continued)

	Forest plot																
Plant species	1	2	3	4	5	6	7	8	9	10	11	12	13	14	15	16	17
Hymenophyllum pectinatum	–	–	–	–	–	–	–	–	–	NM	–	–	–	NM	–	–	–
Hymenophyllum peltatum	–	–	–	–	–	–	–	–	–	–	–	–	–	NM	–	–	–
Hymenophyllum plicatum	NM	–	–	–	–	–	–	–	–	–	–	–	–	NM	–	–	–
Hymenophyllum seselifolium	–	–	–	–	–	–	–	NM	–	–	–	–	–	NM	–	–	–
Hypochaeris tenuifolia	–	–	–	–	–	–	–	–	–	–	–	AM	–	–	–	–	–
Hypolepis poeppigii	–	–	–	AM	–	–	–	–	–	–	–	–	–	–	–	–	–
Juncus microcephalus	–	–	NM	–	–	–	–	–	–	–	–	–	–	–	–	–	–
Juncus pallescens	–	–	NM	NM	–	–	–	–	–	–	–	–	–	–	–	–	–
Juncus planifolius	–	–	–	–	–	–	–	–	–	–	NM	–	–	–	–	–	–
Juncus procerus	–	–	NM	NM	–	–	–	–	–	–	–	–	–	–	–	–	–
Lagenophora hariotti	–	–	–	–	–	–	–	–	–	–	–	–	–	–	–	–	AM
Lapageria rosea	–	–	–	–	AM	–	AM	–	–	–	–	–	–	AM	–	–	–
Lardizabala biternata	–	–	–	–	AM	AM	–	–	–	–	–	–	–	–	–	–	–
Lathyrus subandinus	AM	AM	–	–	–	–	–	–	–	–	–	–	–	–	–	–	–
Latua pubiflora	–	–	–	–	–	–	–	–	AM	–	–	–	–	–	–	–	–
Laurelia sempervirens	–	–	–	–	AM	AM	AM	–	–	–	–	–	–	AM	–	–	–
Laureliopsis philippiana	–	–	–	–	–	–	–	–	AM	AM	–	–	–	AM	AM	–	–
Lepidoceras chilense	–	–	–	NM	NM	–	–	–	–	–	–	–	–	NM	–	–	–
Leptinella scariosa	–	–	–	–	–	–	–	AM	–	–	–	–	–	–	–	–	–
Leptocarpha rivularis	–	–	–	–	AM	AM	–	–	–	–	–	–	–	–	–	–	–
Libertia chilensis	–	–	AM	–	–	–	AM	AM	–	–	–	–	–	–	–	–	–
Lobelia tupa	–	–	–	–	–	–	AM	–	–	–	–	–	–	–	–	–	–
Lomatia dentata	–	–	–	–	NM	–	NM	–	–	–	–	–	–	NM	–	–	–
Lomatia ferruginea	–	–	–	–	–	–	NM	NM	NM	NM	–	NM	–	–	NM	–	–
Lomatia hirsuta	NM	NM	NM	–	–	NM	NM	–	–	–	–	–	–	–	–	–	–

	Forest plot																
Plant species	1	2	3	4	5	6	7	8	9	10	11	12	13	14	15	16	17
Lophosoria quadripinnata	–	–	–	–	–	–	–	–	AM	AM	–	–	–	AM	AM	–	–
Lotus pedunculatus	–	–	AM	–	–	–	–	–	–	–	–	–	–	–	–	–	–
Luma apiculata	–	–	–	AM	AM	AM	AM	–	AM	AM	–	–	–	AM	–	–	–
Luma chequen	–	–	AM	–	–	–	–	–	–	–	–	–	–	–	–	–	–
Luzula racemosa	AM	–	–	–	–	–	–	–	–	–	–	–	–	–	–	–	–
Luzuriaga polyphylla	–	–	–	AM	–	–	AM	–	–	–	–	–	–	AM	AM	–	–
Luzuriaga radicans	–	–	–	AM	AM	–	–	AM	AM	AM	–	–	–	AM	–	–	–
Lycopodium gayanum	–	–	–	–	–	–	–	–	–	–	–	AM	AM	–	–	–	–
Lycopodium magellanicum	–	–	–	–	–	–	–	–	–	–	–	AM	AM	–	–	–	AM
Lycopodium paniculatum	–	–	–	–	–	–	–	–	–	–	–	AM	AM	–	–	AM	AM
Maytenus boaria	–	–	–	AM	AM	AM	–	–	–	–	–	–	–	–	–	–	–
Maytenus disticha	–	AM	–	–	–	–	–	–	–	–	–	–	–	–	–	–	AM
Maytenus magellanica	AM	AM	–	–	–	–	–	–	–	–	–	–	AM	–	–	AM	–
Megalastrum spectabile	–	–	–	–	–	–	–	–	–	–	–	–	–	–	AM	–	–
Misodendrum brachystachium	–	–	–	–	–	–	NM	–	–	–	–	–	–	–	–	–	–
Misodendrum linearifolium	–	–	–	–	NM	–	–	–	–	–	–	–	–	–	–	–	–
Mitraria coccinea	–	–	–	–	–	AM	AM	AM	AM	AM	–	–	–	AM	AM	–	–
Muehlenbeckia hastulata	–	–	–	AM	AM	–	–	–	–	–	–	–	–	–	–	–	–
Mutisia spinosa	AM	AM	–	–	–	–	–	–	–	–	–	–	–	–	–	–	–
Myoschilos oblonga	AM	–	AM	–	–	AM	–	–	–	–	–	AM	AM	–	–	–	–
Myrceugenia chrysocarpa	–	–	–	–	–	–	–	–	–	AM	–	–	–	–	–	–	–
Myrceugenia exsucca	–	–	–	AM	–	–	–	–	–	–	–	–	–	–	–	–	–
Myrceugenia parvifolia	–	–	AM	–	–	–	AM	–	–	–	–	–	–	AM	–	–	–
Myrceugenia planipes	–	–	–	–	–	–	–	AM	AM	–	–	–	–	AM	AM	AM	–

(continued)

Table 16.1 (continued)

Plant species	Forest plot																
	1	2	3	4	5	6	7	8	9	10	11	12	13	14	15	16	17
Nertera granadensis	–	–	AM	–	–	–	–	AM	AM	AM	–	–	–	AM	–	–	–
Notanthera heterophylla	–	–	–	–	–	NM	–	–	–	–	–	–	–	–	–	–	–
Nothofagos obliqua	–	–	–	–	EM	EM	–	EM	–	–	–	–	–	EM	–	–	–
Nothofagus alpina	–	–	–	–	–	–	–	EM	–	–	–	–	–	–	–	–	–
Nothofagus antarctica	–	–	EM	–	–	–	–	–	–	–	EM	–	–	–	–	–	–
Nothofagus betuloides	–	–	–	–	–	–	–	–	–	–	–	–	EM	–	–	EM	–
Nothofagus dombeyi	EM	–	EM	–	–	–	EM	–	EM	–	–	–	–	–	–	EM	–
Nothofagus nitida	–	–	–	–	–	–	–	–	–	EM	–	–	–	–	–	–	–
Nothofagus pumilio	–	–	–	–	–	–	–	–	–	–	–	–	–	–	–	EM	EM
Oreobolus obtusangulus	–	–	–	–	–	–	–	–	–	NM	–	–	–	–	–	–	–
Osmorhiza chilensis	–	AM	–	–	AM	AM	AM	AM	–	–	–	–	–	AM	–	–	–
Ourisia sp.	–	AM	–	–	–	–	–	–	–	–	–	–	–	–	–	–	–
Ovidia andina	–	–	–	–	–	–	–	–	–	–	–	–	–	–	–	AM	AM
Ovidia pillopillo	–	–	AM	–	–	–	AM	AM	AM	AM	–	–	–	AM	–	–	–
Oxalis arenaria	–	–	–	–	AM	AM	–	–	–	–	–	–	–	–	–	–	–
Oxalis dumetorum	–	–	–	–	AM	–	–	AM	–	–	–	–	–	–	–	–	–
Perezia pedicularifolia	AM	–	–	–	–	–	–	–	–	–	–	–	–	–	–	–	–
Persea lingue	–	–	–	–	AM	AM	AM	–	–	–	–	–	–	AM	–	–	–
Peumus boldus	–	–	–	–	–	AM	–	–	–	–	–	–	–	–	–	–	–
Philesia magellanica	–	–	–	–	–	–	–	AM	–	AM	–	AM	–	–	–	–	–
Pilea elliptica	–	–	–	–	–	–	–	–	–	–	–	–	–	–	AM	–	–
Pilgerodendron uviferum	–	–	–	–	–	–	–	–	–	–	AM	–	–	–	–	–	–
Pinguicula antarctica	–	–	–	–	–	–	–	–	–	–	–	NM	–	–	–	–	–
Poa obvallata	AM	AM	–	–	–	–	–	–	–	–	–	–	–	–	–	–	–
Podocarpus nubigenus	–	–	–	–	–	–	–	AM	AM	AM	–	AM	–	–	–	–	–

	Forest plot																
Plant species	1	2	3	4	5	6	7	8	9	10	11	12	13	14	15	16	17
Podocarpus salignus	–	–	–	–	–	–	AM	AM	–	–	–	–	–	AM	–	–	–
Polypodium feuillei	NM	–	–	–	NM	NM	NM	–	NM	–	–	–	–	NM	–	–	–
Polystichum plicatum	–	AM	–	–	–	–	–	–	–	–	–	–	–	–	–	–	–
Polystichum sp.	–	–	–	–	–	–	–	–	–	–	–	–	–	–	AM	–	–
Potentilla sp.	–	–	–	–	–	–	–	–	–	–	–	–	–	AM	–	–	–
Pseudopanax laetevirens	–	–	–	–	–	–	–	AM	AM	AM	–	–	–	AM	–	–	–
Pseudopanax valdiviensis	–	–	–	–	–	–	–	–	AM	–	–	–	–	AM	–	AM	–
Pteris semiadnata	–	–	–	–	–	–	–	–	–	–	–	–	–	–	AM	–	–
Ranunculus peduncularis	–	–	–	–	–	–	–	–	–	–	–	–	–	–	–	–	AM
Rhamnus diffusus	–	–	–	AM	AM	AM	–	–	–	–	–	–	–	AM	–	–	–
Rhaphithamnus spinosus	–	–	–	–	AM	–	AM	AM	AM	–	–	–	–	AM	–	–	–
Ribes magellanicum	–	–	–	–	–	–	–	AM	AM	AM	–	–	–	–	–	AM	AM
Ribes punctatum	–	–	AM	–	AM	–	–	–	–	–	–	–	–	–	–	–	–
Ribes valdivianum	–	AM	–	–	–	–	–	–	–	–	–	–	–	–	–	–	–
Rubus geoides	–	–	–	–	–	–	–	–	–	–	–	–	–	–	–	–	AM
Rubus radicans	–	–	–	–	–	–	–	–	AM	–	–	–	–	–	–	–	–
Rumohra adiantiformis	AM	–	–	–	–	–	–	–	–	–	–	–	–	–	–	–	–
Samolus latifolius	–	–	–	AM	–	–	–	–	–	–	–	–	–	–	–	–	–
Sanicula crassicaulis	–	–	–	–	AM	AM	–	–	–	–	–	–	–	–	–	–	–
Sanicula graveolens	–	–	–	–	–	–	–	–	–	–	–	–	–	AM	–	–	–
Sarmienta repens	–	–	–	NM	NM	NM	–	–	–	–	–	–	–	NM	–	–	–
Saxegothaea conspicua	–	–	–	–	–	–	–	AM	AM	AM	–	–	–	–	–	AM	–
Schinus polygamus	–	–	AM	–	–	AM	–	–	–	–	–	–	–	–	–	–	–
Schizaea fistulosa	–	–	–	–	–	–	–	–	–	–	AM	–	–	–	–	–	–

(continued)

Table 16.1 (continued)

Plant species	Forest plot 1	2	3	4	5	6	7	8	9	10	11	12	13	14	15	16	17
Schoenus rhynchosporoides	–	–	NM	–	–	–	–	–	–	–	NM	–	–	–	–	–	–
Scirpus inundatus	–	–	AM	–	–	–	–	–	–	–	–	–	–	–	–	–	–
Senecio acanthifolius	–	AM	–	–	–	–	–	–	–	–	–	AM	AM	–	–	–	–
Senecio chionophilus	AM	–	–	–	–	–	–	–	–	–	–	–	–	–	–	–	–
Senecio trifurcatus	AM	–	–	–	–	–	–	–	–	–	–	–	–	–	–	–	AM
Sisyrinchium arenarium	–	AM	–	–	–	–	–	–	–	–	–	–	–	–	–	–	–
Solanum krauseanum	–	–	–	–	–	–	–	–	AM	–	–	–	–	–	–	–	–
Solanum valdiviense	–	–	–	–	AM	–	–	–	–	–	–	–	–	–	–	–	–
Sophora cassioides	–	–	–	–	–	AM	–	–	–	–	–	–	–	–	–	–	–
Stellaria arvalis	–	–	–	–	AM	–	–	–	–	–	–	–	–	–	–	–	–
Tepualia stipularis	–	–	AM	AM	–	–	AM	–	–	–	–	–	–	–	–	–	–
Tristerix corymbosus	–	–	–	NM	NM	NM	–	–	–	–	–	–	–	NM	–	–	–
Tristerix verticillatus	–	–	–	–	–	NM	–	–	–	–	–	–	–	–	–	–	–
Ugni candollei	–	–	–	–	–	–	–	–	AM	AM	–	AM	–	–	–	–	–
Ugni molinae	–	–	–	–	–	–	AM	AM	–	–	–	–	–	–	–	–	–
Uncinia phleoides	–	–	–	–	NM	NM	–	–	NM	NM	–	–	–	NM	NM	–	–
Uncinia tenuis	–	–	NM	–	–	–	–	–	NM	–	–	NM	–	–	NM	NM	–
Valeriana lapathifolia	–	–	–	–	–	–	–	–	–	–	–	–	–	–	–	AM	AM
Veronica officinalis	–	–	–	–	–	–	–	NM	NM	–	–	–	–	–	–	–	–
Vicia setifolia	–	AM	–	–	–	–	–	–	–	–	–	–	–	–	–	–	–
Viola buchtienii	–	–	–	–	–	–	–	AM	AM	–	–	–	–	AM	–	–	–
Viola maculata	–	–	–	–	AM	–	–	–	–	–	–	–	–	–	–	–	–
Viola reichei	–	–	–	–	–	–	–	–	–	–	–	–	–	–	–	AM	AM
Viola rubella	–	–	–	–	–	–	AM	–	–	–	–	–	–	–	–	–	–
Weinmannia trichosperma	–	–	–	–	–	–	–	–	AM	AM	–	–	AM	–	AM	–	–

Species list after Rodriguez et al. (2018). Mycorrhizal types: arbuscular mycorrhizal (AM), ectomycorrhizal (EM), ericoid (ER), orchid (OR), and non-mycorrhizal (NM). The plots were located on the Andean mountain range (AR), the Chilean Central Valley (CV), and the Coastal mountain range (CR) and were dominated by: (1) *Austrocedrus chilensis* (AR); (2) *Araucaria araucana* (AR); (3) *Nothofagus antarctica* (CV); (4) *Blepharocalyx cruckshanksii* (CV); (5) *Nothofagus obliqua* (CV); (6) *Peumus boldus* (CV); (7) *Nothofagus dombeyi* and *Eucryphia cordifolia* (CR); (8) *Nothofagus alpina* (CR); (9) *Weinmannia trichosperma* (CR); (10) *Nothofagus nitida* (CR); (11) *Pilgerodendron uviferum* (CR); (12) *Fitzroya cupressoides* (CR); (13) *Nothofagus betuloides* (CR); (14) *Aextoxicon punctatum* (AR); (15) *Luma apiculata* (AR); (16) *Nothofagus dombeyi* (AR); and (17) *Nothofagus pumilio* (AR)

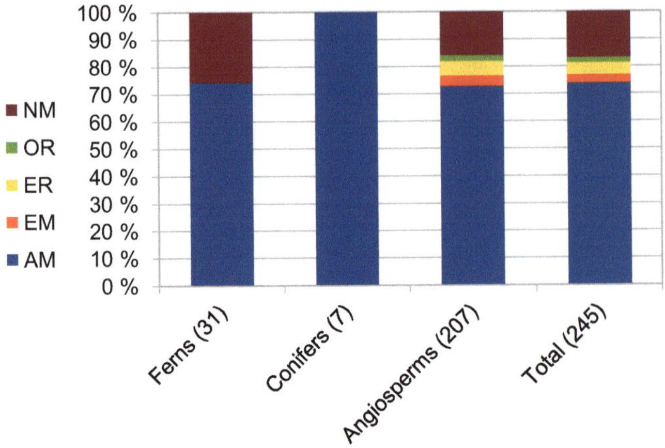

Fig. 16.1 Proportion of mycorrhizal types by different plant groups. Mycorrhizal types: arbuscular mycorrhizal (AM), ectomycorrhizal (EM), ericoid (ER), orchid (OR), and non-mycorrhizal (NM)

OM mycorrhizal associations were found only in four plant species of the family Orchidaceae. Pereira et al. (2018) studied the terrestrial orchid *Codonorchis lessonii*, endemic to southern Chile and Argentina, showing on the plant the presence of fungal binucleate cells and DNA material belonging to the families Ceratobasidiaceae and Tulasnellaceae. Fungal isolates belonging to Ceratobasidiaceae grew at a higher rate than those from Tulasnellaceae (Pereira et al. 2018). Phylogenetic analyses showed that different fungal partners associate with this orchid, suggesting relatively low specificity (Pereira et al. 2018).

A total of 10 vascular plant species presented fungal-bacterial tripartite associations: six species from Fabaceae (AM fungi + *Rhizobium*), two species from Rhamnaceae (AM fungi + *Frankia*), and two plant species from Gunneraceae (AM fungi + Cyanobacteria) (Carú 1993). These plants with tripartite associations, as well as several of the 37 NM plant species, are prevalent as pioneer plants on degraded soil or are known to colonize new substrates (for example after a volcanic event), having the role of ecosystem engineers on degraded ecosystems (Zúñiga-Feest et al. 2010).

The soils of the 17 plots sampled were generally acidic (Table 16.2), with low nitrogen and phosphorous content, and in some plots, extremely high concentrations of aluminum. Under these extremely harsh conditions for plant growth, the mycorrhizal associations play a key role on enhancing plant nutrition (Etcheverría et al. 2009; Marín et al. 2018a), and on giving the plant resistance to phytotoxic aluminum concentrations (Aguilera et al. 2017).

Table 16.2 Soil physicochemical parameters of 17 plots of temperate rainforests in southern Chile. The plot numbers correspond to the same plots as in Table 16.1

Plot No.	pH (KCl)	Cond. (uS/cm^{-1})	TC (%)	TN (%)	C/N	Av. P (ppm)	K (ppm)	Ca (ppm)	Mg (ppm)	Al (ppm)
1	5.60	48	0.87	0.06	14.50	1.9	25	212	25	74
2	4.70	69	9.00	0.43	20.93	4.7	79.00	658	92.00	261
3	4.30	213	13.74	1.31	10.49	2.6	218	698	167	1469
4	4.70	395	9.33	0.94	9.93	6.7	130	3546	901	130
5	5.13	209	11.51	0.98	11.74	3.5	233	4637	716	358
6	6.01	304	10.09	0.76	13.28	52.5	1309	7613	705	14
7	3.24	115	10.98	0.43	25.53	7.4	143	249	71	1790
8	3.98	124	9.1	0.58	15.69	2.7	131	661	113	1002
9	3.61	69	10.99	0.68	16.16	1.0	202	950	139	768
10	3.29	223	14.77	0.92	16.05	8.3	210	285	146	1745
11	3.47	223	7.91	0.50	15.82	20.3	81	254	51	639
12	3.35	126	6.3	0.19	33.16	1.0	90	137	73	752
13	3.15	224	8.71	0.39	22.33	16.7	171	200	93	1355
14	4.55	257	13.47	0.92	15.01	6.0	107	2171	488	721
15	4.27	122	9.53	0.81	12.01	9.0	69	361	84	1043
16	4.61	156	11.97	0.43	28.01	7.0	164	43	48	2230
17	4.06	113	10.85	0.33	33.01	9.0	120	204	61	1375

16.4 Mycorrhizal Bioweathering

In terrestrial ecosystems, crucial nutrients for plant nutrition as phosphorous and base cations, largely come from the *in situ* weathering of the bedrock or from weathering elsewhere and subsequent atmospheric deposition (Boy and Wilcke 2008; Boy et al. 2008). Bioweathering is the physicochemical process by which rocks are degraded by the direct and/or indirect actions of biota (Burford et al. 2003). The role of soil biota in weathering-related processes was largely ignored (Berner 1992). Mycorrhiza represents the most direct connection between photosynthesis and weatherable mineral surfaces, resulting in the lowest energy costs (Hoffland et al. 2004; Gadd 2007). Mycorrhiza, unlike plant roots, can explore small and nutrient-specifc mineral areas, and unlike bacteria, have a direct source of energy. Mycorrhiza degrades minerals by physical mechanisms such as tigmotropism, and by chemical processes such as acidolysis, complexolysis, redox reactions, and metal precipitation (Banfield et al. 1999; Burford et al. 2003; Hoffland et al. 2004; Rosling et al. 2004; Gadd 2007; van Schöll et al. 2008; Finlay et al. 2009; Taylor et al. 2009; Smits and Wallander 2017).

Ecosystem age, which reflects the ecological succession estate, was related to the degree of bioweathering in a mesocosm experiment on temperate rainforests in southern Chile (Marín 2018a), showing also higher bioweathering degree on EM-dominated forests. Furthermore, hyphae channels were seen on phyllosilicate minerals (Fig. 16.2; Marín 2018a). The degree of fungal bioweathering increased

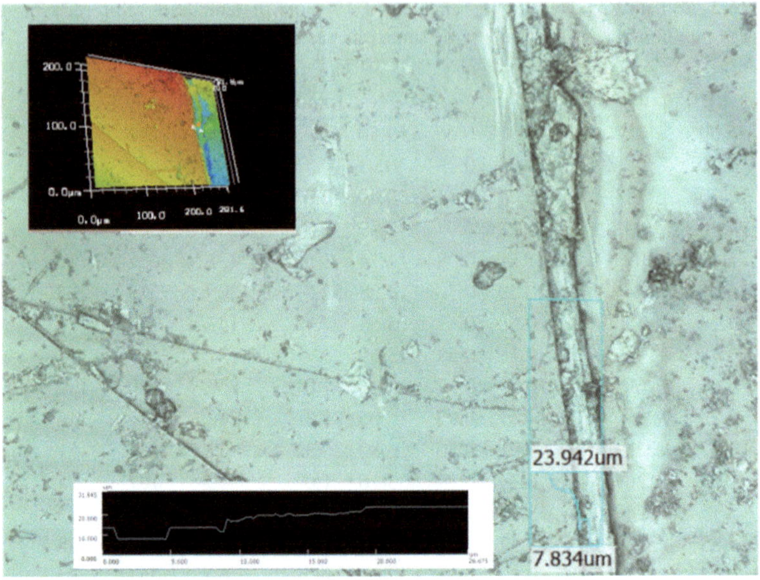

Fig. 16.2 Channels formed by mycorrhizal fungi on Muscovite after one year of exposition in a *Fitzroya cupressoides* forest, Alerce Costero National Park, southern Chile. Photo taken with a Confocal Laser Microscope at 5000×

over time, showing that this is a relevant biogeochemical process on terrestrial ecosystems (Marín 2018a).

16.5 Mycorrhizal Fungi as Ecological Restoration Tools

Up to now forest restoration using native trees has shown limited success (Thomas et al. 2014). Among the potential factors influencing this low success is the lack of integrating underground processes essential for tree survival. Mycorrhizal symbiosis is a key interaction directly affecting plant nutrition and resistance to abiotic and biotic stressors (Godoy et al. 2014), and thus should constitute a fundamental tool for ecological restoration (Lara et al. 2014). Specifically, the whole ´symbiome´ (Tripp et al. 2017) -plants and its symbionts- should be considered in restoration efforts. However, mycorrhizal fungi are rarely subjects of conservation programs, as mycorrhizal biogeography is an incipient area (Tedersoo 2017). In fact, the effects of anthropogenic disturbances on the distribution of plants and its mycorrhizal fungi largely remains to be studied, especially in regions with harsh environmental conditions as the temperate rainforests of southern Chile (Bueno et al. 2017).

AM fungi have been shown to be essential for the survival of the native conifer *Araucaria araucana* after fires, providing access to the remaining available soil-nutrients (Paulino et al. 2009; Lara et al. 2014; Cortés 2016) or even by maintaining glomalin production without mycelia growth (Rivas et al. 2012, 2016). Similarly, endemic EM fungi as the Basidiomycete *Descolea antarctica* seem to be crucial for the recovery of post-fire *Nothofagus alpina* seedlings (Palfner et al. 2008). Despite the conservation threats, restoration plans of Chilean native plant species are still incipient (Lara et al. 2014). However, suggestions as considering systems with native plants growing with their local rhizosphere and AM fungi under greenhouse conditions have emerged (Godoy et al. 2014; Lara et al. 2014; Marín et al. 2018a). It is important to emphasize that endangered plant species should preferably be grown in nurseries with local mycorrhizal fungal symbionts, as local mycorrhizal fungi guarantee greater growth and resistance to environmental stress (Godoy et al. 1994; Marín et al. 2018a). Greenhouse experiments involving mycorrhizal fungal inoculation of both *Nothofagus* spp. (Garrido 1988; Godoy et al. 1995; Marín et al. 2018a), and native conifer (Godoy et al. 1994) species, have shown significant effects on the plant growth rate, biomass, and seedling survival.

16.5.1 Nursery Experiment

In southern Chile, with the objective of producing native flora seedlings, an assay of simple and combined mycorrhizal inoculations with *Pisolithus tinctorius* and *Laccaria laccata* on *Nothofagus alpina* and *N. obliqua* was developed. Simultaneously, the potential of litter applications as a natural source of mycorrhizal incoculum for plant production programs was tested.

Seedlings were obtained under sterile germination conditions in culture chambers, and later transported to the nursery. At the end of the assay (16 weeks), several plant morphometric variables were measured, discerning statistical differences with the Tukey test (p value <0.05).

For *N. alpina*, the treatments involving *P. tinctorius* + *L. laccata* (with and without fertilization) and also litter, resulted on significantly higher values of the morphometric variables (except for the length of the radical systems) (Table 16.3). In contrast, all treatments on *N. obliqua* resulted on significantly higher values of the morphometric variables (Table 16.3). The quality index was significantly higher in respect to the control for the combined inoculations and litter treatments on *N. alpina*, while it was significantly higher for all treatments on *N. obliqua* (Table 16.3).

16.5.2 Reforestation Experiment

A reforestation experiment with the *Nothofagus alpina* and *N. obliqua* seedlings of the previous nursery assay was developed. Plantations were installed on two different sites (Folilco and Riñihue), of the premontane Andean region in southern Chile. After 23 weeks of the plantations, the plants were harvested and transported to the laboratory for the measurement of morphometric parameters and the estimation of

Table 16.3 Inoculation assay (simple and combined) with *Pisolithus tinctorius* and *Laccaria laccata* on two *Nothofagus* species. Treatments: (1) *P. tinctorius* without fertilizer; (2) *P. tinctorius* with fertilizer; (3) *P. tinctorius* and *L. laccata* without fertilizer; (4) *P. tinctorius* and *L. laccata* with fertilizer; (5) litter; (6) control

Treatment	Stem diameter (mm)	Stem length (cm)	Root length (cm)	Fresh stem weight (g)	Fresh root weight (g)	Dry stem weight (g)	Dry root weight (g)
Nothofagus alpina							
1	3.6	14.15	13.13	2.66	1.50	0.73	0.31
2	3.98[a]	19.78[a]	15.65	3.35	1.76[a]	0.91	0.33
3	4.43[a]	25.55[a]	16.2	4.83[a]	1.84[a]	1.25[a]	0.43[a]
4	5.83[a]	36.90[a]	16.68	9.37[a]	3.28[a]	2.49[a]	0.68[a]
5	5.05[a]	29.85[a]	17.50[a]	6.18[a]	2.06[a]	1.68[a]	0.48[a]
6	3.40	13.57	15.8	2.10	1.14	0.58	0.26
Nothofagus obliqua							
1	4.00[a]	33.52[a]	16.40[a]	5.45[a]	1.45[a]	1.40[a]	0.33[a]
2	4.53[a]	41.10[a]	14.55[a]	7.90[a]	2.42[a]	1.94[a]	0.41[a]
3	5.00[a]	51.45[a]	15.95[a]	10.77[a]	2.93[a]	2.73[a]	0.53[a]
4	5.58[a]	57.45[a]	18.45[a]	13.02[a]	3.07[a]	3.46[a]	0.65[a]
5	3.95[a]	34.43[a]	15.37[a]	6.04[a]	1.90[a]	1.41[a]	0.27[a]
6	2.35	17.08	11.63	1.30	0.53	0.32	0.10

Values correspond to the average of 20 plant individuals. [a]denotes statically significant differences between the control and the treatments (Tukey test, p value <0.05)

the quality index (Ritchie 1984). The combined inoculation of *Pisolithus tinctorius* and *L. laccata* in both *Nothofagus* species showed overall excellent results (Table 16.4). According to the quality index, the best treatment for *N. obliqua* was co-inoculation with fertilization, and for *N. alpina* it was co-inoculation without fertilization (Table 16.4). These results show clear advantages of mycorrhizal co-inoculations for the re-establishment of crucial native flora of the temperate rainforests of southern Chile.

16.6 Conclusions and Future Directions

Mycorrhizal fungal communities of the temperate rainforests of Southern Chile are affected by the mountain system in which they are located (Andes and Coastal mountain ranges), the mycorrhizal dominance of the forest (either ectomycorrhizal -EM- or arbuscular mycorrhizal-AM), soil chemistry, and altitude (Marín et al. 2017a; Marín 2018a). Mycorrhizal fungi are important ecosystem components as they play central roles in nutrient cycling, maintenance of biodiversity, and ecosystem productivity (van der Wal et al. 2013; Bardgett and van der Putten 2014; Peay et al. 2016). Mycorrhizal fungal communities can be highly diverse (Tedersoo et al. 2014), and their diversity is affected by edaphic and climatic conditions, as well as

Table 16.4 Reforestation from seedlings of an Inoculation assay (simple and combined) with *Pisolithus tinctorius* and *Laccaria laccata* on two *Nothofagus* species. Treatments: (1) *P. tinctorius* without fertilizer; (2) *P. tinctorius* with fertilizer; (3) *P. tinctorius* and *L. laccata* without fertilizer; (4) *P. tinctorius* and *L. laccata* with fertilizer; (5) litter; (6) control

Treatment	Stem diameter (mm)	Stem length (cm)	Root length (cm)	Fresh stem weight (g)	Fresh root weight (g)	Dry stem weight (g)	Dry root weight (g)	Quality Index
Nothofagus alpina – Folilco								
1	5.00	23.1[a]	13.20	7.40	7.60	7.40	5.50	0.27
2	7.7[a]	35.5[a]	15.5[a]	18.0[a]	13.6[a]	13.6[a]	10.1[a]	0.50
3	9.5[a]	50.3[a]	15[a]	23.1[a]	20.85[a]	16.7[a]	14.3[a]	0.57
4	7.6[a]	36.9[a]	15.1[a]	15.3[a]	15.50	11.0[a]	10.30	0.43
5	9.5[a]	55.4[a]	17.9[a]	23.0[a]	19.3[a]	16.9[a]	15.8[a]	0.55
6	4.10	17.00	12.30	8.00	7.90	6.30	6.20	0.29
Nothofagus alpina – Riñihue								
1	4.90	30.00	12.7[a]	8.6[a]	8.00	6.60	6.1[a]	0.20
2	18.1[a]	40.4[a]	17.8[a]	16.6[a]	14.3[a]	12.3[a]	10.3[a]	0.44
3	8.0[a]	35.3[a]	19.9[a]	14.9[a]	11.8[a]	10.5[a]	9.9[a]	0.45
4	9.9[a]	54.8[a]	19.6[a]	28.9[a]	24.4[a]	17.4[a]	15.2[a]	0.30
5	7.2[a]	38.1[a]	16.2[a]	13.3[a]	13.4[a]	9.8[a]	9.0[a]	0.35
6	4.10	24.60	9.40	5.80	7.70	7.60	4.96	0.20
Nothofagus obliqua – Folilco								
1	7.50	59.20	16.9[a]	19.70	9.60	15.40	8.0[a]	0.29
2	9.3[a]	62.20	21.50	27.70	15.70	21.3[a]	14.70	0.52
3	8.80	77.3[a]	22.00	24.50	19.00	19.00	14.00	0.37
4	11.1[a]	84.9[a]	22.20	37.3[a]	26.7[a]	27.5[a]	23.3[a]	0.65
5	8.9[a]	78.1[a]	22.2[a]	32.6[a]	18.60	27.80	16.60	0.49
6	7.80	55.40	19.70	24.10	17.50	16.70	13.00	0.41
Nothofagus obliqua – Riñihue								
1	7.9[a]	64.1[a]	21.3[a]	25.7[a]	14.30	17.7[a]	9.9[a]	0.33
2	8.5[a]	60.40	17.90	29.0[a]	17.1[a]	18.2[a]	11.1[a]	0.40
3	8.7[a]	80.4[a]	19.00	31.9[a]	19.8[a]	20.5[a]	12.8[a]	0.35
4	11.5[a]	94.3[a]	17.50	51.0[a]	25.7[a]	31.0[a]	16.0[a]	0.56
5	9.4[a]	77.4[a]	21.0[a]	39.4[a]	20.9[a]	24.9[a]	13.5[a]	0.45
6	6.30	53.80	16.80	16.80	10.50	10.90	7.10	0.20

Values correspond to the average of 20 plant individuals. [a]denotes statically significant differences between the control and the treatments (Tukey test, p value <0.05). Quality Index after Ritchie (1984)

biotic factors such as plant diversity (Tedersoo et al. 2014; Davison et al. 2015). However, how these abiotic and biotic factors interact and affect mycorrhizal fungal communities -and the mycorrhizal symbiosis overall- remains to be throughly studied (Truong et al. 2017), especially on the temperate rainforests of South America (Bueno et al. 2017).

The diversity and function of soil biota under scenarios of climate change provides fundamental information on the ecosystem processes that take place over long periods. Such questions cannot be addressed by traditional approaches which are commonly limited to 2–3 years due to funding and logistical restrictions (Amano and Sutherland 2013). Thus, scientific collaboration represents an opportunity to tackle the role of soil biota, particularly mycorrhizal fungi, in future studies of biogeochemical cycles in pristine temperate rainforests of South America (Truong et al. 2017, 2019; Oeser et al. 2018).

Acknowledgments R.G and C.M. were funded by the project Fondecyt 1190642. C.M. was funded by the Universidad de O'Higgins post-doctoral research fund. Special thanks to Dr. Jens Boy for support in the laboratory at the Institute of Soil Science, Leibniz Universität Hannover, Germany.

References

Aguilera-Betti I, Muñoz AA, Stahle D, Figueroa G, Duarte F, González-Reyes Á, Christie D, Lara A, González ME, Sheppard PR, Sauchyn AM, Toledo-Guerrero I, Olea M, Apaz P, Fernandez A (2017) The First Millennium-Age *Araucaria araucana* in Patagonia. Tree-Ring Res 73(1):53–56

Aguilera P, Marín C, Oehl F, Godoy R, Borie F, Cornejo PE (2017) Selection of aluminum tolerant cereal genotypes strongly influences the arbuscular mycorrhizal fungal communities in an acidic Andosol. Agric Ecosyst Environ 246:86–93

Amano T, Sutherland WJ (2013) Four barriers to the global understanding of biodiversity conservation: wealth, language, geographical location and security. Proc R Soc B 280(1756):20122649. https://doi.org/10.1098/rspb.2012.2649

Armesto JJ, Manuschevich D, Mora A, Smith-Ramírez C, Rozzi R, Abarzúa AM, Marquet PA (2010) From the Holocene to the Anthropocene: A historical framework for land cover change in southwestern South America in the past 15,000 years. Land Use Policy 27(2):148–160

Armesto JJ, Smith-Ramírez C, Carmona MR, Celis-Diez JL, Díaz IA, Gaxiola A, Gutiérrez AG, Núñez-Avila MC, Pérez CA, Rozzi R (2009) Old-growth temperate rainforests of South America: conservation, plant–animal interactions, and baseline biogeochemical processes. In: Wirth C, Gleixner G, Heimann M (eds) Old-growth forests. Springer Berlin Heidelberg, Berlin, p 367–390

Banfield JF, Barker WW, Welch SA, Taunton A (1999) Biological impact on mineral dissolution: application of the lichen model to understanding mineral weathering in the rhizosphere. P Nat Acad Sci USA 96(7):3404–3411

Bardgett RD, van der Putten WH (2014) Belowground biodiversity and ecosystem functioning. Nature 515(7528):505–511

Bekessy S, Allnutt T, Premoli A, Lara A, Ennos R, Burgman M, Cortes M, Newton AC (2002) Genetic variation in the vulnerable and endemic Monkey Puzzle tree, detected using RAPDs. Heredity 88:243–249

Bekessy S, Lara A, González M, Cortés M, Premoli A, Newton A (2004) Variación en *Araucaria araucana* (Molina) K. Koch (Araucaria o Pehuén). In: Donoso C, Premoli A, Gallo L, Ipinza R (eds) Variación Intraespecífica en las especies arbóreas de los bosques templados de Chile y Argentina. Editorial Universitaria, Santiago, p. 215–232

Berner RA (1992) Weathering, plants, and the long-term carbon cycle. Geochim Cosmochim Acta 56(8):3225–3231

Bowman DM, Moreira-Muñoz A, Kolden CA, Chávez RO, Muñoz AA, Salinas F, González-Reyes A, Rocco R, de la Barrera F, Williamson GJ, Borchers N, Cifuentes LA, Abatzoglou JT, Johnston FH (2018) Human–environmental drivers and impacts of the globally extreme 2017 Chilean fires. Ambio 48(4):350–362

Boy J, Wilcke W (2008) Tropical Andean forest derives calcium and magnesium from Saharan dust. Global Biogeochem Cycles 22(1):GB1027

Boy J, Godoy R, Guevara G (2014) Transporte de aerosoles, biometeorización y cambio global. In: Donoso C, González M, Lara A (eds) Ecología Forestal: Bases para el Manejo Sustentable y Conservación de los Bosques Nativos de Chile. Ediciones Universidad Austral de Chile, Valdivia, p. 281–295.

Boy J, Rollenbeck R, Valarezo C, Wilcke W (2008) Amazonian biomass burning-derived acid and nutrient deposition in the north Andean montane forest of Ecuador. Global Biogeochemical Cycles 22(4):GB4011

Brundrett MC, Tedersoo L (2018) Evolutionary history of mycorrhizal symbioses and global host plant diversity. New Phytol 220(4):1108–1115

Bueno CG, Marín C, Silva-Flores P, Aguilera P, Godoy R (2017) Think globally, research locally: contrasting patterns of mycorrhizal symbiosis in South America. New Phytol 215(4):1306–1309

Burford EP, Fomina M, Gadd GM (2003) Fungal involvement in bioweathering and biotransformation of rocks and minerals. Mineral Mag 67(6):1127–1155

Carrillo R, Godoy R, Peredo H (1992) Simbiosis micorrícica en comunidades boscosas del Valle Central en el sur de Chile. Bosque 13(2):57–67

Carú M (1993) Characterization of native *Frankia* strains isolated from Chilean shrubs (Rhamnaceae). Plant Soil 157(1):137–145

Castillo C-G, Borie F, Godoy R, Rubio R, Sieverding E (2006) Diversity of mycorrhizal plant species and arbuscular mycorrhizal fungi in evergreen forest, deciduous forest and grassland ecosystems of Southern Chile. J Appl Bot Food Qual 80(1):40–47

Clemmensen KE, Finlay RD, Dahlberg A, Stenlid J, Wardle DA, Lindahl BD (2015) Carbon sequestration is related to mycorrhizal fungal community shifts during long-term succession in boreal forests. New Phytol 205(4):1525–1536

Cortés M (2016) Restauración ecológica de los bosques de *Araucaria araucana* (Molina) K. Koch de la Cordillera de Nahuelbuta en la región de la Araucanía, Chile. Doctoral thesis, Universidad Austral de Chile.

Davison J, Moora M, Öpik M, Adholeya A, Ainsaar L, Bâ A, Burla S, Diedhiou AG, Hiiesalu I, Jairus T, Johnson NC, Kane A, Koorem K, Kochar M, Ndiaye C, Pärtel M, Reier Ü, Saks Ü, Singh R, Vasar M, Zobel M (2015) Global assessment of arbuscular mycorrhizal fungus diversity reveals very low endemism. Science 349(6251):970–973

Etcheverría P, Huygens D, Godoy R, Borie F, Boeckx P (2009) Arbuscular mycorrhizal fungi contribute to 13C and 15N enrichment of soil organic matter in forest soils. Soil Biol Biochem 41(4):858–861

Fernández N, Fontenla S, Messuti MI (2005). Micorrizas en pteridofitas de los bosques templado-lluviosos del Noroeste de Patagonia. II Convención Ambiental Universitaria Patagónica, Universidad Nacional del Comahue, Bariloche, Argentina.

Finlay R, Wallander H, Smits M, Holmstrom S, van Hees P, Lian B, Rosling A (2009) The role of fungi in biogenic weathering in boreal forest soils. Fungal Biol Rev 23(4):101–106

Flores R, Godoy R, Palfner G (1997) Morfo-anatomía de la ectomicorriza *Cenococcum geophilum* Fr. en *Nothofagus alessandrii* Esp. Gayana Bot 54(2):157–162

Fontenla S, Godoy R, Rosso P, Havrylenko M (1998) Root associations in *Austrocedrus* forests and seasonal dynamics of arbuscular mycorrhizas. Mycorrhiza 8(1):29–33

Gadd GM (2007) Geomycology: biogeochemical transformations of rocks, minerals, metals and radionuclides by fungi, bioweathering and bioremediation. Mycol Res 111(1):3–49

Garrido N (1988) Agaricales s. l. und ihre mykorrhizen in den *Nothofagus*-Wäldern mittelchiles. Bibliotheca Mycologica 120. J., Cramer, Berlin.

Godoy R, Carrillo R, Peredo H (1995) Ökologische und experimentelle Arbeiten über Mykorrhiza in Naturwäldern Südchiles. Ver. Ges Ökol 24:619–622.
Godoy R, Valenzuela E, Guevara G, Boy J, Barrientos M, Matus F (2014) Biogeoquímica en bosques templados del sur de Chile. In: Donoso C, González ME, Lara A (eds) Ecología Forestal. Editorial Universidad Austral de Chile, Valdivia, p. 257–280
Godoy R, Mayr R (1989) Caracterización morfológica de micorrizas vesículo-arbusculares en coníferas endémicas del sur de Chile. Bosque 10(2):89–98
Godoy R, Oyarzún C (1998) Water chemistry in forests of southern Chile. Ver Ges Ökol 28:471–474
Godoy R, Oyarzún C, Bahamones J (1999) Flujos hidroquímicos en un bosque de *Nothofagus pumilio* en el Parque Nacional Puyehue, sur de Chile. Rev Chil Hist Nat 72:579–594
Godoy R, Oyarzún C, Gerding V (2001) Precipitation chemistry in deciduous and evergreen *Nothofagus* forests of southern Chile under a low-deposition climate. Basic Appl Ecol 2(1):65–72
Godoy R, Paulino L, Oyarzún C, Boeckx P (2003) Atmospheric N deposition in Central and Southern Chile, an overview. Gayana Bot 60(1):47–53
Godoy R, Paulino L, Valenzuela E, Oyarzún C, Huygens D, Boeckx P (2009) Temperate ecosystems of Chile: characteristic biogeochemical cycles and disturbance regimes. In: Oyarzún C, Verhoest N, Boeckx P, Godoy R (eds) Ecological advances on Chilean temperate forests. Academia Press, Gehnt, p. 31–40
Godoy R, Palfner G (1997) Ectomicorrizas en *Nothofagus alpina* (P. et E.) Oerst y *N. dombeyi* (Mirb.) Oerst. del Sur de Chile. Bol Micol 12:55–61
Godoy R, Romero R, Carrillo R (1994) Estatus micotrófico de la flora vascular en bosques de coníferas nativas. Rev Chil Hist Nat 67:209–220
Hedin LO, Hetherington ED (1996) Atmospheric and geologic constraints on the biogeochemistry of North and South American temperate rainforests. In: Lawford RG, Alaback P, Fuentes E (eds) High-Latitude Rainforests and Associated Ecosystems of the West Coast of the Americas. Springer, New York, p. 55–74
Hoffland E, Kuyper TW, Wallander H, Plassard C, Gorbushina AA, Haselwandter K, Holmström S, Landeweert R, Lundström US, Rosling A, Sen R, Smits MM, van Hees PAW, van Breemen N (2004). The role of fungi in weathering. Front Ecol Environ 2(5):258–264
Kennedy MJ, Hedin LO, Derry LA (2002) Decoupling of unpolluted temperate forests from rock nutrient sources revealed by natural 87Sr/86Sr and 84Sr tracer addition. P Nat Acad Sci USA 99(15):9639–9644
Koske RE, Gemma JN (1989) A modified procedure for staining roots to detect VA mycorrhizas. Mycol Res 92(4):486–488
Lara A, Little C, Cortés M, Cruz E, González ME, Echeverría C, Suárez J, Bahamondez A, Coopman R (2014) Restauración de ecosistemas forestales. In: Donoso C, González ME, Lara A (eds) Ecología Forestal. Editorial Universidad Austral de Chile, Valdivia, p.605–672
Lara A, Villalba R (1993) A 3620-year temperature record from *Fitzroya cupressoides* tree rings in southern South America. Science 260(5111):1104–1106
Marín C (2018a) Fungal biodiversity associated to bioweathering processes in a chronosequence of temperate rainforests in southern Chile. Doctoral Thesis, Universidad Austral de Chile.
Marín C, Aguilera P, Cornejo P, Godoy R, Oehl F, Palfner G, Boy J (2016) Arbuscular mycorrhizal assemblages along contrasting Andean forests of Southern Chile. J Soil Sci Plant Nutr 16(4):916–929
Marín C, Aguilera P, Oehl F, Godoy R (2017a) Factors affecting arbuscular mycorrhizal fungi of Chilean temperate rainforests. J Soil Sci Plant Nutr 17(4):966–984
Marín C, Godoy R, Valenzuela E, Schloter M, Wubet T, Boy J, Gschwendtner S (2017b) Functional land-use change effects on soil fungal communities in Chilean temperate rainforests. J Soil Sci Plant Nutr 17(4):985–1002
Marín C, Valenzuela E, Godoy R, Palfner G (2018a) Diversity and growth-effects of ectomycorrhizal fungi of a *Nothofagus pumilio* forest in the Andes of Southern Chile. Bol Micol 33(1):9–20
Marín C, Torres D, Furci G, Godoy R, Palfner G (2018b) Estado del arte de la conservación del reino Fungi en Chile. Biodiversidata 7:98–115

Nouhra ER, Urcelay C, Longo MS, Fontenla S (2012) Differential hypogeous sporocarp production from *Nothofagus dombeyi* and *N. pumilio* forests in southern Argentina. Mycologia 104(1):45–52

Nouhra E, Urcelay C, Longo S, Tedersoo L (2013) Ectomycorrhizal fungal communities associated to *Nothofagus* species in Northern Patagonia. Mycorrhiza 23(6):487–496

Oeser RA, Stroncik N, Moskwa LM, Bernhard N, Schaller M, Canessa R, van den Brink L, Köster M, Brucker E, Stock S, Fuentes JP, Godoy R, Matus FJ, Oses Pedraza R, Osses McIntyre P, Paulino L, Seguel O, Bader MY, Boy J, Dippold MA, Ehlers TA, Kühn P, Kuzyakov Y, Leinweber P, Scholten T, Spielvogel S, Spohn M, Übernickel K, Tielbörger K, Wagner D, von Blanckenburg F (2018) Chemistry and microbiology of the Critical Zone along a steep climate and vegetation gradient in the Chilean Coastal Cordillera. Catena 170:183–203

Oyarzún CE, Godoy R, Leiva S (2002) Depositación atmosférica de nitrógeno en un transecto Valle Longitudinal-Cordillera de los Andes, Centro-Sur de Chile. Rev Chil Hist Nat 75(1):165–175

Oyarzún CE, Godoy R, de Schrijver A, Staelens J, Lust N (2004) Water chemistry and nutrient budgets in an undisturbed evergreen rainforest of southern Chile. Biogeochemistry 71(1):107–123

Oyarzún CE, Godoy R, Staelens J, Donoso PJ, Verhoest NE (2011) Seasonal and annual throughfall and stemflow in Andean temperate rainforests. Hydrol Process 25(4):623–633

Oyarzún C, Aracena C, Rutherford P, Godoy R, de Schrijver A (2007) Effects of land use conversion from native forests to exotic plantations on nitrogen and phosphorus retention in catchments of southern Chile. Water Air Soil Pollut 179(1–4):341–350

Oyarzún C, Godoy R, Sepúlveda A (1998) Water and nutrient fluxes in a cool temperate rainforest at the Cordillera de la Costa in Southern Chile. Hydrol Process 12(7):1067–1077

Oyarzún C, Verhoest N, Boeckx P, Godoy R (2009) Ecological advances on Chilean temperate forests. Academia Press, Ghent.

Palfner G (2001) Taxonomische studien an Ektomykorrhizen aus den *Nothofagus*-Waldern Mittelsudchiles. Bibliotheca Mycologica 190. J., Cramer, Berlin.

Palfner G (2002) Diversity and community studies on *Fagus/Nothofagus* ectomicorrhizae from European and South American Forest. In: de Schrijver A, Kint V, Lust N (eds) Procceding of the Workshop Comparison of ecosystems functioning and biogeochemical cycles in temperate forests in Southern Chile and Flanders. Academia Press, Ghent, p. 119–129

Palfner G, Godoy R (1996a) "*Nothofagirhiza vinicolor*" + *Nothofagus pumilio* (Poepp. et Endl.) Kraser. Descriptions of Ectomycorrhizae 1:65–70

Palfner G, Godoy R (1996b) *Russula fuegiana* Singer + *Nothofagus pumilio* (Poepp. et Endl.) Krasser. Descriptions of Ectomycorrhizae 1:131–136

Palfner G, Canseco MI, Casanova-Katny A (2008) Post-fire seedlings of *Nothofagus alpina* in Southern Chile show strong dominance of a single ectomycorrhizal fungus and a vertical shift in root architecture. Plant Soil 313(1–2):237–250

Paulino L, Godoy R, Boeckx P (2009) Ecosystem responses of Andean *Araucaria-Nothofagus* communities after a wildfire. In: Oyarzún C, Verhoest N, Boeckx P, Godoy R (eds) Ecological advances on Chilean temperate forests. Academia Press, Gehnt, p. 117–132

Peay KG, Kennedy PG, Talbot JM (2016) Dimensions of biodiversity in the Earth mycobiome. Nat Rev Microbiol 14(7):434–447

Pereira G, Suz LM, Albornoz V, Romero C, Garcia L, Leiva V, Atala C (2018) Mycorrhizal fungi associated with *Codonorchis lessonii* (Brongn.) Lindl., a terrestrial orchid from Chile. Gayana Bot 75(1):447–458

Ritchie GA (1984) Assessing seedling quality. In: Duryea ML, Landis TD (eds) Forestry nursery manual: production of bareroot seedlings. Springer, The Hague, p. 243–259

Rivas Y, Canseco MI, Knicker H, Etcheverría P, Godoy R, Matus F, Valenzuela E, Gallardo R (2016) Variación en el contenido de glomalina relacionada a las proteínas del suelo, después de un incendio forestal en un Andisol en bosques de *Araucaria araucana* del centro-sur de Chile. Bosque 37,409–417

Rivas Y, Huygens D, Knicker H, Godoy R, Matus F, Boeckx P (2012) Soil nitrogen dynamics three years after a severe *Araucaria-Nothofagus* forest fire. Austral Ecol 37:153–163

Rodriguez R, Marticorena C, Alarcón D, Baeza C, Cavieres L, Finot VL, Fuentes N, Kiessling A, Mihoc M, Pauchard A, Ruiz E, Sanchez P, Marticorena A (2018) Catálogo de las plantas vasculares de Chile. Gayana Bot 75(1):1–430

Rosling A, Lindahl BD, Finlay RD (2004) Carbon allocation to ectomycorrhizal roots and mycelium colonising different mineral substrates. New Phytol 162(3):795–802

Sadzawka A, Carrasco MA, Grez R, Mora ML, Flores H, Neaman A (2006) Métodos de análisis recomendados para los suelos de Chile. Revisión 2006. Serie Actas INIA 34.

Simard SW, Beiler KJ, Bingham MA, Deslippe JR, Philip LJ, Teste FP (2012) Mycorrhizal networks: mechanisms, ecology and modelling. Fungal Biol Rev 26(1):39–60

Singer R (1969) Mycoflora Australis. Beihefte Nova Hedwigia 29, J. Cramer, Lehre.

Singer R (1970) *Phaeocollybia* (Cortinariaceae, Basidiomycetes). Flora Neotropica 4:1–13

Singer R, Moser M, Gamundí I, Ellas R, Sarmiento G (1965) Forest mycology and forest communities in South America. Mycopath Mycol Appl 26(2–3):129–191

Singer R, Morello JH (1960) Ectotrophic forest tree mycorrhizae and forest communities. Ecology 41(3):549–551

Smith SE, Read DJ (2008) Mycorrhizal Symbiosis (Third Edition). Academic Press, New York.

Smits MM, Wallander H (2017) Role of mycorrhizal symbiosis in mineral weathering and nutrient mining from soil parent material. In: Johnson N, Gehring C, Jansa J (eds) Mycorrhizal Mediation of Soil: Fertility, Structure, and Carbon Storage. Elsevier, New York, p. 35–46

Staelens J, de Schrijver A, Oyarzún C, Lust N (2003) Comparison of dry deposition and canopy exchange of base cations in temperate hardwood forests in Flanders and Chile. Gayana Bot 60(1):9–16

Staelens J, Godoy R, Oyarzún C, Thibo K, Verheyen K (2005) Nitrogen fluxes in throughfall and litterfall in two *Nothofagus* forests in southern Chile. Gayana Bot 62(2):63–71

Staelens J, Oyarzún C, Almonacid L, Padila E, Verheyen (2009) Aboveground nutrient cycling in temperate forest ecosystems of southern Chile. In: Oyarzún C, Verhoest N, Boeckx P, Godoy R (eds) Ecological advances on Chilean temperate forests. Academia Press, Gehnt, p. 103–106

Taylor LL, Leake JR, Quirk J, Hardy K, Banwart SA, Beerling DJ (2009) Biological weathering and the long-term carbon cycle: integrating mycorrhizal evolution and function into the current paradigm. Geobiology 7(2):171–191

Tedersoo L (2017) Biogeography of mycorrhizal symbiosis. Springer, New York.

Tedersoo L, Bahram M, Põlme S, Kõljalg U, Yorou NS, Wijesundera R, Ruiz L. V, Vasco-Palacios AM, Thu PQ, Suija A, Smith ME, Sharp C, Saluveer E, Saitta A, Rosas M, Riit T, Ratkowsky D, Pritsch K, Põldmaa K, Piepenbring M, Phosri C, Peterson M, Parts K, Pärtel K, Otsing E, Nouhra E, Njouonkou AL, Nilsson RH, Morgado LN, Mayor J, May TW, Majuakim L, Lodge DJ, Lee SS, Larsson K-H, Kohout P, Hosaka K, Hiiesalu I, Henkel TW, Harend H, Guo L, Greslebin A, Grelet G, Geml J, Gates G, Dunstan W, Dunk, Drenkhan R, Dearnaley J, De Kesel A, Dang T, Chen X, Buegger F, Brearley FQ, Bonito G, Anslan S, Abell S, Abarenkov K (2014) Global diversity and geography of soil fungi. Science 346(6213):1256688

Thomas E, Jalonen R, Loo J, Boshier D, Gallo L, Cavers S, Bordács S, Smith P, Bozzano M (2014) Genetic considerations in ecosystem restoration using native tree species. Forest Ecol Manag 333:66–75

Trierveiler-Pereira L, Smith ME, Trappe JM, Nouhra ER (2015) Sequestrate fungi from Patagonian *Nothofagus* forests: *Cystangium* (Russulaceae, Basidiomycota). Mycologia 107(1):90–103

Tripp EA, Zhang N, Schneider H, Huang Y, Mueller GM, Hu Z, Häggblom M, Bhattacharya D (2017) Reshaping Darwin's Tree: Impact of the Symbiome. Trends Ecol Evol 32(8):552–555

Truong C, Mujic AB, Healy R, Kuhar F, Furci G, Torres D, Niskanen T, Sandoval-Leiva PA, Fernández N, Escobar JM, Moretto A, Palfner G, Pfister D, Nouhra E, Swenie R, Sánchez-García M, Matheny PB, Smith ME (2017) How to know the fungi: combining field inventories and DNA-barcoding to document fungal diversity. New Phytol 214(3):913–919

Truong C, Gabbarini LA, Corrales A, Mujic AB, Escobar JM, Moretto A, Smith ME (2019) Ectomycorrhizal fungi and soil enzymes exhibit contrasting patterns along elevation gradients in southern Patagonia. New Phytol. *In press*. https://doi.org/10.1111/nph.15714

Valenzuela E, Leiva S, Godoy R (2001) Variación estacional y potencial enzimático de microhongos asociados con la descomposición de hojarasca de *Nothofagus pumilio*. Rev Chil Hist Nat 74(4):737–749

Valenzuela E, Moreno G, Garnica S, Godoy R, Ramírez C (1999) Mycosociology in native forests of *Nothofagus* of the X Region of Chile, diversity and ecological role. Mycotaxon 72:217–226

Van der Heijden MG, Bardgett RD, van Straalen NM (2008) The unseen majority: soil microbes as drivers of plant diversity and productivity in terrestrial ecosystems. Ecol Lett 11:296–310

van der Wal A, Geydan TD, Kuyper TW, de Boer W (2013) A thready affair: linking fungal diversity and community dynamics to terrestrial decomposition processes. FEMS Microbiol Rev 37(4):477–494

Van Schöll L, Kuyper TW, Smits MM, Landeweert R, Hoffland E, van Breemen N (2008) Rock-eating mycorrhizas: their role in plant nutrition and biogeochemical cycles. Plant Soil 303(1–2):35–47

Weathers KC, Lovett GM, Likens GE, Caraco NF (2000) Cloudwater inputs of nitrogen to forest ecosystems in southern Chile: forms, fluxes, and sources. Ecosystems 3(6):590–595

Zúñiga-Feest A, Delgado M, Alberdi M (2010) The effect of phosphorus on growth and cluster-root formation in the Chilean Proteaceae: *Embothrium coccineum* (R. et J. Forst.). Plant Soil 334(1–2):113–121

Chapter 17
Mycorrhizas in South American Anthropic Environments

Marcela C. Pagano, Newton Falcão, Olmar B. Weber, Eduardo A. Correa, Valeria S. Faggioli, Gabriel Grilli, Fernanda Covacevich, and Marta N. Cabello

17.1 Introduction

The agricultural expansion has leaded to increase the irrigated cropland area and the use of fertilizers, resulting in water degradation, increased energy use, and common pollution (Foley et al. 2011). Of particular concern is the increased interest to reduce the environmental impacts of high quantities of water dedicated to irrigation by agricultural activities (Foley et al. 2011).

M. C. Pagano (✉)
Federal University of Minas Gerais, Belo Horizonte, Brazil

N. Falcão
Instituto Nacional de Pesquisas da Amazonia (INPA), Manaus, Brazil

O. B. Weber
Empresa Brasileira de Pesquisa Agropecuária, Embrapa Agroindústria Tropical, Fortaleza, Brazil

E. A. Correa
Empresa de Pesquisa Agropecuária de Minas Gerais EPAMIG-URECO, Pitanguí, Brazil

V. S. Faggioli
INTA EEA, Marcos Juárez, Argentina

G. Grilli
FCEFyN (CONICET-Universidad Nacional de Córdoba), Córdoba, Argentina

F. Covacevich
CONICET-Unidad Integrada EEA INTA- Facultad de Ciencias Agrárias UNMP, Balcarce, Argentina

M. N. Cabello
Instituto Spegazzini (Facultad de Ciencias Naturales y Museo, UNLP), Comisión de Investigaciones Científicas de la Prov. de Buenos Aires (CICPBA), La Plata, Argentina

M. C. Pagano, M. A. Lugo (eds.), *Mycorrhizal Fungi in South America*, Fungal Biology, https://doi.org/10.1007/978-3-030-15228-4_17

We are now truly recognizing the importance of sustainable measures in agriculture such as conservation of the vegetation cover and management approach to understand surface and deep soil responses to global change (Chaparro et al. 2012). For example, promising plant species can be tested to engineer the cultivable soil microbiome (Ellouze et al. 2013). The new alternatives for the agro-ecosystem management, such as inter-cropping, tillage and organic amendments, affect soils physical and chemical properties, modifying the abundance, diversity and activity of the mycorrhizal communities (Cardoso and Kuyper 2006; Pagano et al. 2011). Thus, the agroecology management based on key processes from natural ecosystems can help to solve some agricultural difficulties. For example, cultural practices (rotation, intercropping and fungal inoculation) that mimic the natural processes can reinforce the mycorrhizal potential in degraded ecosystems (Wahbi et al. 2016).

Increasing studies on the Arbuscular mycorrhizal fungi (AMF) has showed their importance for soil ecology (Bradford 2014) and studies on their biodiversity have spread in some agro-ecosystems such as corn and soybean monocultures (Carrenho et al. 2001; Gomes et al. 2015; Pontes et al. 2017) and coffee plantations (Cogo et al. 2017). Therefore, it is needed to deeply study the mycorrhizal functions under global change. In this chapter, we examine the major developments and advances on mycorrhizal fungi based on recent research from South American countries. New reports on the occurrence of mycorrhizas in Amazonian dark earth, as well as the inoculum production of arbuscular mycorrhizal fungi native of soils under native forest covers (dos Santos et al. 2017), have resulted in a more detailed understanding of the soil biology from South America.

Reports from Amazonian dark earth or "Terra preta do índio" soil has stimulated the use of biochar worldwide as a soil conditioner (Glaser 2007) that can add value to non-harvested agricultural products (Major et al. 2005) and promote plant growth. Few reports from Brazil showed that the addition of inorganic fertilizer, compost and chicken manure resulted in increases in plant cover and plant species richness (Major et al. 2005). In this sense, the biochar/mycorrhizae interactions also can be prioritized for sequestration of carbon in soils to contribute to climate change mitigation (Warnock et al. 2007).

17.2 The Mycorrhizal Symbioses in Agro-Ecosystems

Microorganisms are intensively investigated for novel compounds from saprophytic terrestrial fungi to marine habitats and living plants with their endophytes (Schueffler and Anke 2014). A growing worldwide attention on fungi is noticed, as of 100,000 known fungal species more than one million are predictable to exist (Schueffler and Anke 2014). Among soil fungi, AMF are of special interest for agriculture and increasing investigation from South America is continuously reported (Stürmer and Siqueira 2006; Pagano and Covacevich 2011; Castillo et al. 2016).

More information on indigenous AMF occurrence in agro-ecosystems as well as enough understanding of inoculum persistence, and cover crops that favor the indigenous arbuscular mycorrhizal fungi (AMF) by means of active roots (Douds et al. 2005) is needed. In general, compared to grasslands, conventionally managed fields can present low AMF diversity and low sporulation capacity (Thougnon Islas et al. 2016). Moreover, fruit plants (pineapple, Sapota trees) under organic management systems can also reduce the AMF species richness and abundance in relation to natural vegetation areas (Dantas et al. 2015).

Increasing interest in plant species for forest use as commercial plantations in Brazil has led to studies of response to inoculation of seedlings with AMF at different doses of P, such as for the Australian red cedar (*Toona ciliata* M. Roem var. *australis*) which presented high mycorrhizal root colonization, and thus, high quality seedlings (da Silva et al. 2017). Moreover, other researchers investigated the diversity of mycorrhizal fungi in planted forest in Northeast Brazil (Weber et al. unpublished) providing indication of Brazilian trees for reforestation in the tropical region.

AMF density and distribution vary both spatially and temporally with soil types and with host plant species diversity. Important economic plant species worldwide are being examined for AMF symbioses. Some agroecosystems have high economic interest such as coffee, vineyards and olive plantations, which are in the focus of interest from new technologies for their cultivation including their associated microbiota. Olive trees are mycotrophic species (Roldan-Fajardo and Barea 1986) associated with a high number of AMF species in the rhizosphere of plants growing in Morocco (Chliyeh et al. 2016) and Spain (Porras Soriano et al. 2002). The AMF diversity was studied for sustainable management of vineyards, showing low values in France (Bouffaud et al. 2016) and high values in vineyards from Germany under permanent vegetation cover or not (Oehl and Koch 2018). In Brazil, few studies such as from Rosa et al. (2016) investigated the application of AMF to reduce copper toxicity in young rootstock grapevines, pointing out some fungal species as promoters of great benefit. In the wine-growing regions of Southern Brazil a high humidity increases the susceptibility to foliar pathogens and thus, successive applications of copper fungicides are commonly used.

Much interest is nowadays dedicated in the preparation of inocula suitable for use in nurseries as this symbiosis improve plant performance and resistance to pathogens and water stress after transplantation. In Argentina, robust plants for field cultivation were obtained under greenhouse and nursery conditions by the coinoculation of two AMF strains at the beginning of plant propagation (Bompadre et al. 2014). It is known that the addition of organic amendments to the substrate can improve sporulation avoiding the replacing of nutrient solutions, vermicompost being commonly utilized. In Brazil, inoculated corn presented high number of infective propagules and biomass when inoculated with AMF and amended with vermicompost (Coelho et al. 2014). Peanut also responds positively when inoculated with different AMF species; however, the dependence on phosphorus (P) modified the plant responses (Hippler and Moreira 2013).

With regard to biochar, most reports are from Europe and the USA, and few reports from south American researchers, most from International Conferences, mention its interaction with AMF. One of them showed that biochar from *Eucalyptus* at high temperatures (700 °C) improved plant growth and AMF root colonization of sorghum, besides a higher spore germination (Dela Piccolla et al. 2016). Reports from Chile showed the early effect of the application on wheat in an Andisol and Ultisol improving root colonization by native AMF and glomalin content besides soil properties, thus encouraging implementation of sustainable systems. Biochar also improved sustainable barley grain production in field trials in the Araucanía Region of southern Chile (Curaqueo et al. 2014a, b).

In Brazil, investigating field samples in economic tree plantations and cassava in crop rotation, Pereira et al. (2014) found higher AMF species richness (30 taxa) in rhizospheric soil samples. However, Oehl et al. (2005) stressed that deep soil layers should be included in studies to better know the AMF diversity, especially in agroecosystems, where soil stirring is frequent. At present, research on crops, especially corn (Gomes et al. 2015), have increased and new reports compiled new information on AMF (Table 17.1). Weber (2014) also compiled the importance of biofertilizers and AMF in agriculture (Fig. 17.1).

In Chile, reports compiled during the last 10 years form the Southern-Central zone showed a total of 21 genera (represented by 57 species of AMF) that have been recognized, equivalent to 21% of all AMF species described worldwide (Castillo et al. 2016). Twenty-four AMF species were associated with different cultivars of *Triticum aestivum* and, differently, *Acaulospora* and *Scutellospora* predominate. In that study, AM fungal community structure differed along wheat cultivars: 'Porfiado' and 'Invento', with 19 species in relation to 'Otto' cultivar (15 species) (Aguilera et al. 2014). Castillo et al. (2006) studied the effects of tillage on AMF propagules. They found little differences in spore numbers, however a high root colonization in no-tillage treatments. Moreover, *Scutellospora* was common under no-tillage.

In Argentina, earlier studies have found less management of AMF to increase plant productivity (Covacevich and Echeverría 2009). It is known that soils of the Pampas region present high native AMF that colonize crop plants under different management systems (Covacevich et al. 2006, 2007; Schalamuk et al. 2006; Covacevich and Echeverría 2008); however, they are not yet manipulated. To avoid decreases in the grassland productivity, which leads to decline livestock production, new studies including AMF ecology and on the impact of agricultural practices on AMF symbiosis pointed to a selective decrease of viable spore number with glyphosate applications in native grasslands (in the Flooding Pampa), resulting in altered AMF community structure. However, the use of sublethal doses of the herbicide was more useful contributing to project more sustainable land management agroecosystems (Druille et al. 2015). In this regard, undisturbed (pristine) soils could be considered a reserve pool of diversity of native AMF, showing that spore and large number of propagules (hyphae) can be the main source of inoculum. Thus, low or no relationship between spore number with the root colonization and/or glomalin content can be found (Thougnon Islas et al. 2016).

Investigating the richness of AMF in soybean fields in Argentina (Fig. 17.2), Faggioli et al. (2019) found 95 AMF virtual taxa (VT) belonging to 8 families: Acaulosporaceae, Archaeosporaceae, Claroideoglomeraceae, Diversisporaceae, Gigasporaceae, Glomeraceae (57), Pacisporaceae, and Paraglomeraceae. Among

Table 17.1 Total number of identified species in some agro-ecosystems/anthropic environments from South America

Country/state	Biome/region	Crops/ Vegetation type	AMF species/ genera	Indicator/dominant species	Root colonization by AMF/ECM	Reference
Brazil	Amazonia	Cowpea	Inoculation (13 isolates)	ND	1–82%	Silva et al. (2018)
Brazil	Amazonia	Secondary vegetation	12	ND	NE	Pagano et al. (2016)
Brazil	Amazonia	Secondary vegetation	24	ND	NE	Leal et al. (2009)
Brazil	Amazonia	Babassu palm	16	ND	NE	Nobre et al. (2018)
Brazil, Pernambuco	Atlantic rain forest	Sapodilla, rubber tree, mahogany, eucalyptus plantation and cassava	24–30	Detected	NE	Pereira et al. (2014)
Brazil	Atlantic rain forest/ Caatinga ecotone	Forest trees	15–20	Detected	15–57% AMF / 12–29% ECM	Weber et al. (unpublished)
Brazil	Cerrado (14 sites)	Coffee	70	ND	13–40%	Cogo et al. (2017)†
Brazil	Cerrado	Maize	10 genera	ND	NE	Gomes et al. 2015)
Brazil	Cerrado	Grassland (*Brachiaria brizantha*)	11	ND	40–62%	Ferreira et al. (2012)
Brazil	Atlantic rain forest/ Cerrado ecotone	Native and exotic trees	16	D	NE	Correa et al. (unpublished)
Brazil	Cerrado	Soybean	15–18	D	NE	Pontes et al. (2017)
Brazil / Sao Paulo	Sugarcane cropping region	Sugarcane	22	ND	30–52%	Azevedo et al. (2014)

(continued)

Table 17.1 (continued)

Country/state	Biome/region	Crops/ Vegetation type	AMF species/ genera	Indicator/dominant species	Root colonization by AMF/ECM	Reference
Brazil / Sao Paulo		Leguminous green manure and sunflower in rotation with sugarcane	NE	ND	49–74%	Ambrosano et al. (2010)
Brazil / Londrina, Paraná		Soybean and cotton	*Rhizophagus clarus* inoculation		~20–70%	Cely et al. (2016)
Brazil	Pampa	Various species	NM	NM	Presence	König et al. (2014)†
Brazil	Santa Catarina state/ Experimental Station	Cassava	*Rhizophagus clarus* inoculation	4–9	13–20%	Heberle et al. (2015)
Chile	Agroecosystems of the southern-central zone	Horticultural, wheat managed grasslands, wheat rotation, other crops	5–24	*Glomus* spp.	NM	Castillo et al. (2016)†
Argentina	Pampa	Wheat	NE	ND	NM	Schalamuk et al. (2013)
Argentina	Pampa	Wheat	Inoculation of *Glomus mosseae*	ND	~40%	Schalamuk et al. (2011)
Argentina	Pampa (126 sites)	Soybean	37 species	*Glomus fuegianum* (long term agriculture)		Faggioli (2016)
Argentina	Rainforest of Misiones	*Ilex paraguariensis* (traditional / high technology fertilized crops)	NE	ND	~25–50%	Velázquez et al. (2018)

AMF (spores): species (N° min – N° max); Indicator species (the most characteristic of a site): D (detected) or ND (not detected); NE (not evaluated); NM (not mentioned); †Checklist or review

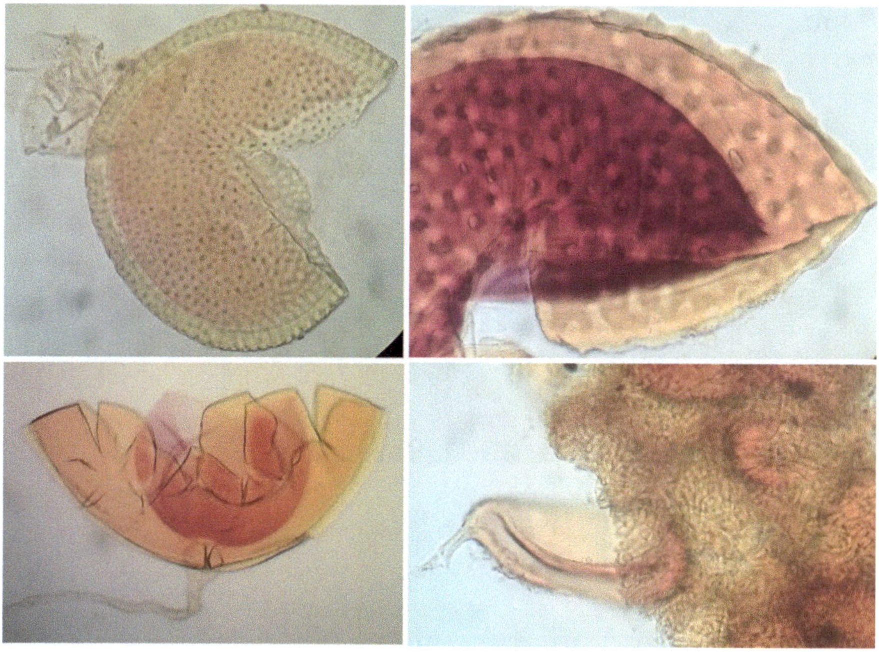

Fig. 17.1 Some AMF spores from cultivated areas in Brazil. Clockwise, from upper left: AMF spores of *Acaulospora* spp., *Glomus* and *Gigasporales* representant isolated from Northeast region (Photo-credit: M. Pagano)

them, Diversisporaceae was the most sensitive to long term Agricultural practices (Fig. 17.3). VT richness per sample did not differ between historical land uses and it could be attributed to the widespread use of no-tillage practices associated with soybean cultivation. This conservative soil management has been well documented as positive in the maintenance of AMF richness (Colombo et al. 2014). Soil textural components (i.e. clay and sand content) appeared as significant determiners of AMF richness (Fig. 17.4). Coarser soils were related to high VT richness in soil but low VT richness in roots. This probably was consequence of different textural preferences of AMF species (Lekberg et al. 2007). However, it is worth to highlight here that sandy soils were located in the driest area. Hence, the effect of drought on plant growth could also negatively affect key stages of AMF colonization resulting in the diminution of VT richness in roots of Livestock sites.

Among crop variables, only plant density was significantly correlated with VT richness (Fig. 17.5). Larger density of plant roots might improve resource availability for AMF because more carbohydrates would be available to support the symbiosis (Lekberg et al. 2010). In addition, roots and the associated fungal network might explore higher soil volume and contact propagules of rare and infrequent AMF species which may result in increases of VT richness. Therefore, our results reveal that appropriate plant density is a promising agronomical parameter for the maintenance of AMF species in agroecosystems.

Fig. 17.2 Historical land uses (HLU) currently cultivated with soybean in Pampas Region (Cordoba, Argentina): (**a**) agricultural, (**b**) Livestock-Agricultural, (**c**) Agricultural after recent deforestation of shrub land area. Each location was approximately 100 km from another one. Ten sampling sites were selected in each situation (Faggioli et al. 2019) (Photo-credit: V Faggioli)

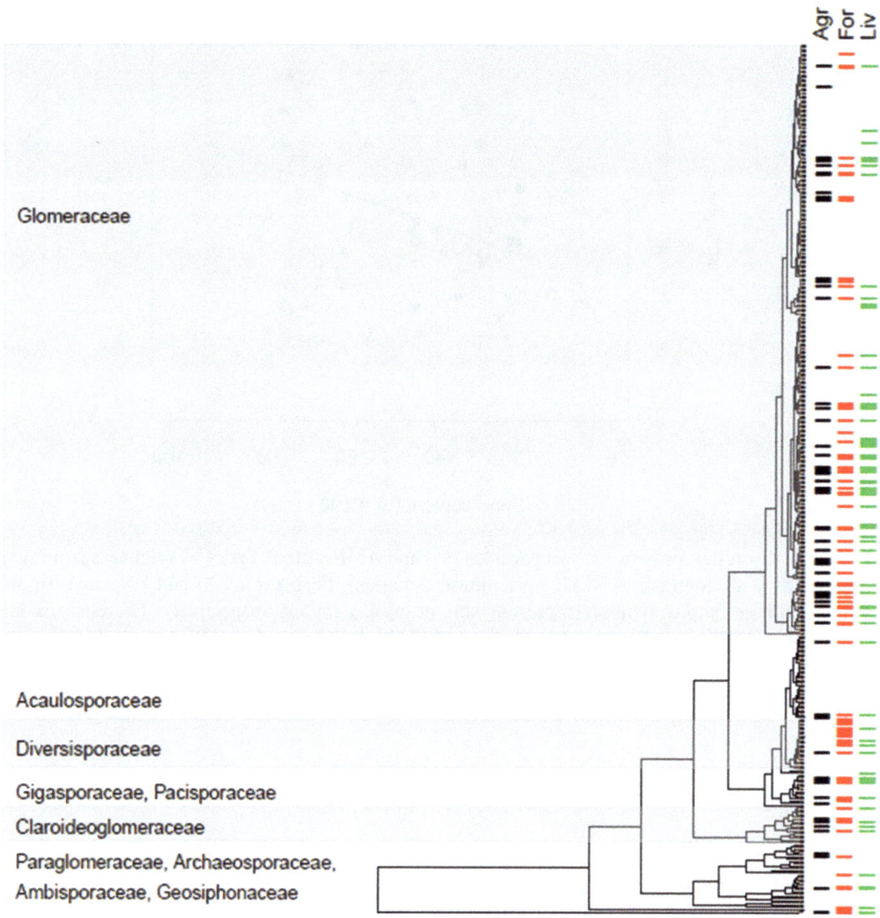

Fig. 17.3 Glomeromycota phylogenetic tree with virtual taxa (VT) recorded in different historical land use (HLU). The tree contains type SSU rRNA gene sequences of VT from the Maarj*AM* database (Öpik et al. 2010). Coloured lines indicate the presence of VT in HLU: Agricultural (Agr., black lines), Forest (For., red lines) and Livestock (Liv., green lines). Molecular study performed by 454 pyrosequencing and taxonomic assignment of sequences against MaarjAM database according to Faggioli et al. (2019)

In South America, the impact of different agricultural practices on AMF in arable fields is still poorly understood. Wheat phenology improved AMF biodiversity during grain filling; however, tilling and fertilization did not decrease spore biodiversity (Schalamuk et al. 2006). Spore populations of AMF communities in arable fields of wheat crop can vary between from just one to 4 spores g−1 soil in conventional tillage, from 3 to 5 in no-tillage (Schalamuk et al. 2013) but it also depends on plant phenological stages. Rarely more than 26 AMF species were reported in field studies (Schalamuk and Cabello 2010a, b). Pioneer studies on propagules in soils (propagule bank) from Argentina showed that different environmental

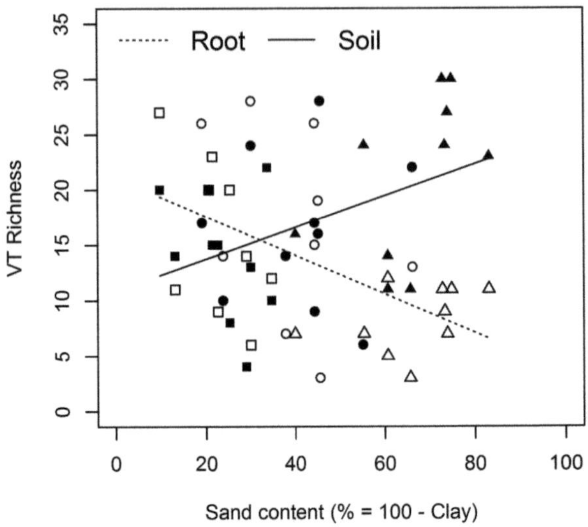

Fig. 17.4 Correlations between soil particles (%) and AMF Virtual Taxa (VT) richness from soybean fields with contrasting HLU: Agricultural (squares), Forest (circles) and Livestock (triangles); solid or empty symbols represent soil or root samples, respectively. Correlations are statistically significant (Spearman Test $p < 0.001$). Molecular study performed by 454 pyrosequencing and taxonomic assignment of sequences against MaarjAM database according to Faggioli et al. (2019)

Fig. 17.5 Correlation between plant density (number or plant per square meter) and VT richness in soil samples in soybean fields with contrasting HLU (p-value < 0.01, Spearman coefficient 0.47). Molecular study performed by 454 pyrosequencing and taxonomic assignment of sequences against MaarjAM database according to Faggioli et al. (2019)

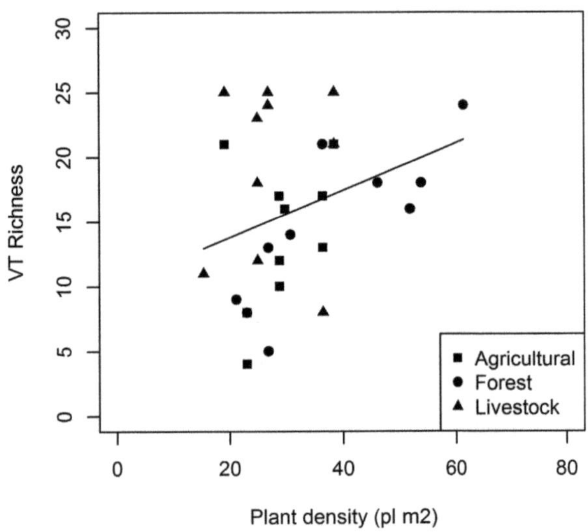

conditions and the effects of tillage and no-tillage modify both the composition of the AMF soil propagule bank and the diversity (Schalamuk and Cabello 2010a, b). Generally, Acaulosporaceae, Gigasporaceae, Glomeraceae can be found in agricultural fields; however, *Glomus* predominate (Schalamuk and Cabello 2010a, b). This can lead to think in different types of AMF inocula based on the proportions of their AMF families (Acaulosporaceae, Gigasporaceae, Glomeraceae) between field and trap cultures. For instance, in the forest garden, Czerniak and Stürmer (2015) tested two AMF species of different families, such as Gigasporaceae and Glomeraceae (*Dentiscutata heterogama* and *Claroideoglomus etunicatum*, respectively) in on farm production of inoculum against residues from the forestry industry (pine bark and pulp sludge).

In the trap cultures from agro-ecosystems more than 90% of AM species belong to Glomeraceae (Schalamuk and Cabello 2010a, b). *Glomus* spp. (Glomeraceae) present more extensive root colonization than other families and lower soil colonization by extraradical hyphae besides rapid colonization of new plants also from colonized roots fragment (Hart and Reader 2002). Thus, in the trap cultures prepared from crop systems generally *Glomus* or *Acaulospora* species are recovered.

In Southern Brazil, increasing studies of AMF in experimental farms and fruit plant orchards have extended the panorama of investigation with this type of soil fungi. Reports on AMF diversity in fruit orchards of Blueberries cultivars showed the prevalence of species of *Glomus* and *Acaulospora* and the potential benefit from inoculated AMF such as *Gigaspora margarita* and *Glomus etunicatum* (Farias 2012). In the semiarid region, Dantas et al. (2015) investigated the AMF occurrence in the establishment of fruits plants (pineapple, Sapota trees) under organic management, detecting *Glomus* spores in all the areas, and corroborated the fact that soil management in organic cropping systems reduce the AMF species richness and abundance in relation to natural vegetation areas.

The AMF occurrence was investigated in an experimental farm in Minas Gerais State (Correa et al. unpublished) under different plant covers. High diversity and abundance were related to adjacent native forest, with 16 AMF species; however, grassland and maize field presented lower values.

Lastly, another anthropic environment is the man-made anthrosoils conformed by Amazonian Dark earth, also called Terra preta do índio (TPI), a highly fertile soil whose processes of formation has not yet been resolved (Hofwegen et al. 2009). In this regard, more recent reports (Tsai et al. 2009; Pagano et al. 2016) on the microbial communities of TPI have pointed to the presence of AMF of varied families/order unlike trends for cultivated field soils, with dominance of Glomeraceae. Black carbon prevalence and its unique physical and chemical characteristics, point it as the chief component conforming recalcitrant biochar with unique microbial communities (Tsai et al. 2009). For example, in TPI samples at different depths: from 0–20 to 100 cm, from Amazonas State, Brazil, Pagano et al. (2016) identified 12 AMF species (*Acaulospora bireticulata, A. mellea, A. rhemii, A. scrobiculata, A. spinosa, Ambispora appendicula, Claroideoglomus etunicatum, Scutellospora calospora, Racocetra castanea, Funneliformes geosporus, Glomus tortuosum, Pacispora franciscana*) and 6 were *Glomus* like species. Glomeromycota were dominated by Diversisporales, followed by Glomerales and Gigasporales.

As seen in previous observations in other soil types most of the AMF species richness and diversity (Shannon index) were concentrated in the topmost soil horizons. The *Scutellospora* species was found only in the deeper strata, in agreement with some previous reports (Oehl et al. 2005). *Scutellospora calospora* was also found in the control soil only at subsurface layer (0–20 cm) in contrast to its occurrence at 60–100 m in TPI soil samples.

With regard to the control adjacent soil samples (oxisol and ultisols), similar AMF species were detected, with 8 species identified and 3 unidentified. *Racocetra castanea* found only in the control soil at subsurface layer (0–20 cm) together with *Glomus tortuosum* (20–40 cm depth) occurred exclusively in adjacent soils and most species (11) were in common between the TPI and adjacent soils. This microbiological analysis showed that the abundance of AMF was greater in TPI than in control soils. AMF richness decreases only at great depth; however, diversity remained similar.

17.3 The Soil Conditioners in Agro-Ecosystems

Similar to methods to potentialize the mycorrhizal fungal inoculation of roots using soil amendments (Smith and Read 2008), no-tillage methods used to apply biochar into the root zone of crop soils and the mycorrhizal responses to biochar addition were amongst the pioneering works in biochar research. Another anthropic environment originated from South America is the ancient man-made anthrosoils conformed by Amazonian Dark earth, also called *Terra preta do índio* (TPI) (Fig. 17.5), a highly fertile soil whose processes of formation has not yet been resolved (Hofwegen et al. 2009). In Brazil, the Amazonian Dark Earth "Terra Preta" is dated about 7000 years being common at the Amazon basin (Falcão et al. 2003; Glaser 2007) and it is a promising subject to help sustainable agriculture, soil C sequestration and thus, climatic change mitigation. The climate at these areas is Koeppen's Af tropical rainforest with an annual average temperature between 25 °C and 35 °C. At the time of sampling the vegetation cover is usually secondary forest. In this regard, more recent reports (Tsai et al. 2009; Pagano et al. 2016) on the microbial communities of TPI have pointed to the presence of AMF of varied families/order unlike trends for cultivated field soils, with dominance of Glomeraceae. Black carbon prevalence and its unique physical and chemical characteristics, point it as the chief component conforming recalcitrant biochar with unique microbial communities (Tsai et al. 2009). For example, in TPI samples at different depths: from 0–20 to 100 cm, from Amazonas State, Brazil, Pagano et al. (2016) identified 11 AMF species (*Acaulospora bireticulata, A. mellea, A. rhemii, A. scrobiculata, A. spinosa, Ambispora appendicula, Claroideoglomus etunicatum, Scutellospora calospora, Racocetra castanea, Funneliformes geosporus, Glomus tortuosum, Pacispora franciscana*) and 6 were *Glomus* like species. Glomeromycota were dominated by Diversisporales, followed by Glomerales and Gigasporales.

As seen in previous observations in other soil types most of the AMF species richness and diversity (Shannon index) were concentrated in the topmost soil horizons. The *Scutellospora* species was found only in the deeper strata, in agreement with some previous reports (Oehl et al. 2005). *Scutellospora calospora* was found in the control soil only at subsurface layer (0–20 cm) in contrast to its occurrence at 60–100 m in TPI soil samples. With regard to the control adjacent soil samples (oxisol and ultisols), similar AMF species were detected, with 8 species identified and 3 unidentified. *Racocetra castanea* was found only in the control soil at subsurface layer (0–20 cm) together with *Glomus tortuosum* (20–40 cm depth) occurred exclusively in adjacent soils and most species (11) were in common between the TPI and adjacent soils. This microbiological analysis showed that the abundance of AMF was greater in TPI than in control soils. AMF richness decreases only at great depth; however, diversity remained similar.

Lastly, there is more nuances in the study of TPI, the Terra preta (very dark, with broken potsherds and highly nutrient content) form under sites of home inhabitation, and the Terra mulata (light brown and with less nutrient content), which is less well documented. Thus, Amazonian dark earths are subdivided into: terra preta and terra mulata (black earths and brown earths respectively) (Kern and Kämpf 1989; Arroyo-Kalin 2008) that associates with respectively, past settlement areas and cultivated fields (Arroyo-Kalin 2010). TPI usually exhibit highly elevated levels of phosphorus (P), calcium (Ca) and other essential minerals for plants (Figs. 17.6 and 17.7). Terra mulata present less nutrient content, light brown, being adjacent to TPI

Fig. 17.6 Sites of Terra Preta de Índio in the Jiquitaia Farm (Lat 2° 37′S, Long 59° 40′W). The vegetation is secondary forest capoeira type, with approximately 40 years of age. Soil samples are usually collected from the 0–20 cm and 20–40 cm depth layers. Clockwise, from upper left: Overview of the area with Latossolo Amarelo with A anthropic horizon (Terra Preta de Indio) at Rio Preto da Eva, AM; Representative profil (Photo-credit: NPS Falcão) and spores of AMF retrieved from soil samples (Photo-credit: M Pagano)

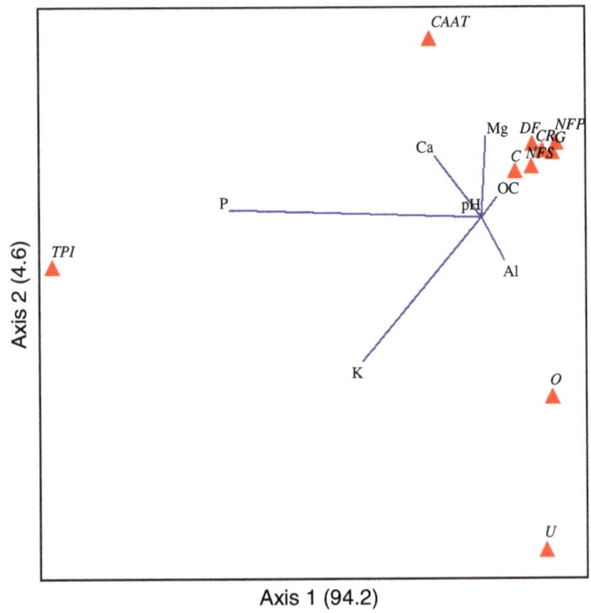

Fig. 17.7 Environmental similarity among some different soil samples (forest, TPI, control soils and cultivated sites) studied in Brazil. The similarity among the geographic areas represented as a nonmetric multidimensional scaling (NMDS). Distance and placement is indicative of similarity among areas TPI = "Terra preta do Indio" soil sample; U = ultisol and O = oxisol (control soils from the Amazon region), NFS = soil sample from Atlantic forest at Minas Gerais state; G = cultivated grassland at Minas Gerais state; NFP = soil sample from a Atlantic forest at Minas Gerais state, DF = soil from a dry forest at Ceará state; CR = soil from dry vegetation type and C = cultivated site

sites. It is believed that it was formed through intensive agriculture involving burning and mulching under low oxygen (Hecht 2003, Fraser et al. 2011).

17.4 AM Inoculation for Agro-Ecosystems

The mycorrhizal inoculation technologies or to manage native arbuscular mycorrhizal fungus communities can serve to replace or reinforce the mycorrhizal potential in degraded ecosystems (Wahbi et al. 2016). For example, to manage AMF soil infectivity in agrosystems it was proposed reductionist and holistic schemes that could be combined: the reductionist pattern aims to improve plant performance in disturbed soils by adding specialized AMF inocula adapted to the environmental conditions and to the target crop. Still, the objectives of the holistic pattern are to preserve and restore the composition of native AMF communities (Wahbi et al. 2016). However, benefits can be obtained from the integration of AMF in agricultural practices through the combination of the "reductionist" and "holistic" approaches (Wahbi et al. 2016).

The management of AMF in the rhizosphere provides an alternative to high inputs of fertilizers and pesticides in sustainable plant production systems (Reviewed by Azcón-Aguilar and Barea 1997). Moreover, crop yield increases showed the potential to be used by farmers (Douds et al. 2005). However, AM inoculation technology is limited by the lack of production of commercial inocula, because a difficult multiplication on artificial growth media without a host (Sieverding 1991).

Some researchers suggested a careful choice of compatible host/mycorrhiza/substrate combination for crop success (Azcón-Aguilar and Barea 1997). Many methods are used to handle AMF, inoculating them on host plants, and replicating large amounts of inoculum. In vivo cultures of AMF species from different regions are preserved in ex-situ collections (Giovannetti and Avio 2002).

Other techniques have been developed to produce large quantities of soil-free inoculum, based on hydroponic and aeroponic cultivation systems (Jarstfer and Sylvia 1995). The roots transformed by *Agrobacterium rhizogenes* are also effective as inocula which generally utilized carrot, but they are generally used as experimental model systems for research purposes (Giovannetti and Avio 2002). But these inoculation procedures are highly expensive and only utilized in agriculture of high value products.

An alternative source of inocula is to use roadsides around crop fields as a repository for the conservation of AMF diversity affected by Land use (Dai et al. 2013). It has become customary to use AM spores as inoculum (Read 2003) and using three representative genera of AMF (mixed inocula) is a common inoculation strategy.

In South America, several works showed the feasibility and importance of AM inoculation in a large number of economic value and fruit plants. The applications of mycorrhizas in agriculture and environmental issues are still incipient. AMF inoculant for farm application requires large-scale multiplication fungi. The expensive technology of inoculum production comprises formation of single cultures of AMF. A cheaper method is the "on farm" system (farmers can produce their inoculum) (Douds et al. 2008, 2010), native AMF being more efficient due to local adaptation to the environment (Sreenivassa 1992). Infective propagules of AMF (spores, hypha and colonized roots) can be used as inoculum (Sieverding 1991).

In fertile soils from Argentina (Pampa Ondulada region), the effects of agronomic practices on the AMF communities, was reported by using pyrosequencing or a morphological approach (Colombo et al. 2014) showing that soil management has a negative effect on AMF community biodiversity. This study greatly improved the knowledge about AM fungi in South America where the molecular diversity of AM fungi was practically unknown.

Maize crop in Argentina is, after soybeans, the second most important crop (with the highest planted area, followed by wheat, citrus, sugarcane, and sunflower (Boix and Zinck 2008). However, non-tillage and contemporary hybrids with high yield that accumulation of crop residues affect the balance of biological and chemical cycles disturbing the P and Zn levels (Ratto and Miguez 2006). In this sense, Astiz et al. (2014) suggested that soil characteristics could be used to select potentially beneficial inoculum to compensate Zn deficiency in maize. The inoculum of indigenous AMF from sites presenting different levels of P and Zn resulted in changes in

Table 17.2 Some book or reviews dealing with AMF and ecological restoration in South America

Reports on AMF and plant restoration	Biome/ Country	References
Restored environments	Argentina, Brazil	Pagano (2012)
Riparian forest	Brazil	Braghirolli et al. (2012)
Arbuscular mycorrhizas in degraded land restoration	Brazil	Soares and Carneiro (2010)
Native species for restoration and conservation of biodiversity in South America	Argentina, Brazil	Pagano et al. (2012), Pagano (2016)

root colonization by AMF and response to inoculation in both Zn uptake and dry matter production. The inoculum indigenous from a site with low P and high Zn content was the lowest efficient. Thus, to compare agricultural fields with high and low soil biota abundance and diversity to assess soil biota potential when soil communities are well developed is urgently needed (Bender and van der Heijden 2014).

Interestingly, in Colombia, the edaphic factors such as Soil pH had a direct relationship with species richness and with the diversity index, but, height above sea level can also affect the AMF community composition. Thus, a heterogeneous distribution in patches with little influence of the type of crop management (mono or polyculture) can be found. This highlight the constraints of developing specific biofertilizers for crops that contain AMF and not including natural adaptations to the different characteristics of the varied agriculture soil types (Mahecha-Vásquez and Sierra 2017). We lack the field studies that are needed to understand with confidence how to do an effective AMF inoculation.

With regard to the ecological restoration of species-rich grasslands that are of priority for conservation of biodiversity, reports have showed many options for that task in South America (Table 17.2). Torrez et al. (2016) determined if plant species recolonization of degraded nutrient-poor grasslands could be increased by adding a local source of AMF inoculum at different distances from intact remnant grasslands. There are effects by the well-dispersed generalist plant species, particularly at 20 m from the intact patches, the role of below-ground processes being crucial for restoration success that can be improved by AMF additions in the short term and at relatively close distances to intact grassland patches (Teste 2016). In Fig. 17.8 we show a protocol to add AM fungi to disturbed ecosystems.

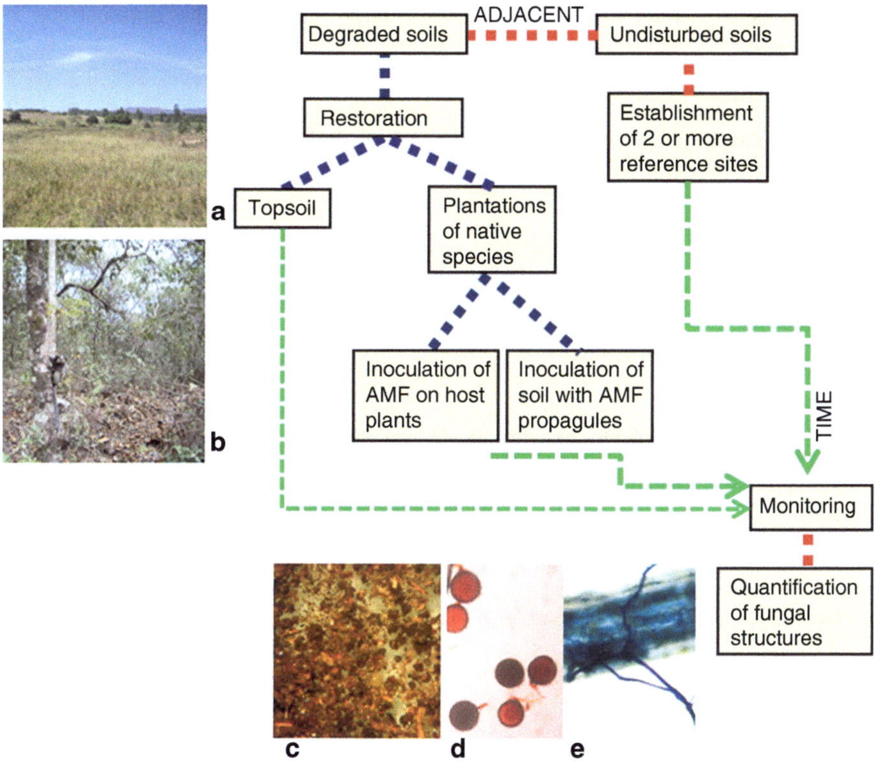

Fig. 17.8 Protocol to add AM fungi to disturbed ecosystems. After evaluation of diagnostic of degraded site (**a**) and establishment of 2 or more undisturbed sites (**b**), restoration can be achieved by introduction of topsoil or plantations. AMF inoculation can be performed on host plants or by inoculation of health soil (**c**). Monitoring the restored sites: determination of infective propagules including spores recovered from rhizospheric soils (**d**) and roots of plants growing in the degraded and reference soils stained for AM colonization (**e**) (Photos by M. Pagano)

17.5 Conclusion

In this chapter, the examination and use of arbuscular mycorrhizas in different crop systems has been mentioned and the needs for more information to understand agro-ecosystems and soils under different management have been highlighted. Throughout the chapter, the study of the occurrence of mycorrhizas in agriculture in South America were showed as still incipient. Morphological identification procedure of AMF continues to be important, although the specific training and experience. Moreover, better technology for commercial mycorrhizal inoculum is needed. Finally, this chapter argues that agro-ecosystems generally present low AMF diversity; however, organically managed fields are more similar to natural ecosystems, Amazonian dark earth being a model of highly fertile soils. Consequently, further research is necessary on this field, especially regarding the applications of mycorrhizas.

Acknowledgements All authors contributed to this chapter. Dr Neimar F. Duarte - Pró-Reitor de Pesquisa, Inovação e Pós-Graduação, Instituto Federal de Minas Gerais, Brazil is gratefully acknowledged. Dr Gabriel Grilli was supported by FCEFyN (CONICET-UNiversidad Nacional de Córdoba).

References

Aguilera P, Cornejo P, Borie F, Barea JM, von Baer E, Oehl F (2014) Diversity of arbuscular mycorrhizal fungi associated with *Triticum aestivum* L. plants growing in an Andosol with high aluminum level. Agriculture, Ecosystems and Environment 186: 178–184

Astiz IP, Barbieri PA, Echeverría HE, Rozas HRS, Covacevich F (2014) Indigenous mycorrhizal fungi from Argentina increase Zn nutrition of maize modulated by Zn fertilization Soil Environ. 33,1: 23–32

Azcón-Aguilar C, Barea JM (1997) Applying mycorrhiza biotechnology to horticulture: significance and potentials. Scientia Horticulturae 68: 1–24

Azevedo LCB, Stürmer SL, Lambais MR (2014) Early changes in arbuscular mycorrhiza development in sugarcane under two harvest management systems. Brazilian Journal of Microbiology 45: 3, 995–1005

Ambrosano EJ, Azcón R, Cantarella H (2010) Crop rotation biomass and arbuscular mycorrhizal fungi effects on sugarcane yield. Sci agric 67:6

Arroyo-Kalin M (2008) Steps towards an ecology of landscape: A geoarchaeological approach to the study of anthropogenic dark earths in the central Amazon region, Brazil. PhD thesis, University of Cambridge

Arroyo-Kalin M (2010) The Amazonian formative: crop domestication and anthropogenic soils. Diversity 2(4):473–504

Boix LR, Zinck JA (2008) Land-Use Planning in the Chaco Plain (Burruyacú, Argentina). Part 1: Evaluating Land-Use Options to Support Crop Diversification in an Agricultural Frontier Area Using Physical Land Evaluation. Environmental Management 42: 1043–1063

Bompadre,MJ Pérgola M, Bidondo LF, Colombo RP, et al. (2014) Evaluation of arbuscular mycorrhizal fungi capacity to alleviate abiotic stress of olive (*Olea europaea* L.) Plants at different transplant conditions. e Scientific World Journal, Article ID 378950, 12 p.

Bouffaud ML, Bernaud E, Colombet A, Van Tuinen D, Wipf D, Redecker D (2016) Regional-scale analysis of arbuscular mycorrhizal fungi: the case of Burgundy vineyards. J. Int. Sci. Vigne Vin 50, 1: 1–8

Bradford MA (2014) Good dirt with good friends. Nature 505: 486–487

Bender SF, Van der Heijden MGA (2014) Soil biota enhance agricultural sustainability by improving crop yield, nutrient uptake and reducing nitrogen leaching losses. Journal of Applied Ecology 52(1)

Braghirolli FL, Sgrott AF, Pescador R, Uhlmann A, Stürmer SL (2012) Arbuscular mycorrhizal fungi in riparian forest restoration and soil carbon fixation. Rev Bras Ciênc Solo 36:3

Cardoso IM, Kuyper TW (2006) Mycorrhizas and tropical soil fertility. Agric. Ecosyst. Environ. 116: 72–84

Castillo CG, Borie F, Oehl F, Sieverding E (2016) Arbuscular mycorrhizal fungi biodiversity: prospecting in Southern-Central zone of Chile. A review. Journal of Soil Science and Plant Nutrition 16 (2): 400–422

Cely MVT, de Oliveira AG, de Freitas VF, de Luca MB et al. (2016) Inoculant of Arbuscular Mycorrhizal Fungi (Rhizophagus clarus) Increase Yield of Soybean and Cotton under Field Conditions. Frontiers in Microbiology 7, Article 720

Chaparro J, Sheflin A, Manter D, Vivanco J (2012) Manipulating the soil microbiome to increase soil health and plant fertility. Biology and Fertility of Soils 48: 489–499

Chliyeh M, Kachkouch W, Zouheir T, Touhami AQ et al. (2016) Evolution of a composite endomycorrhizal inoculums in function of time in the level of the olive plants rhizosphere. IJAPBC 5(1), ISSN: 2277–4688

Coelho IR, Pedone-Bonfim MVL, Silva FSB, Maia LC (2014) Optimization of the production of mycorrhizal inoculum on substrate with organic fertilizer. Brazilian Journal of Microbiology 45, 4: 1173–1178

Cogo FD, Guimarães PTG, Pouyú Rojas E, Saggin Júnior OJ, Siqueira JO, Carneiro MAC (2017) Arbuscular mycorrhiza in *Coffea arabica* l.: Review and meta-analysis. Coffee Science, 12, 3: 419–443

Colombo RP, Bidondo LF, Silvani VA, Carbonetto MB, Rascovan N, Bompadre MJ, Pérgola M, Cuenca G, Godeas AM (2014) Diversity of arbuscular mycorrhizal fungi in soil from the Pampa Ondulada, Argentina, assessed by pyrosequencing and morphological techniques Canadian Journal of Microbiology 60, 12: 819–827

Covacevich F, Echeverría HE (2008) Receptivity of an Argentinean pampas soil to arbuscular mycorrhizal *Glomus* and *Acaulospora* strains. World Journal of Agricultural Sciences 4, 6: 688–698

Covacevich F, Echeverría HE (2009) Mycorrhizal occurrence and responsiveness in tall fescue and wheatgrass are affected by the source of phosphorus fertilizer and fungal inoculation. Journal of Plant Interactions 4,2: 101–112

Covacevich F, Echeverría HE, Aguirrezaba LAN (2007) Soil available phosphorus status determines indigenous mycorrhizal colonization of field and glasshouse-grown spring wheat from Argentina. Applied Soil Ecology 35: 1–9

Covacevich F, Marino MA, Echeverria HE (2006) The phosphorus source determines the arbuscular mycorrhizal potential and the native mycorrhizal colonization of tall fescue and wheat grass in a moderately acidic Argentinean soil. European Journal of Soil Biology 42: 127–138

Curaqueo G, Meier S, Borie F, Navia R (2014a) Biochar and Arbuscular Mycorrhizal Fungi: An Alternative to Contributing to Agroecosystem Sustainability The 20th World Congress of Soil Science June 8~13, Jeju, Korea

Curaqueo G, Meier S, Khan N, Cea M, Navia R (2014b) Use of biochar on two volcanic soils: effects on soil properties and barley yield. Journal of Soil Science and Plant Nutrition 14, 4: 911–924

Czerniak MJ, Stürmer SL (2015) On-farm production of mycorrhizal inoculum using residues from the forestry industry. Revista Brasileira de Ciencia do Solo 38 (6):1712–1721

Carrenho R, Trufem SFB (2001) Bononi V L. R. 2001. Arbuscular mycorrhizal fungi in rhizospheres of three phytobionts established in a riparian area. Acta bot bras 15(1):115–124

Castillo CG, Rubio R, Rouanet L, Borie F (2006) Early effects of tillage and crop rotation on arbuscular mycorrhizal fungal propagules in an Ultisol. Biol Fertil Soils 43:83

da Silva EP, Ferreira PAA, Furtini-Neto AE, Soares CRFS (2017) Arbuscular mycorrhiza and phosphate on growth of Australian Red Cedar seedlings. Ciência Florestal 27, 4: 1269–128

dos Santos RS, Ferreira JS, Scoriza RN (2017) Inoculum production of arbuscular mycorrhizal fungi native to soils under different forest covers. Rev. Ceres: 64, 2: 197–204

Dai M, Bainard LD, Hamel C, Gan Y, Lynch D (2013) Impact of Land Use on Arbuscular Mycorrhizal Fungal Communities in Rural Canada. Appl Environ Microbiol 79, 21: 6719–6729

Dela Piccolla C, Novotny EH, Ryosuke T, Saito M. Growth and Symbiosis of Plants with Arbuscular Mycorrhizal Fungi in Soil Submitted to Biochar Application 2016. In: International Conference of International Humic Substances Society, 18, 2016, Kanazawa. Keystone for future earth: natural organic matter (NOM) in diverse environments: proceedings. Kobe: IHSS: Kobe University

Douds DD, Nagahashi G, Pfeffer PE, Kayser WM, Reider C (2005) On-farm production of AM fungus inoculum in mixtures of compost and vermiculite. Bioresour Technol 97:809–818
Douds DD, Nagahashi G, Reider C, Hepperly PR (2008) Choosing a mixture ratio for the on-farm production of AM fungus inoculum in mixtures of compost and vermiculite. Compost Science and Utilization 16: 52–60
Douds DD, Nagahashi G, Reider C, Hepperly PR (2010) On-farm production of inoculum of indigenous arbuscular mycorrhizal fungi and assessment of diluents of compost for inoculum production. Bioresource Technology 101 (7): 2326–2330
Druille M, Cabello MN, García Parisi PA, Golluscio RA, Omacini M (2015) Glyphosate vulnerability explains changes in root-symbionts propagules viability in pampean grasslands. Agriculture, Ecosystems and Environment 202: 48–55
Dantas BL, Weber OB, Matos EPNB, Neto JPM, Mendes Filho PF, Rossetti AG, Pagano MC (2015) Diversity of arbuscular mycorrhizal fungi in an organic orchard of semi-arid land of Ceará. Ciência Rural 45(8):1480–1486
Ellouze W, Hamel C, Vujanovic V, Gan Y, Bouzid S, St-Arnaud M (2013) Chickpea genotypes shape the soil microbiome and affect the establishment of the subsequent durum wheat crop in the semiarid North American Great Plains. Soil Biology & Biochemistry 63: 129–141
Faggioli VS, Cabello MN, Grilli G, Vasar M, Covacevich F, Öpik M (2019). Root colonizing and soil borne communities of arbuscular mycorrhizal fungi differ among soybean fields with contrasting historical land use. Agriculture, Ecosystems & Environment 269: 174–182
Faggioli VS (2016) Estudio de las comunidades de hongos formadores de micorrizas arbusculares en soja: Relación con la nutrición fosforada en agroecosistemas. Doctoral thesis. Facultad de Ciencias Naturales y Museo. Universidad Nacional de La Plata, 162 p.
Farias DH (2012) Diversity of arbuscular mycorrhizal fungi in orchards and seedling growth of micropropagated blueberry. Dissertation (Master Degree) Post-Graduation Program in Agronomy. Faculdade de Agronomia Eliseu Maciel, Universidade Federal de Pelotas
Ferreira DA, Carneiro MAC, Saggin Junior OJ (2012) Fungos Micorrízicos Arbusculares em um Latossolo Vermelho sob Manejos e Usos no Cerrado. R. Bras. Ci. Solo 36:51–61
Foley JA, Ramankutty N, Brauman KA, Cassidy ES, Gerber JS, Johnston M, Mueller ND, O'Connell C, Ray DK, West PC, Balzer C, Bennett EM, Carpenter SR, Hill J, Monfreda C, Polasky S, Rockström J, Sheehan J, Siebert S, Tilman D, Zaks DPM (2011) Solutions for a cultivated planet. Nature 478: 337–342
Falcão NP, Comerford N, Lehmann J (2003) Determining nutrient bioavailability of Amazonian Dark Earth soils – methodological challenges. In J. Lehmann, DC Kern B Glaser & WI Woods (eds), Amazonian Dark Earths: Origin, Properties, Management (pp. 255–270). The Netherlands: Kluwer Academic Publishers
FRASER JA, TEIXEIRA W, FALCÃO N et al (2011) Anthropogenic soils in the Central Amazon: from categories to a continuum. Area 43(3):264–273
Giovannetti M, Avio L (2002) Biotechnology of Arbuscular Mycorrhizas. Applied Mycology and Biotechnology 2: 275–310
Gomes EA, Oliveira CA, Lana UGP, Noda RW, Marriel IE, de Souza FA (2015) Arbuscular mycorrhizal fungal communities in the roots of maize lines contrasting for Al tolerance grown in limed and non-limed Brazilian Oxisoil. J. Microbiol. Biotechnol. 25(7): 978–987
GLASER B (2007) Prehistorically modified soils of central Amazonia: a model for sustainable agriculture in the twenty-first century. Phililosophical Transactions of the Royal Society B 362(1478):187–196
Hart MM, Reader RJ (2002) Taxonomic basis for variation in the colonization strategy of arbuscular mycorrhizal fungi. New Phytologist 153 (2): 335–344
Heberle ES, Armas RD, Heberle DA, Stürmer SL et al. (2015) Mycorrhizal Fungal Communities in Cassava after Cultivation of Cover Crops as Observed by the "PCR-DGGE" Technique. R. Bras. Ci. Solo 39:1292–1301
Hippler FWR, Moreira M (2013) Dependência micorrízica do amendoinzeiro sob doses de fósforo. Bragantia 72:184–191

Hofwegen G, Kuyper TW, Hoffland E, Broek JA, et al. Opening the black box: deciphering carbon and nutrient flows in Terra Preta (2009). In: Amazonian Dark Earths: Wim Sombroek's Vision. Woods WI, Teixeira WG, Lehmann J, Steiner C, WinklerPrins AMGA, Rebellato L (Eds.)

HECHT SBISM (2003) Implications of Kayapó practices. In: LEHMANN, J.; KERN, D. C.; GLASER, B.; WOODS, W. I. (Orgs.). Amazonian Dark Earths: Origin, properties, and management. Kluwer Academic Publishers, Dordrecht

Jarstfer AG, Sylvia DM (1995) Aeroponic culture of VAM fungi. In: Varma A and Hock B (eds) Mycorrhiza. Springer-Verlag, Berlin

König F, Gonçalves CEP, Aguiar AR, Silva ACF (2014) Pampa Biome: Interactions between microorganisms and native plant species. Revista de Ciências Agrárias 37(1): 3–9 3

Kern DC, Kämpf N (1989) Efeitos de antigos assentamentos indígenas na formação de solos com Terra Preta Arqueológicas na região de Oriximiná-PA. Revista Brasileira de Cienca do Solo 13:219–225

Leal PL, Stürmer SL, Siqueira JO (2009) Occurrence and diversity of arbuscular mycorrhizal fungi in trap cultures from soils under different land use systems in the Amazon, Brazil. Brazilian Journal of Microbiology 40: 111–121

Lekberg Y, Hammer EC, Olsson PA (2010) Plants as resource islands and storage units–adopting the mycocentric view of arbuscular mycorrhizal networks. FEMS Microbiology Ecology 74: 336–345

Lekberg Y, Koide RT, Rohr JR, Aldrich-Wolfe L, Morton JB (2007) Role of niche restrictions and dispersal in the composition of arbuscular mycorrhizal fungal communities. Journal of Ecology 95: 95–105

Mahecha-Vásquez G, Sierra S, Posada R (2017) Diversity indices using arbuscular mycorrhizal fungi to evaluate the soil state in banana crops in Colombia Applied Soil Ecology 109: 32–39

Major J, Steiner C, Ditommaso A, Falcão NP, Lehmann J (2005) Weed composition and cover after three years of soil fertility management in the central Brazilian Amazon: compost, fertilizer, manure and charcoal applications. Weed Biol Manag 5: 69–76

Nobre CP, Costa MG, Goto BT, Gehring C (2018) Arbuscular mycorrhizal fungi associated with the babassu palm (Attalea speciosa) in the eastern periphery of Amazonia, Brazil. Acta Amazonica 48: 321–329

Oehl F, Koch B (2018) Diversity of arbuscular mycorrhizal fungi in no-till and conventionally tilled vineyards. Journal of Applied Botany and Food Quality 91: 56–60

Oehl F, Sieverding E, Ineichen K, Ris EA, Boller T, Wiemken A (2005) Community structure of arbuscular mycorrhizal fungi at different soil depths in extensively and intensively managed agroecosystems. New Phytol. 165: 273–283

Öpik M, Vanatoa A, Vanatoa E, Moora M, Davison J, Kalwij JM, Reier U, Zobel M (2010) The online database Maarj*AM* reveals global and ecosystemic distribution patterns in arbuscular mycorrhizal fungi (Glomeromycota). New Phytologist 188: 223–241

Pagano MC, MA Lugo, FS Araújo, MA Ferrero, E Menoyo, Diego Steinaker (2012) Native Species for Restoration and Conservation of Biodiversity in South America in: Native Species: Identification, Conservation and Restoration. Nova Science Publishers Editors: Lluvia Marín, Dimos Kovač (eds)

Pagano MC (2012) Mycorrhiza: occurrence in natural and restored environments. Nova Science Publishers

Pagano MC (2016) Recent Advances on Mycorrhizal Fungi, Springer, Fungal Biology book series

Pagano MC, Covacevich F (2011) Arbuscular Mycorrhizas in Agroecosystems. In: Fulton SM (ed) Mycorrhizal Fungi: Soil, Agriculture and Environmental Implications. Nova Science Publishers, New York

Pagano MC, Ribeiro-Soares J, Cançado LG, Falcão NPS, et al. (2016) Depth dependence of black carbon structure, elemental and microbiological composition in anthropic Amazonian dark soil. Soil and Tillage Research 155, 298–307

Pagano MC, Utida MK, Gomes EA, Marriel IE, Cabello MN, Scotti MR (2011) Plant-type dependent changes in arbuscular mycorrhizal communities as soil quality indicator in semi-arid Brazil. Ecological Indicators 11 (2): 643–650

Pereira CMR, da Silva DKA, Ferreira ACA, Goto BT, Maia LC (2014) Diversity of arbuscular mycorrhizal fungi in Atlantic forest areasunder different land uses. Agriculture, Ecosystems and Environment 185 (2014) 245–252

Pontes JS, Oehl F, Pereira CD, de Toledo Machado CT, Coyne D, da Silva DKA, Maia LC (2017) Diversity of arbuscular mycorrhizal fungi in the Brazilian's Cerrado and in soybean under conservation and conventional tillage. Applied Soil Ecology 117–118 178–189

Porras Soriano A, Domench Menor B, Castillo Rubio J, Soraino Martin ML, Porras Piedra A (2002) Influence of vesicular arbuscular mycorrhizae on the growth of olive cuttings multiplied under nebulation. Olivae 92: 33–37.

Ratto SE, Miguez FH (2006) Zinc en el cultivo de maíz, deficiencia de oportunidade. Informaciones agronómicas 31: 11–14

Read DJ (2003) Towards Ecological Relevance – Progress and Pitfalls in the Path Towards an understanding of mycorrhizal functions in nature. In: van der Heijden MGA, Sanders IR (eds), Mycorrhizal Ecology. Springer, Berlin

Roldan-Fajardo BE, Barea JM. Mycorrhizal dependency in the olive tree (*Olea europaea* L.), Physiology and genetics aspects of mycorrhizae, Paris, 1986; 323–326

Rosa DJ, Ambrosini VG, Brunetto G, Soares CRFS et al. (2016) Physiological parameters in vines 'Paulsen 1103' (*Vitis berlandieri* x *Vitis rupestris*) inoculated with mycorrhizal fungi in soil contaminated with copper. Ciência Téc. Vitiv. 31(1): 14–23

Schalamuk S and Cabello MN (2010a) Effect of Tillage Systems on the Arbuscular Mycorrhizal Fungi (AMF) Propagule Bank in Soils. CAB International. Management of Fungal Plant Pathogens 162 (eds) A. Arya and A.E. Perelló.

Schalamuk S, Cabello M (2010b) Arbuscular mycorrhizal fungal propagules from tillage and no-tillage systems: possible effects on Glomeromycota diversity. Mycologia 102(2): 261–268

Schalamuk S, Cabello MN, Chidichimo H, Golik S (2011) Effects of Inoculation with *Glomus mosseae* in conventionally tilled and nontilled soils with different levels of nitrogen fertilization on wheat growth, Arbuscular Mycorrhizal Colonization, and nitrogen nutrition. Communications in Soil Science and Plant Analysis 42(5):586–598

Schalamuk S, Velázquez S, Cabello M (2013) Dynamics of arbuscular mycorrhizal fungi spore populations and their viability under contrasting tillage systems in wheat at different phenological stages. Biological Agriculture & Horticulture 29: 38–45

Schalamuk, S. Velazquez, H. Chidichimo, M. Cabello (2006) Fungal spore diversity of arbuscular mycorrhizal fungi associated with spring wheat: effects of tillage. Mycologia 98:1, 16–22, DOI: https://doi.org/10.1080/15572536.2006.11832708

Schueffler A, Anke T (2014) Fungal natural products in research and development. Nat. Prod. Rep. 31: 1425–1448

Sieverding E (1991) Vesicular-arbuscular mycorrhiza management in tropical agrosystems. Eschborn: Deutsche Gesellschaft für Technische Zusammenarbeit.(GTZ) GmnBH, Esch boran

Silva GAE, Siqueira JO, Stürmer SL, Moreira FMS (2018) Effectiveness of Arbuscular Mycorrhizal Fungal Isolates from the Land Uses of Amazon Region in Symbiosis with Cowpea. Anais da Academia Brasileira de Ciências 90, 1: 357–371

Sreenivassa MN (1992) Selection of an efficient vesicular–arbuscular mycorrhizal fungus for Chili (Capsicum annuum L.). Sci Hortic 50:53–58

Stürmer SL, Siqueira JO (2006) Diversity of Arbuscular Mycorrhizal Fungi in Brazilian Ecosystems. In: Moreira FMS, Siqueira JO, Brussaard L (eds) Soil Biodiversity in Amazonian and Other Brazilian Ecosystems. CABI Publishing, London, 206–236

Smith SE, Read DJ (2008) Mycorrhizal symbiosis, 3rd edn. Academic Press, London, UK

Soares CRFS, Carneiro MAC (2010) Micorrizas arbusculares na recuperação de áreas degradadas. In: Siqueira JO, Souza FA, Cardoso EJBN, Tsai SM (eds) Micorrizas: 30 anos de Pesquisa no Brasil. Lavras: Editora UFLA, 441–474 p.

Thougnon Islas AJ, Hernandez Guijarro K, Eyherabide M, Sainz Rozas HR, Echeverría HE, Covacevich F (2016) Can soil properties and agricultural land use affect arbuscular mycorrhizal fungal communities indigenous from the Argentinean Pampas soils? Applied Soil Ecology 101: 47–56

Tsai SM, O'neill B, Cannavan FS, Saito D, Falcão NPS, Kern DC, Grossman J, Thies J et al. (2009) The Microbial World of Terra Preta. In: Woods WI, Teixeira WG, Lehmann J, Steiner C, WinklerPrins A, Rebellato L (eds) Amazonian Dark Earths: Wim Sombroek's Vision. Springer, Dordrecht

Teste FP (2016) Restoring grasslands with arbuscular mycorrhizal fungi around remnant patches. Applied Vegetation Science 19, 1.

Torrez V, Ceulemans T, Mergeay J, de Meester L, Honnay O (2016) Effects of adding an arbuscular mycorrhizal fungi inoculum and of distance to donor sites on plant species recolonization following topsoil removal. Applied Vegetation Science 19:7–19

Velázquez MS, Fabisik JC, Abarca CL, Allegrucci N, Cabello M (2018) Colonization dynamics of arbuscular mycorrhizal fungi (AMF) in *Ilex paraguariensis* crops: Seasonality and influence of management practices. Journal of King Saud University – Science https://doi.org/10.1016/j.jksus.2018.03.017

Wahbi S, Sanguin H, Baudoin E, Tournier E, Maghraoui T, Prin Y, Hafidi M, Duponnois R (2016) Managing the Soil Mycorrhizal Infectivity to Improve the Agronomic Efficiency of Key Processes from Natural Ecosystems Integrated in Agricultural Management Systems. Springer International Publishing Switzerland, 17 K.R. Hakeem et al. (eds.), Plant, Soil and Microbes, https://doi.org/10.1007/978-3-319-27455-3_2

Warnock DD, Lehmann J, Kuyper TW, Rillig MC (2007) Mycorrhizal responses to biochar in soil – concepts and mechanisms.

Weber OB (2014) Biofertilizers with arbuscular mycorrhizal fungi in agriculture. In: Solaiman ZM, Abbott LK, Varma A (eds). Mycorrhizal fungi: use in sustainable agriculture and restoration. Soil biology: 41. Springer-Verlag, Berlin

Index

M. C. Pagano, M. A. Lugo (eds.), *Mycorrhizal Fungi in South America*, Fungal Biology, https://doi.org/10.1007/978-3-030-15228-4

B

N

T

V

W

X

MIX
Papier aus verantwortungsvollen Quellen
Paper from responsible sources
FSC® C105338

If you have any concerns about our products,
you can contact us on
ProductSafety@springernature.com

In case Publisher is established outside the EU,
the EU authorized representative is:
Springer Nature Customer Service Center GmbH
Europaplatz 3, 69115 Heidelberg, Germany

Printed by Libri Plureos GmbH
in Hamburg, Germany